Principles
and Applications of
SEMICONDUCTORS
AND
CIRCUITS

PRENTICE-HALL
SERIES IN ELECTRONIC TECHNOLOGY

Dr. IRVING L. KOSOW, Editor

*Consulting editors*
Charles M. Thomson
Joseph J. Gershon
Joseph A. Labok

PRENTICE-HALL INTERNATIONAL, INC., *London*
PRENTICE-HALL OF AUSTRALIA PTY. LTD., *Sydney*
PRENTICE-HALL OF CANADA, LTD., *Toronto*
PRENTICE-HALL OF INDIA PRIVATE LIMITED, *New Delhi*
PRENTICE-HALL OF JAPAN, INC., *Tokyo*

# Principles and Applications of SEMICONDUCTORS AND CIRCUITS

ROBERT F. COUGHLIN

*Master Instructor,*
*Dept. of Electronic Engineering Technology*
*Wentworth Institute, Boston, Massachusetts*

PRENTICE-HALL, INC., ENGLEWOOD CLIFFS, NEW JERSEY

© 1971 by
PRENTICE-HALL, INC.
Englewood Cliffs, N.J. 07632


All rights reserved. No part of this
book may be reproduced in any form
or by any means without permission
in writing from the publisher.


Current printing (last digit):
10 9 8 7 6 5 4 3 2 1


13-700971-2
Library of Congress catalog card number: 78-149974


Printed in the United States of America

**1617184**

to
*Barbara Ann, Shaun, and Susan*

# Preface

The purpose of this book is to present the basic principles of semiconductors so that their behavior in fundamental circuit applications can be analyzed and predicted. The essentials of the physics of each device are presented in sufficient detail to explain the shape of their characteristic curves. An electric circuit model for the device is then given, based on either measurements from the characteristic curves or the physics of the device. Experience with the circuit model is gained through studying typical applications, supplemented by analysis or design examples.

Extensive use of sweep techniques is utilized throughout the text to show how characteristic curves of any device can be displayed accurately on a cathode ray oscilloscope using a low cost transformer and resistor. Input and output characteristics of a bipolar or field-effect transistor, dc or ac load lines, diode current-voltage curves including zener or avalanche break down, voltage gain of an amplifier, frequency response of a tuned circuit, UJT valley voltage, SCR forward breakover voltage and almost any other device parameter can be drawn by the CRO. Furthermore, once the sweep circuit is set up, parameter variations between the same type of device or parameter variations due to environmental factors can be easily and quickly demonstrated and measured.

Advanced concepts are taught through circuit application. For example, in Chapter 10, a basic differential amplifier circuit is analyzed for methods of controlling gain by voltage. This is developed into an automatic gain control circuit. By a shift of viewpoint, using the same circuit, a change of gain at a signal frequency yields amplitude modulation that in turn illustrates frequency shifting.

Methods of constructing and simplifying relatively complex circuit equations quickly and efficiently, have been developed by an extension of current or voltage divider relations and the superposition theorem. For example, in Chapter 4, equations for base or collector currents, including leakage currents are developed *directly from an inspection* of the circuit without recourse to calculus, simultaneous equations or memory aids. In Chapters 5 and 6 expressions for input resistance, current and voltage gain are developed in a standard format so that they too can be expressed directly from an inspection of the circuit.

I gratefully thank: Dean Charles M. Thomson, for constant encouragement; Professors Campbell L. Searle and Bruce B. Wedlock of Massachusetts Institute of Technology, for teaching me how to teach; Dr. Irving I. Kosow, Matthew Fox, Nicholas C. Loomos, William F. Norton and Dominic Giampetro for their many suggestions and assistance; Miss Mary Kelleher, Mrs. Pauline Campbell and Miss Janice Prevett for their skillful typing.

ROBERT F. COUGHLIN

# Contents

## 1. Introduction to Semiconductors — 2

- 1-0  Introduction .................................................... 3
- 1-1  Intrinsic Semiconductors ........................................ 3
- 1-2  Doping .......................................................... 6
- 1-3  Current Flow in Semiconductor Material .......................... 7
- 1-4  The *pn* Junction ............................................... 10
- 1-5  Diode Characteristics and Models ................................ 12
- 1-6  Sweep Measurements of the Diode ................................. 21
- 1-7  Sweep Measurement of Diode Incremental Resistance ............... 27
- 1-8  Diode Applications in Meter Circuits ............................ 30
- 1-9  The Zener Diode ................................................. 32

## 2. Characteristics of Bipolar Transistors and Their Measurements — 38

- 2-0  Introduction .................................................... 39
- 2-1  Standard Letter Symbols ......................................... 41
- 2-2  Operating Modes and Circuit Configurations ...................... 43
- 2-3  Transistor Action in the Common-Base Configuration .............. 46
- 2-4  Transistor Action in the Common-Emitter Configuration ........... 50
- 2-5  Measurement and Display of Common-Emitter Characteristic Curves ........................................................... 56
- 2-6  Measurement of Common-Emitter Current Gain ...................... 59
- 2-7  Cutoff and Saturation of the Characteristic Curves .............. 61
- 2-8  Measurement of Common-Base Characteristics ...................... 64
- 2-9  Measurement of Other Useful Characteristic Curves ............... 66

## 3. Graphical Analysis of a Transistor Circuit — 74

- 3–0 Introduction — 75
- 3–1 The DC Load Line — 75
- 3–2 The Operating Point — 79
- 3–3 Measuring the Operating Point — 82
- 3–4 Input Resistance by Graphical Analysis — 83
- 3–5 Voltage Amplification by Graphical Analysis — 85
- 3–6 The AC Load Line — 88
- 3–7 AC–DC Load Line Demonstration — 92

## 4. Biasing and Stability — 96

- 4–0 Introduction — 97
- 4–1 Biasing a Basic Common-Emitter Amplifier — 98
- 4–2 Bias Calculations with an Emitter Resistor — 103
- 4–3 Factors Affecting Operating Point Stability — 108
- 4–4 Stability Analysis by Superposition — 112
- 4–5 Stabilizing the Operating Point with an Emitter Resistor — 115
- 4–6 Stabilization with a Collector-Base Resistor — 123
- 4–7 Emitter and Collector-Base Resistor Stabilization — 126
- 4–8 Other Basic Biasing Circuits — 130
- 4–9 Biasing with Diodes for Temperature Compensation — 134
- 4–10 Transistor Stabilized Biasing — 138
- 4–11 Demonstration of Operating Point Stability — 140

## 5. Small-Signal Low-Frequency Amplifiers — 140

- 5–0 Introduction — 141
- 5–1 Basic Low-Frequency Hybrid-Pi Model — 142
- 5–2 Common-Emitter Amplifier — 145
- 5–3 Common-Emitter Amplifier with Emitter Resistance — 153
- 5–4 Common-Collector Amplifier — 159
- 5–5 Output Resistance and Resistance Transformations — 161
- 5–6 Common-Base Amplifier — 166
- 5–7 Measurement of Output and Input Resistance — 170
- 5–8 Gain in Decibels — 171

## 6. Frequency Limitation of Voltage Gain — 176

- 6–0 Introduction — 177
- 6–1 Identification of Cut-Off Frequency — 177
- 6–2 Low-Frequency Cut-Off by the Coupling Capacitor — 179

Contents    xi

6–3   Low-Frequency Cut-Off by the Bypass Capacitor .................... 182
6–4   Cut-off Frequency with Both Coupling and Bypass Capacitors ........ 184
6–5   High-Frequency Model of a BPT ............................... 187
6–6   Dependence of $\beta$ on Frequency ................................. 189
6–7   Common-emitter High Frequency Cut-Off ......................... 191
6–8   Derivation of Miller Effect ...................................... 195
6–9   Emitter Resistance and High-Frequency Cut-Off ................... 197
6–10  Common-base and Common-Collector Upper Cut-Off
      Frequency .................................................... 199
6–11  Measurement of the Basic Hybrid-Pi parameters ................... 204

## 7. Selected Applications for Analysis and Design    206

7–0   Introduction .................................................. 207
7–1   Bootstrapping the Emitter-Follower ............................. 207
7–2   Bootstrapping the Common-Emitter ............................. 210
7–3   Single Stage Collector Feedback ................................ 211
7–4   Control of Upper Cut-off Frequency by Collector Feedback ........ 216
7–5   Single Stage Collector and Emitter Feedback ..................... 219
7–6   The Darlington Pair ........................................... 221
7–7   The Inverted Darlington Amplifier .............................. 225
7–8   Common-emitter to Common-base Cascode ....................... 227
7–9   Linear Mixing ................................................ 231

## 8. Power Amplifiers    234

8–0   Introduction .................................................. 235
8–1   Current and Voltage Limitations ................................ 236
8–2   Thermal and Power Dissipation Limitations ...................... 239
8–3   Thermal Resistance ........................................... 240
8–4   Heat Sinking ................................................. 242
8–5   Review of Power Fundamentals ................................. 246
8–6   Power Supply to a Transistor Circuit ............................ 247
8–7   Class A Direct-Coupled Amplifier ............................... 248
8–8   Class A Transformer-Coupled Power Amplifier ................... 253
8–9   Collector-Emitter Voltage Considerations with Transformer-
      coupling ..................................................... 258
8–10  Class A Power Relationships under Non-Sinusoidal Drive .......... 260
8–11  Class B Amplifier ............................................. 261

## 9. Complementary and Multistage Operation    268

9–0   Introduction .................................................. 269
9–1   The Basic Complementary Power Amplifier ....................... 269

xii    *Contents*

9–2   Biasing the Complementary Amplifier ............................. 270
9–3   Qualitative Analysis of the Complementary Amplifier ............. 273
9–4   Power Relationships ............................................. 274
9–5   Quantitative Analysis of the Complementary Output Stage ......... 276
9–6   Bootstrapping the Complementary Output Stage .................... 279
9–7   The Driver Stage ................................................ 282
9–8   Design Example for a Multistage Audio Amplifier ................. 285
9–9   Operation of a Quasi-Complementary Amplifier .................... 290
9–10  A Practical Quasi-Complementary Multistage Amplifier ............ 293

## 10. Selected Communication and Control Circuits    298

10–0   Introduction ................................................... 299
10–1   Basic Differential Amplifier Circuit Analysis .................. 300
10–2   Common Mode Rejection Ratio .................................... 304
10–3   Differential Output Voltage .................................... 308
10–4   Constant Current Biasing ....................................... 309
10–5   Phase Inverter ................................................. 310
10–6   Voltage Gain Control ........................................... 311
10–7   Gain Control by Offset Voltage ................................. 313
10–8   Linear Multiplier .............................................. 314
10–9   Amplitude Modulation ........................................... 316
10–10  Balanced and Single Side Band Modulators ....................... 320
10–11  Frequency Shifting ............................................. 321
10–12  The Comparator ................................................. 321

## 11. Tuned Circuits    328

11–0   Introduction ................................................... 329
11–1   Tuned Circuit Review ........................................... 329
11–2   Inductance Measurement and Quality Factor ...................... 334
11–3   Transformer Review ............................................. 336
11–4   The Tapped Inductor ............................................ 338
11–5   The Tuned Amplifier ............................................ 342
11–6   Bandwidth Reduction in Cascade ................................. 348
11–7   Multiple-Tuned-Circuit Amplifier ............................... 350

## 12. Negative and Positive Feedback    358

12–0   Introduction ................................................... 359
12–1   Negative Feedback and Voltage Gain ............................. 360
12–2   Fundamental Concepts of Negative Feedback ...................... 363

| | | |
|---|---|---|
| 12–3 | Single-Stage Amplifiers with Negative Feedback | 365 |
| 12–4 | Reduction of Distortion by Negative Feedback | 368 |
| 12–5 | Multistage Parallel-Output to Series-Input Negative Feedback | 371 |
| 12–6 | Multistage Series-Output to Parallel-Input Negative Feedback | 375 |
| 12–7 | Parallel-Output to Parallel-Input Negative Feedback | 381 |
| 12–8 | Multistage Series-Output to Series-Input Negative Feedback | 385 |
| 12–9 | Effect of Negative Feedback on Resistance Levels | 389 |
| 12–10 | Negative Feedback and Frequency Response | 390 |
| 12–11 | Design and Analysis Examples | 391 |
| 12–12 | Oscillation in Amplifiers | 396 |
| 12–13 | Oscillators | 398 |
| 12–14 | Phase-Shift Oscillators | 399 |
| 12–15 | Tuned-Circuit Oscillators | 403 |

## 13. Vacuum Tubes — 412

| | | |
|---|---|---|
| 13–0 | Introduction | 413 |
| 13–1 | Characteristics of the Vacuum Diode | 414 |
| 13–2 | Triode Characteristics and Model | 416 |
| 13–3 | Biasing | 423 |
| 13–4 | Voltage Gain and Resistance Levels | 424 |
| 13–5 | Low-Frequency Response | 429 |
| 13–6 | High-Frequency Response | 431 |
| 13–7 | Pentode Vacuum Tubes | 433 |
| 13–8 | Vacuum Tube Phase-Inverters | 437 |
| 13–9 | Push-Pull Operation | 439 |

## 14. Field-Effect Transistors and Applications — 448

| | | |
|---|---|---|
| 14–0 | Introduction | 449 |
| 14–1 | Physical Model of the JFET | 449 |
| 14–2 | Characteristic Curves and Parameter Measurements of the JFET | 452 |
| 14–3 | Measurement of Pinch-off Voltage for the JFET | 456 |
| 14–4 | Physical Model of the IGFET | 458 |
| 14–5 | Characteristic Curves and Parameter Measurement of the IGFET | 461 |
| 14–6 | Maximum Ratings | 464 |
| 14–7 | Biasing for Depletion Mode Operation | 465 |
| 14–8 | Biasing for Enhancement Mode Operation | 467 |
| 14–9 | Zero Bias Shift of the Operating Point | 471 |
| 14–10 | Low-Frequency Circuit Models for the FET | 473 |

xiv  *Contents*

14–11  Voltage Gain ................................................... 475
14–12  Resistance Levels .............................................. 480
14–13  High Frequency Dependence ..................................... 482
14–14  Low-Frequency Voltage Gain..................................... 484
14–15  Hybrid and Two-Stage Amplifiers ............................... 486
14–16  Chopper Amplifiers ............................................ 488
14–17  Analog Switching and Commutation .............................. 493

## 15. The Unijunction Transistor and Silicon Controlled Rectifier     498

15–0   Introduction .................................................. 499
15–1   Characteristic Curves of the UJT .............................. 500
15–2   Parameter Measurements of the UJT ............................. 505
15–3   Temperature Dependence of $V_P$ ............................... 508
15–4   The Unijunction as a Relaxation Oscillator .................... 510
15–5   Sawtooth Generator ............................................ 516
15–6   Timing and Voltage Level Sensing with the UJT ................. 518
15–7   Introduction to the Silicon Controlled Rectifier SCR .......... 520
15–8   Pulsed Gate Operation.......................................... 523
15–9   Phase Control ................................................. 525
15–10  SCR Applications .............................................. 529

Appendix 1 .........................................................535
Appendix 2 .........................................................538
Appendix 3 .........................................................544
Appendix 4 .........................................................548
Appendix 5 .........................................................550

Bibliography .......................................................533

Index ..............................................................553

Principles
and Applications of
SEMICONDUCTORS
AND
CIRCUITS

# Chapter 1

**1-0** INTRODUCTION ................................. 3
**1-1** INTRINSIC SEMICONDUCTORS ...................... 3
**1-2** DOPING ........................................ 6
**1-3** CURRENT FLOW IN SEMICONDUCTOR MATERIAL ...... 7
**1-4** THE *pn* JUNCTION .............................10
**1-5** DIODE CHARACTERISTICS AND MODELS .............12
**1-6** SWEEP MEASUREMENT OF THE DIODE ...............21
**1-7** SWEEP MEASUREMENT OF DIODE INCREMENTAL
     RESISTANCE ....................................27
**1-8** DIODE APPLICATIONS IN METER CIRCUITS ..........30
**1-9** THE ZENER DIODE ..............................32
     PROBLEMS ......................................35

# Introduction to Semiconductors

## 1-0 Introduction

The study of semiconductors will begin by examining the purest or *intrinsic* form of semiconductor material. We shall consider this material as an environment containing free charges that can move to constitute a conduction of current. The intrinsic semiconductor environment produces two types of free charges, the negatively charged electron and the positively charged hole. We shall consider how temperature controls the number of free charges and how the intrinsic environment can be modified to increase their number so that enough current can be conducted to do useful work. We shall see how charges move within the semiconductor and in an external circuit connected to the semiconductor. These basic factors are the prelude to understanding the electrical behavior of the semiconductor junctions contained in most of the semiconductor devices.

The simplest form of semiconductor junction device is the junction diode. Its electrical behavior will be studied by (1) measurements of the relationships between current through and voltage across the junction in the form of a characteristic curve, (2) forming a circuit model of the diode so that its behavior in a circuit can be calculated with the tools of basic circuit analysis, (3) graphical analysis on the characteristic curve for a pictorial representation of diode and circuit operation, and (4) examples of useful basic applications. We shall conclude with methods of measuring the maximum current or voltage ratings of a conventional diode and the Zener diode.

## 1-1 Intrinsic Semiconductors

Intrinsic semiconductor material is a very pure crystalline form of silicon, germanium, gallium arsenide, and occasionally diamond. All of these elements have four electrons, designated as *valence* electrons, orbiting in their outer atomic shell. Chemical and electrical behavior of any element is determined largely by the number of valence electrons and how they interact between atoms.

When a pure sample of germanium or silicon is melted and then cooled slowly around a seed crystal, the atoms usually arrange themselves into a uniform, symmetrical structure. The structure formed by *tetravalent* or four-valence electron material, when so treated, is described as a *cubic lattice*; a two-dimensional picture is shown in Fig. 1-1.

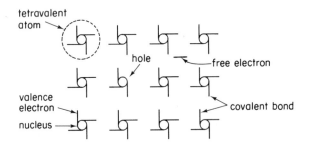

**Fig. 1-1.** Two-dimensional pictorial representation for the cubic lattice of a semiconductor crystal.

Each silicon atom in Fig. 1-1 shares one of its valence electrons with one of a neighboring atom, forming a *covalent bond.* Each atom has four of its own electrons, shares four with its neighbors, and effectively has a full outer shell with eight electrons. Chemically and electrically an outer shell with eight electrons is an extremely stable arrangement since there is no room for additional electrons in an eight-electron shell.

At absolute zero all of the valence electrons would be bound to a parent atom and there would be no free electrons available to move and constitute a current flow. But energy can be added to the semiconductor sample as (1) thermal energy in the form of heat or (2) electromagnetic energy in the form of light or radiation from radioactive sources. For example, at room temperature sufficient energy will be absorbed by the lattice and converted into kinetic energy. We can visualize the atoms vibrating faster, the electrons orbiting farther away from the nucleus, and occasionally an electron obtaining enough energy to escape from the influence of its parent atom.

There are two consequences of an electron escape: (1) A free electron can now wander through the predominately empty space of the crystal environment, becoming a charge carrier that can contribute to a conduction current; (2) a covalent bond was broken when the electron departed, leaving a vacancy or *hole* in the parent atom. A valence electron from a neighboring atom can move easily into the vacancy so that a hole appears to have moved in a direction opposite to the valence electron. For our purposes we shall consider the hole as a free charge carrier with a positive charge equal in magnitude but opposite in sign to that of the electron. Thus two charge carriers are created by one broken covalent bond. Hole and electron flow will be discussed further in Section 1-3.

At a stable temperature such as room temperature, energy is constantly added to the semiconductor so that covalent bonds are broken at a constant rate. However, free electrons can lose energy and combine with a hole to reform a covalent bond in a process called *recombination*. When the number of recombinations per second equals the number of covalent bonds broken per second, a state of dynamic equilibrium exists where the net number of free charges is stable and constant. The number of free charges thus depends on the material. It takes more energy to free an electron from a silicon atom than it does from a germanium atom, so for a given temperature a germanium sample would contain more free charges than a silicon sample. The number of free charges also depends on temperature because an increase in temperature increases the rate of bond rupture and the aggregate number of free electrons must increase until the chances of recombination have increased to meet the rate of bond breakage. To appreciate how few charges are available for current carriers, the number of broken bonds at room temperature and number of atoms in a cubic centimeter of silicon and germanium are compared in Table 1-1.

Table 1-1

|  | Germanium | Silicon |
|---|---|---|
| Atoms/cm$^3$ | $4.5 \times 10^{22}$ | $5.0 \times 10^{22}$ |
| Free holes, electrons, or broken bonds | $2.4 \times 10^{13}$ | $1.5 \times 10^{10}$ |
| Atoms/free electron | $1.9 \times 10^9$ | $3.3 \times 10^{12}$ |

Since the resistance of a material depends on the number of free charges available for conduction the semiconductor is, as its name implies, neither a conductor nor an insulator but has resistance characteristics between these two extremes, as shown in Table 1-2.

Table 1-2

| Material | Classification | Resistance ($\Omega$/cm$^3$) |
|---|---|---|
| silver | conductor | $10^{-5}$ |
| pure silicon | semiconductor | $5 \times 10^4$ |
| pure germanium | semiconductor | 50 |
| mica | insulator | $10^{12}$ |

## 1-2 Doping

The resistance of intrinsic semiconductor material is too high for most applications and must be lowered by adding precise amounts of nontetravalent atoms by a process called *doping*. The additives are either trivalent or pentavalent atoms. Trivalent atoms have only three electrons in their outer shell and are found in group 3 of the periodic table. Pentavalent atoms have five electrons in their outer shells and are found in group 5 of the periodic table. Commonly used doping materials are listed in Table 1-3.

Table 1-3

| Acceptor (Trivalent) | Donor (Pentavalent) |
|---|---|
| boron | phosphorous |
| aluminum | arsenic |
| gallium | antimony |
| indium | |

In the doping process, each of the added doping atoms *replaces* a silicon or germanium atom in the crystalline structure. The pentavalent atom is called a *donor* atom because four of its valence electrons lock into covalent bonds with, for example, four silicon atoms. The fifth electron is bound very loosely to its parent atom and escapes to become a free charge carrier. The donor atom not only contributes one free electron or negative charge carrier to participate in the conduction process but also becomes an ionized atom with a local positive charge. The donor is locked in the lattice, unable to move. The positive donor ion is *not* a hole because the local absence of the fifth electron is not a covalent bond waiting to be filled, and is *not* considered to influence *directly* the recombination process.

Trivalent doping atoms are known as *acceptor* atoms because the absence of the fourth valence electron is an unfilled covalent bond. A bound electron from a neighboring silicon atom may move into the vacancy, leaving a hole. The term acceptor atom has been derived from the idea that it accepts an electron from its neighbors. However, it really contributes a hole. Once it has contributed a positive charge carrier to the sample, the acceptor atom is ionized. This negative ion is locked in the lattice and cannot move. The acceptor negative ion is not a free electron in the same sense that the positive donor ion is not a hole.

One other fundamental concept must now be stated. A sample of pure silicon is electrically neutral. It contains exactly as many positive charges as it does negative charges. A phosphorous or aluminum atom is also neutral. Adding any amount of either to silicon will not change the electrical neutrality of the result. Even though we shall find many free electrons, free holes, locked positive ions, or locked negative ions in the same material, it will be electrically neutral because all the negative charges will be balanced exactly by the positive charges.

A typical doping pricess would add $10^{15}$ donor atoms, which in effect provides $10^{15}$ free electrons to a cubic cemtimeter of pure silicon, or one doping atom for every 50 million silicon atoms. The crystal structure remains essentially unchanged but the $1.5 \times 10^{10}$ electrons from thermally broken bonds have been increased through doping to $10^{15}$ so resistance has been reduced by a factor of approximately $10^{15}/10^{10}$, or about 100,000! Of course the additional electrons will increase the recombination rate to reduce the original number of free holes.

Since two types of doping material are used, there are two types of resulting semiconductor material. Doping with donor atoms results in *n*-type semiconductor material, which contains a majority of free electrons or negative charge carriers and a minority of holes. Doping with acceptor atoms yields *p-type* semiconductor material, which contains a majority of free hole carriers and a minority of electron charge carriers. A pictorial aid to visualize the two classifications of semiconductor material is given in Fig. 1-2.

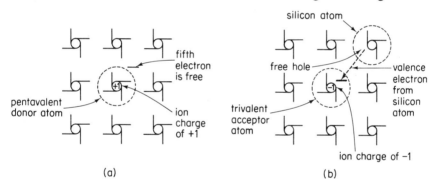

**Fig. 1-2.** Representation of *n-type* semiconductor material, (b) *p-type* semiconductor material.

## 1-3 Current Flow in Semiconductor Material

One basic method of picturing current flow in a conductor is to visualize a cross-sectional area and count the total number of electrons which pass through the area in 1 sec. The number of electrons per second is expressed in coulombs per second to give current flow in amperes. In a semiconductor, holes are available as charge carriers in addition to electrons, and the movement of both types of charge carriers contributes to the total current We must consider how both types of charge carriers move within the semiconductor crystal before we can understand current flow in the external circuit.

The free electron exists from the moment it gains enough energy to escape from the influence of its parent atom. Once free, the electron's movement will be influenced by the crystal's atoms. We can visualize these atoms as vibrating continually and the electron may approach a particular atom as it is either moving with or against the electron's direction. The electron can rebound with more energy (and more speed) if the atom is moving toward it at the time of impact or with less energy (and consequently less speed) if the atom is moving away from it at the time of impact. We therefore visualize the free electrons as (1) moving continuously and (2) changing direction and speed periodically and in a random fashion. This type of electron velocity is called *thermal velocity* since it depends on the energy of the atoms, which depends in turn on the semiconductor's temperature. Analogy with a pinball machine would be useful except that there is no friction or gravity acting on the electron as acts on the pinball.

What has been said for the electron is valid for the hole provided we think of the hole as a positively charged particle. Thus if a valence electron travels left from atom A to occupy a hole in atom B, it is equivalent to saying that a hole moved right from atom B to atom A. Holes move closer to the nucleus in the valence shell while free electrons travel in the open space of the lattice. Consequently hole velocity is slower than electron velocity.

Thermal charge motion does *not* constitute current flow because, over a period of time, as many charges will pass through a cross-sectional area in one direction as in the other. *Drift* and *diffusion* are the two mechanisms by which current will flow through a semiconductor.

*Drift current* is the current which flows in a semiconductor due to an applied electric field. In Fig. 1-3 we show one electron and one hole of an

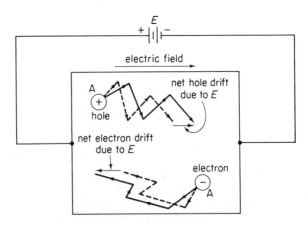

**Fig. 1-3.** Drift current in an intrinsic semiconductor.

intrinsic semiconductor. An electric field (direction taken by a positive charge) that results from the battery $E$ is shown. Dashed lines show hole or electron thermal motion without the electric field and solid lines show how motion will be changed with an electric field. The locked atoms cannot move and are not shown.

If we watch a cross section of the semiconductor sample in Fig. 1-3, more holes will pass to the right than to the left and more electrons will pass to the left than to the right. The net number of holes moving right and the net number of electrons moving left add to form a drift current.

There is no mechanism to support hole flow in the wires or battery of the external circuit where only electrons can flow. Special contacts are attached between the semiconductor material and the lead wires such as gold-antimony for $p$-type material or gold-gallium for $n$-type material to provide sites for the exchange of holes and electrons. These contacts must exhibit negligible voltage drop and are known as *ohmic* contacts. In Fig. 1-4 electron

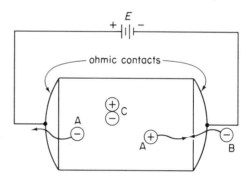

**Fig. 1-4.** External circuit electron current.

A must leave the semiconductor sample and electron B must enter simultaneously to maintain the sample's neutrality. Hole A will combine with electron B. Another hole-electron pair C will be generated and the cycle in Figs. 1-3 and 1-4 will repeat. Thus only electrons flow in the external circuit while both electrons and holes flow in the semiconductor.

*Diffusion current* results from the thermal motion of free charge carriers. We could suddenly break a large number of covalent bonds by shining a high-intensity light on the center of the semiconductor sample in Fig. 1-3. (Assume battery $E$ is removed to eliminate drift current.) A large concentration of holes and electrons would be generated initially at the sample's center. Random thermal motion would force some holes and electrons to diffuse right and left, away from the concentration. This action is analogous to the diffusion of perfume molecules away from a concentration of perfume. The heavy perfume molecules move randomly because they absorb energy from the vibrating air molecules and perfume sprayed in one corner of a room will therefore diffuse evenly throughout the room.

Diffusion current can be demonstrated by the hot-probe experiment in Fig. 1-5, which also determines if a semiconductor is *p*-type or *n*-type. In Fig. 1-5(a) energy is imparted to holes near the hot probe and some move by

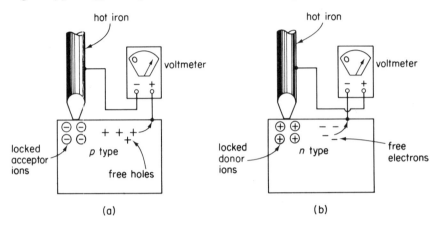

**Fig. 1-5.** Hot-probe demonstration of diffusion and type of doping.

diffusion to the positive terminal of a high-impedance voltmeter to cause an upscale deflection. In Fig. 1-5(b) meter connections must be reversed for an upscale reading to signify an *n*-type sample.

## 1-4 The *pn* Junction

Assume that a predominantly *p*-type semiconductor and a predominantly *n*-type semiconductor are suddenly brought together to form a *pn* junction at their interface. Free majority electrons from the *n*-type would diffuse across the junction and some would fill the fourth covalent bond in acceptor atoms, near the junction, in the *p*-type material. In similar fashion, majority holes from the *p*-type material diffuse across the junction and recombine with free electrons in the *n* side.

We might expect that electrons from the *n* side and holes from the *p* side would continue to diffuse until both electrons and holes were distributed uniformly through both *p*- and *n*-type samples. This would be the case if it were *not* for the action of the *pn* junction. Each hole crossing the junction into the *n*-type material leaves behind a negatively charged donor ion locked in the lattice. More holes cross the junction and a wall of immobile negative ions is formed in the *p*-type material adjacent to the junction, as shown in Fig. 1-6. Likewise electrons diffusing into the *p* side leave behind a wall of positively charged donor ions locked in the lattice adjacent to the junction's *n* side.

Sec. 1-4                                            The pn Junction    11

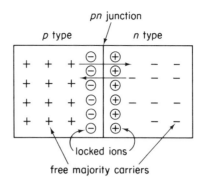

**Fig. 1-6.** Diffusing majority carriers leave locked ions around the *pn* junction.

Whenever a group of stationary positive charges (donor ions) are near another group of stationary negative charges (acceptor ions) there is an electric field line, shown conventionally as an arrow which begins on a positive charge and terminates with its head on a negative charge. This electric field not only signifies attraction between unlike charges but also tells us what direction a positive charge will take when it finds itself between the oppositely charged ions.

Recall that a charged capacitor can be represented by stationary negative charges on one plate and stationary positive charges on the other plate with electric field lines through the insulation as in Fig. 1-7. The capa-

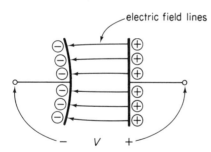

**Fig. 1-7.** Voltage and electric field of a charged capacitor.

citor analogy is introduced for two reasons. First, it shows that the *pn* junction will have properties of capacitance. Second, and more important, the region of locked ions around the junction establishes a *contact potential*.

We can depict these properties of a *pn* junction more clearly by redrawing the junction as in Fig. 1-8. We replace the immobile ions by a region extending on either side of the junction and show their action simply as an electric field line and a *potential barrier* voltage $V_j$. Potential barrier rather than contact potential is used for the junction voltage because it it more descriptive of the electric field's effect on majority carriers. Any majority hole in the *p* side can diffuse into the electric field but will be repelled and prevented from crossing to the *n* side. Electrons from the *n* side are also barred from crossing the junction. Thus the contact voltage acts as a *poten-*

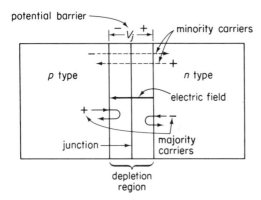

Fig. 1-8. Potential barrier and depletion region of a *pn* junction.

*tial barrier* that prevents *majority* carriers from crossing the junction and diffusing uniformly through both *p*-type and *n*-type samples. Because few majority carriers will be found in the region occupied by the locked ion walls it is designated as the *depletion region* to signify the depletion of free charge carriers. Because this region can be compared with a charged capacitor it is also called a *space-charge region*. Any minority electrons from the *p*-type material which diffuse into the depletion region will be accelerated across the junction to the *n* side. Minority holes from the *n* side can diffuse into the depletion region and be accelerated to the *p* side. This action is shown in Fig. 1-8.

Velocities or energy levels of the majority electrons on the *n* side differ widely. Some will travel fast enough to overcome the electric field and cross into the *p* side, where they become indistinguishable from the minority electrons. Corresponding behavior is exhibited by majority holes from the *p* side. A dynamic equilibrium exists where (1) a few high-energy majority carriers hurdle the potential barrier, (2) become minority carriers, (3) diffuse back to the junction, and (4) are accelerated back to where they began. This dynamic equilibrium persists until it is upset by applying an external voltage across the *pn* structure. This external voltage is called a *bias voltage;* its polarity is extremely important in that the polarity determines whether a large current or a small current flows through the junction.

## 1-5  Diode Characteristics and Models

The junction diode is a semiconductor device whose operating principles depend on the action of a *pn* junction. The diode's electrical symbol is drawn to correspond with its structure in Fig. 1-9. The positive terminal of a bias voltage $V$ is connected to the *p* side or anode of the diode. The negative side of $V$ is connected to the *n* side or cathode of the diode. This bias arrangement is known as *forward bias* because external voltage $V$ acts to reduce the barrier potential. Resistance of the heavily doped *n*-and *p*-type

Sec. 1-5                    Diode Characteristics and Models         13

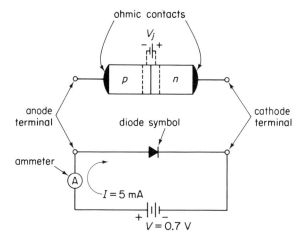

**Fig. 1-9.** Forward-biased diode; $I$ represents conventional current flow.

material is low and resistance of the ohmic contacts is low so that $V$ appears across the junction. Increasing $V$ lowers the potential barrier and allows a greater number of majority carriers to cross the junction. In fact when $V$ exceeds 0.7 V for a silicon junction diode, or 0.3 V for a germanium diode, current flow must be limited by external resistance or the diode will burn out. In Fig. 1-9 only electrons flow in the external circuit from the cathode through the battery to the anode. However, the direction of current $I$ is shown according to conventional current flow to match the diode symbol whose arrowhead also points in the direction of conventional current flow.

In Fig. 1-10 the diode is *reverse biased* because external voltage $V$ acts in place of $V_j$ to raise the potential barrier. None of the majority carriers have enough energy to cross the depletion region. However, all of the minority carriers can cross the junction with ease. For example, a minority

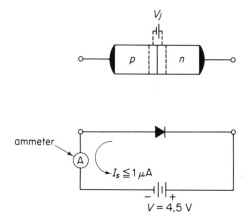

**Fig. 1-10.** Reverse-biased diode.

electron from the *p* region will diffuse to the depletion region, be accelerated to the *n* region, diffuse to the anode terminal, and return through the battery to the *p* side. In fact, all of the minority carriers contribute to this reverse current and since the number of minority carriers is small the reverse current is small. The symbol $I_s$ stands for reverse diode current. The subscript *s* stands for *saturation*, since all available minority carriers are being conducted and consequently $I_s$ will not depend on the magnitude of reverse bias. The value of $I_s$ will depend on the number of minority carriers and, since the number of minority carriers depends on the number of broken covalent bonds, $I_s$ will be temperature dependent.

The reverse-bias voltage causes majority carriers to move away from the junction and widens the depletion region because more ions are uncovered on either side of the junction. With more uncovered field lines the electric field across the junction increases. There are two significant results of this action; (1) minority carriers will be accelerated faster through the space-charge region; and (2) using a capacitor analogy, the depletion region's capacitance will decrease with increasing reverse bias because widening the depletion region corresponds to spreading the plates of a capacitor.

In summary, a reverse-biased diode is characterized by a small reverse current $I_s$, which is practically independent of applied voltage. A forward-biased diode is characterized by a large forward current and a small forward voltage drop. Forward biasing is accomplished by applying a more positive potential to the *p* side or anode terminal than is connected to the *n* side or cathode terminal. Ideally the diode acts as a closed switch when forward biased and as an open switch when reverse biased.

If we know the size of a *linear* element such as a resistor, we can predict all current-voltage relations that it can possess. For example a 10-$\Omega$ resistor will have a 10-V drop across it when it conducts 1 A, and a $-20$-V drop when conducting $-2$ A. From Ohm's law, current through a resistance equals voltage divided by resistance, regardless of voltage polarity or magnitude. The diode is a *nonlinear* element because current through it depends on bias voltage polarity and exponentially upon the bias voltage magnitude. It is often necessary to have a record of all possible current-voltage relations for a particular type of diode. Such a record is called a *characteristic curve* and is drawn by plotting a point for each value of current through the diode and the corresponding voltage across the diode.

A typical current-voltage (*I-V*) characteristic curve is shown in Fig. 1-11. The diode's forward characteristic is located in the first quadrant and its reverse characteristic is located in the third quadrant. It can be shown that the diode's curve in Fig. 1-11 may be expressed by the *diode equation*

$$I = I_s \left( \exp \frac{qV}{kT} - 1 \right) \tag{1-1}$$

Sec. 1-5             Diode Characteristics and Models     15

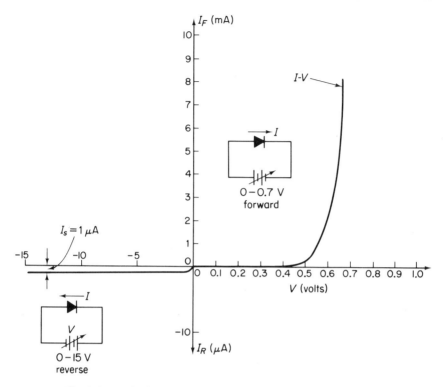

**Fig. 1-11.** Typical current-voltage characteristic curve of a silicon junction diode. Note differences in current scales.

where $I$ = current through the diode (in either direction)
$I_s$ = reverse saturation current,
$q = 1.6 \times 10^{-19}$ C (electron charge),
$k$ = Boltzmann's constant = $1.38 \times 10^{-23}$ J/°K,
$T$ = temperature (°K),
$V$ = voltage applied across the terminals of the diode (substitute positive values for $V$ when the diode is forward biased and negative values when it is reverse biased), and
exp = $e$ raised to the power of $qV/kT$ (where $e = 2.713\ldots$).

While we will refer to the *diode equation* [Eq. (1-1)] there is a more practical method used to approximate the diode's behavior in an actual circuit: a process called *modeling*. For dc and large-signal applications we construct an approximate *circuit model* for the diode when it will be forward biased and another for when it will be reverse biased. The circuit model allows us to represent the real diode by linear elements so that we may use regular circuit-analysis techniques for analysis or design.

A forward-biased dc model is constructed from the forward char-

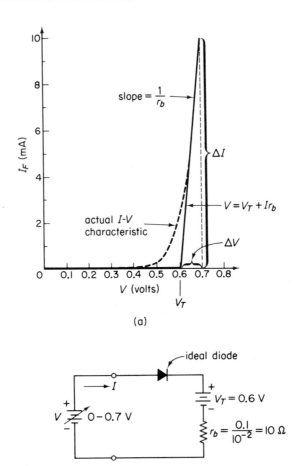

**Fig. 1-12.** The forward diode characteristic of Fig. 1-11 is approximated in (a) and its dc model for forward-bias operation is shown in (b).

acteristic of Fig. 1-11 as shown in Fig. 1-12(a). In Fig. 1-12(b) an *ideal-diode* symbol represents an ideal diode that behaves as a short circuit during forward bias and as an open circuit during reverse bias. We extend the linear portion of the actual characteristic to the voltage axis to locate the *threshold voltage* $V_T$. $V_T$ is shown as a 0.6-V battery connected to oppose $V$ and indicates that $V$ must be increased beyond 0.6 V before the ideal diode will conduct. Once $V$ is greater than 0.6 V, forward current $I$ is limited in the approximate characteristic by the small *bulk resistance* $r_b$ of the diode. Bulk resistance $r_b$ is evaluated from the slope as shown in Fig. 1-12(a) to be

$$r_b = \frac{\Delta V}{\Delta I} = \frac{0.1}{10 \times 10^{-3}} = 10\ \Omega \tag{1-2}$$

and the approximate forward *I-V* characteristic curve is simplified to the expression

$$V = V_T + Ir_b \tag{1-3}$$

These approximations hold only for dc or large-signal operation where current through the diode will exceed 5 mA. *Usually external circuit resistance will be much larger than $r_b$ and $r_b$ can be removed from the model.* Manufacturers often give a point on the diode characteristic at a high forward current $I = I_F$ and the corresponding forward voltage $V = V_F$; bulk resistance $r_b$ may then be calculated from Eq. (1-3).

For *small-signal forward operation* of the diode at a point on the curved portion of the diode's characteristic we find that the diode's behavior depends predominantly on the junction. In Fig. 1-13(a), tangents are drawn to points *A*, *B*, and *C* at diode currents of 1.0, 0.5, and 0.1 mA, respectively. The slopes of these tangents are not equal and are extended as dashed lines to intersect the voltage axis at threshold voltages that differ only slightly. The reciprocal of the slope at each point represnts *junction resistance $r_j$*. The slope is valid for only small changes of voltage and current around the particular point of interest. For example, in Fig. 1-13(a), $r_j$ equals 50 $\Omega$ when *I* decreases to 0.5 mA. Rather than evaluate $r_j$ graphically from the technique shown in Fig. 1-12 we can find a simple expression for the slope $\Delta V/\Delta I$ from Eq. (1-1); that is,

$$\frac{\Delta V}{\Delta I} = \frac{\frac{kT}{q}}{I} = \frac{25\ \text{mV}}{I} = r_j \tag{1-4}$$

Equation (1-4) shows that $r_j$ depends on the dc forward current through the diode. The combination $kT/q$ has a value of 25 mV at room temperature. The resulting model is shown in Fig. 1-13(b) to consist of an ideal diode in series with a threshold voltage and junction resistance $r_j$. If a diode characteristic is not available it is common practice to assume $V_T = 0.2$–0.3 V for germanium diodes and $V_T = 0.6$–0.7 V for silicon diodes. Junction resistance will be considered further in Section 1-6.

A reverse-biased diode is modeled by a *reverse resistance $R_r$* shunting an ideal-diode symbol in Fig. 1-14. Manufacturer's data sheets often specify one point on the reverse-bias characteristic (third quadrant in Fig. 1-11) at a high value of *reverse voltage* $V = V_R$ with the corresponding value of *reverse current* $I = I_R$. The reverse resistance is found from

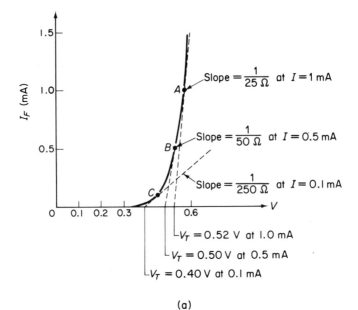

(a)

(b)

**Fig. 1-13.** The forward diode characteristic of Fig. 1-11 is expanded in (a) to amplify the low current region and its circuit model is shown in (b).

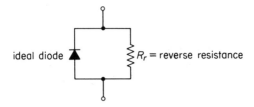

**Fig. 1-14.** Circuit model for a reverse-biased diode.

$$R_r = \frac{V_R}{I_R} \tag{1-5}$$

An ohmmeter can be used as a rapid, although inaccurate, diode tester. The ohmmeter will read approximately $R_R$ when connected to reverse bias the diode. With a forward-bias connection the ohmmeter will indicate a value approximately equal to $r_b$.

*Example 1-1.* A diode parameter listing shows the standard format for a point on the forward characteristic as $I_F/V_F$ in units of milliamperes per volt. (a) Find the large-signal model under forward bias for the general-purpose germanium diode 1N567 when given $I_F/V_F = 150/1$. (b) Repeat for the silicon diode 1N676 with $I_F/V_F = 200/1$.

*Solution*

(a) From Eq. (1-3) and assuming $V_T = 0.3$ V,

$$r_b = \frac{V_F - V_T}{I_F} = \frac{1 - 0.3}{150 \times 10^{-3}} = 4.7 \ \Omega$$

(b) Assuming $V_T = 0.7$ V,

$$r_b = \frac{V_F - V_T}{I_F} = \frac{1 - 0.7}{200 \times 10^{-3}} = 1.5 \ \Omega$$

*Example 1-2.* A diode parameter listing shows $I_R/V_R$ in microamperes per volt. Find the circuit model under reverse-bias conditions for (a) the 1N567 with $I_R/V_R = 150/100$ and (b) the 1N676 with $I_R/V_R = 1/100$.

*Solution*

(a) From Eq. (1-5),

$$R_r = \frac{100}{150 \times 10^{-6}} = 660 \ \text{k}\Omega$$

(b) From Eq. (1-5),

$$R_r = \frac{100}{1 \times 10^{-6}} = 100 \ \text{M}\Omega$$

*Example 1-3.* Given $I_s = 1 \ \mu\text{A}$ for a germanium diode and $I_s = 1$ nA for a silicon diode. Both diodes obey the diode equation. Find the diode current if each diode is forward biased to (a) $V = 0.075$ V, (b) $V = 0.3$ V. Assume the diodes are at room temperature and $kT/q = 25$ mV.

*Solution.* The power of $e$ in Eq. (1-1) can be expressed as

$$\frac{V}{\frac{kt}{q}} = \frac{V}{0.025}$$

20   Introduction to Semiconductors                                Chap. 1

We can then tabulate the calculations as in Table 1-4.

Table 1-4

|     | V (volts) | V/0.025 | exp (V/0.025) | Ge I | Si I |
|-----|-----------|---------|---------------|------|------|
| (a) | 0.075 | 3 | 20 | 19.0 μA | 19.0 nA |
| (b) | 0.30 | 12 | 163 × 10³ | 163 mA | 163 μA |

*Example 1-4.* For the same diodes as in Example 1-3, find the diode current if each is reverse biased to (a) $V = -0.100$ V, (b) $V = -0.125$ V.

*Solution.* Tabulate calculations from Eq. (1-1) from the data in Table 1-5.

Table 1-5

|     | V (volts) | V/0.025 | exp (V/0.025) | Ge I | Si I |
|-----|-----------|---------|---------------|------|------|
| (a) | −0.100 | −4 | 0.018 | −0.98 μA | −0.98 nA |
| (b) | −0.125 | −5 | 0.006 | −1 μA | −1 nA |

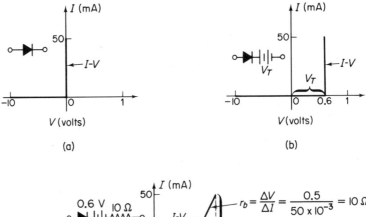

Fig. 1-15. Solutions to Example 1-5.

*Sec. 1-6*  *Sweep Measurement of the Diode*

We conclude from Example 1-4 that reverse-bias current is constant and equal to $I_s$ for reverse-bias voltages greater than 0.1 V. In Example 1-3 we see that forward current is related exponentially to forward-bias voltage for voltages exceeding $3kT/q$. Finally, in Examples 1-1 and 1-2, the diode's forward bulk resistances are very low as compared to their reverse resistances, which are very high.

*Example 1-5.* Draw the circuit model and characteristic curve for (a) an ideal diode, (b) an ideal diode with $V_T = 0.6$ V and $r_b = 0$, and (c) an ideal diode with $V_T = 0.6$ V and $r_b = 10$ Ω.

*Solution.* The solutions are shown in Fig. 1-15.

## 1-6 Sweep Measurement of the Diode

It is assumed that the student has had an introduction to the cathode-ray oscilloscope (CRO) and understands that the CRO is most commonly used to measure and display voltage. An electron beam is passed between a pair of vertical and a pair of horizontal plates to cause a bright spot on the screen of the CRO. The spot may be moved to any point on the screen by positioning controls to establish a reference or zero position.

Vertical plates are controlled by a $y$ amplifier that has adjustable sensitivity and two input terminals designated $+y$ and $-y$, respectively. Assume that 100 mV are applied to make the $+y$ terminal more positive than the $-y$ terminal. If sensitivity of the $y$ amplifier is set at 10 mV per division (10 mV/div) the spot on the CRO face will deflect upwards 10 divisions from its zero-voltage position, to point $A$ in Fig. 1-16. We mentally calibrate the $y$ axis at 10 mV/div and read the 10 division deflection of point $A$ as $V = 100$ mV. The CRO may also be used as an ammeter because if the value of

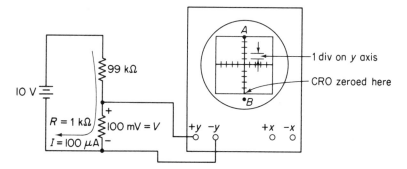

**Fig. 1-16.** Voltage and current measurement with a cathode-ray oscilloscope (CRO).

external circuit resistance across the *y* terminals is known, we can calculate the current. For example, a 1-div deflection with a sensitivity of 10 mV/div from $R = 1000\,\Omega$ means a current sensitivity of 10 mV ÷ 1000 $\Omega$ = 10 $\mu$A/div. Thus the *y* axis may also be calibrated as 10 $\mu$A/div and the deflection to *A* is read as 10 div × 10 $\mu$A/div = 100 $\mu$A. If connections to the *y* input terminals are reversed, and sensitivity is decreased to 100 mV/div the spot will deflect down by 1 div to point *B* in Fig. 1-16, and the *y* axis may be calibrated as 100 mV/1000 $\Omega$ or 0.1 mA/div.

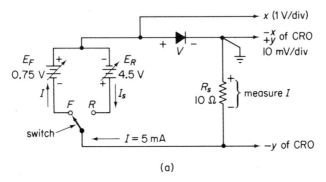

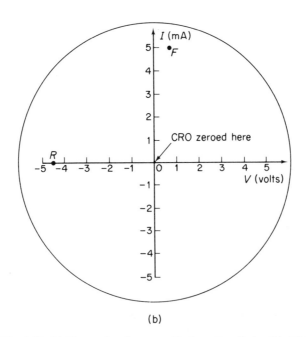

**Fig. 1-17.** (a) Forward and reverse-biasing of a diode. (b) CRO measurements of the operating points of (a).

The same general remarks hold for the $x$ amplifier except that making the $+x$ input positive with respect to the $-x$ input deflects the spot to the right. Usually the $x$ amplifier is arranged to be controlled either externally by the $+x$ and $-x$ inputs or can be controlled by an internal sweep, which sweeps the spot from left to right at a constant rate so the $x$ axis can be cali-

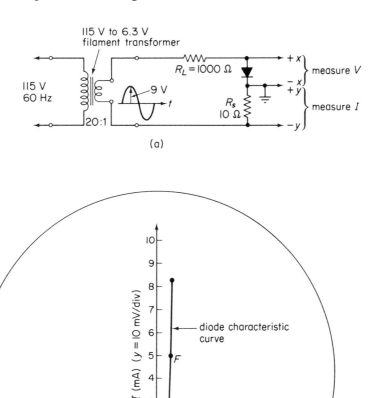

Fig. 1-18. The sweep circuit in (a) generates a characteristic curve for the diode on the CRO screen shown in (b).

brated by a time base. When a varying voltage is applied to the $y$ inputs and the $x$ amplifier is switched to internal sweep, the CRO displays the voltage wave as it varies with time. *In all our sweep measurements the $x$ amplifier will be switched to external sweep.*

We employ the CRO to measure, simultaneously, current through a diode and corresponding voltage across the diode by the test circuit of Fig. 1-17(a). $R_s$ is a current-sensing resistor, so a $y$ amplifier sensitivity of 10 mV/div allows us to calibrate the $y$ axis as $10 \text{ mV}/10 \text{ }\Omega = 1$ mA/div in Fig. 1-17(b). The $x$ amplifier is set to 1 V/div and the CRO is zeroed (with both inputs short circuited) to the center of the screen. With the switch on forward bias ($F$), to apply a forward-bias voltage $E = 0.75$ V, the spot deflects to point $F$ in Fig. 1-17(b). The diode is operating at 5 mA, with 0.70 V across it, and point $F$ is therefore called the *operating point*. With the switch on reverse bias ($R$), the CRO locates operating point $R$ at $I = 0$ mA, $V = 4.5$ V. The polarity of $V$ is opposite to that shown in Fig. 1-17(a) so the spot deflects left. $I_s$ would deflect the spot down except that the reverse saturation current is of the order of 1 $\mu A$ and is insufficient to be seen.

If we reduced $E_R$ to 0 with the switch on point $R$ in Fig. 1-17(a), the spot would move from point $R$ to the origin in Fig. 1-17(b). Throwing the switch from $R$ to $F$ and increasing $E_F$ from 0 to 0.75 V would move the spot from the origin to point $F$. This procedure would have traced all possible operating points for the diode between points $R$ and $F$. There is a simple arrangement of commonly available parts that will perform this procedure continuously so that the spot traces all possible operating points to display the diode's characteristic curve on the CRO.

In Fig. 1-18(a) an ordinary filament transformer provides a sweep voltage that varies sinusoidally and reverses polarity 60 times each second. During the forward-voltage half-cycle, resistor $R_L$ is required to limit current through the diode because the filament transformer's rms rating of 6.3 V will sweep to a peak of $6.3 \times 1.41 \simeq 9$ V. The resulting display on the CRO is shown in Fig. 1-18(b) and represents the diode's *I-V* characteristic curve. Amplifier $y$ and $x$ sensitivities of 10 mV/div and 1 V/div, respectively are also shown in Fig. 1-18(b), with the $y$-axis calibrated at $10 \text{ mV}/10 \text{ }\Omega = 1$ mA/div.

Considerable inprovement can be made in the sweep circuit of Fig. 1-18. For example, the reverse characteristic extends off the screen and in some critical applications we may need to expand the forward characteristic for better accuracy in reading $V$. We employ the diode's rectifying properties in Fig. 1-19 to allow measurement of another diode's forward characteristic by eliminating the reverse-bias sweep. In Fig. 1-19(a) diode $D_s$ blocks the reverse-bias half-sinusoid and allows only a variable forward-bias voltage to sweep the diode's forward characteristic. We gain an advantage whereby we can expand the sensitivity of the CRO's $x$ amplifier to 0.1 V/div

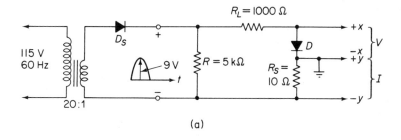

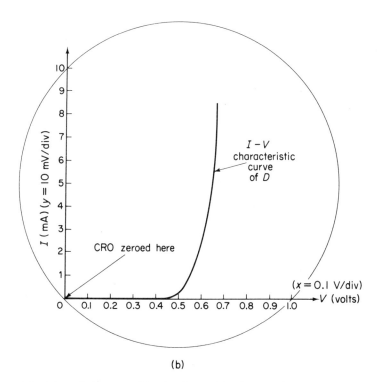

**Fig. 1-19.** Diode $D_s$ rectifies the sinusoidal voltage in (a) to apply positive-going sweeps to diode $D$. The forward characteristic is displayed on the CRO screen in (b).

and expand the display area by zeroing the CRO at the left bottom corner. Resistor $R$ insures that most of the reverse-sweep voltage is developed across $D_s$ so that we do not get a negative deflection on the $x$ axis. Resistor $R_L$ is again required or the diode will burn out.

Rectifying properties of diode $D_s$ are applied in the sweep circuit of Fig. 1-20(a) to observe the maximum allowable reverse-bias or *reverse-breakdown* voltage $V_{BD}$. The $V_{BD}$ of diode $D_s$ should be larger than the $V_{BD}$

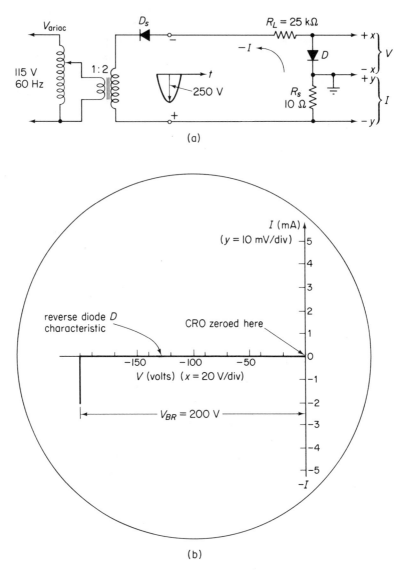

**Fig. 1-20.** $D_s$ allows only the negative sweep voltage in (a) to be applied to diode $D$. In (b) we measure a reverse-breakdown voltage of $V_{BR} = 200$ V.

of diode $D$. We see from the CRO trace in Fig. 1-20(b) that diode current is negligible at $I_s \lesssim 1$ μA until the applied voltage reaches 200 V. Most of this 200 V exists across the depletion region, creating a strong electric field which accelerates minority carriers to a velocity sufficient to rupture covalent bonds by collision with atoms. Electrons and holes freed from the broken covalent bonds can be accelerated to free more carriers. This regenerative process is known as *avalanching* and is accompanied by an abrupt increase in reverse current.

*Example 1-6.* In the circuit of Fig. 1-19(a), what value of current will flow through diode $D$ when the instantaneous sweep voltage equals 9 V across resistor $R$?

*Solution.* $R_s$ is negligible with respect to $R_L$, so $R_L$ would limit diode current to $9/1000 = 9$ mA if the diode voltage drop were 0. However, from Fig. 1-19(b) we see that the diode drop above $I \simeq 5$ mA is approximately 0.7 V. Therefore the net current through $R_L$ is $(9 - 0.7)/1000 = 8.3$ mA, and is verified at the tip of the diode curve.

*Example 1-7.* When the sweep voltage in Fig. 1-20(a) is at its peak of 250 V, what reverse current flows through diode $D$?

*Solution.* From the characteristic curve in Fig. 1-20(b) we see that voltage across diode $D$ is 200 V when it is in the breakdown condition. At this instant, voltage across $D_s$ will be negligible at a few tenths of a volt. Since $R_s$ is negligible with respect to $R_L$ there will be 250 V on one side of $R_L$ and 200 V on the other. Therefore current through $R_L$ and diode $D$ will be $(250 - 200)/25,000 = 2$ mA, and is verified at the tip of the reverse characteristic curve in Fig. 1-20(b).

## 1-7 Sweep Measurement of Diode Incremental Resistance

In many applications of semiconductor devices we encounter a situation where an incremental voltage is superimposed on a dc voltage. By incremental we think of a change in voltage or ac, and often employ a sinusoidal voltage as an incremental voltage generator. For example, in Fig. 1-21 an ac voltage $E_g$ with a peak value of 1 V and a frequency of 100 Hz is coupled through a 1:1 transformer to appear unchanged, in series with $E$, $R_L$, $R_s$ and the diode.

The amount of incremental current flow due to $E_g$ will be dependent on the total series ac or incremental resistance of the circuit. The internal ac or dc resistance is assumed to be negligible for battery $E$ and the transformer's secondary. Under normal circumstances $R_L$ and $R_s$ present 1000 Ω and 10 Ω, respectively, to either $E_g$ or $E$. In calculations involving dc current flow we can estimate the dc voltage drop across the diode and therefore do not need a dc resistance concept for the diode. The ac diode resistance has already been introduced, as junction resistance $r_j$ and bulk resistance $r_b$, in Eqs. (1-2) and (1-4). At low forward current $r_j$ dominates and at high forward current $r_b$ dominates; the diode's incremental resistance may be expressed *generally* as $r_d$, where

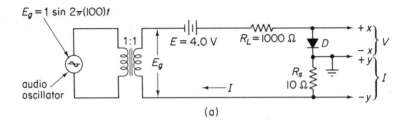

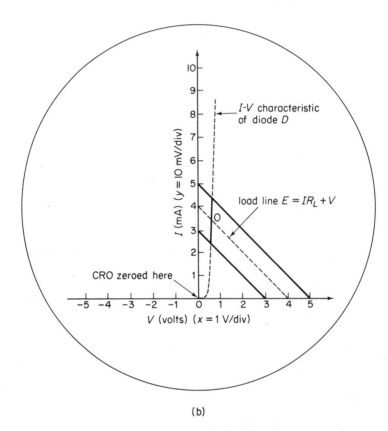

**Fig. 1-21.** Battery $E$ and $R$ in (a) establish dc operating point O in (b). Sinusoid $E_g$ sweeps an operating path in (b) whose slope is the reciprocal of the diode's incremental resistance.

$$r_d = r_j + r_b \tag{1-6}$$

By connecting the CRO to plot total instantaneous current against total instantaneous voltage in the circuit of Fig. 1-21(a), not only can we locate the dc operating point but we can also measure incremental resistance at the operating point.

Assume that the signal voltage $E_g$ is adjusted to 0 in Fig. 1-21(a). We can estimate the diode's operating point, O, by reasoning that the diode is forward biased and will carry a current of approximately $E/R_L = 4.0$ mA. The diode's voltage drop will be approximately 0.7 V at this current and we can get more accuracy by including the diode drop to find $I = (4.0 - 0.7)/1000 = 3.3$ mA.

There is a graphical method of finding the operating point current by writing the dc loop equation for Fig. 1-21(a) (resistance of $R_s$ and transformer secondary are negligible).

$$E = IR_L + V \tag{1-7}$$

Equation (1-7) is known as the *load line* and by plotting it as a dashed line on Fig. 1-21(b) we superimpose restrictions due to the external circuit on the diode's characteristic. Intersection of the load line and characteristic curve locates operating point O, and the coordinates agree with these calculated above.

When $E_g$ is adjusted to a peak value of 1 V, the total series voltage will vary between $E + E_g$ and $E - E_g$ or between 3.0 and 5.0 V. Drawing solid load lines between these two extremes in Fig. 1-21(b) locates the extreme excursions of operating point O. The locus of all possible operating points is traced out by the CRO as an *operating path* shown by the heavy line in Fig. 1-21(b). The slope of the operating path is the reciprocal of the diode's incremental resistance $r_d$. The CRO has traced the operating path easily for us but the slope is too steep for accurate measurement. We could switch the $x$ amplifier to a sensitivity of 0.1 V/div as in Fig. 1-19 and measure the slope by a tangent to the diode characteristic at $I = 3.3$ mA to obtain a more accurate measurement of incremental resistance.

If $E$ is increased in Fig. 1-21 the load line will shift right with slope unchanged and operating point O will shift up along the diode's characteristic. Decreasing $R_L$ will pivot the load line around its $x$-axis intercept at $V = E$, to increase its slope and shift point O up along the load line.

*Example 1-8.* Calculate the ac current resulting from $E_g$ in Fig. 1-21(a) and the peak-to-peak (p/p) voltage swing across the diode.

*Solution.* Since dc operating point current is low, $r_d = r_j$ and from Eq. (1-4), $r_j = (0.025)/(0.0033) = 7.6 \, \Omega$. Peak ac current $I_{gp}$ is found from

$$I_{gp} = \frac{E_g}{r_j + R_L + R_s} = \frac{1}{7.6 + 1000 + 10} = \cong 1 \text{ mA}$$

and p/p diode voltage $V_{p/p}$ is

$$V_{p/p} = 2I_{gp}r_d = 2 \times 10^{-3} \times 7.6 = 15 \text{ mV}$$

From Example 1-8 we verify the p/p incremental current swing of 2 mA shown in Fig. 1-21 and observe that the p/p voltage swing is too small to be measured accurately. An important lesson is to be learned from Example 1-8 and Fig. 1-21. We have employed a principle of superposition in analyzing a circuit having both dc and ac sources; that is, the dc operating-point current was evaluated independent of the ac signal voltage. Then we evaluated the ac current independently of the dc voltage. One evaluation can then be superimposed on the other if it is necessary to know the total current due to both sources at any instant. In most of our semiconductor work we shall apply this superposition principle by (1) replacing the ac source by its internal resistance and solving the dc problem, and (2) replacing the dc source by its internal resistance and solving the ac problem. Thus in calculating $I_{gp}$ in Example 1-8 we would model the diode simply by the incremental resistance $r_d$.

*Example 1-9.* In the circuit of Fig. 1-21(a), $R_L$ is changed to 5000 $\Omega$, $E$ to 5.0 V and the characteristic of $D$ is given by Fig. 1-19(b). (a) Find the operating-point current. (b) Describe a circuit to model the incremental resistance presented to $E_g$.

*Solution.* (a) Without the diode, dc current would be approximately $E/R_L = 5/5000 = 1$ mA. From Fig. 1-19(b) we read $V = 0.55$ V at $I = 1$ mA. Therefore the operating-point current will be $I = (E - V)/(R_L + R_s) = (5 - 0.55)/5010 = 0.89$ mA.
(b) Evaluate $r_j$ from Eq. (1-4), $r_j = (25 \times 10^{-3})/(89 \times 10^{-5}) = 28\ \Omega$. Draw the ac signal voltage in series with 5000 $\Omega$, 10 $\Omega$, and 28 $\Omega$ to model the ac problem completely.

The model in Example 1-9(b) would be useful, for example, in determining the ac voltage drop across the diode directly from the voltage divider made up of $R_L$, $R_s$, and $r_j$. Thus peak incremental diode voltage $V_p$ would be $V_p = E_g(r_j)/(r_j + R_L + R_s) = 1(28)/5038 = 5.6$ mV.

## 1-8 Diode Applications in Meter Circuits

The forward *I-V* characteristic of a silicon diode makes it particularly effective for protection of meter movements. In Fig. 1-22 two silicon diodes are connected across a basic 50-$\mu$A meter movement. At full deflection of

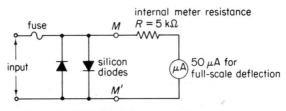

**Fig. 1-22.** Meter protection.

the meter a voltage drop of $(50 \times 10^{-6})(5 \times 10^3) = 250$ mV is developed across terminals $MM'$. This magnitude of voltage is insufficient to cause the forward-biased diode to conduct a current which is significant with respect to 50 $\mu$A. When an overvoltage is applied to the input terminals one of the diodes will turn on and limit voltage across $MM'$ to 0.7 V and limit meter current to $0.7/5000 = 140$ $\mu$A. The conducting diode will continue to shunt current around the meter movement until the fuse opens.

Rectifying properties of the diode are employed to construct a peak-reading voltmeter in Fig. 1-23. When terminal $A$ is positive with respect to

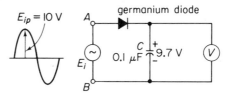

**Fig. 1-23.** Peak-reading voltmeter.

$B$ the diode conducts and capacitor $C$ charges to the peak of the input voltage. Since voltmeter $V$ is a large resistance, the capacitor holds its charge between peaks, and $V$ reads peak voltage minus the 0.3-V diode drop. The voltmeter face can be calibrated to 10 V to minimize the diode voltage difference.

Two diodes are employed in the circuit of Fig. 1-24 to construct a peak-to-peak reading voltmeter. During the negative-going half-cycle of

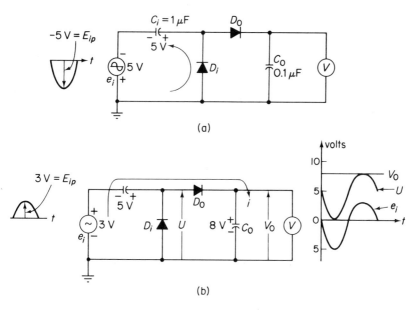

**Fig. 1-24.** Voltage doubler or peak-to-peak voltmeter. Negative and positive input voltages are shown in (a) and (b), respectively.

32  Introduction to Semiconductors  Chap. 1

$E_i$ in Fig. 1-24(a), diode $D_i$ charges $C_i$ to the positive peak input voltage of 5 V. During the positive-going half-cycle of $E_i$ in Fig. 1-24(b), $E_i$ and the voltage across $C_i$ are series-aiding and charge capacitor $C_0$ to 8 V. The resistance of voltmeter $V$ is so high that $C_0$ will hold its charge for a long period, and the voltmeter reads the peak-to-peak voltage of 8 V. If $E_i$ is a symmetrical sine wave the dc voltage across $C_0$ is double the peak input voltage. Consequently this diode-capacitor combination is called a *voltage doubler* and is frequently used as an ac-dc rectifier when small load currents are required. It is an economical circuit because voltage can be stepped up without using an expensive transformer.

Wave shapes are also shown in Fig. 1-24(b) for voltages measured with respect to ground. Observe that voltage $V$ is clamped to ground by diode $D_i$. That is, if $D_i$ is a germanium diode, its cathode can never become more negative than 0.3 V or essentially ground. We conclude that $D_i$ clamps the negative peak of $E_i$ to ground potential so that $V$ varies between 0 and $+8$ V. For this reason the $D_i$-$C_i$ arrangement is called a *clamp circuit* or *dc-restorer*.

## 1-9  The Zener Diode

A *Zener* diode is designed to operate continuously in the reverse-breakdown condition. By increasing doping levels the potential barrier is raised so that the electric field in the depletion region can easily reach a strength of $1 \times 10^6$ V/cm. A few volts of reverse bias will then increase the field strength sufficiently to rupture covalent bonds directly within the depletion region, without the aid of high-speed minority carriers. This direct rupture of covalent bonds is the Zener breakdown mechanism. The value of reverse breakdown at which direct rupture occurs is called the *Zener voltage* $V_Z$. Technically, the Zener region extends from 0 to 5 V of reverse bias. Between 5 and approximately 8 V the breakdown voltage is due to a combination of both Zener action and avalanching while above 8 V the breakdown is due predominantly to avalanching. Under forward bias the Zener has the same $I$-$V$ characteristic as a normal diode.

A typical characteristic of a 6-V Zener diode is shown in Fig. 1-25(a) and may be obtained from the sweep circuit in Fig. 1-25(b). No appreciable current flows through the reverse-biased Zener until the Zener voltage $V_Z$ is reached. As $V$ is increased beyond $V_Z$ an appreciable reverse current $I_Z$ flows and since $V$ increases slightly with increasing $I_Z$ we see that the Zener has a small incremental resistance $r_Z$. This Zener resistance may be evaluated graphically from the slope of the line between points $AA'$ in Fig. 1-25(a) or from

$$r_Z = \frac{\Delta V}{\Delta I} = \frac{0.25}{90 \times 10^{-3}} = 2.8 \, \Omega \tag{1-8}$$

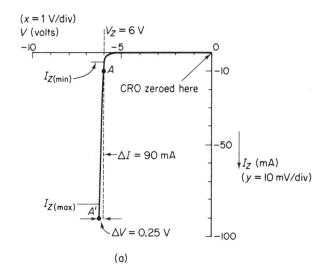

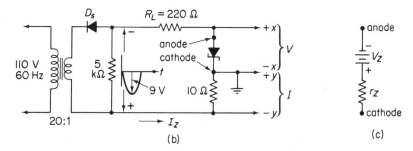

**Fig. 1-25.** The reverse characteristic of a Zener diode in (a) is swept by the circuit of (b) and the Zener is modeled in (c).

A model is constructed from our measured characteristic and in Fig. 1-25(c) is shown to consist of a battery, to symbolize Zener breakdown voltage $V_Z$, in series with the small incremental Zener resistance $r_Z$.

The Zener diode is used primarily in applications that require regulation of voltage to a constant value and Zener diodes are available with $V_Z$ ratings from a few volts to tens of volts. Circuit applications for the Zener must ensure that it is always furnished with enough current to maintain Zener operation at $I_{Z(\min)}$, which is typically equal to a few miliamperes. The maximum current rating $I_{Z(\max)}$ will be determined by heat dissipation of the Zener and is specified by the manufacturer in terms of maximum power dissipation $P_{d(\max)}$. A typical value of $P_{d(\max)}$ for a small Zener diode would be 0.5 W so that in Fig. 1-25(a) we could evaluate $I_{Z(\max)}$ from

$$I_{Z(\max)} = \frac{P_{d(\max)}}{V_Z} = \frac{0.5}{6} = 83 \text{ mA} \qquad (1\text{-}9)$$

Regulating action by the Zener is introduced in the basic circuit of Fig. 1-26(a) where the Zener symbol is replaced by its circuit model. For purposes of analysis it is convenient to show the Thévenin equivalent circuit presented to the Zener branch in Fig. 1-26(b), where

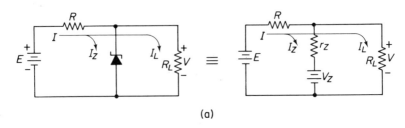

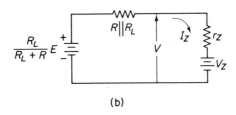

**Fig. 1-26.** The Zener diode and its model are shown in the voltage regulator circuit of (a) and the Thévenin equivalent circuit is shown in (b).

$$\text{Thévenin voltage} = \frac{R_L}{R_L + R} E \qquad (1\text{-}10)$$

$$\text{Thévenin resistance} = R \| R_L = \frac{R R_L}{R + R_L} \qquad (1\text{-}11)$$

Usually load current is specified at some maximum value of $I_L = I_{L(\max)}$. Zener current will be at its minimum value $I_{Z(\min)}$ when maximum load current is flowing.

If we can approximate $I_{Z(\min)} = 0$, then from Fig. 1-26(b) we can maintain regulation if

$$E \frac{R'_L}{R'_L + R} \geq V_Z \qquad (1\text{-}12)$$

where $R'_L = V/I_{L(\max)}$ and $V \cong V_Z$.

Often we cannot make this approximation and must write the inside loop equation from the model in Fig. 1-26(a). Assuming $I_{Z(\min)} r_Z$ is negligible we can maintain regulation if

$$E \geq (I_{Z(\min)} + I_{L(\max)})R + V_Z \qquad (1\text{-}13)$$

Maximum Zener current $I_{Z(\max)}$ will flow through the Zener when the load resistor $R_L$ is removed in Fig. 1-26(a). $E$ and $R$ must then restrict Zener current to or below this value from the loop equation

$$E \leq (I_{Z(\max)})(R + r_Z) + V_Z \qquad (1\text{-}14)$$

*Example 1-10.* Given $I_{Z(\min)} = 10$ mA, $I_{Z(\max)} = 100$ mA, $E = 9$ V, $r_Z = 3\,\Omega$, and $V_Z = 6$ V in Fig. 1-26. For what range of load current variation will this circuit maintain regulation at $V \cong 6$ V?

*Solution.* From Eq. (1-14), $R$ must have a minimum value of

$$9 \leq (100 \times 10^{-3})(R + 3) + 6$$
$$R \geq 27\,\Omega$$

Solve for $I_{L(\max)}$ from Eq. (1-13):

$$9 = (0.010 + I_{L(\max)})27 + 6$$
$$I_{L(\max)} = 101 \text{ mA}$$

The Zener will regulate $V$ to approximately 6 V for all values of $I_L$ between 0 and 101 mA.

# PROBLEMS

*1-1* A diode has a threshold voltage of 0.6 V and has a voltage drop of 0.7 V when it conducts 50 mA. (a) Is the diode made from silicon or germanium? (b) Show the polarity of voltage across the diode when it conducts 70 mA. (c) Is the diode forward or reverse biased in (b)? (d) What is its bulk resistance?

*1-2* A diode conducts 2 $\mu$A with either 200 V or 100 V across it. (a) Is the diode forward or reverse biased? (b) Sketch the polarity of voltage drop across the diode symbol and indicate the $p$ and $n$ sides. (c) What electrical element could model the diode and what would be its value for each voltage?

*1-3* In Example 1-3 both diodes are forward biased by 25 mV. What are the forward currents?

*1-4* In Example 1-4, both diodes are reverse biased by 10 V. What are the reverse currents?

36   *Introduction to Semiconductors*   Chap. 1

*1–5* Draw a circuit model and characteristic curve for an ideal diode with $V_T = 0.7$ V, $r_b = 2\,\Omega$. What is the voltage drop across this diode when it conducts 100 mA?

*1–6* What is the junction resistance of a diode carrying (a) 1 mA; (b) 100 $\mu$A; (c) 10 $\mu$A?

*1–7* Sketch the CRO trace in Fig. 1-17 if (a) $E_F$ is gradually reduced to 0; (b) if the switch is thrown to $R$ and $E_R$ is gradually reduced to 0. (c) Repeat (a) and (b) with the $x$ connection removed. (d) Repeat (a) and (b) with the $y$ connection removed.

*1–8* Sketch the CRO display for Fig. 1-18 if (a) the diode is replaced by a 100-$\Omega$ resistor and (b) the diode is reversed.

*1–9* In Example 1-6, what value of current will flow through diode $D$ when the instantaneous sweep voltage equals 10 V; 0.1 V?

*1–10* In Example 1-7, what current flows through $D$ when (a) the instantaneous sweep voltage equals 300 V; (b) when $R_L = 50\,\mathrm{k}\Omega$ at 250 V?

*1–11* The value of $E$ is changed to 3 V in Fig. 1-21. (a) What is the dc value of $I$? (b) Find the diode's small-signal resistance, ac current, and ac voltage drop in p/p values.

*1–12* Repeat Example 1-9 with $R_L$ and $E$ doubled to 10 k$\Omega$ and 10 V. Find the signal voltage drops across $R_L$, $R_S$, and $r_d$.

*1–13* Sketch the wave shapes for $V$ and $V_0$ in Fig. 1-24 if $e_i$ is a sinusoid with peak values of 10 V.

*1–14* Sketch the wave shapes for Fig. 1-25 if (a) $D_s$ is short circuited; (b) $D_s$ is reveresd.

*1–15* Sketch the characteristic for a Zener diode with $V_Z = 5$ V and $r_z = 5\,\Omega$.

*1–16* What is the maximum Zener current in Problem 1-15 if $P_{d(\max)} = 1$ W.

*1–17* What is the Zener current in Example 1-10 ($R = 27\,\Omega$) for load currents of 0 and 100 mA? What power is dissipated by the Zener for each condition?

*1–18* If $R_L = 100\,\Omega$ in Fig. 1-26, $V_Z = 6$ V, $r_Z = 2\,\Omega$, $P_{D(\max)} = 1$ W, $R = 30\,\Omega$ and $I_{Z(\min)} = 5$ mA, (a) What is the highest value of $E$ that will not exceed $P_{D(\max)}$? (b) What is the lowest value of $E$ that will maintain regulation?

*1–19* A sinusoidal voltage with peak value of 1 V and frequency of 100 Hz is applied to the input of Fig. 1-22. (a) If the meter is removed, sketch the voltage wave appearing between terminals $MM'$ if $V_T = 0.6$ V.

(b) On same drawing show the wave shape at $MM'$ if germanium diodes are substituted with $V_T = 0.3$ V.

*1–20* If a second *identical* diode is connected in parallel and with the same polarity as the diode in Fig. 1-18, show that the current through $R_s$ will remain approximately unchanged for forward bias and doubles for reverse bias.

*1–21* In Problem. 1-20 show that (a) forward voltage across the diodes is almost the same although slightly lower than the case with one diode, and (b) reverse voltages are the same whether one diode or two diodes are present.

*1–22* In Problem. 1-20, show that forward current through each diode is one-half and reverse current is equal to the single diode values.

*1–23* If a second identical diode is connected in parallel with but opposite in polarity to the diode of Fig. 1-18, draw the resulting composite *I-V* characteristic.

*1–24* If a second identical diode is connected in series-aiding (cathode to anode) with the diode in Fig. 1-8, draw the resulting composite characteristic curve.

# Chapter 2

**2-0** INTRODUCTION .................................39
**2-1** STANDARD LETTER SYMBOLS .....................41
**2-2** OPERATING MODES AND CIRCUIT CONFIGURATIONS ...43
**2-3** TRANSISTOR ACTION IN THE COMMON-BASE
  CONFIGURATION ................................46
**2-4** TRANSISTOR ACTION IN THE COMMON-EMITTER
  CONFIGURATION ................................50
**2-5** MEASUREMENT AND DISPLAY OF COMMON-EMITTER
  CHARACTERISTIC CURVES.........................56
**2-6** MEASUREMENT OF COMMON-EMITTER CURRENT GAIN ..59
**2-7** CUTOFF AND SATURATION ON THE CHARACTERISTIC
  CURVES .......................................61
**2-8** MEASUREMENT OF COMMON-BASE CHARACTERISTICS ..64
**2-9** MEASUREMENT OF OTHER USEFUL CHARACTERISTIC
  CURVES .......................................66
  PROBLEMS .....................................72

# Characteristics of Bipolar Transistors and Their Measurements

## 2-0 Introduction

There are two major types of transistors, *field-effect* transistors (FETs) and *bipolar* junction transistors (BJTs). Operation of FETs depends on the control of one type of carrier and will be considered in Chapter 14. Most of this text will be concerned with the bipolar junction transistor, or BJT; so called because its operation depends on the motion of *both* types of charge carriers and the action of two *pn* junctions. BJTs are formed by many processes but each has the same general configuration: a thin region of doped semiconductor material is sandwiched between two regions of oppositely doped semiconductor material. An *n*-type layer sandwiched between two *p*-type layers is designated as a *pnp* transistor, while two *n*-type layers with a *p*-type layer in the middle is an *npn* transistor. The middle layer is called the *base* because early point-contact transistors consisted of two rectifying contacts applied to a base wafer. Point-contact transistors are no longer important and will not be considered. The two outside layers are designated separately as the *emitter* and *collector*. Both are either *p* type or *n* type but usually have different conductivities or doping for reasons which will be considered subsequently. In Fig. 2-1, the basic construction of a *pnp* transistor shows two *pn* junctions, one between the emitter and base and another between the base and collector. At each junction there is a *space-charge* region which controls the passage of charge carriers. Since it is possible to reverse bias or forward bias each junction, there are four possible combinations. Each combination is referred to as a region of operation or *operating mode* and will be discussed in Section 2-2. Operating the emitter junction with a forward bias and the collector junction with a reverse bias is the most common mode of operation. With this type of biasing the transistor can act as an amplifier and is said to operate in the *active mode*.

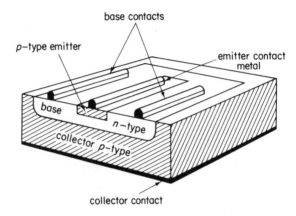

**Fig. 2-1.** Construction of a *pnp* planar-type bipolar junction transistor (not to scale).

In talking about BJTs we shall need a pictorial representation for a physical model and then a circuit symbol. The pictorial representations in Fig. 2-2(a) are not drawn to scale, for the base region is actually very thin.

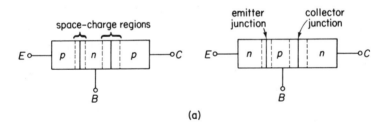

(a)

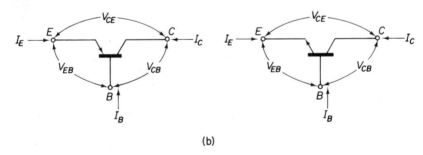

(b)

**Fig. 2-2.** (a) Pictorial and (b) circuit symbols for bipolar transistors.

Often, the space-charge regions are not shown, although they are present. Figure 2-2(b) shows the circuit symbol for a transistor and identifies the emitter by an arrowhead. When the arrowhead points into the base, it indicates a *pn* junction with an *n*-type base. When the arrowhead points into the emitter it indicates an *n*-type emitter. A useful memory aid is to think of the arrowhead as pointing at the *n*-type material, and in the direction of conventional current flow. In this chapter we shall be introduced to the transistor by studing its action from circuit models of the transistor and measurements of its characteristic curves. An understanding of transistor action is a necessary foundation for the study of biasing and small-signal operation in the following chapters.

## 2-1 Standard Letter Symbols

In discussing transistor operation it is helpful if the symbol which represents a variable will tell whether the variable is an ac or dc quantity. Further, if the variable is ac we should know if it is the rms, peak, or instantaneous component. Also it would be desirable if the symbol for a resistance indicated a resistor in the circuit or a transistor parameter. IEEE standard letter symbols (Table 2-1) will be used for the most part.

Table 2-1

| Physical Quantity or variable | Symbol Letter, subscript | Example |
|---|---|---|
| dc | capital, capital | $I_E$, $V_{BE}$ |
| rms | capital, lowercase | $I_e$, $V_{be}$ |
| ac instantaneous | lowercase, lowercase | $i_e$, $v_{be}$ |
| total instantaneous | lowercase, capital | $i_E$, $v_{BE}$ |
| peak or maximum | capital, lowercase with additional subscript $p$ or $m$ | $I_{em}$, $V_{bem}$ |
| circuit elements | capital | $R$, $C_1$ |
| transistor parameters | lowercase or capital | $r_\pi$, $C_\mu$ |

Standard letter symbols are shown applied to an emitter current in Fig. 2-3(a). A discussion of the current directions in Fig. 2-2(b) and the double-subscript notation for the voltage symbols is necessary. First, all dc currents (flow of positive charge) are assumed to be positive when they flow into the device, in both the *pnp* and *npn* transistors. This assumption has nothing to do with actual transistor operation and confuses the student, especially when he tries to connect a milliammeter to measure a current. The standard assumption is shown in Fig. 2-2(b) for the sole purpose of explaining the signs given by manufacturers on their characteristic curve plots. Refer to the circuit symbol of a *pnp* transistor in Fig. 2-2(b). The emitter current is correct as shown, but the dc collector current $I_C$ and the dc base current $I_B$ actually flow out of the transistor. Therefore, if the manufacturer

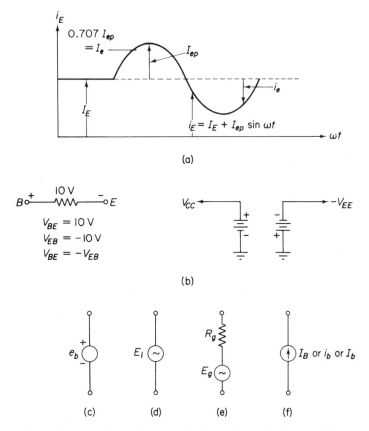

**Fig. 2-3.** Standard notation of variables is shown in (a) and (b). Constant voltage sources in (c), (d), and (e) represent general instantaneous and sinusoidal sources, respectively. Constant-current sources have one symbol in (f).

gives a plot of the dc base current versus base-emitter voltage, he must show values of the base current with a minus sign. For the *npn* transistor, the actual emitter current is reversed from that shown. In this text, the *actual current directions*, as would be measured by a milliammeter, will be shown on the circuit schematic and the loop equations written accordingly. As will be shown, $I_E = I_B + I_C$ for both the *pnp* and *npn* transistors. We shall *not* write $I_E + I_C + I_B = 0$ and then debate about which variables should be assigned minus signs.

The terminal voltage variables $V_{CB}$, $V_{EB}$, and $V_{CE}$ of Fig. 2-2(b) are formed by a capital letter with *two* capital subscripts to signify not only a dc variable but also to tell something about the polarity of voltage between the terminals designated by the subscripts. The first subscript represents the terminal which is assumed to be more positive, so that the equality $V_{CE} = 10$ V means that the collector is more positive than the emitter by 10 V.

Fig. 2-3(b) shows this convention. To read this value, the red-positive terminal of a voltmeter should be connected to the collector and the black-negative terminal of a voltmeter to the emitter for an upscale reading. On the other hand, to say $V_{CE} = -10$ V means that the collector is more negative than the emitter by 10 V and the voltmeter leads must be reversed for an upscale reading. However, once we get beyond the study of characteristic curves, it is far less confusing to the student if we (a) always draw the circuit schematic, (b) show the *actual* conventional current directions, (c) show the expected polarity of voltage drops on the drawing, and (d) write the loop equations accordingly. The voltage between terminals, for example $V_{BE}$ or $V_{CE}$, will indicate magnitude only; the sign will be obtained from the schematic.

Constant current sources and constant voltage sources are shown in Fig. 2-3. In (c) and (d), the voltage sources are ideal with zero internal resistance. This means they will supply as much current as required with no change in terminal voltage. Where the internal resistance of the source is important, it is shown by (e). Only one symbol is used to represent a current source in (f), with the arrowhead pointing in the direction of conventional current flow. The nature of the current source is specified by the letter symbol. Polarities for the instant under consideration are indicated by the plus and minus signs. All other current directions and voltage drops are related to and caused by the instantaneous polarities shown for the voltage sources or current sources.

## 2-2 Operating Modes and Circuit Configurations

Before introducing transistor characteristics we must consider the operating modes and circuit configurations. As already described there are two junctions and each may be either forward or reverse biased for a particular application. There are four possible modes of operation and each has a name (see Table 2-2).

Table 2-2

| Operating Mode | Bias on the Emitter Junction | Bias on the Collector Junction |
|---|---|---|
| active | forward | reverse |
| saturated | forward | forward |
| cutoff | reverse | reverse |
| inverse | reverse | forward |

Switching circuits operate in either the saturated or cutoff mode. That is, the collector-emitter of the transistor acts as a short circuit or as an open circuit when the transistor is saturated or cut off, respectively. Amplifying circuits require operation in the active mode and we shall be concerned primarily with this type of operation. The inverse region is not commonly found.

There are three terminals on the transistor. One lead is selected as a common terminal. An input signal is applied between a second lead and the common terminal. Output is taken between the remaining lead and the common terminal. (Of course the output signal could be extracted from the second lead and the input signal could be injected into the remaining lead.) Once the common terminal is established, two loops are formed linking the common terminal. Either loop may be connected as the input. Thus there are six possible ways of injecting an input signal into the transistor, two for each of three possible selections of a common terminal. It turns out that injection of a signal into a collector terminal or extraction of a signal from a base terminal is never encountered, so there remain only three standard circuit configurations as shown in Fig. 2-4, namely common emitter (*CE*), common base (*CB*), and common collector (*CC*).

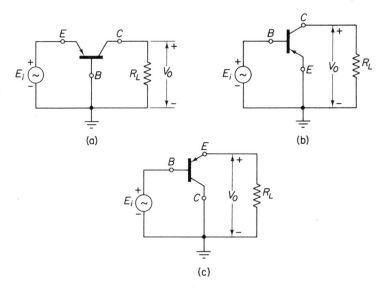

**Fig. 2-4.** Standard circuit configurations are indicated by (a) common base, (b) common emitter, and (c) common collector. Supply voltages are not shown.

The input voltage for all three circuits consists of the sinusoidal signal voltage $E_i$. Voltage $E_i$ is the input signal voltage and an output voltage is developed across load resistor $R_L$. The simplest rule to remember for classifying a circuit configuration is (1) identify the input terminal, (2) identify the output terminal, and (3) the remaining terminal is common and names the classification. Often, the common terminal is grounded and the circuits are classified as grounded emitter or grounded collector. There will be instances of circuits where this classification is meaningless. For example, a common-emitter circuit, where outputs are taken from equal resistors inserted in both the collector and emitter leads, is used as *a phase inverter*. It gives equal output signals which are 180° out of phase. By the time the student has progressed to this level he does not need to think in terms of circuit classification.

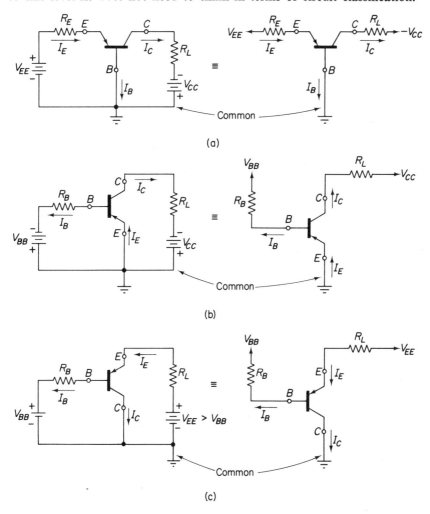

**Fig. 2-5.** Supply-voltage polarities for operation of a *pnp* transistor in the active region in the circuit configuration of (a) common base, (b) common emitter, and (c) common collector.

The base-emitter junction and collector-base junction must be forward and reverse biased, respectively, for operation in the active mode. Biasing voltages are furnished by power supplies and must establish the proper operating conditions for the transistor so that it can handle the input and output signals of Fig. 2-4. Basic supply potentials are connected in Fig. 2-5 to show the required polarity to bias a *pnp* transistor in the active region for all three circuit configurations. For an *npn* transistor, reverse all the supply voltage polarities and all the current directions.

## 2-3 Transistor Action in the Common-Base Configuration

When a transistor is used as an amplifier or control device the emitter junction is forward biased and the collector junction is reverse biased for operation in the active mode. The emitter is doped with a greater density of doping atoms than the base. Therefore most of the current crossing the emitter space-charge region is due to majority carriers from the emitter. For a *pnp* BJT, forward biasing the emitter junction injects a steady stream of holes across the emitter space-charge region, from emitter to base, where they become minority holes in the base region. Practically all of these holes then diffuse across the very thin base region toward the collector junction. As soon as holes from the base diffuse into the collector space-charge region they are accelerated into the collector body. This action is shown by the majority carrier current $I_{EC}$ in Fig. 2-6. Of course holes from the base enter-

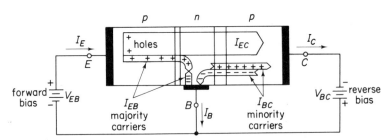

**Fig. 2-6.** Common-base action for a *pnp* transistor in the active region.

ing the emitter space-charge region will be swept back to the emitter, but the diffusion pressure from the large concentration of holes in the emitter maintains a net constant injection of holes into the base. Not quite all of the net number of holes per second injected into the base reach the collector. Less than 1% of the holes combine with electrons in the base. To feed these recombinations, a relatively small current of electrons must be fed into the base from the base contact since there are not enough electrons available in the collector. The recombination current is represented by $I_{EB}$ in Fig. 2-6. (Some of the majority electrons in the base will cross into the emitter and their loss must be made up for by other electrons fed from the base contact. The base is purposely doped lighter than the emitter to minimize the flow of electrons from base to emitter and this normally unimportant component will not be considered further.)

There is another component of current through the reverse-biased collector junction consisting of base minority holes and collector minority electrons. These electrons flowing to the base must be removed from the base through the base contact. This reverse-saturation component is represented by $I_{BC}$ in Fig. 2-6. This minority carrier current is temperature dependent because electrons in the collector are thermally generated from broken covalent bonds. Doping the collector heavily with respect to the base will minimize the number of collector electrons and, therefore, this component of electron current.

Finally, when collector holes reach the collector's ohmic contact they combine with electrons supplied to the collector by the negative battery terminal. Electrons being fed into the collector from the negative terminal of battery $V_{BC}$ correspond to conventional current flowing in the opposite direction so $I_C$ is shown leaving the transistor at the collector terminal. At the emitter ohmic contact, electrons are furnished to leave the emitter and holes are furnished to maintain the hole (majority carrier) density level in the emitter. This corresponds to conventional current entering the transistor at the emitter terminal. For an *npn* transistor, voltage polarities and current directions are reversed in Fig. 2-6 and the signs of the carriers are interchanged.

A circuit model will be developed so that we may learn the relationships between transistor terminal currents. The model will account for the majority carrier currents $I_{EC}$ and $I_{EB}$ as well as for minority carrier current $I_{BC}$.

$I_{BC}$ is the leakage current through the reverse-biased collector junction. It may be identified and measured independently of the majority carrier current by the circuit of Fig. 2-7, where the emitter terminal is left open to eliminate any action by the emitter junction. Collector current equals, in this special case only, the *reverse saturation* or *leakage current* $I_{CBO}$. The

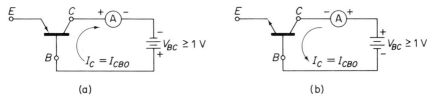

**Fig. 2-7.** Circuit to measure reverse leakage current $I_{CBO}$ for (a) a *pnp* transistor, and (b) an *npn* transistor.

subscripts $C$ and $B$ indicate those terminals of the transistor between which current flows. The subscript $O$ is intended to indicate that the unidentified terminal is open circuited. In a *pnp* transistor $I_{CBO}$ consists of minority base holes crossing into the collector and minority collector electrons crossing into the base. In the external ciruit, $I_{CBO}$ is depicted by conventional current flow as shown by Fig. 2-7. $I_{CBO}$ is highly temperature dependent, approximately doubling for every 6°C temperature increase in a silicon transistor and for every 10°C temperature increase in a germanium transistor. As in the diode, $I_{CBO}$ for germanium transistors will be a few microamperes, and about a thousand times less for silicon transistors. Usually $I_{CBO}$ is negligible for silicon transistors. We shall include its effects in the development of our dc model, for completeness, but shall usually discard it whenever silicon transistors are used.

In our discussion of a reverse-biased diode it was emphasized that the saturation current remained constant, small, and independent of the magnitude of the reverse-bias voltage, until avalanche or Zener breakdown occurred. $I_{CBO}$ is also constant as long as the collector junction is reverse biased and therefore may be modeled by a current source placed between the base and collector. A maximum value for $I_{CBO}$ may be given in the "Electrical Characteristics" section of a data sheet for a particular value of $V_{BC}$. If no value is given it is assumed to be negligible.

Since the emitter-base diode is forward biased, it will be modeled by an ideal diode with battery $V_{BE}$ representing the threshold voltage. Forward resistance for the slope of the emitter-diode characteristic is eliminated from the model because external circuit resistance is usually large with respect to the forward resistance. $V_{BE}$ is typically about 0.7 V in low-power silicon and 0.3 V in germanium transistors.

Once the emitter current $I_E$ is found from the input circuit, the dc *forward current-transfer ratio* $\alpha_F$ tells what portion of the emitter current reaches the collector terminal. That is, the collector current is determined primarily by the component $I_{EC} = \alpha_F I_E$. This component, $\alpha_F I_E$, may be represented by a *dependent, constant-current generator*, located between the base and collector. The remainder of the emitter current $(1 - \alpha_F)I_E$ feeds recombinations in the base. Typical values of $\alpha_F$ are 0.91 to over 0.99 and may be given in manufacturers' data sheets under the symbol $h_{FB}$. A dc model for the transistor in the common-base circuit of Fig. 2-5(a) is given in Fig. 2-8(a). It consists of the following:

1. An ideal diode and battery to model the base-emitter diode. Temperature dependence may be included in the threshold voltage, $V_{BE}$.
2. A dependent current generator in the collector, $\alpha_F I_E$, which is a function of the emitter current.

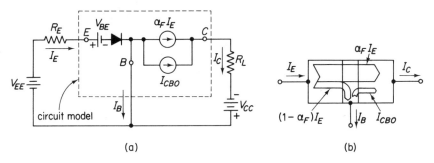

**Fig. 2-8.** The *pnp* transistor in Fig. 2-5(a) is replaced by its dc model in Fig. 2-8(a). Internal transistor currents are shown in (b).

3. A constant current generator $I_{CBO}$, shunting $\alpha_F I_E$, whose value is independent of $V_{BC}$ but is temperature dependent.
4. Element (3) often may be eliminated in the dc circuit model for a *silicon* transistor. The temperature dependences are approximately:

|  | Silicon | Germanium |
|---|---|---|
| $I_{CBO}$ doubles for every temperature rise of | 6°C | 10°C |
| $V_{BE}$ decreases for every 1°C temperature rise by | 2 mV | 2 mV |

Internal transistor currents in Fig. 2-6 are represented in terms of $\alpha_F$ and $I_{CBO}$ in Fig. 2-8(b). Terminal current relationships are developed from Fig. 2-8 by observing that current entering the transistor equals the sum of the currents leaving the transistor:

$$I_E = I_B + I_C \qquad (2\text{-}1)$$

A dependent current generator and a constant current generator make up the collector current

$$I_C = \alpha_F I_E + I_{CBO} \qquad (2\text{-}2)$$

Terminal base current is the difference between the recombination current $(1 - \alpha_F)I_E$ and leakage current $I_{CBO}$, or

$$I_B = (1 - \alpha_F)I_E - I_{CBO} \qquad (2\text{-}3)$$

***Example 2-1.*** Given: $V_{EE} = 1.7$ V, $R_E = 1$ kΩ, $R_L = 10$ kΩ, $V_{CC} = 20$ V, $\alpha_F = 0.990$, and the transistor is silicon in Fig. 2-8. Find (a) $I_E$, (b) $I_C$, and (c) $I_B$.

*Solution.* (a) Write the input loop equation from Fig. 2-8(a):

$$V_{EE} = I_E R_E + V_{BE}$$

Substituting for $V_{BE} = 0.7$ V because of the silicon transistor,

$$1.7 = I_E(1000) + 0.7$$
$$I_E = 1 \text{ mA}$$

(b) $I_{CBO}$ is negligible for a silicon transistor and from Eq. (2-2),

$$I_C = \alpha_F I_E = (0.990)(1 \text{ mA}) = 990 \text{ } \mu\text{A}$$

(c) From Eq. (2-3)

$$I_B = (1 - \alpha_F)I_E = (1 - 0.990)(1 \text{ mA}) = 10 \text{ } \mu\text{A}$$

Check from Eq. (2-1): $I_E = 1000 \text{ } \mu\text{A} = I_C + I_B = 990 + 10$

***Example 2-2.*** Given the same conditions as Example 2-2 except that the transistor is germanium, $V_{EE} = 1.3$ V, and $I_{CBO} = 5$ $\mu$A. Find (a) $I_E$, (b) $I_C$, and (c) $I_B$.

*Solution.* (a) Assuming $V_{BE} = 0.3$ V for the germanium transistor and using the input loop equation from Example 2-1(a),

$$1.3 = I_E(1000) + 0.3$$
$$I_E = 1 \text{ mA}$$

(b) From Eq. (2-2),

$$I_C = \alpha_F I_E + I_{CBO} = (0.990)(1000 \text{ } \mu\text{A}) + 5 \text{ } \mu\text{A} = 995 \text{ } \mu\text{A}$$

(c) From Eq. (2-3),

$$I_B = (1 - \alpha_F)I_E - I_{CBO} = (0.01)(1000 \text{ } \mu\text{A}) - 5 = 5 \text{ } \mu\text{A}$$

Check from Eq. (2-1): $I_E = 1000 \text{ } \mu\text{A} = I_C + I_B = 995 \text{ } \mu\text{A} + 5 \text{ } \mu\text{A}$.

## 2-4 Transistor Action in the Common-Emitter Configuration

The common-emitter circuit of Fig. 2-5(b) is represented by the dc circuit model in Fig. 2-9(a). This model is useful but it is advantageous to develop the equivalent model in Fig. 2-9(b), where the dependent current generator is controlled by $I_B$ rather than by $I_E$. Refer to Figs. 2-6 and 2-8(b) and define a new dc *forward current-transfer ratio* $\beta_F$ where (ignoring $I_{BC}$ or $I_{CBO}$)

$$\beta_F = \frac{I_{EC}}{I_{EB}} = \frac{\alpha_F I_E}{(1 - \alpha_F)I_E} = \frac{\alpha_F}{1 - \alpha_F} \tag{2-4}$$

Sec. 2-4  Common-Emitter Configuration  51

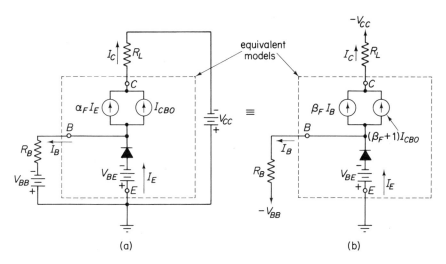

**Fig. 2-9.** An emitter-current controlled model in (a) and base-current controlled model in (b) are both equivalent circuit models for the transistor in the common-emitter circuit of Fig. 2-5(b).

Solving Eq. (2-4) for $\alpha_F$ gives $\alpha_F$ in terms of $\beta_F$:

$$\alpha_F = \frac{\beta_F}{\beta_F + 1} \tag{2-5}$$

Another useful relationship is found by adding 1 to both sides of Eq. (2-4) to yield

$$\beta_F + 1 = \frac{\alpha_F}{1 - \alpha_F} + 1 = \frac{1}{1 - \alpha_F} \tag{2-6}$$

Now we can obtain an expression for the collector-base current relationship in terms of $\alpha_F$ by substituting Eq. (2-1) for $I_E$ in Eq. (2-2).

$$I_C = \alpha_F(I_B + I_C) + I_{CBO} \tag{2-7}$$

Solving for $I_C$:

$$I_C = \frac{\alpha_F}{1 - \alpha_F} I_B + \frac{I_{CBO}}{1 - \alpha_F} \tag{2-8}$$

Equation (2-8) is more useful in terms of $\beta_F$. Therefore we substitute Eqs. (2-4) and (2-6) to give

$$I_C = \beta_F I_B + (\beta_F + 1)I_{CBO} \tag{2-9}$$

Combing Eq. (2-9) with Eq. (2-1) gives

$$I_E = (\beta_F + 1)I_B + (\beta_F + 1)I_{CBO} \qquad (2\text{-}10)$$

Equation (2-9) says that $I_C$ can be modeled by a dependent current generator, $\beta_F I_B$, which is dependent on the base current, plus a leakage current which is much larger than $I_{CBO}$, since typical values of $\beta_F$ vary from 30 to 600. A dc model based on Eqs. (2-9) and (2-10) is given in Fig. 2-9(b), and it is evident that opening the base terminals should make $I_B$ go to zero and force the dependent current generator $\beta_F I_B$ to go to zero. The resulting collector current will be $(\beta_F + 1)I_{CBO}$. Let us reason out what is happening in the transistor when $I_C = (\beta_F + 1)I_{CBO}$.

A test circuit to measure $I_C$ with the base terminal open is shown in Fig. 2-10(a). The single power supply $V_{CC}$ will reverse bias the collector junction and forward bias the emitter junction. Since the collector junction is reverse biased, a constant current $I_{CBO}$ will be generated at the collector junction and can be represented by the conventional current direction shown in Fig. 2-10(b). Because the base terminal is open, $I_{CBO}$ must flow around the loop through the battery and emitter to make a complete circuit. If it were not for transistor action, our analysis would stop here. But is is evident that $I_{EB}$ in Fig. 2-6 corresponds to $I_{CBO}$ crossing from emitter to base. The emitter sees this current as holes fed into the base and recombining with electrons just as if the electrons were furnished from the base terminal rather than minority electrons from the collector.

Therefore, from Eq. (2-4) the component $I_{EB} = I_{CBO}$ in Fig. 2-10(b) will cause a current of $\beta_F I_{EB} = \beta_F I_{CBO}$ to be generated in the collector, and of course, furnished by the emitter as shown in Fig. 2-10(c). Alternatively we can reason that $I_{EB}$ in Fig. 2-10(b) corresponds to the recombination current $(1 - \alpha_F)I_E$ in Fig. 2-8(b) or $(1 - \alpha_F)I_E = I_{CBO}$ as shown in Fig. 2-10(d). In either case the emitter must furnish:

For Fig. 2-10(c),

1. a recombination current equal to $I_{CBO}$ plus
2. a current component to the collector equal to $\beta_F I_{CBO}$ for a total of $(\beta_F + 1)I_{CBO} = I_E$;

For Fig. 2-10(d)

1. a recombination current equal to $(1 - \alpha_F)I_E = I_{CBO}$ plus
2. a current to the collector of $\alpha_F I_E = I_{CBO}/(1 - \alpha_F)$, so that $\alpha_F I_E = [\alpha_F/(1 - \alpha_F)]I_{CBO} = \beta_F I_{CBO}$ and

$$I_E = \frac{I_{CBO}}{1 - \alpha_F} = (\beta_F + 1)I_{CBO} = I_C \qquad (2\text{-}11)$$

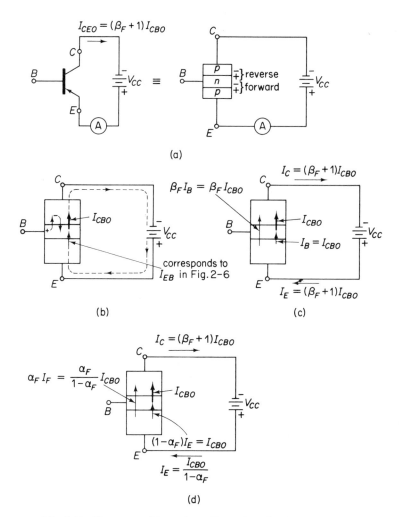

**Fig. 2-10.** $V_{CC}$ reverse biases the collector junction in (a) to cause leakage current $I_{CBO}$ in (b) which crosses the forward-biased emitter junction in (c) to excite the generator, causing $\beta_F$. An alternative viewpoint in (d) considers $I_{CBO}$ to cause $(1 - \alpha_F)I_E$ and therefore $I_E = I_{CBO}/(1 - \alpha_F)$.

Collector and emitter currents are equal in Fig. 2-10 because the base terminal is open; the triple-subscript notation for this leakage current is $I_{CEO}$, so Eq. (2-11) may be expressed as

$$I_{CEO} = \frac{I_{CBO}}{1 - \alpha_F} = (\beta_F + 1)I_{CBO} \tag{2-12}$$

From this point on we shall use loop currents to analyze transistor action because it is easy to separate majority carrier or injection currents from minority carrier or leakage currents. If the transistors are made of silicon the leakage currents are negligible and are not shown, so it is necessary to show only the majority carrier currents. For germanium transistors we add or superimpose the leakage currents.

In Fig. 2-11(a) we show the majority current loops for a common-emitter circuit by drawing the base-current loop and the resultant collector-current component loop $\beta_F I_B$. This completes the transistor currents for a silicon transistor. Observe that we can write the expressions for $I_C$ and $I_E$ directly from inspection of Fig. 2-11(a). The results correspond to Eqs. (2-9) and (2-10) for $I_{CBO} = 0$.

We add the leakage current loops to the majority current loops in Fig. 2-11(b) for a germanium transistor by proceeding as follows. (1) Draw

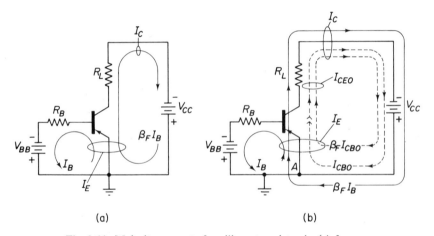

**Fig. 2-11.** Majority currents for silicon transistor in (a) for common-emitter operation. Leakage currents are added in dashed lines in (b) for a germanium transistor.

the majority current loops exactly as we did for the silicon transistor. (2) Draw $I_{CBO}$ as a current source flowing from collector junction through $R_L$ and $V_{CC}$ to point $A$. (Remember that a constant current source causes a current that is unaffected by any series resistance or voltage source. The current can divide only when it encounters parallel paths.) At point $A$, $I_{CBO}$ sees one path via a short circuit through the emitter back to the base and a second path through $R_B$ to the base. Therefore $I_{CBO}$ does not divide at point $A$ but all of $I_{CBO}$ crosses the emitter junction completing the loop. (3) Because $I_{CBO}$ crosses the emitter junction, it acts as a base current to excite the $\beta_F$ generator and causes the loop current given by $\beta_F I_{CBO}$ in Fig. 2-11(b).

Figure 2-11 is a different presentation of the model in Fig. 2-9(b), from which it was developed. But observe that all of the equations from Eq. (2-9) to Eq. (2-12) can be written *by inspection* of Fig. 2-11, which was constructed by a straightforward procedure. This procedure is really an application of superposition where currents caused by the dependent voltage sources are superimposed on currents caused by the independent current sources. Recall that $\alpha_F I_E$ and $\beta_F I_B$ are *dependent current* sources which are excited by currents crossing the emitter junction. $I_{CBO}$ is considered to be an independent current source because it is on as long as the collector junction is reverse biased.

***Example 2-3.*** Given $R_B = 100 \text{ k}\Omega$, $V_{BB} = 1.7 \text{ V}$, $R_L = 10 \text{ k}\Omega$, $V_{CC} = 20 \text{ V}$, $\alpha_F = 0.990$, and the transistor is silicon in Fig. 2-11(a). Find (a) $I_B$, (b) $I_C$, and (c) $I_E$.

*Solution.* (a) Evaluate $\beta_F$ from Eq. (2-4):

$$\beta_F = \frac{\alpha_F}{1 - \alpha_F} = \frac{0.990}{0.010} = 99$$

Find $I_B$ from the input loop equation of Fig. 2-11(a):

$$V_{BB} = I_B R_B + V_{BE}$$
$$1.7 = I_B(10^5) + 0.7$$
$$I_B = 10 \text{ }\mu\text{A}$$

(b) Assuming $I_{CBO} = 0$, evaluate $I_C$ from Eq. (2-9):

$$I_C = \beta_F I_B = 99(10 \text{ }\mu\text{A}) = 990 \text{ }\mu\text{A}$$

(c) From Eq. (2-10),

$$I_E = (\beta_F + 1)I_B = 100(10 \text{ }\mu\text{A}) = 1000 \text{ }\mu\text{A}$$

Check from Eq. (2-1): $I_E = 1000 \text{ }\mu\text{A} = I_C + I_B = 990 + 10 \text{ }\mu\text{A}$. Compare this example with Example 2-1 to see the similarities and differences between the common-emitter and common-base circuits.

***Example 2-4.*** Given the same conditions as Example 2-3 except that the transistor is germanium, $V_{BB} = 1.3 \text{ V}$, and $I_{CBO} = 5 \text{ }\mu\text{A}$. Find (a) $I_B$, (b) $I_C$, and (c) $I_E$. (d) How much were $I_C$ and $I_E$ increased because of $I_{CBO}$ in comparison to Example 2-3?

*Solution.* (a) Using the same value of $\beta_F$ and the input loop equation of Example 2-3, find $I_B$ for Fig. 2-11(b):

$$V_{BB} = I_B(10^5) + 0.3$$
$$I_B = 10 \text{ }\mu\text{A}$$

(b) Evaluate $I_C$ from Eq. (2-9):

$$I_C = \beta_F I_B + (\beta_F + 1)I_{CBO} = 99(10 \ \mu A) + (100)(5 \ \mu A)$$
$$= 1.49 \text{ mA}$$

(c) From Eq. (2-10),

$$I_E = (\beta_F + 1)I_B + (\beta_F + 1)I_{CBO} = 100(10 \ \mu A) + 100(5 \ \mu A)$$
$$= 1.5 \text{ mA}$$

(d) Both $I_C$ and $I_E$ were increased by $I_{CEO}$ in an amount calculated from Eq. (2-12):

$$I_{CEO} = (\beta_F + 1)I_{CBO} = 100(5 \ \mu A) = 0.5 \text{ mA}$$

*Example 2-5.* Draw majority carrier-current and leakage-current loops for the common-base circuit of Figs. 2-8 and 2-5(a) with the values given in Example 2-2, but with an *npn* germanium transistor.

*Solution.* Power-supply batteries are reversed with the *npn* transistor for the solution in Fig. 2-12. Solid loops are drawn for $\alpha_F I_E$ and $(1 - \alpha_F)I_E$ to

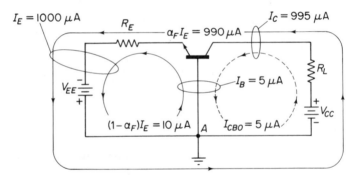

**Fig. 2-12.** Solution to Example 2-5.

account for majority carrier currents. $I_{CBO}$ begins at the reverse-biased collector junction and we trace it backwards through $R_L$ and $V_{CC}$ to junction point $A$, where it all goes through the short circuit to the base rather than through the high-resistance parallel path offered by $R_E$.

## 2-5 Measurement and Display of Common-Emitter Characteristic Curves

Our study of transistor action was based on the assumption that we knew values of input currents $I_B$ for the common emitter and $I_E$ for the common base. Compare Figs. 2-5(a) and (b) to see that removing the output loop $R_L$ and $V_{CC}$ results in identical input circuit loops because the base-emitter diode is forward biased with a battery and series resistance. Input current will flow according to the diode equation and $I_B$ or $I_E$ will depend on the diode voltage $V_{BE}$. Consequently $I_B$ or $I_E$ will depend exponentially on input voltage unless we insert a large series resistance in the input circuit so that input current is controlled primarily by input voltage and series resistance rather than the diode equation. For example, if we forward bias the emitter-base diode in Fig. 2-5(a) with $R_B$ and $V_{BB}$ chosen to set $I_B = 10\ \mu\text{A} = I_E$ and then connect $R_L$ and $V_{CC}$, there will be no change in input current but a collector current of 990 $\mu$A will flow if $\beta_F = 99$. The student may suggest that setting $R_B$ and $V_{BB}$ to 1 mA and then connecting $R_L$ and $V_{CC}$ would yield the same result. However, connecting the collector circuit would cause $I_E$ to split into 1 part for the base and $\beta_F$ parts for the collector. The base current would try to decrease but the $R_B - V_{BB}$ combination would hold $I_B$ to 1 mA, forcing $I_C$ to rise to 99 mA and $I_E$ to 100 mA.

For the reasons outlined above, the transistor is considered to be a current-controlled device. If we halve the input control voltage the input current will be halved. Since output collector current is related to input current by a constant ($\alpha_F$ for common base and $\beta_F$ for common emitter) the output current will be halved. We conclude that input current is our best choice for an independent variable to be controlled at will. We can then analyze its effect on the transistor's output. The transistor's output characteristic is defined as the relation between output current and output voltage where (1) output current is defined as the current flowing between the output terminal and the common terminal and (2) output voltage is the voltage between these two terminals.

Thus we shall measure output characteristics by (1) establishing a constant input current and (2) varying output voltage to measure corresponding values of output current. The result is plotted on a graph of output current versus output voltage and yields one characteristic curve. We then establish a different constant input current in step (1) and repeat step (2) to obtain another characteristic curve. We repeat steps (1) and (2) until a family of output characteristic curves is obtained.

Commercially available curve plotters can display a family of output characteristic curves quickly and conveniently on a cathode-ray tube. It is highly desirable to gain experience in their use but it is even more valuable for the student to construct his own curve plotter by combining sweep techniques with the CRO to obtain the same results.

A sweep circuit to display common-emitter output characteristics is given in Fig. 2-13. $V_{BB}$ and $R_B$ can adjust input current $I_B$ for any desired value. The sweep circuit automatically varies output voltage $V_{CE}$ from 0–9 V and holds the collector negative with respect to the emitter. $V_{CE}$ is displayed on the x axis of the CRO. $R_S$ samples collector current $I_C$ and connections to the y amplifier display output collector current on the vertical axis.

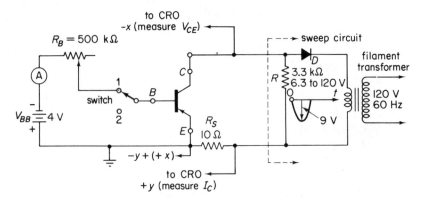

**Fig. 2-13.** Circuit to display common-emitter collector characteristics of $I_C$ versus $V_{CE}$ in the first quadrant. The $x$-amplifier is set to 1 V/div and $y$ to 10 mV/div = 1 mA/div.

With no power on the circuit of Fig. 2-13 and with the switch on point 2, the CRO is zeroed by positioning controls to place the spot at an origin located at the lower left corner of the screen as in Fig. 2-14. When power is applied to the sweep circuit, $V_{CE}$ is varied from 0 to $-9$ V at a rate of 60 times per second, and the CRO displays the characteristic curve labeled $I_B = 0$ in Fig. 2-14. Conditions correspond to Fig. 2-10(a) so that we measure the relatively constant leakage current of $I_{CEO} \simeq 0.15$ mA. Observe that $I_{CEO}$ is independent of $V_{CE}$ for negative values of $V_{CE}$ greater than $-0.2$ V. The area under $I_B = 0$ in Fig. 2-14 depicts *cutoff* because no majority carrier currents flow and the transistor is essentially cut off from conducting substantial output current $I_C$.

Throw the switch to point 1 in Fig. 2-13 to connect the base biasing network and adjust $R_B$ until ammeter A reads 20 μA. The CRO will display the characteristic curve labeled $-20$ μA in Fig. 2-14. As base current is increased the curve rises on the CRO and is shown for $I_B = -40, -60$, and $-80$ μA. Of course, the transistor can operate at any point between the curves of $I_B = -80$ μA and $I_B = 0$. To locate an operating point on the transistor's output characteristic curve, it is necessary to know only two out of the three variables $I_B$, $I_C$, and $V_{CE}$. The remaining variable is specified automatically For example, if we measured 7 V between collector and emitter, and a collector current of 3 mA, this would locate point $A$. It follows automatically that $I_B$ is approximately $-30$ μA, since point $A$ lies halfway between $I_B = -40$ μA and $I_B = -20$ μA.

Directing attention to the curve $I_B = -40$ μA in Fig. 2-14, we see that the collector current is roughly $-4.1$ mA for all values of $V_{CE}$ greater than $-0.6$ V showing that once the collector junction is reverse biased, the magnitude of $V_{CE}$ has practically no effect on $I_C$. That region of the graph where $I_C$ does not change appreciably with changes in $V_{CE}$ is called the

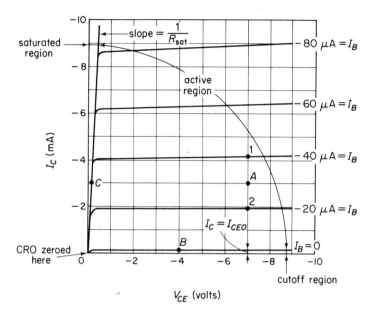

**Fig. 2-14.** Common-emitter collector characteristic curves obtained by the circuit of Fig. 2-13 for a germanium transistor.

*active region.* If the transistor is operated at any point within this region its emitter junction is forward biased and its collector junction is reverse biased. The remaining area of the graph is called the *saturated region;* in Fig. 2-14 it lies between the vertical axis and the line labeled $R_{\text{sat}}$. Both saturated and cutoff regions will be examined further in Section 2-7.

## 2-6 Measurement of Common-Emitter Current Gain

Common-emitter characteristic curves in Fig. 2-14 are approximated by Eq. (2-9) and repeated here for convenience:

$$I_C = \beta_F I_B + I_{CEO}$$

Since output current $I_C$ is larger than input current $I_B$ there is a dc current gain represented by $\beta_F$; solving Eq. (2-9) for $\beta_F$,

$$\beta_F = \frac{I_C - I_{CEO}}{I_B} \tag{2-13}$$

One of the primary uses of the collector characteristic curves is to measure $\beta_F$. For example, in Fig. 2-14 the value of $\beta_F$ is calculated for point $A$ to be at:

$$\beta_F = \frac{(3.0 - 0.15) \times 10^{-3}}{30 \times 10^{-6}} = 95$$

Spacing between the curves increases with increasing values of collector current. That is, along a constant vertical line, at $V_{CE} = 7$ V, when base current $I_B = 20$ $\mu$A, collector current $I_C$ will equal 4.2 mA, and when $I_B = 80$ $\mu$A, $I_C = 19$ mA. We summarize calculations of $\beta_F$ in Table 2-3 to show that $\beta_F$ increases somewhat with increasing $I_C$ and also with increasing $V_{CE}$ for the characterisitcs in Fig. 2-14.

Table 2-3

| $I_B$ ($\mu$A) | $V_{CE} = -1$ V | | $V_{CE} = -7$ V | |
|---|---|---|---|---|
| | $I_C$ (mA) | $\beta_F$ | $I_C$ (mA) | $\beta_F$ |
| 20 | 1.9 | 87 | 1.9 | 87 |
| 40 | 4.0 | 96 | 4.2 | 101 |
| 60 | 6.2 | 101 | 6.4 | 104 |
| 80 | 8.7 | 107 | 9.0 | 111 |

Small-signal current gain $\beta_0$ can also be measured from the characteristic curves. Suppose $I_B$ was changed by a small-signal $\Delta I_B$ to vary between 20 $\mu$A and 40 $\mu$A as shown by points 1 and 2 in Fig. 2-14. $V_{CE}$ is held constant at 7 V during the change. We want to know the small-signal current gain $\beta_0$ that relates $\Delta I_B = 20$ $\mu$A to the resulting change in collector current $\Delta I_C$. Write the expression for $I_C$ at point 1 and point 2, then subtract the equations:

$$\begin{aligned} I_{C1} &= I_{B1}\beta_{F1} + I_{CEO} \\ -(I_{C2} &= I_{B2}\beta_{F2} + I_{CEO}) \\ \hline \Delta I_C = I_{C1} - I_{C2} &= I_{B1}\beta_{F1} - I_{B2}\beta_{F2} \end{aligned} \quad (2\text{-}14)$$

For small enough signal currents such that $\beta_{F1} = \beta_{F2}$, the small-signal current gain is defined by $\beta_0 = \beta_{F1} = \beta_{F2}$. Substituting in Eq. (2-14) gives

$$\Delta I_C = \beta_0(I_{B1} - I_{B2}) = \beta_0 \, \Delta I_B$$

Solving for $\beta_0$ and stipulating that operating points must be measured at the same $V_{CE}$,

$$\beta_0 = \left.\frac{\Delta I_C}{\Delta I_B}\right|_{V_{CE}=\text{const}, \Delta I_B \to 0} \quad (2\text{-}15)$$

*Example 2-6.* $\beta_F$ was calculated to equal 95 for point $A$ in Fig. 2-14. Calculate $\beta_0$ for a small symetrical base-current signal swing of 10 $\mu$A peak value, around point $A$.

**Solution.** Graphically we work with peak-to-peak (p/p) values so that $\Delta I_B = 20\ \mu\text{A}$ and corresponds to points 1 and 2. At $V_{CE} = -7$ V we read $\Delta I_C = 4.2 - 1.9 = 2.2$ mA so that

$$\beta_0 = \left.\frac{\Delta I_C}{\Delta I_B}\right|_{V_{CE}=-7\text{V}} = \frac{2.2 \times 10^{-3}}{20 \times 10^{-6}} = 110$$

From Eq. (2-14) it is evident that $\beta_0$ does not depend on $I_{CEO}$ and from Fig. 2-14 $\beta_0$ will vary with operating-point location because of the unequal spacing between curves. If $I_{CEO} = 0$ and characteristic curves are equally spaced and parallel, then $\beta_0 = \beta_F$. A commonly accepted symbol for $\beta_0$ is $h_{fe}$.

## 2-7 Cutoff and Saturation on the Characteristic Curves

When the transistor is operated at point $B$ on the curve $I_B = 0$ in Fig. 2-14, it is said to be *cut off*. Base current is zero, collector current is about 0.15 mA, and $V_{CE}$ is equal to 4 V. The area under the curve of $I_B = 0$ is often referred to as the *cutoff region*, and normally is to be avoided. Look at point 1 in Fig. 2-15(a) for an example of a circuit operating exactly at point $B$.

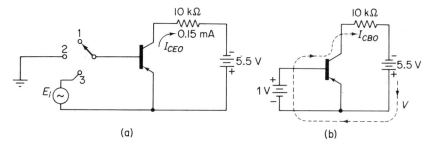

**Fig. 2-15.** Circuits to demonstrate operation near cutoff in (a) and at actual cutoff in (b).

Strictly speaking, the transistor is not entirely cut off, in the sense that $I_{CEO}$ is not the minimum collector current that can flow in the common-emitter circuit. Decreasing the emitter-junction voltage below the threshold voltage to a value less than about 3 $KT/q = 75$ mV is the true measure of cutoff, for it is the base-emitter voltage which controls the emitter junction. Grounding the base by switching to point 2 in Fig. 2-15(a) will not cut off the emitter junction because there is resistance in the inactive portion of the base material and much of $I_{CBO}$ will still cross the emitter junction to excite the $\beta_F$ generator. The emitter junction is still forward biased, even though base and emitter terminals are connected together. By placing a resistor of a few kilohms between the emitter and base it is possible to ensure that all of $I_{CBO}$ crosses the emitter junction and makes the collector current return to a magnitude of $I_{CEO}$. To cut off the emitter junction completely, a small reverse bias must be applied, as in Fig. 2-15(b) where $I_C$ is minimum at $I_{CBO}$.

This leads to the interesting circuit of Fig. 2-15(a), where a common-emitter circuit is operated without a dc bias network by switching to point 3. A small sinusoidal input voltage can cause an approximately sinusoidal base current to flow provided that the peak base-current change is held at or below $I_{CBO}$. This means the collector current of $I_{CEO}$ (with no signal) will be varied sinusoidally by a peak value of roughly $\beta_F I_{CBO}$. This circuit is inexpensive but has poor operating-point stability and is rarely used. However, it can illustrate the true meaning of cutoff and help us to interpret the characteristic curves in terms of transistor action.

Operation in the *saturated mode* is exemplified by biasing the transistor to operate at point C in Fig. 2-14. Here the transistor is passing a collector current of 3 mA and the voltage across it is $V_{CE} = -0.15$ V. However, the base current may have any value greater than about $-30$ $\mu$A. For example, if $I_B$ were either $-40$ $\mu$A or $-80$ $\mu$A, the transistor would still be operated at point C. In saturation, both collector and emitter junctions are forward biased. To illustrate this fact we assume the germanium emitter junction must be forward biased at about 0.3 V because it is conducting a base current of at least 30 $\mu$A. This operating-point information is shown in

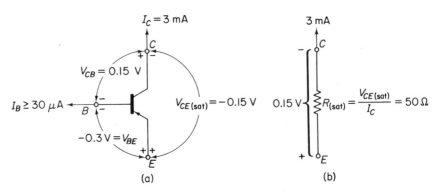

**Fig. 2-16.** Operating-point currents and voltages in (a) are given for saturated operation at point C in Fig. 2-12. The saturated transistor may be modeled by the resitor in (b).

Fig. 2-16(a), where forward bias of both junctions is proved by the fact that the *n*-type base is negative with respect to both emitter and collector.

In Fig. 2-14, the straight portion of the near-vertical line joining the origin and point C and continuing, has a slope of rise over run of $I_C/V_{CE}$ equal to the reciprocal of the saturation resistance $R_{sat}$. Operating the transistor on or to the left of this line defines the saturated mode. $R_{sat}$ is found graphically from the output characteristics by its definition of

## Sec. 2-7  Cutoff and Saturation on the Characteristic Curves  63

$$R_{sat} = \frac{V_{CE(sat)}}{I_C} \quad (2\text{-}16)$$

and is modeled by a resistor between the collector and emitter in Fig. 2-16(b). There is an ill-defined but small region where a particular curve of $I_C$ versus $V_{CE}$ makes the transition from a slope of $1/R_{sat}$ to a horizontal line of almost constant current. Here the transistor is proceeding from the saturated mode to the active mode as the collector junction goes from forward to reverse

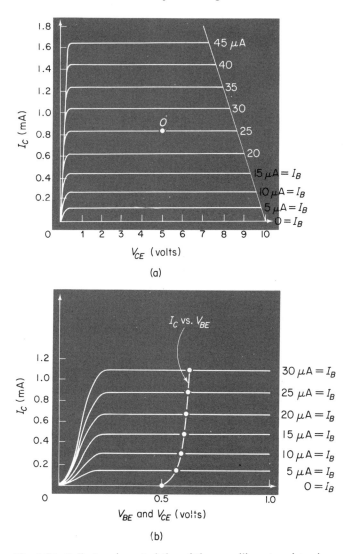

Fig. 2-17. Collector characteristics of the *npn* silicon transistor in (a) are expanded in (b).

bias. A practical definition follows by considering *saturation to occur when a change in base current produces no significant change in collector current.*

**Example 2-7.** Collector characteristic curves are given for an *npn* silicon transistor in Fig. 2-17(a). Note that $I_{CEO}$ is negligible since the curve of $I_B = 0$ coincides with the horizontal axis. To investigate the saturation region we expand the horizontal sensitivity by a factor of 10 in Fig. 2-17(b) and see that the curves do not coincide in the saturation region but have different slopes and consequently different values of $R_{sat}$. Superimposed on the collector characteristic curves of Fig. 2-17(b) is a plot of $I_C$ versus $V_{BE}$. This plot was obtained by holding $I_B$ constant at values of 5 μA and multiples thereof while $V_{CE}$ was swept between 0 and approximately 1 V. Thus when $I_B = 20$ μA, $V_{BE} = 0.6$ V regardless of the value of $I_C$ or $V_{CE}$. (a) What collector current flows when $V_{CE} = 100$ mV and $I_B = 20$ μA? (b) At this operating point, what is the value of $R_{sat}$, and (c) what is the bias voltage applied to each junction?

**Solution.** (a) From Fig. 2-17(b), locate the intersection of $I_B = 20$ μA and $V_{CE} = 0.1$ V to read $I_C = 0.3$ mA. (b) Calculate $R_{sat}$ from Eq. (2-16) to find $R_{sat} = 0.1/(0.3 \times 10^{-3}) = 333$ Ω. (c) Since $V_{BE} = 0.6$ V at $I_B = 20$ μA (base positive with respect to emitter) and $V_{CE} = 0.1$ V (collector positive with respect to emitter), $V_{BC} = V_{BE} - V_{CE} = 0.5$ V and the base is more positive than the collector. Polarity signs are reversed from those in Fig. 2-16(a).

## 2-8 Measurement of Common-Base Characteristics

A sweep circuit is given in Fig. 2-18(a) that will display output collector current versus output voltage $V_{CB}$ for constant values of input emitter current in the common-base configuration. With $V_{EE}$ set to 0 the collector sweep circuit traces out the *x* axis on the CRO since $V_{CB}$ is too small to cause any vertical deflection. This curve locates the cutoff region in Fig. 2-18(b), where it is indentified as $I_E = 0$.

$V_{EE}$ and $R_E$ are next adjusted to set $I_E = 0.2$ mA and the curve labeled $I_E = 0.2$ mA is swept out in Fig. 2-18(b). Observe that $I_C \cong I_E$ for all values of $I_C$ where the collector is negative with respect to the emitter. Note also that *during the half-cycle when the sweep voltage is zero*, emitter current continues to inject current into the base which diffuses to the collector. This $I_C = 0.2$ mA causes a small drop across the 1-kΩ resistor so that $V_{CB}$ goes positive for a 0.2-V forward bias on the collector junction. An interesting point is encountered when $I_E = 1.0$ mA and the sweep voltage is 0 in Fig. 2-18(b). $I_C$ continues to flow at 1 mA until $V_{CB}$ builds up to approximately +0.6 V, where the collector junction becomes heavily forward biased and begins injecting holes into the base. $I_C$ decreases rapidly to a value which will stabilize $V_{CB}$ at approximately 0.6 V and is determined by the 1-kΩ

## Sec. 2-8 Measurement of Common-Base Characteristics

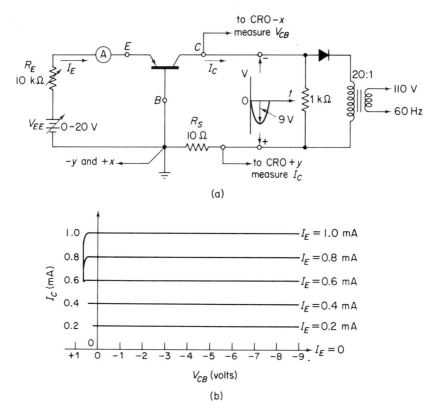

**Fig. 2-18.** Common-base characteristic curves in (b) are measured by the sweep circuit in (a) for a *pnp* transistor.

resistor to be $I_C = 0.6/1000 = 0.6$ mA. The saturation region is located in the second quadrant of Fig. 2-18(b).

Common-base characteristic curves exhibit uniform spacing so that current gain $\alpha_F$ is uniform throughout the active region in the first quadrant of Fig. 2-18(b). We could attempt a graphical measurement of $\alpha_F$ by reading values for $I_E$, $I_C$, and $I_{CBO}$ at an operating point and substituting into the following form of Eq. (2-2):

$$\alpha_F = \frac{I_C - I_{CBO}}{I_E} \qquad (2\text{-}17)$$

The result would be inaccurate, however, since $I_C$ is approximately equal to $I_E$ and $I_{CBO}$ is often negligible. It is preferable to measure $\beta_F$ at the same operating point in a common-emitter configuration and calculate $\alpha_F$ from Eq. (2-5). It follows that small-signal current gain $\alpha_0$ will be equal to $\alpha_F$, due to uniform curve spacing, and should be evaluated from Eq. (2-18a) by a measurement of $\beta_0$ at the same operating point.

where
$$\alpha_0 = \frac{\beta_0}{\beta_0 + 1} \quad (2\text{-}18a)$$

$$\alpha_0 = \frac{\Delta I_C}{\Delta I_E}\bigg|_{V_{CB}=\text{const},\,\Delta I_E \to 0} \quad (2\text{-}18b)$$

Characteristic curves for the common-collector configuration are almost identical with those for the common-emitter configuration. Input base current is held constant while $I_E$ is measured on the vertical axis and $V_{CE}$ is swept and measured on the horizontal axis. Since $I_C \cong I_E$ the similarity between the two configurations is apparent and will not be considered further.

*Example 2-8.* Given $I_{CBO} = 2\ \mu\text{A}$ and $\beta_F = 100$ for a *pnp* germanium transistor. Assume that $\beta_F$ is uniform and sketch output common-base characteristics in the range of $I_C = 0$ to $-2$ mA, $V_{CB} = 0$ to $-10$ V.

*Solution.* Calculate $\alpha_F$ from Eq. (2-5) to obtain

$$\alpha_F = \frac{100}{101} = 0.99$$

Assume $I_{E1} = 1$ mA and $I_{E2} = 2$ mA. The corresponding output curves are plotted in Fig. 2-19 and are expressed from Eq. (2-2) by

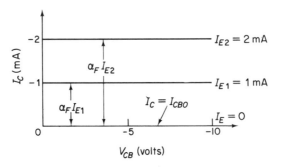

**Fig. 2-19.** Solution to Example 2-8.

$$I_{C1} = 0.99\ \text{mA} + 2\ \mu\text{A} \qquad I_{C2} = 1.98\ \text{mA} + 2\ \mu\text{A}$$

## 2-9 Measurement of Other Useful Characteristic Curves

Transistor characteristics are by no means limited to a description of the relationships between output current and output voltage for various values of input current. A curve tracer is employed in Fig. 2-20 to portray

Sec. 2-9        Measurement of Other Useful Characteristic Curves        67

common-emitter input characteristics of a silicon transistor. This is the same transistor whose output characteristics are given in Fig. 2-17. Input current $I_B$ is plotted against input voltage $V_{BE}$ in Fig. 2-20(a), where it can

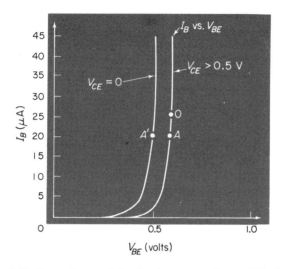

**Fig. 2-20.** Input characteristics for the *npn* transistor of Fig. 2-15.

be seen that their relationship does not depend strongly on $V_{CE}$. For example, locate the point $A'$ at $I_B = 20\ \mu\text{A}$, $V_{BE} = 0.5$ V. It corresponds to a collector-to-emitter voltage of 0 V. The curve tracer maintains $I_B$ constant at 20 $\mu$A for an interval while $V_{CE}$ is swept to a peak value of 10 V. While $V_{BE}$ increased from 0 to about 0.5 V, the value of $V_{BE}$ increased from 0.5 to 0.6 V to the dot at point $A$. Then as $V_{CE}$ is increased further, the base current remains constant at 20 $\mu$A and $V_{BE}$ stays constant at 0.6 V. The value of $V_{CE}$ is less than $V_{BE}$ between the points $A'$ and $A$ so that we may think of the area between the two curves as identifying operation in the saturated region. For operation in the active region we shall select points on the curve of $I_B$ versus $V_{BE}$ along the line of $V_{CE} > 0.5$ V, as illustrated in Fig. 2-20. The curve labeled $V_{CE} = 0$ is the characteristic of the base-emitter diode. One possible use for this curve is to allow a graphical measurement of incremental input resistance $h_{ie}$.

Collector current versus base-emitter voltage is plotted for the same *npn* transistor in Fig. 2-21. This curve is called a transfer characteristic because it shows the relationship between an output quantity $I_C$ and an input quantity $V_{BE}$. Its slope at an operating point may be used to evaluate incremental *transconductance* $g_m$.

Another easily obtainable and useful transfer characteristic is a plot of $I_C$ versus $I_B$ in Fig. 2-22 for the same transistor. The slope of a line from

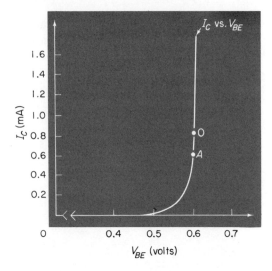

**Fig. 2-21.** Transfer characteristic curve of $I_C$ versus $V_{BE}$, common emitter.

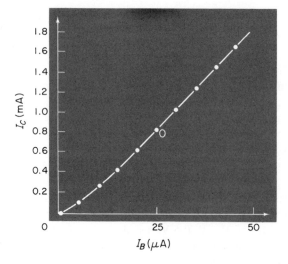

**Fig. 2-22.** Transfer characteristic curve of $I_C$ versus $I_B$, common emitter.

origin to operating point is a measurement of $\beta_F$. The slope of a line drawn tangent to the curve at the operating point is a measurement of $\beta_0$.

Input and output characteristics are the basic curves because all data concerning one operating point may be obtained from them. For example, in Fig. 2-17(a), operating point 0 is described by $I_B = 25$ $\mu A$, $I_C = 0.84$ mA, and $V_{CE} = 5$ V. From Fig. 2-20, point 0 is located from knowlege of $I_B$

Sec. 2-9    Measurement of Other Useful Characteristic Curves    69

and $V_{CE}$ and we read the final bit of data as $V_{BE} = 0.6$ V. From knowledge of $I_C$ we can verify $V_{BE}$ in Fig. 2-21 and $I_B$ in Fig. 2-22 (within the accuracy limits of the curve plotter and operator) to locate point 0 on both figures.

*Example 2-9.* Given point $A$ on the input characteristic of Fig. 2-20. Wherever possible, locate $A$ on Figs. 2-22, 2-21, and 2-17.

*Solution.* From Fig. 2-20 we read operating-point data of point $A$ as $I_B = 20$ $\mu$A, $V_{BE} = 0.6$ V, and $V_{CE} > 0.5$ V. On the collector characteristic in Fig. 2-17(a) we can enter at the base-current line of $I_B = 20$ $\mu$A. No specific value of $V_{CE}$ is obtainable from the input characteristic, except that $V_{CE}$ exceeds 0.5 V. Therefore, point $A$ on the input characteristic is represented as the entire horizontal portion of the $I_B = 20$ $\mu$A line on the output characteristic and the coordinates of $A$ are given as $I_C = 0.62$ mA, $V_{CE} \geq 0.5$ V. Point $A$ is located on Fig. 2-22 by entering with the base current of 20 $\mu$A and reading $I_C = 0.62$ mA from the curve of $I_C$ versus $I_B$. The fourth dot from the origin is point $A$. Finally, $A$ is located in Fig. 2-21 by entering the value of $I_C = 0.62$ mA, calling this value of $I_C$ one coordinate and $V_{BE} \cong 0.6$ V the other.

*Example 2-10.* Refer to Fig. 2-17. (a) Evaluate $\beta_F$ at point 0. (b) Use the resulting value and Eq. (2-9) to approximate the transistor's output characteristics. (c) Plot the results of (b) for $I_B$ in 10-$\mu$A intervals and compare them with the actual output characteristics.

*Solution.* (a) $I_{CEO}$ is negligible and $\beta_F$ is calculated to be $I_C/I_B = 0.84 \times$

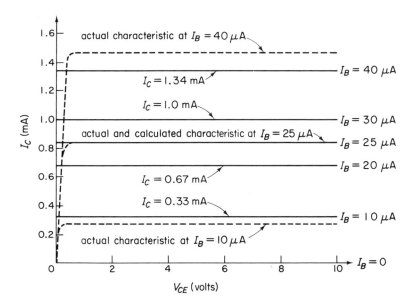

**Fig. 2-23.** Solution to Example 2-10.

$10^{-3}/(25 \times 10^{-6}) = 33.6$. (b) Transistor characteristics are expressed by Eq. (2-9) as

$$I_C = 33.6 I_B + 0$$

This equation is plotted in Fig. 2-23 and compared with the actual characteristics for $I_B = 10\ \mu A$ and $40\ \mu A$. It is seen there is reasonable agreement between the actual and approximate characteristics.

*Example 2-11.* Common-emitter output characteristics are given in Fig. 2-24,

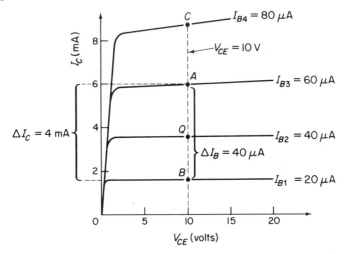

**Fig. 2-24.** Measurement of $\beta_0$ around operating point $Q$ along $V_{CE} = 10\ V$ for Example 2-11.

where measurement of $\beta_0$ is shown to be $\Delta I_C/\Delta I_B = 100$, at point $Q$. (a) Plot the transfer characteristic of $I_C$ versus $I_B$ for $V_{CE} = 10\ V$. (b) From the transfer characteristic, illustrate the difference between $\beta_0$ and $\beta_F$.

*Solution.* (a) From point $B$ in Fig. 2-24 we read $I_C = 1.6\ mA$, $I_B = 20\ \mu A$. Plot this data as shown by point $B$ in Fig. 2-25. Repeat this process for points $Q$, $A$, and $C$ to complete the $I_C - I_B$ curve in Fig. 2-25. (b) Calculate $\beta_F$ to be 100 at point $A$ as shown in Fig. 2-25. Draw a tangent (shown dashed) to point $A$ and calculate $\beta_0 = 126$ from its slope.

Manufacturers supply correction curves for transistor parameters on their data sheets. For example $\beta_0$ may be listed as $h_{fe}$ and have a large spread in values. The example in Table 2-4 is not unusual. The differences in $\beta_0$ between transistors of the same number may be larger than those between different transistors. Collector characteristics are given on the data sheet and usually correspond to typical values of the transistor's parameters. Test condi-

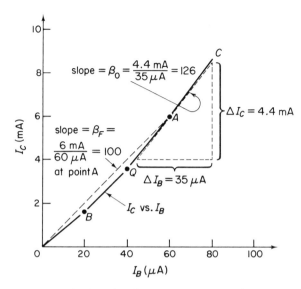

**Fig. 2-25.** $I_C$ is plotted against $I_B$ at $V_{CE} = 10$ V in Fig. 2-24 to illustrate the difference between $\beta_0$ and $\beta_F$.

Table 2-4

|  | Min. | Typ. | Max. | Tested at | |
| --- | --- | --- | --- | --- | --- |
| $h_{fe}$ | 100 | 300 | 600 | $I_C = 1$ mA | $V_{CE} = -5$ V |

tions are listed by the manufacturer to show the operating point at which the $\beta_0$ was measured. This means that a typical value must be corrected if your operating point is different. The manufacturer supplies correction curves; one for different collector currents, one for different collector voltages, and a third for different temperatures. Two common systems of supplying correction data are shown together in Fig. 2-26. Correction factors may be normalized to the test-point data as shown on the right-hand vertical axes. Otherwise actual values of $\beta_0$ are given as shown on the left-hand axis and present no problem. But if the data sheet is in the normalized system, it is necessary to multiply the test-point data by the correction factor. The procedure is illustrated by Example 2-12.

*Example 2-12.* An operating point is selected at $I_C = 0.1$ mA, $V_{CE} = 10$ V. Find the value of $\beta_0$ from Fig. 2-26, assuming that the normalized system is presented by the manufacturer.

*Solution.* (1) Enter the abscissa of Fig. 2-26(a) at $I_C = 0.1$ mA and read the

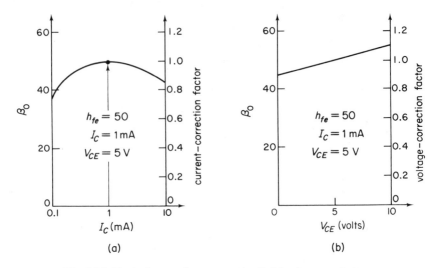

**Fig. 2-26.** Typical correction curves supplied by the manufacturer for $\beta_0$ or $h_{fe}$. Current-correction and voltage-correction factors are shown for two systems of data presentation in (a) and (b), respectively.

current-correction factor of 0.9. (2) Enter the abscissa of Fig. 2-26(b) at $V_{CE} = 10$ V and read the voltage-correction factor of 1.1. (3) Multiply the test-point value of $\beta_0$ by both correction factors:

$$\beta_0 = (50)(0.9)(1.1) \cong 49$$

In this example the correction factors offset one another.

## PROBLEMS

2-1  Sketch the circuit symbols and pictorial representations for a *pnp* and an *npn* bipolar transistor.

2-2  Draw the supply-voltage polarities for an *npn* transistor intended to operate in the active region as (a) a common-base, (b) a common-emitter, and (c) a common-collector amplifier.

2-3  Repeat Problem 2-2 for a *pnp* transistor.

2-4  Sketch the internal transistor currents, identifiying majority and minority carriers for an *npn* transistor in the active region. Include bias batteries.

2-5  If $I_{CBO}$ measured 1 μA at $V_{BC} = 5$ V in Fig. 2-7, what would $I_{CBO}$ probably measure at $V_{BC} = 10$ V?

2-6  Redraw Fig. 2-8 for an *npn* transistor.

2-7  Redraw Fig. 2-9 for an *npn* transistor. Does $I_E = I_B + I_C$ for both *npn* and *pnp*?

2-8  Recalculate $I_C$ and $I_B$ in Example 2-1 if $\alpha_F$ is changed to 0.995.

2-9  What is $I_C$ and $I_B$ in Example 2-2 if $\alpha_F = 0.995$?

2-10  Redraw Fig. 2-11 for *npn* silicon and germanium transistors.

2-11  How do the currents change in Example 2-3 if $\alpha_F = 0.98$?

2-12  How much will $I_C$ increase in Problem 2-11 if a germanium transistor is substituted and $V_{BB} = 1.3$ V, $I_{CBO} = 5$ μA?

2-13  If $I_{CBO} = 5$ μA at 25°C and $\beta_F = 200$ in both Figs. 2-11 and 2-12, what change will occur in the collector currents of each circuit if $I_{CBO}$ is quadrupled to 20 μA by a temperature rise?

2-14  Modify Fig. 2-13 to display a common-emitter collector characteristic of an *npn* transistor in the first quadrant.

2-15  Calculate $\beta_F$ for each base-current curve in Fig. 2-14 at $V_{CE} = -4$ V (except $I_B = 0$).

2-16  Calculate the corresponding values of $\alpha_F$ in Problem 2-15.

2-17  Evaluate $\beta_0$ and $\alpha_0$ for each point in Problem 2-15.

2-18  Assume $I_{CBO} = 2$ μA in Fig. 2-15. What is $\beta_F$?

2-19  From Fig. 2-17(b), determine $R_{sat}$ at $V_{CE} = 0.1$ V for each base-current curve.

2-20  Given $R_{sat} = 200$ Ω, $I_{CBO} \cong 0$, and $\beta_F = 50$. Sketch the collector characteristic curves for this transistor at $I_B = 5, 10, 15,$ and 20 μA, for $V_{CE}$ between 0 and 10 V, in the active region.

2-21  Plot $I_C$ versus $I_B$ from Fig. 2-24 for $V_{CE} = 15$ V. Illustrate the difference between $\beta_0$ and $\beta_F$ at $I_{B2} = 40$ μA, $V_{CE} = 15$ V.

2-22  What is the value of $\beta_0$ at 10 mA in Fig. 2-26?

# Chapter 3

**3-0** INTRODUCTION ................................. 75
**3-1** THE DC LOAD LINE ........................... 75
**3-2** THE OPERATING POINT ........................ 79
**3-3** MEASURING THE OPERATING POINT ............... 82
**3-4** INPUT RESISTANCE BY GRAPHICAL ANALYSIS ........ 83
**3-5** VOLTAGE AMPLIFICATION BY GRAPHICAL ANALYSIS .. 85
**3-6** THE AC LOAD LINE ............................. 88
**3-7** AC-DC LOAD LINE DEMONSTRATION ............... 92
      PROBLEMS ....................................... 93

# Graphical Analysis of a Transistor Circuit

## 3-0 Introduction

Graphical analysis of a basic transistor circuit is studied to give a pictorial representation of amplification resulting from the interaction of a transistor and its external circuit. Graphical analysis is a visual aid that gives a picture of (1) the concept of amplification, (2) the concept of maximum possible output voltage swing, (3) the dependence of operating point on both transistor characteristics and external circuit elements, and (4) relates operating-point measurements for purposes of verification and interpretation.

The transistor does not function alone. A low-energy input signal controls the transistor and the transistor in turn controls the delivery of energy from a high-energy power supply to a load. Resistors are the simplest form of load, and since many other loads may often be approximated by a resistance we shall use the resistor to introduce a graphical method of solving basic transistor circuits.

## 3-1 The DC Load Line

The graphical procedure which superimposes circuit characteristics over device characteristics is known as drawing the *load line*. In Fig. 3-1, $V_{CC}$ and $R_L$ are the external circuit and their characteristics are to be drawn as a load line on the characteristics of transistor $Q$. Of course, an equilibrium or *quiescent* condition will occur when transistor, resistor, and power supply are connected together and the resulting circuit equilibrium is fully described by the *operating* or *quiescent point*. Since the three are connected in series, each carries the same current. Knowing the current and load resistor value will give the voltage drop across the resistor. Usually the power supply has a negligible internal resistance so the supply voltage does not depend on the

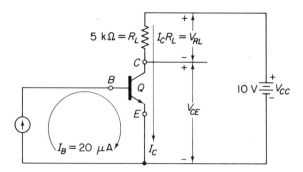

**Fig. 3-1.** $V_{CC}$, $R_L$, and $Q$ interact to produce equilibrium at some value of $I_C$, $I_B$, and $V_{CE}$ whose values define a transistor operating point.

magnitude of the current being drawn from it. Therefore the supply voltage furnishes the voltage drop across the load resistor and also the drop across the transistor.

There are two ways we can tell what the voltage drop across the transistor must be: (1) Subtract the drop across the load resistor from the supply voltage, and the remainder is what is left across the transistor. (2) Knowing the value of current through the transistor, obtain its voltage drop from the characteristic curve.

Figure 3-1 illustrates a basic common-emitter circuit. It shows $V_{CC}$ supplying both $V_{RL}$ and $V_{CE}$, with their relationship expressed by a loop equation of the output circuit as

$$V_{CC} = V_{RL} + V_{CE} \tag{3-1}$$

$V_{RL}$ may be replaced by its equivalent, $I_C R_L$, so that Eq. (3-1) can be written as

$$V_{CC} = I_C R_L + V_{CE} \tag{3-2}$$

Equation (3-2) is the *load-line equation*. That is, plotting Eq. (3-2) on a graph of $I_C$ versus $V_{CE}$ superimposes $V_{CC}$ and $R_L$ on top of the output characteristics.

A load line is plotted on a graph, without regard for other curves that may already be displayed on the same graph. To emphasis this important point, the same load line and supply voltage will be plotted on three graphs made up of an $I_C$ ordinate and $V_{CE}$ abscissa, as in Fig. 3-2. In all three examples $R_L$ is equal to 5 kΩ and $V_{CC} = 10$ V so that Eq. (3-2) may be plotted as $10 = 5000 I_C + V_{CE}$. There are several shortcuts to plotting a load line other than picking any value of $I_C$ and solving for $V_{CE}$, while hoping it falls on the graph. The fastest and easiest method is shown in Fig. 3-2 and the procedure is as follows (this procedure is general and valid for circuits with BJTs, FETs, or vacuum tubes):

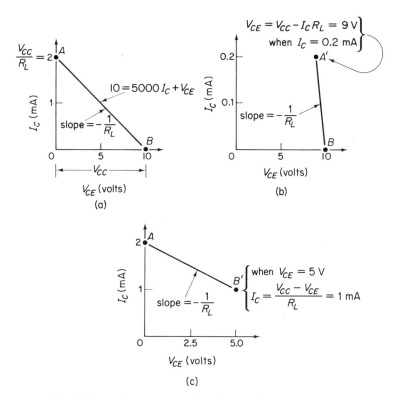

**Fig. 3-2.** Load-line equation (3-2) is plotted on all three graphs for $R_L = 5\ \text{k}\Omega$, $V_{CC} = 10$ V. Both axis intercepts can be plotted in (a). Only the voltage-axis intercept and current-axis intercept can be plotted in (b) and (c), respectively.

1. Find the current-axis intercept of $I_C = V_{CC}/R_L$, by setting $V_{CE} = 0$ in Eq. (3-2) and solving for $I_C$. Point $A$ locates this intercept in Fig. 3-2(a) with coordinates at $I_C = V_{CC}/R_L = 2$ mA, $V_{CE} = 0$.
2. Find the voltage-axis intercept of $V_{CC} = V_{CE}$, by setting $I_C = 0$ in Eq. (3-2) and solving for $V_{CE}$. Point $B$ locates this intercept with coordinates at $I_C = 0$, $V_{CE} = V_{CC} = 10$ V.
3. Since $A$ and $B$ are the locations of two points on a straight-line curve, connect them to obtain the load line. In many cases our problem of drawing the load line is completed at this point. However in some problems the coordinates of the current intercept will be located off the graph so that $A$ cannot be plotted, as in Fig. 3-2(b). However, point $B$ can be plotted so we continue to step (4a).
4. (a) Knowing the location $B$, we want to find another point on the load line which falls on the graph. While any current values may be chosen on a trial-and-error basis, it is best to look at the current

axis and pick the *largest* value of $I_C$. This value is 0.2 mA in Fig. 3-2(b). Substitute in Eq. (3-2) for $I_C$ and solve

$$10 = (0.2 \times 10^{-3})(5000) + V_{CE}$$

for $V_{CE}$ to obtain $V_{CE} = 9$ V. Point $A'$ lies on the load line and is located at $I_C = 0.2$ mA, $V_{CE} = 9$ V. Connecting $A'$ and $B$ will yield the load line. In other problems it may be possible to plot the current intercept of the load line on the given graph, but impossible to plot the voltage intercept as in Fig. 3-2(c). In this event perform steps (1) and (2), then go to step (4b).

(b) Knowing the location of point $A$, we go for the easiest second point on the load line which will (i) fall on the graph and (ii) be easy to calculate and plot. Pick the largest value of $V_{CE}$ available on the voltage axis. This value is 5 V in Fig. 3-2(c). Substitute in Eq. (3-2) and solve

$$10 = 5000 I_C + 5$$

for $I_C$ to obtain $I_C = 1$ mA. Point $B'$ is located at $I_C = 1$ mA, $V_{CE} = 5$ V. Connecting $A$ and $B'$ will yield the load line.

It should be noted that the situation may arise where no point on the load line can be found on the graphs. For example, if the current axis had a maximum value of 0.2 mA and $V_{CE}$ of 5 V, then we could not locate any point on a load line of $R_L = 5$ kΩ and a supply of $V_{CC} = 10$ V.

This situation merely means we have tried something that is physically impossible. The procedure informs us of this fact and we must change any or all of $V_{CC}$, $R_L$, or the transistor region of operation (corresponding to the graph's limits) until a realizable circuit is made.

There is a second common method of analyzing load lines which the student must know. It is based on the standard formula for a straight line as studied under analytic geometry and is given by

$$y = mx + b$$

where $y$ = dependent variable plotted on the vertical axis or ordinate,
$x$ = independent variable (the one you choose to vary) plotted on the horizontal axis or abscissa.
$m$ = slope of the curve, and
$b$ = intercept on the $y$ axis.

Compare this notation with Eq. (3-2) by solving for $I_C$:

$$I_C = -\frac{1}{R_L} V_{CE} + \frac{V_{CC}}{R_L} \qquad (3\text{-}3)$$

The direct comparison shows that

$y$ corresponds to $I_C$,
$x$ corresponds to $V_{CE}$,
$m$ corresponds to $-1/R_L$, and
$b$ corresponds to the current-axis intercept of $V_{CC}/R_L$.

Thus we can also use the techniques of analytic geometry and draw the load line by (1) locating the current-axis intercept and (2) laying out a line with slope of $-1/R_L$ from the intercept. Of course, once any point on the load line is plotted, we can pass through it a line with slope of $-1/R_L$ to obtain the load line.

## 3-2 The Operating Point

Thus far, references to the transistor's relationships with the load line have been purposely minimized. But now we look below the load line at the transistor's characteristic curves to learn what is going on in the circuit and how to interpret a meter measurement in terms of the circuit's action. Measurement is the crucial test of our understanding, for we can make some simple measurements and infer from them what is going on in the circuit in terms of the load-line picture.

Characteristic curves of a transistor are shown together with a load line for the upper circuit in Fig. 3-3, to assemble (1) Fig. 3-1—the circuit—with (2) Fig. 3-2(a)—the load line—and (3) the transistor's characteristics. Look for the line of $I_B = 20 \ \mu A$, which is one restriction imposed by the transistor. The load line is the remaining restriction imposed by the external circuit. Intersection of load line and transistor characteristic locate the only point that will satisfy both restrictions simultaneously. By reading the coordinates of point $Q$ from the axes we see that $Q$ identifies one specific combination of currents and voltages allowed by this particular transistor as determined from the characteristic curves. Voltage coordinate $V_{CEQ}$ shows that 5 V exist across the transistor's output terminals. $I_{CQ}$ shows that the transistor is passing 1 mA. Voltage $V_{RL}$ is the difference between $V_{CC}$ and $V_{CE}$, or 5 V, and may also be found by calculating the product of $I_C$ and $R_L$.

Another example of matching the graphical solution to the circuit condition is seen by considering the same circuit in Fig. 3-3 at a different operating point $Q'$. From the graph we locate $Q'$ from the intersection of $I_B = 10 \ \mu A$ and the load line and read directly that $V_{CC}$ is divided to develop $V'_{CEQ} = 7.5$ V across the transistor and $V'_{RL} = 2.5$ V across the load. The new operating-point current is read from the graph to be $I'_{CQ} = 0.5$ mA. Multiplying $I'_{CQ}$ by $R_L$ should equal $V'_{RL}$ and provide a check.

Adding the transistor's characteristics to the graph gives a vital bit of information about the input circuit's requirement to establish the $Q$ point.

80    Graphical Analysis of a Transistor Circuit    Chap. 3

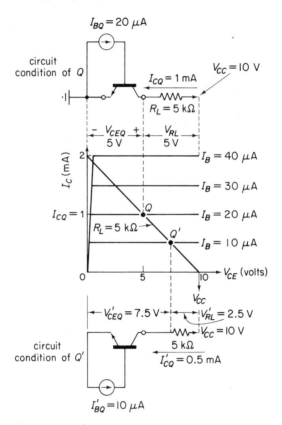

**Fig. 3-3.** Operating points $Q$ and $Q'$ describe two separate equilibrium conditions of the same circuit and define current and voltages in the top and bottom schematics, respectively.

The base current value at the $Q$ point is taken from the transistor's characteristic curves. Visualize many different characteristics for different transistors, all given on different graphs with axes identical to those of Fig. 3-3. Connecting each transistor in series with identical $V_{CC}$ and $R_L$ is the same as drawing the same load line on every graph. At the same operating point on every graph, the only difference will be in the magnitude of the base current. When the base-current value at the operating point is identified, we must proceed to design an input circuit to deliver this required base current.

*Biasing* is the procedure by which the proper value of base current is delivered to the transistor to establish the operating point. Note that the biasing issue has been avoided temporarily by showing the base current emanating from a noncommittal current source. This issue is mentioned now but will be dealt with squarely in Chapter 4. Two examples will be undertaken to examine the effect of varying circuit elements $V_{CC}$ and $R_L$ on the operating point.

*Example 3-1.* A constant-current bias circuit delivers $I_B = 20\ \mu A$ to the circuit of Fig. 3-3. (a) If $V_{CC}$ is changed from 10 V to 7.5 V, find the new operating point and describe how the 5-k$\Omega$ load line shifts. (b) What is the value of $\beta_F$ at the operating point? (c) At approximately what value of $V_{CC}$ will the transistor enter saturation?

*Solution.* (a) As $V_{CC}$ decreases, the load line shifts to the left, maintaining the same slope in Fig. 3-4. Original operating point $Q$ moves left along

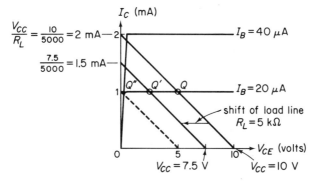

**Fig. 3-4.** Changing $V_{CC}$ moves the operating point along the characteristic curve of $I_B = 20\ \mu A$.

the characteristic curve of $I_B = 20\ \mu A$ to the new operating point $Q'$. The coordinates of $Q'$ are $V_{CE} = 2.5$ V, $I_C = 1$ mA.
(b) The value of $\beta_F$ at both $Q$ and $Q'$ is

$$\beta_F = \frac{I_C}{I_B} = \frac{10^{-3}}{20 \times 10^{-6}} = 50$$

(c) By inspection of the dashed load line in Fig. 3-4, point $Q''$ locates saturation when $V_{CC} = 5$ V. Alternatively, saturation is reached when $V_{CC} \cong I_C R_L$ (since $V_{CE(sat)} \cong 0$) or $V_{CC} = 10^{-3} \times 5000 = 5$ V.

*Example 3-2.* A constant-base bias current of 20 $\mu A$ is delivered to the circuit of Fig. 3-3. (a) If $V_{CC}$ is held at 10 V and $R_L$ is changed from 5 to 7.5 k$\Omega$, find the new operating point and describe the motion of the load line. (b) At approximately what value of $R_L$ will the transistor enter saturation? (c) Apply Eq. (3-2) to operating point $Q'$ in Fig. 3-5.

*Solution.* (a) As $R_L$ increases, the $I_C$ intercept moves down the $I_C$ axis, rotating the load line around pivot point $I_C = 0$, $V_{CE} = V_{CC}$. Operating point $Q$ moves to a new location at $Q'$, where $I_C = 1$ mA, $V_{CE} = 2.5$ V. (b) By inspection of Fig. 3-5 we draw a dashed load line through saturation point $Q''$ and read a current-axis intercept of 1 mA, to find $R_L = 10/10^{-3} = 10$ k$\Omega$. (c) Substituting $Q'$ values in Eq. (3-2),

$$V_{CC} = V_{CE} + I_{CQ}R_L$$
$$10 = 2.5 + (10^{-3})(7.5 \times 10^3) = 10$$

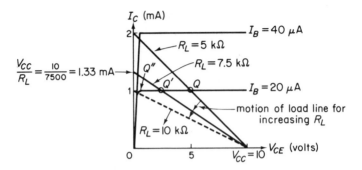

**Fig. 3-5.** Changing $R_L$ rotates the load line around the voltage-axis intercept.

## 3-3 Measuring the Operating Point

It is most important to restate the load-line discussion in terms of electrical measurement. Perhaps the single most important measurement to verify an operating point is a dc voltage measurement between collector terminal and common terminal. It is inconvenient to insert ammeters for two reasons: (a) transistor or integrated circuit terminals are often soldered directly to a printed circuit board, and (b) inserting a microammeter can add 5 k$\Omega$ or more of series resistance and change the $Q$ point. Therefore, it is sound practice to look for a way to measure voltage across a known resistance and calculate the current. It is apparent from a restudy of Fig. 3-3 that if $R_L$ and $V_{CC}$ are known the entire condition of the circuit can be deduced accurately from one measurement of $V_{CE}$. This concept is pursued in Example 3-3.

*Example 3-3.* An input circuit is designed to deliver 20 $\mu$A to the base of a transistor with characteristics supposedly like those of Fig. 3-3. $R_L$ is measured to be 5 k$\Omega$ and $V_{CC}$ is set at 10 V, with intentions of operating the circuit at point $Q$. The circuit is built and, as is quite common, we do not measure the desired $V_{CE}$ but measure, for example, a $V_{CE}$ of 2.5 V. From the measurement of $V_{CE}$, (a) locate the actual operating point; (b) state the most probable action required to relocate the operating point to $Q$.

*Solution.* (a) Since we know values of $R_L$, $V_{CC}$, and $V_{CE}$, the value of $I_C$ can be calculated by Eq. (3-2). However, it is better to reason that the voltage across $R_L$ is $V_{CC} - V_{CE}$ or 7.5 V, and $I_C$ must be this voltage divided by $R_L$ or $7.5/5000 = 1.5$ mA. The actual operating point is located on the load line, halfway between $Q$ and the current-axis intercept, at $V_{CE} = 2.5$ V, $I_C = 1.5$ mA. (b) To restore the $Q$ point we (1) cannot replace the transistor or (2) change $V_{CC}$, or (3) change $R_L$ because it is this combination that must work satisfactorily at the predicted point. The problem is that the base current is too high, and corrective action must be taken to lower it at the input circuit. (Note: Adjustments to the output circuit are not the solution.)

Since real loads may be encountered which vary from very low to very high magnitudes, the limits of load-line plots are of interest. Obviously the lower limit of any load is 0 Ω. A load line of 0 Ω has a slope of $-1/0 = \infty$ and goes straight up from the voltage-axis intercept. The other extreme is an arbitrarily large load with a slope of $-1/\infty = 0$ and appears as a horizontal line coinciding with the horizontal axis. Operating points are indeterminate with infinite loads.

## 3-4 Input Resistance by Graphical Analysis

A pictorial concept of input resistance and voltage gain is introduced by performing graphical analysis on the common-emitter circuit of Fig. 3-6.

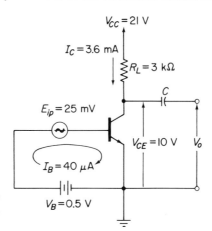

Fig. 3-6. Common-emitter circuit to introduce voltage gain $V_o/E_i$.

The desired dc operating-point currents and voltages are shown and in addition we see a sinusoidal signal input voltage with a peak value of $E_{ip}$ in series with the base loop. $E_{ip}$ will cause dc bias current $I_B$ to vary sinusoidally. The sinusoidal variation of $I_B$ constitutes a signal current and will have some peak value $I_{bp}$, determined by ac resistance presented to $E_{ip}$. From Fig. 3-6 we see that the only ac resistance in series with $E_{ip}$ is the transistor's input resistance, since we assume that the internal resistance of $V_B$ is negligible.

In order to evaluate $I_{bp}$ we must refer to the transistor's input characteristic in Fig. 3-7 and locate operating point $Q$ from the dc values of $I_B = 40 \ \mu A$ and $V_B = V_{BE} = 0.5$ V. Since $E_{ip}$ is in series with $V_B$ we can draw $E_{ip} \sin \omega t$ superimposed on $V_B$ to show that $E_{ip}$ aids and opposes $V_B$ once during each cycle in Fig. 3-7. $V_{BE}$ must swing between $0.5 - 0.025 = 475$ mV and $0.5 + 0.025 = 525$ mV.

These peak excursions of $V_{BE}$ are projected upwards to locate points $D$ and $C$ on the input characteristic and we can read the corresponding peak base currents at $I_{bp} = 20 \ \mu A$ and $I_{bp} = 60 \ \mu A$, respectively. Intermediate

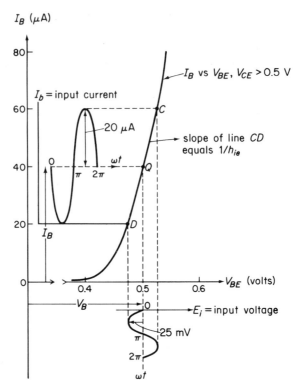

**Fig. 3-7.** Input signal current-voltage relationship is obtained from the input characteristic and defines input resistance $h_{ie}$.

values of $E_i$ will cause the operating point to be located either between $Q$ and $D$ or between $Q$ and $C$. The operating point must therefore always swing along the characteristic curve between $D$ and $C$ a line known as the *operating path*. It is clear from Fig. 3-7 that the slope of line $CD$ is the reciprocal of *incremental input resistance* $h_{ie}$, where

$$h_{ie} = \frac{\Delta V_{BE}}{\Delta I_B} \tag{3-4}$$

We conclude that $E_{ip}$ in Fig. 3-6 sees the transistor's input resistance as $h_{ie}$, where $h_{ie}$ is calculated from peak-to-peak values of $\Delta I_B = 2I_{bp} = 40\ \mu A$ and $\Delta V_{BE} = 2E_{ip} = 50$ mV:

$$h_{ie} = \frac{50 \times 10^{-3}}{40 \times 10^{-6}} = 1250\ \Omega$$

Finally, as with the diode, input resistance to incremental or ac voltages varies with operating-point location because the slope of the $I_B - V_{BE}$ characteristic differs, particularly at low values of base current.

## 3-5 Voltage Amplification by Graphical Analysis

From the results of Section 3-4 we see that $E_{ip}$ in Fig. 3-6 causes an ac base current to ride on the dc base current. The transistor provides an ac current gain $\beta_0$ which amplifies the ac base current and causes an ac collector current to vary around the dc collector current. This amplified ac or incremental collector current passes through load resistor $R_L$ and develops an incremental output voltage $V_o$. Capacitor $C$ allows only the ac output voltage to be measured at $V_o$.

To find $V_o$, begin by drawing the load line for Fig. 3-6 on the output characterisitcs in Fig. 3-8. Locate operating point $Q$ at the intersection of base bias current $I_{B2} = 40$ μA and the load line to complete the dc portion of our problem.

We have already established from Section 3-4 that $E_{ip}$ will vary base current by $I_{bp} = 20$ μA. The sinusoidal input current $20 \sin \omega t$ is drawn in Fig. 3-8. When $E_{ip}$ decreases base current by $I_{bp}$ to a value of $I_{B1} = 20$ μA,

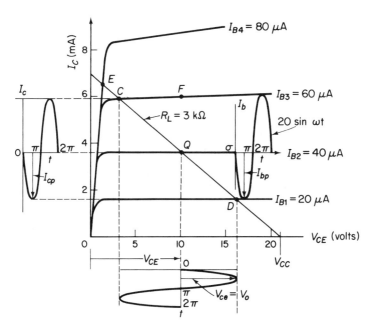

**Fig. 3-8.** Signal input current $I_{bp}$ from Fig. 3-7 causes signal output current $I_{cp}$ and voltage $V_o$.

the operating point must move along the load line to point $D$. Furthermore, if $E_i$ drives the base negative with respect to the emitter, sinusoidally, (1) base signal current will decrease sinusoidally causing (2) the $Q$ point to move sinusoidally toward point $D$, indicating (3) collector current changes sinusoidally from 3.6 to 1.6 mA, and (4) $V_{CE}$ changes from 10 V to approximately 16 V. This sequence of cause and effect is shown in the time interval $0$–$\pi$ in Fig. 3-8. During the interval $\pi$–$2\pi$, $E_i$ drives the base *positive* with respect to the emitter, increasing $I_B$ by $I_{bp}$, increasing $I_C$ by $I_{cp}$ and resulting in $V_{CE}$ going *negative* from 10 to 3.5 V.

The peak-to-peak change in $V_{CE}$ is measured from Fig. 3-8 to be $16 - 3.5 = 12.5$ V. This change is coupled through capacitor $C$ in Fig. 3-6 to give the same peak-to-peak output voltage swing for $V_o$. Voltage gain $A_V$ is defined as the ratio of output voltage $V_o$ to input voltage $E_i$, and is evaluated from

$$A_V = -\frac{V_o}{E_i} = -\frac{12.5}{50 \times 10^{-3}} = -250 \qquad (3\text{-}5)$$

To summarize: Graphical analysis has shown in Fig. 3-8 that the base signal current moves the operating point along *operating path CD* and from points $C$ and $D$ we can measure the peak excursions of collector current and voltage.

1. When input signal voltage $E_i$ drives the base-input terminal negative, the collector-output terminal goes positive. Therefore $V_o$ is 180° out of phase with $E_i$ and this relationship is accounted for by a minus sign in the voltage-gain equation $V_o = -A_V E_i$.

2. When the peak value of $I_{bp}$ is raised much above 20 $\mu$A, severe distoration will be introduced by the transistor. For example, if signal current $I_b$ raises base current to 65 $\mu$A in Fig. 3-8, collector current will increase to saturation at point $E$. Any further increase in $I_{bp}$ will not change collector current so that the signal input information is not reproduced at the output. Tops of the collector-current wave and output-voltage wave will remain flat (clipping) for all values of instantaneous base current greater than 65 $\mu$A or signal base-current peaks greater than 25 $\mu$A. Thus for maximum output voltage swing we would bias the transistor halfway up the load line at $V_{CE} = V_{CC}/2$.

3. Current gain of the circuit is slightly less than $\beta_0$ because point $C$ in Fig. 3-8 represents a smaller collector current than point $F$. Current $QF$ is a current change due to $\beta_0$ and $I_{bp}$, while $QC$ is the actual current gain. This is because the characteristic curve of $I_{B3} = 60$ $\mu$A is not exactly horizontal and, as we shall see in Chapter 5, is due to a decrease in width of the collector-base junction with increased $V_{CE}$.

4. From Fig. 3-7 we see that $E_i$ must be very small or one half-cycle

of base current will differ in magnitude from the other because of the input characteristic. Furthermore, in Fig. 3-8 equal changes in $I_B$ produce unequal changes in $I_C$ because the spacing between characteristic curves increases with increasing $I_C$ and $V_{CE}$.

All of the above principles indicate that signals must be limited to small variations of bias currents or bias voltages so that distortion will be minimized.

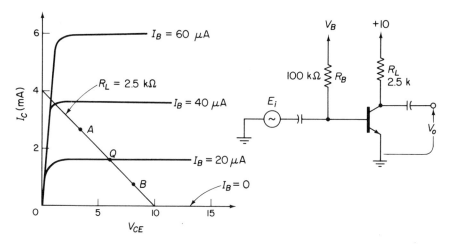

Fig. 3-9. Circuit and solution to Example 3-4.

*Example 3-4.* Input characteristics of the transistor in Fig. 3-9 are given in Fig. 3-7. (a) Using graphical techniques, find the value of $V_B$ to establish operating point $Q$ on the load line of Fig. 3-9. (b) Find the approximate peak-to-peak signal base current change for a peak-to-peak $E_i$ of 50 mV. (c) Find the approximate peak-to-peak output voltage $V_o$. (d) Find the voltage gain.

*Solution.* (a) For a $Q$-point base current of 20 $\mu$A, $V_B$ must furnish 0.47 V for the transistor plus $(20 \times 10^{-6})(10^5) = 2$ V for the drop across $R_B$ or approximately 2.5 V. (b) The reciprocal of the slope at point $D$ in Fig. 3-7 is estimated to be 2.5 k$\Omega = h_{ie}$ so that $I_{b(p/p)}$ is $E_i/h_{ie} = 50 \times 10^{-3}/2500 = 20$ $\mu$A. (c) Points $A$ and $B$ in Fig. 3-9 approximate the operating path caused by $I_{b(p/p)} = 20$ $\mu$A and specify a collector current change of $2.6 - 0.8 = 1.8$ mA. Extrapolating down from $A$ and $B$, locate a $V_{CE}$ variation, or $V_o$ of $8 - 3.5 = 4.5$ V. (d) Voltage gain is $A_V = -V_o/E_i = -4.5/(50 \times 10^{-3}) = -90$.

Example 3-4 illuminates some of the difficulties inherent in graphical analysis. Measurement of the input characteristic's slope is quite inaccurate, because of curvature of the characteristic. Small-signal changes are impossible to plot. In short, graphical analysis is useful for relatively large signal swings and to introduce basic circuit analysis pictorially for very simple circuits.

## 3-6 The AC Load Line

When enough gain is not available from one transistor, its output is connected to the input of another transistor. The second transistor stage amplifies the output of the first so that the overall gain is the product of the individual stages. Graphical analysis is useful in such cases, provided it is possible to represent the input of the second stage as a load resistor. In Fig. 3-10, $R_L$

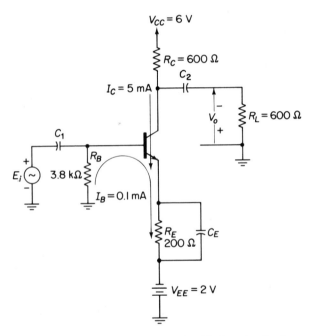

Fig. 3-10. Common-emitter circuit with dc and ac load.

represents the equivalent input resistance of the second stage and could be the base emitter of another transistor. The usable ac signal output current flows through $R_L$ as the base signal current of the other transistor. Capacitors $C_1$ and $C_2$ block dc currents from flowing through the signal generator and load $R_L$. Capacitor $C_E$ bypasses any signal current component around $R_E$. It is standard practice to consider the capacitors as open circuits to dc currents and as short circuits to ac currents. This practice allows a simpler solution

Sec. 3-6　　　　　　　　　　　　　　　　　The AC Load Line　　89

by separating the circuit of Fig. 3-10 into a dc problem and an ac problem for graphical solution.

In the dc solution, three complicating elements, $V_{EE}$, $R_E$, and $R_L$, are in the circuit which were not present in previous considerations of the dc load line. To draw the dc load line, we look for the *total* dc circuit resistance which will limit the dc collector current, and find it in the sum of $R_E$ and $R_C$. The *total* dc voltage supplying the transistor is the sum of $V_{CC}$ and $V_{EE}$. Assuming that emitter current and collector current are approximately equal, load-line Eq. (3-2) is modified for Fig. 3-10 by writing the dc output loop to obtain

$$V_{CC} + V_{EE} = I_C(R_C + R_E) + V_{CE} \qquad (3\text{-}6)$$

The voltage-axis intercept occurs when $I_C = 0$ at $V_{CE} = V_{CC} + V_{EE}$. The dc load line rises up and to the left from this point at a slope of $-1/(R_C + R_E)$ to the current-axis intercept at

$$V_{CE} = 0; \quad I_C = \frac{V_{CC} + V_{EE}}{R_C + R_E}$$

After the dc load line is plotted on the output characteristic curves, base bias current is found to locate the quiescent operating point and complete the dc solution.

An intuitive approach is adopted to analyze the ac operation. Let $E_i$ be a small sinusoidal voltage in Fig. 3-10 which causes a small sinusoidal base current to flow. The signal component of base current is limited only by $h_{ie}$ of the transistor and not by any contribution due to $R_E$, because capacitor $C_E$ appears as a short circuit to the emitter signal current and shunts all signal current increments around $R_E$. The signal base current is multiplied by $\beta_0$ because of transistor action, causing a collector signal current of $\beta_0 I_b = I_c$. This incremental collector current sees a path through both $R_C$ and $R_L$ in parallel or an *ac load resistance* of $R_{ac} = R_C \| R_L$. This is because $C_2$ appears as a short circuit to signal current variations. We cannot expect the operating point to move along the dc load line because the collector signal current sees a different load resistance than the dc collector current. The peak signal current $I_{cp}$ will develop a peak signal voltage across the ac load resistance equal to $I_{cp} R_{ac}$. The transistor's collector-to-emitter voltage must vary to deliver this peak value so that its peak magnitude is

$$V_{cep} = I_{cp} R_{ac} \qquad (3\text{-}7)$$

This means that the operating point will move from the quiescent location a peak distance of $I_{cp}$ and $V_{cep}$; that is, the operating point moves on an operating path along the *ac load line*. The operating point always returns to the quiescent point when there is no signal, and especially when the signal is passing through its zero value so that the ac load line always passes through the dc quiescent operating point. Slope of the ac load line is $-1/R_{ac}$ or $-I_{cp}/V_{cep}$.

This discussion is illustrated in Example 3-5.

*Example 3-5.* In the circuit of Fig. 3-10, $R_E$, $R_B$, and $V_{EE}$ furnish a bias current of $I_B = 0.10$ mA. Find the dc operating point $Q$ on the transistor's output characteristics in Fig. 3-11.

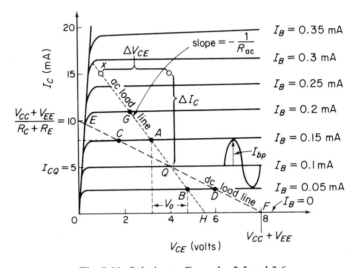

**Fig. 3-11.** Solution to Examples 3-5 and 3-6.

*Solution.* The dc load line is drawn with a slope of $-1/(R_C + R_E) = -1/800$ from the voltage-axis intercept at $I_C = 0$, $V_{CE} = V_{CC} + V_{EE} = 8$ V. In Fig. 3.11 the dc load line is shown as a dashed line. Its intersection with $I_B = 0.1$ mA locates operating point $Q$, where the dc quiescent values of $I_C$ and $V_{CE}$ are 5 mA and 4 V, respectively. *With respect to ground*, the dc potentials at the transistor terminals are

| | |
|---|---|
| Emitter | $-1$ V |
| Base | $-0.38$ V |
| Collector | $+3$ V |

*Example 3-6.* Using the results of Example 3-5,
 (a) plot the ac load line;
 (b) evaluate the transistor's signal current gain graphically, by assuming a base signal current with peak value of 0.05 mA.

*Solution.* (a) By inspection of Fig. 3-10, the ac load is $R_L \| R_C = 600 \| 600 = 300 \, \Omega = R_{ac}$. Pick a convenient $\Delta I_C$ of 10 mA and solve for the corresponding $\Delta V_{CE}$ by Eq. (3-7):

## Sec. 3-6  The AC Load Line

$$V_{cep} = \Delta V_{CE} = \Delta I_C R_{ac} = 10 \times 10^{-3} \times 300 = 3 \text{ V}$$

Since $Q$ is one point on the ac load line, proceed from point $Q$ up on a rise of 10 mA and left on a run of 3 V to locate point $x$ on Fig. 3-11. Extend the ac load line beyond points $x$-$A$ as shown by the dotted line. For a base-current p/p change between $I_B = 0.15$ mA and $I_B = 0.05$ mA, locate points $A$ and $B$ on the ac load line. Point $A$ corresponds to $I_C = 7.8$ mA and point $B$ to 2.5 mA. Thus the transistor's current gain $A_i$, using p/p values from input-base to collector-output terminal is

$$A_i = \frac{\Delta I_C}{\Delta I_B} = \frac{(7.8 - 2.5)}{(0.15 - 0.05)} = 53$$

Note that since the lines of $I_B = 0.15$ mA and $I_B = 0.05$ mA are practically horizontal, $A_i \cong \beta_0$.

**Example 3-7.** Given $h_{ie} = 250 \, \Omega$ in Examples 3-5 and 3-6.
(a) What p/p value of $E_i$ is necessary to cause a p/p base-signal current of 0.1 mA?
(b) Evaluate the circuit's voltage gain $A_V = V_0/E_i$.
(c) What p/p signal current flows through $R_L$ and from $E_i$?
(d) What is the circuit's current gain from source $E_i$ to load $R_L$?

*Solution.* (a) Since $C_E$ and $V_{EE}$ are short circuits to signal currents,

$$E_{i(p/p)} = [I_{b(p/p)}]h_{ie} = (0.10 \times 10^{-3})250 = 25 \text{ mV}$$

(b) From the $V_{CE}$ coordinates of points $B = 4.7$ V and $A = 3.2$ V in Fig. 3-11 we measure $V_{o(p/p)} = 4.7 - 3.2 = 1.5$ V. Using p/p values,

$$A_V = \frac{V_{o(p/p)}}{E_{i(p/p)}} = \frac{-1.5}{25 \times 10^{-3}} = -60$$

(c) The p/p signal current through $R_L$ is $I_{L(p/p)}$, where

$$I_{L(p/p)} = \frac{V_{o(p/p)}}{R_L} = \frac{1.5}{600} = 2.5 \text{ mA}$$

Of course an equal signal current $I_{RC}$ flows through $R_C$, since $R_C = R_L$ and both signal currents should add to $\Delta I_C$ on our graph. We check and ascribe any discrepancies to the unavoidable graphical inaccuracies:

$$I_{L(p/p)} + I_{RC(p/p)} \cong \Delta I_C$$
$$2.5 + 2.5 \cong 5.2$$

(d) Using p/p values, $E_i$ furnishes 0.1 mA to the base and a current of $E_i/R_B = 25 \times 10^{-3}/(3.8 \times 10^3) = 6.5 \, \mu A$ to $R_B$ for a total source current of 106 $\mu A = I_i$. Current gain of the circuit in p/p values is

$$A_i = \frac{I_L}{I_i} = \frac{2.5 \times 10^{-3}}{106 \times 10^{-6}} = 23.6$$

Examples 3-5 through 3-7 demonstrate how graphical techniques and basic circuit theory are combined to analyze a simple transistor amplifier. The most important use of graphical analysis has been reserved until last.

Graphical analysis demonstrates more vividly than any other technique the effect of an ac load on the maximum available signal output voltage. With just the dc load we could increase $I_{b_p}$ to 0.1 mA, driving the operating point from point $E$ to $F$ in Fig. 3-11 for a maximum p/p signal output voltage almost equal to the entire supply voltage. With the same base-current drive we switch in the ac load and drive the operating point from points $G$ to $H$. This cuts our maximum p/p signal output voltage in half. We conclude that when other techniques of circuit analysis are employed, we may have to return to graphical analysis to see what maximum output swing can be obtained.

## 3-7 AC-DC Load Line Demonstration

All of the basic principles of amplification can be demonstrated by the circuit of Fig. 3-12. A CRO is connected (with direct coupling for both $x$

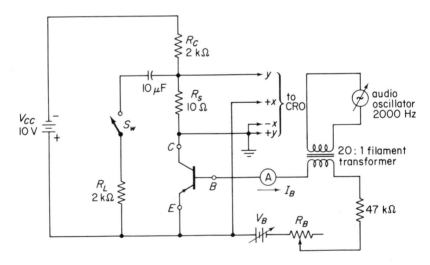

Fig. 3-12. Circuit to demonstrate dc and ac load lines.

and $y$ amplifiers) to display collector current on the vertical axis versus $V_{CE}$ on the horizontal axis. $R_s$ samples collector current with the $y$ amplifier set to a sensitivity of 10 mV/div. The vertical axis is calibrated to display 10 mV/10 = 1 mA/div, and we set the $x$ amplifier to 1 V/div.

For display of an operating point, set the audio oscillator to zero output level, and with switch $Sw$ open, adjust $R_B$ and $V_B$ for the desired $I_B$. As $V_B$ is varied the CRO shows an operating point moving along the dc load line (assuming the CRO has been zeroed at the lower left-hand corner of the screen). Place the operating point halfway up the load line due to $R_C$ and increase the audio oscillator output level to display the dc load line.

When $Sw$ is closed the load-line trace is seen to rotate clockwise around the operating point to a new slope, corresponding to an ac load line due to $R_L \| R_C = 1$ kΩ. The circuit in Fig. 3-12 provides a vivid demonstration of load lines. The CRO can be switched to internal sweep to display time variation of plate current. Moving the operating point with $V_B$ and adjusting signal amplitude with the audio oscillator gives a graphic picture of saturation, cutoff, and maximum output voltage swing.

## PROBLEMS

3-1 Assume $R_L$ is increased to 10 kΩ in Fig. 3-1. Plot the new load lines in Fig. 3-2.

3-2 Assume $V_{CC}$ is changed to 5 V in Fig. 3-1. Plot the new load lines in Fig. 3-2.

3-3 Let $I_B$ be held constant at 10 μA in Example 3-1 and show the resulting changes.

3-4 Compare the effects of holding $I_B = 10$ μA in Example 3-2.

3-5 Assume $V_{CE}$ is measured to be 6.0 V in Example 3-3. What action is required to relocate the operating point to point $Q$?

3-6 Plot load lines for $R_L = \infty$ and $R_L = 0$ in Fig. 3-3 and locate the operating points.

3-7 If $E_{ip}$ were reduced to 10 mV in Fig. 3-7, (a) would $h_{ie}$ change? (b) What is the resultant peak signal base current?

3-8 Estimate $h_{ie}$ at operating points $D$ and $C$ in Fig. 3-7 from a graphical measurement of the slope at those points.

3-9 In Figs. 3-6, 3-7, and 3-8, bias current is adjusted to $I_B = 20$ μA. (a) What is the value of $V_B$? (b) locate the resulting operating-point collector current and voltage. (c) What is the voltage drop across $R_L$. (d) If a peak base current signal of 20 μA is introduced, what are the operating path, peak collector current and voltage values? (e) Evaluate $\beta_F$ and $\beta_0$ at the operating point.

3-10 What happens to voltage gain in Example 3-4 when $R_L$ is increased to 3.3 kΩ.

**3–11** An *npn* transistor is to be operated at $I_C = 2$ mA, $V_{CE} = 4$ V in Fig. 3-8. (a) What value of $V_{CC}$ is required for load resistors of 2 k$\Omega$, 4 k$\Omega$, or 8 k$\Omega$? (b) Assuming the collector bias current is varied by a peak of 0.5 mA, what is the p/p output voltage $V_o$ for each load resistor and the instantaneous values of $V_{CE}$ at each positive-and negative-going peaks?

**3–12** An ac load of 2000 $\Omega$ is coupled to the collector resistors in Problem 3-11. With the same collector signal current peak of 0.5 mA what are the new p/p output voltages and instantaneous values of $V_{CE}$ for each load resistor?

**3–13** Plot dc load lines for $R_L = 1$ k$\Omega$, 2 k$\Omega$, 3.3 k$\Omega$, 5 k$\Omega$, and 10 k$\Omega$ in Fig. 3-9. What happens to the operating point $V_{CE}$ as $R_L$ increases?

**3–14** Plot load lines for $V_{CC} = 15$ V, 10 V, 5 V, 2.5 V, and 0 V in Fig. 3-9. What happens to the operating point $V_{CE}$ as $R_L$ increases?

**3–15** (a) Plot the ac load line of Figs. 3-10 and 3-11 if $R_L$ is reduced to 300 $\Omega$. (b) Assuming the same ac and dc input conditions as in Example 3-7, what are the new values of $V_{o(p/p)}$? (c) What are the p/p signal output currents through $R_L$ and $R_C$? (d) Evaluate graphically the collector signal current swing and verify that it equals the sum of the currents in (c).

**3–16** With switch $Sw$ open in Fig. 3-12, describe the resulting CRO display if (a) the audio oscillator is removed and $V_B$ or $R_B$ is varied, (b) $V_B$ and $R_B$ are adjusted to center the operating point and the oscillator volume is increased.

**3–17** With the dc load line displayed in Fig. 3-12, what happens to the display if $R_C$ is increased?

# Chapter 4

**4-0** INTRODUCTION ............................... 97
**4-1** BIASING A BASIC COMMON-EMITTER AMPLIFIER ...... 98
**4-2** BIAS CALCULATIONS WITH AN EMITTER RESISTOR ....103
**4-3** FACTORS AFFECTING OPERATING-POINT STABILITY ..108
**4-4** STABILITY ANALYSIS BY SUPERPOSITION ...........112
**4-5** STABILIZING THE OPERATING POINT WITH
    AN EMITTER RESISTOR ........................115
**4-6** STABILIZATION WITH A COLLECTOR-BASE RESISTOR ..123
**4-7** EMITTER AND COLLECTOR-BASE RESISTOR
    STABILIZATION .............................126
**4-8** OTHER BASIC BIASING CIRUITS ...................130
**4-9** BIASING WITH DIODES FOR TEMPERATURE
    COMPENSATION .............................134
**4-10** TRANSISTOR-STABILIZED BIASING .................138
**4-11** DEMONSTRATION OF OPERATING-POINT STABILITY ....140
    PROBLEMS .................................141

# Biasing and Stability

## 4-0 Introduction

*Biasing* encompasses the procedures which establish an operating point. *Stability* is concerned with maintaining the operating-point location within required limits despite variations in environmental factors or device characteristics. Biasing circuits vary from simple resistor combinations to complex feedback networks, which may contain nonlinear elements such as temperature compensating thermistors, diodes, or even other transistors.

In practice, bias networks are often selected experimentally. However, it is necessary to master the basic ideas concerned with the analysis or design of a bias network in order to experiment intelligently. Not only must we understand how to establish an operating point and minimize its shift due to environmental factors, but we must also become aware of the bias network's affect on input resistance. We shall rely on a biasing procedure that uses Thévenin's theorem because one of the calculations in the procedure simplifies the evaluation of ac input resistance.

In our biasing procedure we shall generally follow the practice of locating the operating point midway between saturation and cutoff to allow for maximum output swing. This corresponds to a location halfway up the dc load line. Locating the operating point high on the load line will result in signal clipping by transistor saturation when the amplifier is overdriven. An operating point which is low on the load line will cause clipping due to cutoff when the amplifier is overdriven. Also, as we will see when we consider stability, the operating point will rise along the dc load line as temperature increases or shift down as temperature decreases. Our biasing studies will concentrate on the common-emitter configuration. The biasing procedure is identical for the common-base and common-collector configurations.

## 4-1 Biasing a Basic Common-Emitter Amplifier

The operating point of a circuit is selected after careful consideration of many possible requirements, some of which are usually in competition. One ends up with a location for an operating point which specifies (1) the dc collector current $I_C$, (2) the dc voltage across the transistor $V_{CE}$, and (3) the dc base current $I_B$, which will hereafter be referred to as the *bias* current. Bias current is related to $I_C$ by the forward current transfer ratio $h_{FE}$ or $\beta_F$ in the common-emitter configuration.

Thus the biasing procedure is to determine the required value of bias current $I_{BQ}$ and then to design an input biasing network to deliver this current to the transistor. An introductory biasing network consists of a dc voltage supply $V_B$ in series with a single resistor, as shown in Fig. 4-1(a). Voltage suppy $V_B$ may be a separate supply but usually is taken from the same supply,

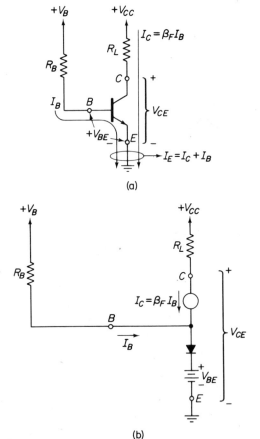

Fig. 4-1. The single-battery bias network for the common-emitter circuit in (a) is modeled in (b).

$V_{CC}$, that feeds the collector terminal. The symbol $V_B$ is chosen in this first analysis so that it may be associated directly with the Thévenin equivalent voltage of our next biasing example.

If $V_B$ is given, then $R_B$ and the dc voltage drop $V_{BE}$ across the base emitter will determine $I_B$. Resistance $R_B$ is the only element that can be adjusted to deliver the proper value of $I_B$. Voltage $V_{BE}$ is taken from the input characteristic curves to correspond with the value of $I_B$. However, if this data is not available we approximate $V_{BE}$ by 0.7 V for a silicon transistor and 0.3 V for a germanium transistor. This approximation introduces an uncertainty which is often not significant, because if $V_B$ is 10 times greater than $V_{BE}$, the base current will be determined primarily by $V_B$ and $R_B$. However, if $V_B$ is less than 10 times $V_{BE}$, we must strive to obtain an accurate value for $V_{BE}$ and also consider its temperature dependence.

A model of the common-emitter circuit is given in Fig. 4-1(b), where $I_{CBO}$ is omitted under the assumption that the transistor is made of silicon. We obtain an expression for $I_B$ in terms of the circuit elements from the input loop equation of Fig. 4-1 as

$$I_B = \frac{V_B - V_{BE}}{R_B} \tag{4-1}$$

Collector current is expressed in terms of circuit elements by

$$I_C = \beta_F I_B = \beta_F \frac{V_B - V_{BE}}{R_B} \tag{4-2}$$

Once we have calculated $I_B$ and $I_C$ from Eqs. (4-1) and (4-2), $I_E$ can be calculated readily from $I_C + I_B$, and $V_{CE}$ can be calculated from knowledge of $V_{CC}$ and $R_L$. Should the circuit be built we would test the operating point by a measurement of $V_{CE}$. If $V_{CE}$ is higher then the value required for the operating point, it is because $I_C$ is too low. The base current also must be too low and $R_B$ should be decreased to compensate.

*Example 4-1.* In the circuit of Fig. 4-1, $V_{CC} = 8$ V, $V_B = V_{CC}$, and $R_L = 1$ kΩ. The transistor is made of silicon, and has a $\beta_F = 100$. Find the value of $R_B$ to bias the circuit in the center of the load line.

*Solution.* At the load line's center, $V_{CE} = V_{CC}/2 = V_{RL} = 4$ V. Current $I_C$ must equal $V_{RL}/R_L = 4/1000 = 4$ mA. Operating point $Q$ is at $I_C = 4.0$ mA and $I_B = I_C/\beta_F = 40$ µA. In the absence of an input characteristic we select 0.7 V for $V_{BE}$. Resistor $R_B$ is calculated from Eq. (4-1) to be

$$R_B = \frac{V_B - V_{BE}}{I_{BQ}} = \frac{8 - 0.7}{40 \times 10^{-6}} = 182 \text{ k}\Omega$$

Had $V_{BE}$ been neglected, $R_B$ would be calculated at 200 kΩ. If resistors of

10% tolerance are used, the difference in calculated values is less than the tolerance.

*Example 4-2.* A single power supply with a value of 20 V is available to feed a load resistor of 2 kΩ. Collector current at the operating point is $I_C = 2$ mA. The silicon transistor has a $\beta_F$ of 100. (a) What is the required value of $R_B$ to establish the proper value of bias current in a circuit like Fig. 4-1(a)? (b) What is the value of $V_{CE}$? (c) If the operating point is not at the midpoint of the load line, is it closer to saturation or to cutoff?

*Solution.* Because the leakage current is negligible in a silicon transistor, assume $I_{CEO}$ is zero and from Eq. (4-2),

$$I_B = \frac{I_C}{\beta_F} = \frac{2000 \ \mu A}{100} = 20 \ \mu A$$

Solve for $R_B$ from Eq. (4-1) to yield

$$R_B = \frac{V_B - V_{BE}}{I_B} = \frac{20 - 0.7}{20 \times 10^{-6}} \cong 1 \ M\Omega$$

(b) $V_{CE}$ is found by subtracting the dc voltage drop across the load resistance from the supply voltage, or

$$V_{CE} = V_{CC} - I_C R_L = 20 - (2 \times 10^{-3})(2 \times 10^3) = 16 \ V$$

(c) The collector voltage is higher than $V_{CC}/2$ so we are lower on the load line than the midpoint and within 4 V of cutoff. The maximum possible output signal swing, without clipping, therefore, will be a peak value of 4 V.

A second basic biasing circuit is given in Fig. 4-2(a), where collector-, base-, and bleeder-current loops are shown for a silicon transistor. We simplify the circuit by application of Thévenin's theorem. Take the Thévenin equivalent circuit of Fig. 4-2(a) as indicated by the section line $A$-$A'$. The Thévenin voltage is found by breaking the connection to the base and calculating the open-circuit voltage across $R_2$ which, according to the voltage divider expression, will be

$$\text{Thévenin voltage} = V_B = \frac{R_2}{R_1 + R_2} V_{CC} \tag{4-3}$$

Thévenin resistance $R_B$ is found by assuming voltage source $V_{CC}$ is a short circuit so that, looking between $A$ and $A'$ we see $R_2 \| R_1$, or

$$\text{Thévenin resistance} = R_B = \frac{R_1 R_2}{R_1 + R_2} \tag{4-4}$$

By comparing the Thévenin equivalent circuit in Fig. 4-2(b) with the bias circuit of Fig. 4-1(b) it is apparent that the circuits are identical. A point of

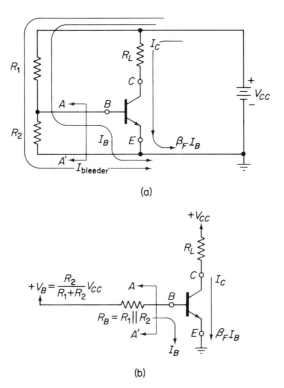

**Fig. 4-2.** The single-battery biasing circuit in (a) is simplified to its Thévenin equivalent in (b).

caution is that the Thévenin voltage *cannot* be measured at any point in the actual circuit. One can measure $V_B$ in Fig. 4-2(a), only by breaking the base connection to $R_2$ and measuring the voltage across $R_2$. Voltage $V_B$ is not $V_{BE}$. The value of $V_B$ differs from that of $V_{BE}$ by the voltage drop across $R_B$. This point is now readily apparent, but when an emitter resistor is introduced in our next biasing network, the student may lose sight of this fact in the laboratory.

In the application of Eqs. (4-3) and (4-4) we can choose a value for $V_B$ and, knowing the required bias current, solve for $R_B$. However, when considering the effect of $R_B$ on the ac input resistance, we may select $R_B$ and, knowing the required bias current, solve for $V_B$. In either case we must be able to solve for $R_1$ and $R_2$ from a knowledge of $V_B$ and $R_B$. Equations (4-3) and (4-4) are solved simultaneously to accomplish this and yield

$$R_1 = \frac{V_{CC}}{V_B} R_B \tag{4-5}$$

$$R_2 = \frac{V_{CC}}{V_{CC} - V_B} R_B \tag{4-6}$$

It is useful to develop an expression for the dc collector current in terms of the external elements for the circuits in Fig. 4-2. (Leakage current is considered to be negligible.) Substituting Eqs. (4-3) and (4-4) into Eq. (4-2) gives

$$I_C = \left(\frac{V_{CC}}{R_1} - \frac{V_{BE}}{R_1 \| R_2}\right)\beta_F \qquad (4\text{-}7a)$$

However, the preferred method of finding $I_C$ is to solve for the value of $I_B$ from Eq. (4-1) and multiply the result by $\beta_F$:

$$I_C = \frac{V_B - V_{BE}}{R_B}\beta_F \qquad (4\text{-}7b)$$

*Example 4-3.* In Example 4-1 the value of $R_B$ was found to be 182 k$\Omega$ and $V_B$ to be 8 V. It is desired to replace $R_B$ with two resistors, $R_1$ and $R_2$, but to maintain the same bias current of $I_B = 40~\mu$A. (a) In choosing $V_B$, what are the maximum and minimum values possible? (b) Choose two intermediate values of $V_B$ and find the values of $R_1$ and $R_2$ to deliver the proper bias current. (c) Calculate the bleeder current for each case and relate its relative size back to the original choice of $V_B$.

*Solution.* (a) The maximum possible value of $V_B$ is equal to $V_{CC}$, so $R_1 = R_B$ and $R_2$ is an open circuit. The minimum value of $V_B$ is equal to $V_{BE} = 0.7$ V and corresponds to a circuit where the battery is connected directly to the base, with no resistors in the bias network.
(b) Pick $V_B = 6.7$ V and $V_B = 1.7$ V and solve for $R_B$:

$V_B = 6.7$ V $\qquad\qquad\qquad\qquad\qquad V_B = 1.7$ V

$$R_B = \frac{V_B - V_{BE}}{I_B} = \frac{6}{40 \times 10^{-6}} = 150~\text{k}\Omega \qquad R_B = \frac{1}{40 \times 10^{-6}} = 25~\text{k}\Omega$$

Solve for $R_1$ from Eq. (4-5):

$$R_1 = \frac{V_{CC}}{V_B}R_B = \frac{8}{6.7}(1.5 \times 10^5) = 179~\text{k}\Omega,\ V_B = 6.7~\text{V}$$

$$R_1 = \frac{8}{1.7}(25 \times 10^3) = 118~\text{k}\Omega,\ V_B = 1.7~\text{V}$$

Solve for $R_2$ from Eq. (4-6):

$$R_2 = \frac{V_{CC}}{V_{CC} - V_B}R_B = \frac{8}{8 - 6.7}(1.5 \times 10^5) = 992~\text{k}\Omega,\ V_B = 6.7~\text{V}$$

$$R_2 = \frac{8}{8 - 1.7}(25 \times 10^3) = 31.8~\text{k}\Omega,\ V_B = 1.7~\text{V}$$

(c) The voltage across $R_2$ is identical in either circuit at $V_{BE} = 0.7$ V. Current $I_{BL}$ is found easily from:

$$I_{BL} = \frac{V_{BE}}{R_2} = \frac{0.7}{922 \times 10^3} = 0.76 \, \mu A \qquad I_{BL} = \frac{0.7}{31.8 \times 10^3} = 22 \, \mu A$$

We conclude that the lower choice of $V_B$ results in a higher value of bleeder current, as is predicted by the lower value of $R_B$.

## 4-2 Bias Calculations with an Emitter Resistor

In Fig. 4-3(a) an emitter resistor $R_E$ has been added. It will be shown in Section 4-5 that this helps stabilize the operating point against temperature variations. But for now consider that leakage currents are negligible so that we may focus attention on the resistance-transformation properties of the transistor. Assuming that the operating point is known, we wish to choose values for $R_E$, $R_B$, and $V_B$ to establish a desired value for the bias current $I_B$. Writing the input loop for the Thévenin equivalent circuit gives

$$V_B = I_B R_B + V_{BE} + I_E R_E \qquad (4\text{-}8)$$

By substituting $I_B = I_E/(\beta_F + 1)$ or $I_E = (\beta_F + 1)I_B$ we can write Eq. (4-8) in two forms:

1. In terms of base current only:

$$V_B = I_B R_B + V_{BE} + I_B(\beta_F + 1)R_E \qquad (4\text{-}9)$$

2. In terms of emitter current only:

$$V_B = \frac{R_B}{\beta_F + 1} I_E + V_{BE} + I_E R_E \qquad (4\text{-}10)$$

Equations (4-9) and (4-10) represent the circuit of Fig. 4-3(a), but now we can draw a circuit which is valid for each of these equations in Fig. 4-3(b) and (c).

First, the circuit of Fig. 4-3(b) represents Eq. (4-9). Base current $I_B$ sees $R_B$ unchanged but sees $R_E$ apparently $(\beta_F + 1)$ times as large. If we divided Eq. (4-9) through by $I_B$ the result would be an expression for resistance:

$$\frac{V_B}{I_B} = R_B + (\beta_F + 1)R_E + \frac{V_{BE}}{I_B} = R_{in} \qquad (4\text{-}11)$$

The resistance $R_{in}$ is that seen looking into the base terminal by $I_B$, and is $R_B + (\beta_F + 1)R_E$ plus the term $V_{BE}/I_B$. What has happened is that by looking into the base with $I_B$ we see $I_B$ going through $R_B$ and $R_E$. But $I_B$ excites the $\beta_F$ generator, causing another component $\beta_F I_B$ to pass through $R_E$ and

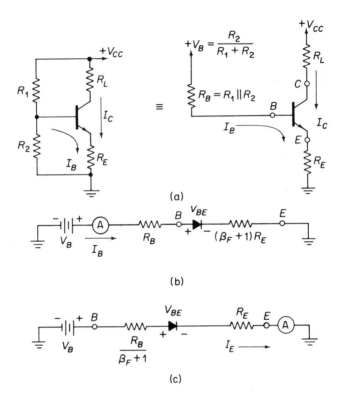

**Fig. 4-3.** The circuit and its Thévenin equivalent in (a) contain emitter resistor $R_E$. In (b) $R_E$ is transformed into the base leg and in (c) $R_B$ is transformed into the emitter leg.

set up a voltage drop. *With respect to the base current* this larger voltage drop may be accounted for by thinking that $R_E$ was increased by $\beta_F$ so that $R_E$ has been transformed out of the emitter leg and into the base leg. Thus the transistor exhibits resistance-transformation properties. From now on we shall use this principle to determine input resistance almost by inspection. That is, Eq. (4-9) will be written directly from Fig. 4-3(a).

In Fig. 4-3(c) the ammeter in the emitter circuit indicates that the ratio of $V_B$ to $I_E$ is the resistance seen looking into the emitter terminal of the transistor. $I_E$ sees $R_E$ unchanged, but if we think of the actual voltage drop across $R_B$ in terms of the larger current $I_E$, then $R_B$ must be divided by $(\beta_F + 1)$. This follows from the fact that the voltage drop across $R_B$ can be expressed in two ways by the equality

$$I_B R_B = \frac{I_E}{\beta_F + 1} R_B$$

Sec. 4-2    *Bias Calculations with an Emitter Resistor*    105

Thus, resistance in the base leg is transformed into the emitter leg by *dividing* any base-leg resistance with the factor $(\beta_F + 1)$. Employing this principle, we can write Eq. (4-10) directly from the circuit of Fig. 4-3(a).

The student has not been given enough background at this point to make an intelligent choice of $R_E$ or $R_B$, since such a choice requires a compromise between requirements of stability, input resistance, available output-voltage swing, and supply voltage. However, the following general guide will prove useful in designing a basic circuit with emitter resistance:

1. Select an operating point to specify $I_C$, $I_B$, and $V_{CE}$.
2. Choose $R_E$ to give a dc voltage drop of no more than about 10% $V_{CC}$.
3. Make $R_B$ about 10–20 times greater than $R_E$.
4. Solve for $V_B$ from Eq. (4-9).
5. Using Eqs. (4-5) and (4-6), solve for $R_1$ and $R_2$.

To *analyze* or find an operating point, knowing $R_1$, $R_2$, $R_E$, $\beta_F$, and $V_{CC}$:

1. Solve for $V_B$ from Eq. (4-3).
2. Solve for $R_B$ from Eq. (4-4).
3. Find $I_B$ from the following form of Eq. (4-9):

$$I_B = \frac{V_B - V_{BE}}{R_B + (\beta_F + 1)R_E} \quad (4\text{-}12)$$

4. Find $V_{CE}$ from $V_{CE} = V_{CC} - (I_C R_L + I_E R_E)$ and $I_C$ from $\beta_F I_B$.

*Example 4-4.* In Fig. 4-4(a), $\beta_F = 50$ for the silicon transistor at an operating point of $I_C = 2$ mA, $V_{CE} = 6$ V. Resistance $R_E$ is given as 1 k$\Omega$. Calculate $R_1$ and $R_2$ to deliver the proper base bias current.

*Solution.* The value of $I_B$ is found from:

$$I_B = \frac{I_C}{\beta_F} = \frac{2 \text{ mA}}{50} = 40 \text{ }\mu\text{A}$$

The actual value of emitter current is $I_C + I_B = 2.04$ mA. In Fig. 4-4(b), $R_E$ is transformed into the base leg, so we may employ Eq. (4-9) to find $V_B$:

$$V_B = I_B[R_B + (\beta_F + 1)R_E] + V_{BE} \quad (4\text{-}9)$$

Substituting,

$$V_B = 40 \times 10^{-6}[R_B + (51 \times 10^3)] + 0.7$$

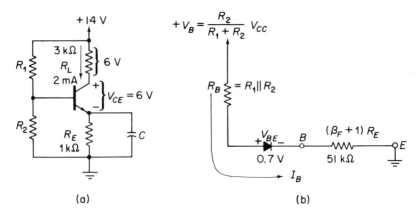

**Fig. 4-4.** Common-emitter circuit in (a) has its equivalent base bias circuit modeled in (b), for Example 4-4.

There are two unknowns in this equation: $R_B$ and $V_B$. Choose $R_B = 10R_E = 10\ \text{k}\Omega$ and solve for $V_B$:

$$V_B = 40 \times 10^{-6}\,(10 + 51) \times 10^3 + 0.7 = 2.44 + 0.7 \cong 3.1\ \text{V}$$

Calculate $R_1$ from Eq. (4-5):

$$R_1 = \frac{V_{CC}}{V_B} R_B = \frac{14}{3.1} \times 10^4 = 45\ \text{k}\Omega$$

Calculate $R_2$ from Eq. (4-6):

$$R_2 = \frac{V_{CC}}{V_{CC} - V_B} R_B = \frac{14 \times 10^4}{14 - 3.1} = 12.8\ \text{k}\Omega$$

**Example 4-5.** Given $V_{BE} = 0.6\ \text{V}$ and $\beta_F = 50$ in the circuit of Fig. 4-5(a). Find (a) $I_B$, (b) $I_C$, (c) $V_{CE}$, and (d) $V_{\text{base}}$.

**Solution.** (a) Evaluate $V_B$, $R_B$, and $(\beta_F + 1)R_E$ for the input loop in Fig. 4-5(b) as follows:

$$V_B = \frac{R_2}{R_1 + R_2} V_{CC} = \frac{40 \times 10^3}{(40 + 80) \times 10^3}(15) = 5\ \text{V}$$

$$R_B = R_1 \parallel R_2 = \frac{(40)(80) \times 10^6}{(40 + 80) \times 10^3} = 26.7\ \text{k}\Omega$$

$$(\beta_F + 1)R_E = (50 + 1) \times 10^3 = 51\ \text{k}\Omega$$

$I_B$ may now be calculated directly by Eq. (4-12):

$$I_B = \frac{V_B - V_{BE}}{R_B + (\beta_F + 1)R_E} = \frac{5 - 0.6}{(26.7 + 51) \times 10^3} = 57\ \mu\text{A}$$

(b) $I_C = \beta_F I_B = 50(57)\ \mu A = 2.85$ mA
(c) Emitter current is $I_C + I_B = I_E = 2.9$ mA. To find $V_{CE}$ we obtain the voltage drops across $R_E$ and $R_L$ and subtract their sum from $V_{CC}$:

$$V_{RE} = I_E R_E = 2.9 \times 10^{-3} \times 10^3 = 2.9\ V$$
$$V_{RL} = I_C R_L = (2.85 \times 10^{-3})(3 \times 10^3) = 8.6\ V$$

Finally:

$$V_{CE} = V_{CC} - (V_{RL} + V_{RE}) = 15 - (8.6 + 2.9)$$
$$= 3.5\ V$$

(d) The voltage at the base is determined by adding the voltage drop across $R_E$ to $V_{BE}$ to give

$$V_{base} = V_{RE} + V_{BE} = 2.9 + 0.6 = 3.5\ V$$

*Example 4-6.* To illustrate a circuit analysis technique find (a) all the circuit currents, (b) $V_{CE}$, and (c) $\beta_F$ from the two voltage measurements $V_{collector}$ and $V_{base}$ in Fig. 4-5(a).

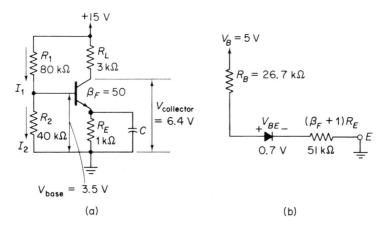

Fig. 4-5. Common-emitter circuit and input model for Example 4-5.

*Solution.* (a) From the collector measurement determine $I_C$:

$$I_C = \frac{V_{CC} - V_{collector}}{R_L} = \frac{15 - 6.4}{3 \times 10^3} = 2.85\ mA$$

From the base voltage measurement we find

(1) $I_2 = \dfrac{V_{base}}{R_2} = \dfrac{3.5}{40 \times 10^{-3}} = 87\ \mu A$

(2) $I_1 = \dfrac{V_{CC} - V_{\text{base}}}{R_1} = \dfrac{15 - 3.5}{80 \times 10^{-3}} = 144 \ \mu\text{A}$

(3) $I_B = I_1 - I_2 = 144 - 87 = 57 \ \mu\text{A}$

(b) $V_{CE}$ is found by subtracting the voltage drop across $R_E$ from the collector voltage measurement:

$$V_{RE} = V_{\text{base}} - V_{BE} = 3.5 - 0.6 = 2.9 \text{ V}$$
$$V_{CE} = V_{\text{collector}} - V_{RE} = 6.4 - 2.9 = 3.5 \text{ V}$$

(c) Finally, $\beta_F =$ is calculated from

$$\beta_F = \dfrac{I_C}{I_B} = \dfrac{2.85 \text{ mA}}{57 \ \mu\text{A}} = 50$$

## 4-3  Factors Affecting Operating-Point Stability

Once an operating point is chosen, the collector current requirement is known and a bias network is designed to deliver it. However, environmental factors, such as temperature variation or radiation exposure, may cause the operating point to move. This section is concerned with thermal effects only, so operating-point stability is defined as determining how collector bias current will be changed by a change in temperature Discussion will be concerned mainly with the common-emitter circuit. Collector bias current increases or decreases directly with a temperature increase or decrease, respectively. To understand how much collector current can be allowed to shift we consider the p/p requirement of output voltage. In Fig. 4-6, $V_o$ is the maximum expected output voltage. Point $L$ locates the lowest allowable operating point and point $H$ locates the highest allowable operating point. If the operating point shifts above the high-temperature extreme (point $H$) or below the low-temperature extreme (point $L$) then serious distortion results because of saturation and cutoff, respectively. Point $Q$ is the best mid-temperature operating-point location and the biasing circuit must be designed to limit temperature-induced excursions of the operating point to $\Delta I_C$.

Three transistor parameters are temperature dependent: $I_{CBO}$, $V_{BE}$, and $\beta_F$. Temperature dependence of each parameter can be summarized by

|  | Si | Ge |
|---|---|---|
| 1. $I_{CBO}$ doubles for a temperature increase of | 6°C | 10°C |
| At room temperature $I_{CBO}$ is roughly | $10^{-3} \mu\text{A}$ | $1 \ \mu\text{A}$ |

2. $V_{BE}$ decreases by 2 mV for each 1-°C temperature rise for both silicon and germanium BJTs.

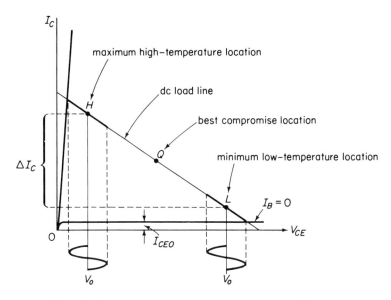

**Fig. 4-6.** Operating point $Q$ moves along the dc load line because temperature variations of $I_C$ are at low frequency.

3. $\beta_F$ is proportional to the spacing between curves of constant base current which spread apart with increasing temperature. $\beta_F$ increases with increasing temperature because higher temperature increases charge-carrier velocity through the base region.

Compare the collector characteristics for a germanium transistor at 25°C in Fig. 4-7(a) with those at approximately 100°C in Fig. 4-7(b). First compare $I_{CEO}$ at $V_{CE} = 10$ V. From the curve of $I_B = 0$, $I_{CEO} \cong 3$ mA at 25°C in Fig. 4-7(a) and $I_{CEO} \cong 11$ mA at 100°C in Fig. 4-7(b). Compare $\beta_F$ at $I_C \cong 18$ mA, $V_{CE} = 10$ V in Fig. 4-7(a), [from Eq. (2-9)],

$$\beta_F = \frac{I_C - I_{CEO}}{I_B} = \frac{(18-3) \times 10^{-3}}{150 \times 10^{-6}} \cong 100 \text{ at } 25°C$$

with $\beta_F$ at $I_C \cong 17$ mA, $V_{CE} = 10$ V in Fig. 4-7(b),

$$\beta_F = \frac{(17-11) \times 10^{-3}}{50 \times 10^{-6}} \cong 120 \text{ at } 100°C$$

Collector characteristics for a silicon transistor at 25°C are shown in Fig. 4-7(c), and at approximately 100°C in Fig. 4-7(d). We compare the values of $\beta_F$ at $V_{CE} = 10$ V and $I_C \cong 8$ mA as follows:

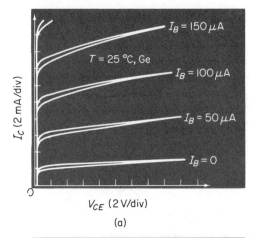

(a)

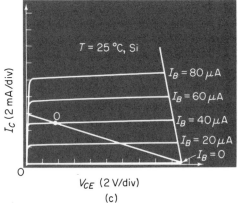

$V_{CE}$ (2 V/div)

(c)

At 25°C,

$$\beta_F = \frac{I_C}{I_B} = \frac{8 \times 10^{-3}}{62 \times 10^{-6}} \cong 130$$

At 100°C,

$$\beta_F = \frac{8 \times 10^{-3}}{35 \times 10^{-6}} \cong 230$$

Observe that $I_{CEO}$, and consequently $I_{CBO}$, are negligible for the silicon transistor since the line $I_B = 0$ almost coincides with the horizontal axis in Figs. 4-7(c) and (d). However, $I_{CBO}$ is most significant in a germanium transistor because when $I_{CEO}$ increases it raises all of the collector curves above $I_B = 0$ by a large amount. The inherent superiority of silicon over germanium with respect to a stable operating point stems directly from the differences in $I_{CBO}$.

Sec. 4-3  Factors Affecting Operating-Point Stability  111

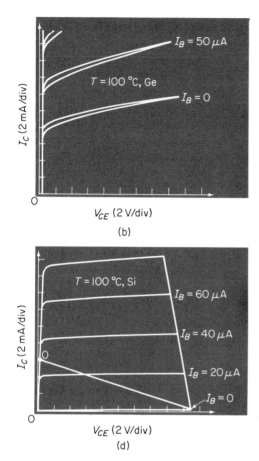

Fig. 4-7. Output characteristics for a germanium transistor are shown at room and elevated temperatures in (a) and (b), respectively. Characteristics for a silicon transistor are given in (c) and (d) for room and elevated temperatures, respectively.

Refer to operating point 0 in Fig. 4-7(c), which indicates the transistor operates at $I_C = 5$ mA, $I_B = 40$ μA, and $V_{CE} = 3$ V at 25°C. Assume the transistor is biased to this operating point by the biasing networks of Fig. 4-1 or 4-2, which hold the base current constant. If the transistor is heated, the curve of $I_B = 40$ μA moves up to a new position in Fig. 4-7(d). Operating point 0 has moved up the dc load line to saturation. $I_C$ has increased to 6 mA and $V_{CE}$ is less than 1 V. Clearly any biasing network which holds base current constant will give poor operating-point stability.

Temperature dependence of $V_{BE}$ for a germanium BJT is illustrated by the input characteristics in Fig. 4-8(a). As the transistor is heated, the $I_B - V_{BE}$ curve moves to the left. The temperature change experienced by the silicon transistor in Fig. 4-8(b) is three times larger and its $I_B - V_{BE}$ characteristic shifts three times as much. In both examples, at a constant base current, $V_{BE}$ decreases as temperature increases by approximately 2 mV/°C.

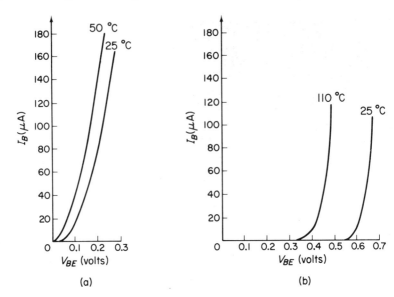

**Fig. 4-8.** Temperature dependence of $V_{BE}$ for (a) germanium and (b) silicon BJT.

We conclude that the temperature dependence of $V_{BE}$ and $\beta_F$ will affect the operating-point stability of both silicon and germanium BJTs. $I_{CBO}$ is not significant for a silicon BJT but it is for a germanium BJT. It is therefore logical to analyze the stability problem by investigating operating-point shift as a consequence of $V_{BE}$ and $\beta_F$ variation with temperature. Any conclusions drawn will be applicable to both types of BJTs. Effects on the operating point due to $I_{CBO}$ will be segregated and will apply to germanium transistors only. This approach leads to treating $I_{CBO}$ in our circuit models as a current source that can be turned on if the transistor is germanium or eliminated if the transistor is silicon.

## 4-4 Stability Analysis by Superposition

A superposition method of analysis is developed here whereby one can examine clearly and simply the individual affects of $I_{CBO}$, $V_{BE}$, and $\beta_F$ on operating-point stability. An incidental benefit is gained by being able to write stability coefficients, *by inspection*, for many circuits, without reference to calculus. In the laboratory we would monitor the dc voltage between collector and ground of a transistor circuit to see it *increase* toward cutoff for *decreasing* collector current and decreasing temperature. Conversely, the transistor operating point shifts toward saturation for increasing temperature. It is therefore necessary to begin our analysis of how to mitigate operating-point shift by deriving expressions for collector current in terms of the temperature-dependent transistor parameters and external circuit elements. Next the expression will be manipulated into a form suitable for *examination of the effect of each parameter separately on the operating-point shift*. Finally we shall develop a method of writing the complicated expression for $I_C$ by inspection. This is accomplished by taking the contribution of each parameter to collector current, then *superimposing* them to give the final result. In the resulting expression we delete the $I_{CBO}$ factors for silicon transistors and retain them for germanium transistors.

Temperature dependence of the simple single-battery bias circuit is examined by reference to Fig. 4-9(a), which models Fig. 4-1(a), and includes $I_{CBO}$. We obtain an expression for $I_C$ in standard fashion by writing the general expression for $I_C$ in terms of $I_B$:

$$I_C = \beta_F I_B + (\beta_F + 1)I_{CBO} \tag{4-13}$$

$I_B$ is written by inspection of the input loop,

$$I_B = \frac{V_B - V_{BE}}{R_B} \tag{4-14}$$

and substituted back into the general expression for $I_C$:

$$I_C = \beta_F \left(\frac{V_B - V_{BE}}{R_B}\right) + (\beta_F + 1)I_{CBO} \tag{4-15}$$

Compare the difference between Eqs. (4-15) and (4-7b) due to leakage current $I_{CBO}$.

From Eq. (4-15) we can predict the affect of each transistor parameter on $I_C$. As temperature increases, $I_{CBO}$ increases, and the increase is multiplied by $(\beta_F + 1)$. Collector current increases and we cannot mitigate against leakage current with this circuit. If $V_B$ is made large with respect to $V_{BE}$ then variations in $V_{BE}$ are unimportant.

The ratio of $V_B$ to $R_B$ is fixed by the required valued of $I_B$ (assuming $V_B > V_{BE}$), so we cannot mitigate against changes in $\beta_F$.

It is possible to pick a low value of $V_B$ so that the decrease in $(V_B - V_{BE})$ with increasing temperature will partially offset the increase in $\beta_F$ and $I_{CBO}$. However, the resulting value of $R_B$ is so low that it will parallel the transistor input and shunt most of the signal current away from the base. As indicated by Eq. (4-14) this circuit tends to keep base current constant but allows collector current to change.

To write Eq. (4-15) by inspection and to introduce the superposition approach, place the equation into the form *collector current equals (1) $\beta_F$ times the base current crossing the emitter junction, plus (2) $I_{CBO}$ (which is*

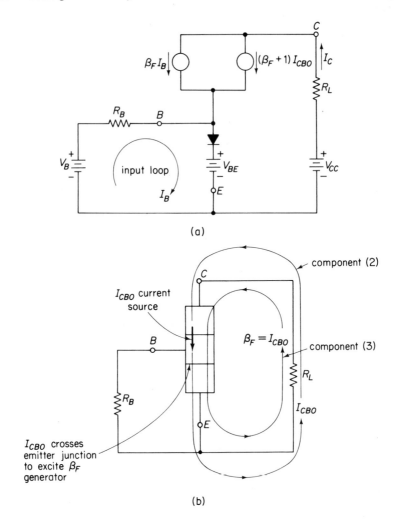

**Fig. 4-9.** Leakage-current components may be superimposed on currents caused by voltage sources. (a) circuit model of Fig. 4-1 (a) with leakage currents. (b) Collector current components due to current generator $I_{CBO}$.

always present) plus (3) $\beta_F$ times that part of $I_{CBO}$ which crosses the emitter junction.

$$I_C = \beta_F\left(\frac{V_B - V_{BE}}{R_B}\right) + I_{CBO} + \beta_F I_{CBO} \qquad (4\text{-}16)$$
$$= \quad (1) \quad + (2) + \quad (3)$$

Component (1) is due only to the voltage sources ($V_{BE}$ is considered a voltage source), and has been illustrated in Fig. 4-2(b). $I_B$ is written by inspec-

*Sec. 4-5*  *Stabilizing the Operating Point with an Emitter Resistor*

tion, ignoring current sources, and multiplied by $\beta_F$ to obtain collector component (1) in Eq. (4-16).

Component (2) is illustrated in Fig. 4-9(b). The constant-current source $I_{CBO}$ cannot be attenuated but it can divide between branches. At the base region $I_{CBO}$ can divide between the base leg and emitter leg. But we assume that $R_B$ is large enough to divert it all through the emitter junction. Since we are examing only $I_{CBO}$ there is no need to show $V_B$ or $V_{BE}$, presumably they are of the proper value to estabilish $I_{CBO}$.

Component (3) results from $I_{CBO}$ crossing the emitter junction and adds $\beta_F I_{CBO}$ to the collector current. Remember, $I_{CBO}$ is dependent principally on temperature, and is considered to be an independent source. On the other hand, $\beta_F I_{CBO}$ is a dependent generator so don't try to multiply the $\beta_F I_{CBO}$ component again by $\beta_F$.

For silicon transistors, components (2) and (3) in Eq. (4-16) may be neglected, while for germanium transistors all three components must be evaluated.

To summarize, we should now be able to write Eq. (4-16) directly from the circuit of Fig. 4-1 by constructing component (1) from $\beta_F$ times $I_B$ solved from the base loop, plus $I_{CBO}$, plus $\beta_F$ times $I_{CBO}$ crossing the emitter junction.

## 4-5 Stabilizing the Operating Point with an Emitter Resistor

A measure of stability may be added to the circuit of Fig. 4-9(a) by adding an emitter resistor, as shown in Fig. 4-10(a). Any action that will tend to hold the collector current constant will stabilize the operating point. A physical explanation of why the emitter resistor acts to stabilize collector current is deferred for the moment. Our problem now is to derive an expression for collector current in terms of the circuit elements and temperature dependent transistor parameters, as given by Eq. (4-17):

$$I_C = \beta_F \left[ \frac{V_B + V_{EE} - V_{BE}}{R_B + (\beta_F + 1)R_E} \right] + I_{CBO} + \beta_F \left[ \frac{R_B}{R_B + (\beta_F + 1)R_E} \right] I_{CBO} \quad (4\text{-}17)$$

$$= \qquad (1) \qquad + \quad (2) \quad + \qquad (3)$$

The result of our derivations will prove that we can write Eq. (4-17) *by inspection* if we realize that component (1) is $\beta_F$ times the base current component plus component (2)—the leakage current generator (always present) —plus component (3)—the portion of $I_{CBO}$ that crosses the emitter junction multiplied by $\beta_F$. The derivation proceeds as follows:

1. Derive an expression for $I_B$ in terms of the circuit elements by writing the input loop from the circuit model of Fig. 4-10(b):

$$V_B - I_B R_B - V_{BE} - I_E R_E + V_{EE} = 0$$

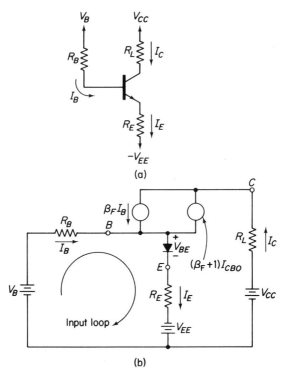

**Fig. 4-10.** The common-emitter circuit in (a) and its model in (b) are stabilized somewhat against operating point variation by $R_E$.

Substitute for emitter current its expression in terms of base current,

$$I_E = (\beta_F + 1)I_B + (\beta_F + 1)I_{CBO}$$

to obtain

$$V_B + V_{EE} - V_{BE} = I_B[R_B + (\beta_F + 1)R_E] + I_{CBO}(\beta_F + 1)R_E$$

and solve for $I_B$:

$$I_B = \frac{V_B + V_{EE} - V_{BE}}{R_B + (\beta_F + 1)R_E} - \frac{(\beta_F + 1)R_E}{R_B + (\beta_F + 1)R_E}I_{CBO} \quad (4\text{-}18)$$
$$= \quad\quad (1) \quad\quad - \quad\quad (2)$$

2. Derive an expression for $I_C$, beginning with

$$I_C = \beta_F I_B + (\beta_F + 1)I_{CBO}$$

Substitute Eq. (4-18) for $I_B$ and factor the term $(\beta_F + 1)I_{CBO}$:

$$I_C = \beta_F \left[ \frac{V_B + V_{EE} - V_{BE}}{R_B + (\beta_F + 1)R_E} \right] + (\beta_F + 1)I_{CBO} \left[ 1 - \frac{\beta_F R_E}{R_B + (\beta_F + 1)R_E} \right]$$

The right-hand term is then manipulated into

$$(\beta_F + 1)I_{CBO}\left[\frac{R_B + R_E}{R_B + (\beta_F + 1)R_E}\right] = I_{CBO}\left\{\frac{[R_B + (\beta_F + 1)R_E] + \beta_F R_B}{R_B + (\beta_F + 1)R_E}\right\}$$

$$= I_{CBO}\left[1 + \frac{\beta_F R_B}{R_B + (\beta_F + 1)R_E}\right]$$

to result in Eq. (4-17).

The interpretation of component (1) is relatively simple. We ignore the leakage-current components, write the expression for the base current, and multiply the result by $\beta_F$. This is exactly the same procedure as led to Eq. (4-12), except that $V_{EE}$ is series-aiding with $V_B$. The circuit model in Fig. 4-11(a) shows how this may be done by inspection. With practice, the student can work directly from the schematic and reason that $I_B$ is forced by the sum of $V_B$ and $V_{EE}$ less the drop across $V_{BE}$, and limited by $R_B$ plus $(\beta_F + 1)R_E$. This gives component (1) in Eq. (4-18) and multiplying by $\beta_F$ gives component (1) in Eq. (4-17). (With a silicon transistor our analysis would stop here.)

The interpretation of component (2) is still straightforward, but as depicted in Fig. 4-11(b), $I_{CBO}$ now has two possible paths from the base region. Some fraction $k$ of $I_{CBO}$ will go through the emitter lead. But since there is now a resistance in the emitter leg, the rest of $I_{CBO}$ will leave the base via $R_B$. We treat $I_{CBO}$ as a current generator whose current divides between two branches proportionally to their resistances. But regardless of how $I_{CBO}$ divides it returns through the collector and accounts for component (2).

Component (3) requires a little discussion. The key to its makeup is contained in the current-dividing term $R_B/[R_B + (\beta_F + 1)R_E]$. This infers that $I_{CBO}$ divides between a resistance $R_B$ and a resistance $(\beta_F + 1)R_E$ and takes the branch represented by $(\beta_F + 1)R_E$. The fraction $k$ is therefore the current-divider term. But to interpret this conclusion physically, refer to Fig. 4-11(b). We see the dependent generator causes $\beta_F k I_{CBO}$ to flow in the collector circuit (it does not split at the base). Thus from the base region $I_{CBO}$ sees $R_E$ apparently multiplied by $(\beta_F + 1)$ and $R_B$ unchanged. We can develop a proof and solve for $k$ by reasoning that the voltage drop between base and emitter via $R_B$ is equal to that via $R_E$, or

$$(1 - k)I_{CBO}R_B = (kI_{CBO} + \beta_F k I_{CBO})R_E$$

and

118   Biasing and Stability   Chap. 4

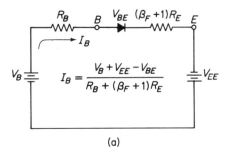

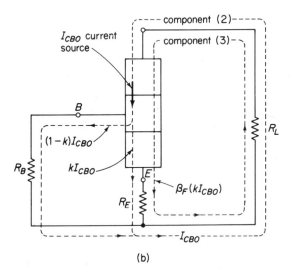

(b)

**Fig. 4-11.** To construct Eq. (4-17) find component (1) from voltage-source currents in (a). Add irreducible component (2) in (b). Fraction $k$ is the current divider between $R_B$ and $(\beta_F + 1) R_E$ to complete component 3.

$$(1 - k)R_B = kR_E(\beta_F + 1)$$

Solving for $k$:

$$\frac{R_B}{R_B + (\beta_F + 1)R_E} = k \qquad (4\text{-}19)$$

This technique clearly identifies the split of $I_{CBO}$ at the base. *Adding emitter resistance diverts part of* $\mathrm{I}_{CBO}$ *away from the emitter junction and prevents that part from being multiplied by* $\beta_F$. This *reduces* considerably the increase in $I_C$ due to an increase in $I_{CBO}$. There is nothing that can be done to the external circuit to reduce component (2). The addition of an emitter battery merely allows us to put in a larger $R_E$ and divert even more of $I_{CBO}$

## Sec. 4-5  Stabilizing the Operating Point with an Emitter Resistor

away from the emitter junction. Note that $R_L$ has no effect on the operating-point stability.

The effect of $V_{BE}$ as it varies with temperature is contained in component (1) of Eq. (4-17). If $V_{EE}$ and $V_B$ are large with respect to $V_{BE}$, then its variation is swamped out. The introduction of $R_E$ requires a higher value of $V_B$ and therefore acts indirectly to reduce the effect of variation in $V_{BE}$.

Analyzing separately, the effect of $\beta_F$ is complicated unless we introduce a simplification by stipulating that $R_B$ is small with respect to $(\beta_F + 1)R_E$; we can approximate Eq. (4-17) by

$$I_C \cong \beta_F \left[\frac{V_B + V_{EE} - V_{BE}}{(\beta_F + 1)R_E}\right] + I_{CBO} + \beta_F \left[\frac{R_B}{(\beta_F + 1)R_E}\right] I_{CBO} \qquad (4\text{-}20)$$

for $R_B \ll (\beta_F + 1)R_E$. Since $\beta_F \cong (\beta_F + 1)$, Eq. (4-20) may be further simplified to

$$I_C = \frac{V_B + V_{EE} - V_{BE}}{R_E} + I_{CBO} + \frac{R_B}{R_E} I_{CBO} \qquad (4\text{-}21)$$

Equation (4-21) indicates that the $\beta_F$ terms disappear and so we conclude that the emitter resistor also stabilizes the collector current against changes in $\beta_F$. In practical circuits, $R_B$ is 10–15 times $R_E$ and $\beta_F$ is usually over 100 for small-signal transistors. Thus comparing $R_B = 15 R_E$ with $(\beta_F + 1)R_E = 101 R_E$ we see that the stipulation is not unreasonable.

As a matter of interest, *stability factors* are identified in Eq. (4-17) as the *coefficients* of the temperature-dependent parameters. They are listed in Table 4-1.

Table 4-1

| Stability Factor or Coefficient | Symbol | Expression |
|---|---|---|
| $I_{CBO}$ | S | $1 + \beta_F \dfrac{R_B}{R_B + (\beta_F + 1)R_E} \cong \dfrac{R_B}{R_E}$ |
| $V_B, V_{BE}, V_{EE}$ | M | $\dfrac{\beta_F}{R_B + (\beta_F + 1)R_E} \cong \dfrac{1}{R_E}$ |

A feeling for the circuit's stabilizing action may be developed by two approaches. First, an examination of Eq. (4-18) shows that as $I_{CBO}$ increases, the external base current decreases and may be driven to zero. In fact, if $R_B$ is made up of resistors $R_1$ and $R_2$, then $R_2$ can pass enough of the high leakage current to reverse $I_B$. [Base-current reversal may be demonstrated conclusively by designing a bias circuit for Fig. 4-4(a) to deliver a low base current of a few microamperes at room temperature to a germanium transistor. Place a zero-center milliammeter in series with the base. $V_B$ should be low, at a value of about $\frac{1}{15}V_{CC}$ to give a low value of $R_2$, and $R_E$ should be about 1 kΩ.] In the collector circuit the temperature-dependent parameters will tend to increase collector current with increasing temperature according to $I_C = \beta_F I_B + (\beta_F + 1)I_{CBO}$. But if $I_B$ is decreasing with increasing temperature it tends to offset the increase due to $\beta_F$ and $I_{CBO}$. Thus the collector current is stabilized.

A second and intuitive approach is based on a feedback concept. Assume the circuit of Fig. 4-4 is operating at a reasonable temperature. $I_B$ is determined by $V_B$, $R_B$, and the voltage across $R_E$. Any disturbance, including temperature, which tends to increase the collector or emitter current will raise the voltage drop across $R_E$. Since $R_B$ and $V_B$ do not change, the higher voltage across $R_E$ will decrease the base current, which, in turn, will decrease the emitter current. Therefore this circuit acts to stabilize the emitter current as well as the collector current.

*Example 4-7.* Assume $R_E = 0$ in Fig. 4-12, $R_B = 10$ kΩ, and $V_{BE}$ and $\beta_F$

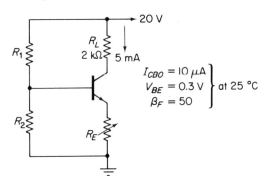

Fig. 4-12. Circuit for Examples 4-7 and 4-8.

do not change with temperature. (a) Calculate $I_B$, $R_1$, and $R_2$ to set $I_C$ at 5 mA. (b) Calculate $I_B$ and $I_C$ when $I_{CBO}$ doubles to 20 μA at 35°C. (c) Repeat (b) for $I_{CBO} = 80$ μA at 55°C.

*Solution.* (a) From Eq. (4-13),

$$I_B = \frac{I_C - (\beta_F + 1)I_{CBO}}{\beta_F} = \frac{5000 - (51)(10)}{50} = 89.8 \ \mu A$$

From Eq. (4-14), $V_B = I_B R_B + V_{BE} = (89.8 \times 10^{-6})(10^4) + 0.3 = 1.19$ V. From Eqs (4-5) and (4-6),

$$R_1 = \frac{V_{CC}}{V_B}(R_B) = \frac{20}{1.19}(10^4) = 168 \text{ k}\Omega$$

$$R_2 = \frac{V_{CC}}{V_{CC} - V_B}(R_B) = \frac{20}{20 - 1.19}(10^4) = 10.6 \text{ k}\Omega$$

Sec. 4-5   Stabilizing the Operating Point with an Emitter Resistor   121

(b) Since $R_E = 0$, $I_{CBO}$ will not divide and $I_B$ is constant at 89.8 μA. The new value of $I_C$ due to $I_{CBO}$ is

$$I_C = \beta_F I_B + (\beta_F + 1)I_{CBO} = 50(89.7 \text{ μA}) + 51(20) \text{ μA} = 5.5 \text{ mA}$$
$$\phantom{I_C} = \phantom{\beta_F I_B + (\beta_F + 1)I_{CBO} = 50} (1) \phantom{\text{ μA}} + (2+3)$$

Note that we should be able to predict this answer from Fig. 4-9 by reasoning (a) component (1) will not be changed by temperature variation, but (b) component (2) in Fig. 4-9(b) is doubled, adding 10 μA to $I_C$, and (c) component (3) is also doubled, adding 50(10) μA = 500 μA for a total increase in $I_C$ of 510 μA or 0.5 mA.

(c) Let's do this part by inspection. Component (2) is increased by 70 μA. Component (3) is increased by 70(50) μA = 3500 μA. $I_C$ goes up by 3.57 mA to 8.57 mA. A check follows for nonbelievers:

$$I_C = \beta_F I_B + (\beta_F + 1)I_{CBO} = 50(89.7) + 51(80) = 4490 + 4080$$
$$= 8.57 \text{ mA}$$

*Summary:*

$$R_E = 0, \quad R_B = 10 \text{ kΩ}, \quad I_{CBO} = 10 \text{ μA}$$

| $I_B$ (μA) | $I_C$ (mA) | $T$ (°C) |
|---|---|---|
| 89.7 | 5 | 25 |
| 89.7 | 5.51 | 35 |
| 89.7 | 8.57 | 55 |

*Example 4-8.* Let $R_E$ be 1 kΩ in Example 4-7. (a) Keep the same value of $R_B$ and calculate $I_B$, $R_1$, and $R_2$. (b) Assume $I_{CBO}$ doubles to 20 μA and find the effect on $I_B$ and $I_C$. (c) Repeat (b) for $I_{CBO} = 80$ μA. Assume $\beta_F$ and $V_{BE}$ remain constant.

*Solution.* (a) The required value of $I_B$ is 89.7 μA. Get $V_B$ from Eq. (4-18):

$$V_B = I_B[R_B + (\beta_F + 1)R_E] + I_{CBO}(\beta_F + 1)R_E + V_{BE}$$
$$= (89.7 \times 10^{-6})(61 \times 10^3)$$
$$+ (10 \times 10^{-6})(51 \times 10^3) + 0.3 = 6.27 \text{ V}$$

From Eqs. (4-5) and (4-6),

$$R_1 = \frac{20 \times 10^4}{6.27} = 32 \text{ kΩ} \qquad R_2 = \frac{20 \times 10^4}{13.73} = 14.5 \text{ kΩ}$$

(b) At 25°C, from Eq. (4-18) $I_B$ was made of two components evaluated by

$$I_B = \frac{V_B - V_{BE}}{R_B + (\beta_F + 1)R_E} - \frac{(\beta_F + 1)R_E}{R_B + (\beta_F + 1)R_E} I_{CBO}$$
$$= \frac{6.27 - 0.3}{61 \times 10^3} - \frac{(51 \times 10^3)}{(61 \times 10^3)}(10 \times 10^{-6})$$
$$= 98.0 \qquad -8.3 = 89.7 \text{ μA}$$
$$\phantom{=} (1) \phantom{98.0 ABC} (2)$$

Observe that $\frac{51}{61}$ of $I_{CBO}$ [component (2)] flows against component (1). If $I_{CBO}$ increases to 20 μA, component (1) remains unchanged but component (2) increases to $\frac{51}{61}20$ μA = 16.6 μA, so $I_B$ decreases to 98.0 − 16.6 = 81.4 μA.

Evaluate components of $I_C$ from Eq. (4-17) at 25°C. Component (1) is $\beta_F$ times component (1) of $I_B$ or 50(98.0) = 4.9 mA. Component (2) is $I_{CBO} = 10$ μA. Component (3) is

$$k\beta_F I_{CBO} = \frac{(10,000)(50)(10)}{61,000} = 82 \text{ μA}$$

When $I_{CBO}$ increases to 20 μA, component (3) changes by twice its former value to 164 μA, so $I_C$ increases to 4.9 + 0.02 + 0.16 = 5.08 mA. (c) When $I_{CBO}$ increases to 80 μA, $\frac{51}{61}I_{CBO}$ enters the base leg, reducing $I_B$ to 98.0 − 67 = 32 μA. Component (3) of $I_C$ increases to

$$\frac{10}{61}(50)(80) = 655 \text{ μA}$$

and $I_C$ increases to 4.9 + 0.08 + 0.65 = 5.63 mA.

Compare Examples 4-7 and 4-8 to observe the change in base current due to $R_E$. By comparison an improvement in collector-current stability is obvious (see Table 4-2).

Table 4-2

|  | $I_C$ at 25°C | $I_C$ at 35°C | $I_C$ at 55°C |
|---|---|---|---|
| Without $R_E$ | 5 mA | 5.51 mA | 8.57 mA |
| With $R_E$ | 5 mA | 5.08 mA | 5.63 mA |

To review the action of $R_E$ on operating-point stability and also to simplify the concept, assume we know from a sketch similar to Fig. 4-6 that $I_C$ of a germanium BJT should be held between 2 mA and 4 mA in an environment where the temperature will vary between 25 and 55°C. We could estimate the variation in $I_{CBO}$ from the transistor's data sheet and might, for example, expect a variation of 10–80 μA over the same temperature range with perhaps a variation in $\beta_F$ of 50–100. Initially assume $\beta_F$ is unchanged by temperature and remains at 50.

With $R_E = 0$, components (2) and (3) of $I_C$ would add to 10 + 500 μA at 25°C and 80 + 100 (80) = 8.0 mA at 80°C for a shift of 7.5 mA. In order to reduce component (3) to a maximum shift of 2 mA we would have to reduce 8080 μA by a fraction of at least $\frac{1}{4}$ and choose $\frac{1}{6}$ for a margin of safety.

In order to obtain $k = \frac{1}{6}$ with $\beta_F = 50$, let $10R_E = R_B$ so that from Eq. (4-19)

$$k = \frac{10R_E}{10R_E + 51R_E} \cong \frac{1}{6} \quad \text{and} \quad k\beta_F = \frac{50}{6} \cong 8$$

And then if $\beta_F$ increases to 100, $k$ changes to

$$k = \frac{10R_E}{10R_E + 100R_E} = \frac{1}{11}$$

But $k\beta_F$ remains approximately constant:

$$k\beta_F = \frac{100}{11} \cong 9$$

Thus component (3) is held to approximately the same increase of 9(80) $\mu$A = 0.72 mA due to our choice of $k$, whether $\beta_F$ doubles or not.

We would accordingly design our bias network to establish $I_C = 2$ mA at 25°C and make $R_B = 10R_E$. If the operating-point shift must be held to 0.5 mA, halve the size of $k$ to $\frac{1}{12}$ and make $R_B \cong 5R_E$. Check that $k\beta_F$ will be approximately 5 for $\beta_F = 50$ and $\beta_F = 100$ and component (3) will change by 5(80) $\mu$A = 0.4 mA.

## 4-6 Stabilization with a Collector-Base Resistor

The basic stabilizing circuit in Fig. 4-13 has a feedback resistor between collector and base. Although this circuit differs from that of Fig. 4-12,

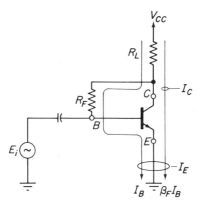

Fig. 4-13. Common-emitter stabilization with collector-base resistor $R_F$. Voltage-source currents are shown for a silicon transistor.

the stabilizing action is similar because base current is automatically adjusted to resist changes in collector current. Goals in the following analysis are also identical: (1) to write the equations for base and collector currents by inspection, and (2) to display the equations so that effects of variations in $I_{CBO}$, $V_{BE}$, and $\beta_F$ may be studied separately.

For a silicon transistor, only the voltage-source components are important and from Fig. 4-13 we can write an expression for the base current component as

$$I_B = \frac{V_{CC} - V_{BE}}{R_F + (\beta_F + 1)R_L} \qquad (4\text{-}22)$$

The collector current component is $\beta_F I_B$, or

$$I_C = \beta_F I_B = \beta_F \frac{V_{CC} - V_{BE}}{R_F + (\beta_F + 1)R_L} \qquad (4\text{-}23)$$

Temperature dependence of $V_{BE}$ will be negligible if $V_{CC}$ is greater than $10V_{BE}$ or approximately 6 V. If $R_F \ll (\beta_F + 1)R_L$, Eq. (4-23) reduces to

$$I_C \simeq \frac{\beta_F(V_{CC} - V_{BE})}{(\beta_F + 1)R_L} \simeq \frac{V_{CC} - V_{BE}}{R_L} \qquad (4\text{-}24)$$

If a germanium transistor is substituted in Fig. 4-13 we must account for leakage-current components and superimpose them upon those already derived. Refer to the current-source model in Fig. 4-14.

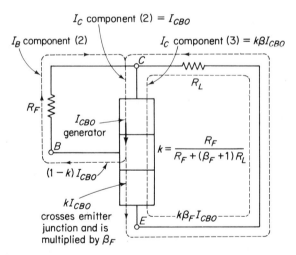

Fig. 4-14. Leakage-current components to be considered when a germanium transistor is substituted in the circuit of Fig. 4-13.

A fraction of $I_{CBO}$ will flow through $R_F$. This fraction is $(1-k)$. The remainder of $kI_{CBO}$ crosses the emitter junction and is multiplied by $\beta_F$. This remainder is $k\beta_F I_{CBO}$, and it traverses $R_L$. Since $V_{BE}$ is accounted for in Fig. 4-13, the voltage drop from $C$ to $B$ is equal to that from $C$ to $E$ in Fig. 4-14 and

Sec. 4-6    Stabilization with a Collector-Base Resistor    125

$$(1 - k)I_{CBO}R_F = (k + k\beta_F)I_{CBO}R_L$$

$$k = \frac{R_F}{R_F + (\beta_F + 1)R_L} \quad (4\text{-}25)$$

Actually Eq. (4-25) can be deduced by reasoning that any current crossing the emitter junction will see an apparent resistance $(\beta_F + 1)$ times as large due to transistor action. If there were an emitter resistor, it would also appear to $kI_{CBO}$ as if it were $(\beta_F + 1)R_E$.

The same principle applies in Fig. 4-13. Base current $I_B$ sees $R_F$ in series with $R_L$, $V_{CC}$, and $V_{BE}$ but because it crosses the emitter junction it turns on $I_B\beta_F$. $I_B\beta_F$ is conducted through $R_L$, increasing the voltage drop across it by a factor of $\beta_F$ so that $R_L$ looks $\beta_F$ times as big to the original $I_B$. Thus $I_B$ sees $R_F + (\beta_F + 1)R_L$.

It is now possible to construct the collector-current equation by adding all components. One component is given in Eq. (4-23). The second component of collector current in Fig. 4-14 is due to leakage current and is seen to be the unreducible $I_{CBO}$. The final component is $k\beta_F I_{CBO}$. Applying superposition to all three components gives the general expression

$$I_C = \beta_F \frac{V_{CC} - V_{BE}}{R_F + (\beta_F + 1)R_L} + I_{CBO} + \beta_F \frac{I_{CBO}R_F}{R_F + (\beta_F + 1)R_L} \quad (4\text{-}26a)$$

$$= \quad (1) \quad\quad + (2) + \quad\quad (3)$$

The companion base-current equation is

$$I_B = \frac{V_{CC} - V_{BE}}{R_F + (\beta_F + 1)R_L} - I_{CBO}\frac{(\beta_F + 1)R_L}{R_F + (\beta_F + 1)R_L} \quad (4\text{-}26b)$$

$$= \quad (1) \quad\quad - \quad (2)$$

The coefficient of $I_{CBO}$ in Eq. (4-26a) is the stability factor for leakage current. *But it is much more informative to consider how* $I_C$ *will shift as a function of* $I_{CBO}$ *by studying the fraction* k. *Any action to reduce* k *reduces the amount of* $I_{CBO}$ *being multiplied by* $\beta_F$. Increasing $R_L$ or $\beta_F$ and decreasing $R_F$ will divert more of $I_{CBO}$ from crossing the emitter junction.

Collector-current shift is explained qualitatively by noting that $I_B$ is established by $V_{CE}$ in Fig. 4-13. Any increase in collector current will increase the voltage drop across $R_L$, resulting in a decrease in $V_{CE}$. This decrease reduces $I_B$, reducing the tendency of $I_C$ to increase.

*Example 4-9.* (a) In the circuit of Fig. 4-13, calculate $R_F$ to establish $I_C = 1$ mA. Assume $\beta_F = 50$, $V_{BE} = 0.6$ V, and neglect $I_{CBO}$. $R_L = 10$ k$\Omega$ and $V_{CC} = 20$ V. (b) Calculate $I_C$ for a transistor with $\beta_F = 100$ to see if $I_C$ is stabilized by $R_F$.

*Solution.* (a) The required value of $I_B$ is $I_B = I_C/\beta_F = 10^{-3}/50 = 20$ $\mu$A. From Eq. (4-22),

$$R_F = \frac{V_{CC} - V_{BE}}{I_B} - (\beta_F + 1)R_L = \frac{20 - 0.6}{20 \times 10^{-6}} - 51 \times 10^4 = 460 \text{ k}\Omega$$

(b) With no stabilization, doubling $\beta_F$ would double $I_C$ to 2 mA, but with $R_F$, $I_C$ is found from Eq. (4-22):

$$I_C = \beta_F \left[ \frac{V_{CC} - V_{BE}}{R_F + (\beta_F + 1)R_L} \right] = 100 \left[ \frac{19.4}{[460 + (101)(10)]10^3} \right] = 1.3 \text{ mA}$$

*Example 4-10.* (a) A germanium transistor is substituted in Example 4-9. Given $I_{CBO} = 2$ μA, $V_{BE} = 0.2$ V, and $\beta_F = 50$, calculate $R_F$ to establish $I_C$ at 1 mA. (b) What fraction of $I_{CBO}$ will cross the emitter junction? (c) Assuming the temperature changes of $V_{BE}$ and $\beta_F$ are negligible, how much will collector current increase if $I_{CBO}$ is increased by 10 μA?

*Solution.* (a) Solving Eq. (4-26a) for $R_F$ gives

$$R_F = \frac{\beta_F(V_{CC} - V_{BE}) - (\beta_F + 1)R_L(I_C - I_{CBO})}{I_C - (\beta_F + 1)I_{CBO}}$$

$$= \frac{50(19.8) - (51)(10^4)(998)(10^{-6})}{10^{-3} - (51)(2)(10^{-6})}$$

$$= 536 \text{ k}\Omega$$

(b) From Eq. (4-25),

$$k = \frac{R_F}{R_F + (\beta_F + 1)R_L} = 0.51$$

(c) Without stabilization, collector current would be increased by $I_{CBO}(\beta_F + 1) = (10 \text{ μA})(51) = 0.51$ mA. The stabilizing circuit allows an increase in component (2) of $I_{CBO} = 10$ μA, plus $k\beta_F I_{CBO} = (0.51)(50)(10) = 225$ μA or a total increase in $I_C$ of 0.26 mA.

## 4-7 Emitter and Collector-Base Resistor Stabilization

Feedback resistors $R_F$ and $R_E$ may both be introduced into a common-emitter circuit for additional inprovement of stability in Fig. 4-15. Assume $I_{CBO} = 0$ to analyze this circuit with respect to silicon transistors. We can then obtain an expression for $I_B$ from the loop equation

$$V_{CC} = (\beta_F + 1)I_B R_L + I_B R_F + (\beta_F + 1)I_B R_E + V_{BE}$$

Solving for $I_B$,

$$I_B = \frac{V_{CC} - V_{BE}}{R_F + (\beta_F + 1)(R_L + R_E)}, \quad I_{CBO} = 0 \quad (4\text{-}27)$$

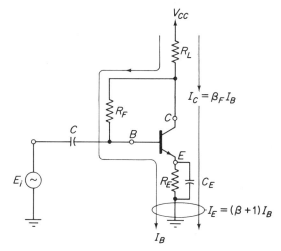

**Fig. 4-15.** Common-emitter silicon transistor circuit with stabilization by both $R_E$ and $R_F$. Voltage-source components of current.

Equation (4-27) shows that $I_B$ sees resistors $R_L$ and $R_E$ apparently multiplied by $(\beta_F + 1)$ in series with $R_F$. This is because the voltage drop across $R_L$ is due to current $(\beta_F + 1)I_B$ and therefore in terms of $I_B$ only, the voltage drop seems to be across $(\beta_F + 1)R_L$. With this viewpoint of the resistance-transforming properties of the transistor, Eq. (4-27) can be written by inspection. Still discounting $I_{CBO}$, the collector current is $\beta_F I_B$, or

$$I_B = \beta_F \frac{V_{CC} - V_{BE}}{R_F + (\beta_F + 1)(R_L + R_E)}, \quad I_{CBO} = 0 \quad (4\text{-}28)$$

The temperature dependence of $\beta_F$ will affect collector current. But if $R_F$ is small with respect to $(\beta_F + 1)(R_L + R_E)$, then Eq. (4-28) may be approximated by

$$I_C \simeq \frac{\beta_F(V_{CC} - V_{BE})}{(\beta_F + 1)(R_L + R_E)} \simeq \frac{V_{CC} - V_{BE}}{R_L + R_E}, \quad R_F \ll (\beta_F + 1)(R_L + R_E) \quad (4\text{-}29)$$

Now, $I_C$ is not dependent on $\beta_F$, and if $V_{CC} > V_{BE}$, the operating point will be quite stable. Physically we can see from Eq. (4-27) that increasing $\beta_F$ will decrease $I_B$. The smaller value of $I_B$ is multiplied by the increased $\beta_F$. These two effects compensate one another and reduce the shift in $I_C$.

The effect of $I_{CBO}$ on operating-point stability will be investigated by examining how it will split at the base rather than by a straight derivation. The superposition principle will be applied by adding the currents due to $I_{CBO}$ with the transistor-current component (1) in Eq. (4-28).

In Fig. 4-16, $I_{CBO}$ will split between the base leg and the emitter leg. A fraction of $I_{CBO}$ equal to $kI_{CBO}$ will cross the emitter junction and be mul-

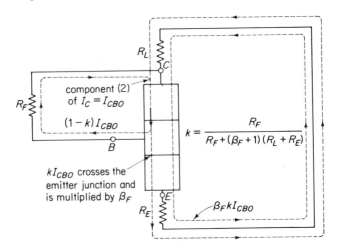

**Fig. 4-16.** Current-source model to account for leakage currents when a germanium transistor is used in Fig. 4-15.

tiplied by $\beta_F$ so that a total of $k(\beta_F + 1)I_{CBO}$ flows throw both $R_L$ and $R_E$. Thus $R_L$ and $R_E$ are apparently increased by a factor of $(\beta_F + 1)$ to the current $kI_{CBO}$. Therefore $I_{CBO}$ sees $R_F$ in the base leg and $(\beta_F + 1)(R_L + R_E)$ in the emitter leg and divides accordingly. The current division in the base leg is, by inspection,

$$(1 - k)I_{CBO} = \frac{(\beta_F + 1)(R_L + R_E)}{R_F + (\beta_F + 1)(R_L + R_E)} I_{CBO} \quad (4\text{-}30)$$

Current division into the emitter leg is

$$kI_{CBO} = \frac{R_F}{R_F + (\beta_F + 1)(R_L + R_E)} I_{CBO} \quad (4\text{-}31)$$

For good stability we want the fraction $k$ to be small, so that the smallest possible part of $I_{CBO}$ will be multiplied by $\beta_F$. From Eq. (4-31) this can be accomplished by making $R_F$ small and $R_L$, $R_E$, and $\beta_F$ large. The effect of $I_{CBO}$ on collector current may be defined by two components:

$$\text{Component (2)} = I_{CBO}$$

$$\text{Component (3)} = \beta_F \frac{R_F}{R_F + (\beta_F + 1)(R_L + R_E)} I_{CBO} \quad (4\text{-}32)$$

These two components may be added to Eq. (4-28) to construct a final equation for $I_C$ in terms of all the parameters. This equation will not be written here because it is identical with Eq. (4-26a) if $R_L$ is replaced by $(R_L + R_E)$. We shall, however, write the total base-current equation by adding the component in Eq. (4-27) to the opposing component in Fig. 4-16 of $(1 - k)I_{CBO}$ to obtain

$$I_B = \frac{V_{CC} - V_{BE}}{R_F + (\beta_F + 1)(R_L + R_E)} - \frac{(\beta_F + 1)(R_L + R_E)}{R_F + (\beta_F + 1)(R_L + R_E)} I_{CBO} \quad (4\text{-}33)$$

Compare Eqs. (4-32) and (4-26b) to see how $R_E$ helps to adjust the base current so that the shift in $I_C$, due to a change in $I_{CBO}$, will be minimized.

*Example 4-11.* (a) Find the values of $I_B$, $I_C$, and $V_{CE}$ for the circuit of Fig. 4-15. Assume $I_{CBO}$ is negligible, $\beta_F = 99$, $R_L = 5\,k\Omega = R_E$, $R_F = 1\,M\Omega$, $V_{CC} = 20$ V. (b) Assume $\beta_F$ is changed to 199. What is the shift in $I_C$? (c) If a leakage current of $I_{CBO} = 5\,\mu A$ were introduced in (a), what would the new values be for $I_B$ and $I_C$?

*Solution.* (a) From Eq. (4.27),

$$I_B = \frac{V_{CC} - V_{BE}}{R_F + (\beta_F + 1)(R_L + R_E)} = \frac{20 - 0.6}{[1000 + (100)(10)](10^3)} = 9.7\,\mu A$$

$$I_C = \beta_F I_B = 99(9.7 \times 10^{-6}) = 0.96 \text{ mA}$$

$$V_{RL} = I_E R_L = (9.7 + 960)(10^{-6})(5000) = 4.85\,V = V_{RE}$$

$$V_{CE} = V_{CC} - V_{RL} - V_{RE} = 20 - 9.7 = 10.3\,V$$

(b) $$I_B = \frac{20 - 0.6}{[1000 + (200)(10)]10^3} = \frac{19.4}{3 \times 10^6} = 6.5\,\mu A$$

$$I_C = \beta_F I_B = 199(6.5)\,\mu A = 1.23 \text{ mA}$$

$I_C$ increases by about 25% for a 100% increase in $\beta_F$.

(c) $I_B$ would be reduced by the fraction of $I_{CBO}$ traversing $R_F$, or

$$I_{CBO} \frac{(\beta_F + 1)(R_L + R_E)}{R_F + (\beta_F + 1)(R_L + R_E)} = 5 \frac{100 \times 10 \times 10^3}{1000 + 100 \times 10^4} = 2.5\,\mu A$$

The new value of $I_B$ will be $9.7 - 2.5 = 7.2\,\mu A$ and $I_C$ will be made up of the original component (1) of 0.96 mA, plus the unchangeable component $I_{CBO} = 5\,\mu A$, plus $\beta_F$ times the fraction crossing the emitter junction, or

$$\beta_F I_{CBO} \frac{R_F}{R_F + (\beta_F + 1)(R_L + R_E)} = (99)(5) \frac{1000}{2000} = 248\,\mu A$$

The new value of $I_C$ is therefore $I_C = 0.96 + 0.005 + 0.25 = 1.21$ mA.

## 4-8 Other Basic Biasing Circuits

When two power supplies are available, *emitter biasing* results in the extremely stable circuit arrangement of Fig. 4-17. The emitter terminal is fairly close to ground potential, separated from ground by $V_{BE}$ and the small

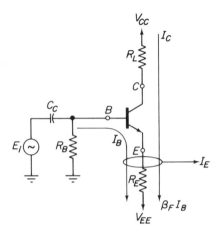

Fig. 4-17. Common-emitter circuit with emitter biasing (silicon transistor).

drop across $R_B$. Emitter current will then depend primarily on $V_{EE}$ and $R_E$, and will be held constant. $R_B$ shunts the ac input signal voltage and cannot be made too small. Neglecting $I_{CBO}$ in Fig. 4-17, $I_B$ sees $R_E$ as $(\beta_F + 1)R_E$ and can be written by

$$I_B = \frac{V_{EE} - V_{BE}}{R_B + (\beta_F + 1)R_E} \qquad (4\text{-}34)$$

$I_E$ is written by inspection as

$$I_E = \frac{V_{EE} - V_{BE}}{\frac{R_B}{\beta_F + 1} + R_E} \qquad (4\text{-}35\text{a})$$

If $V_{EE} \gg V_{BE}$, $R_E$ will be large with respect to $R_B/(\beta_F + 1)$ and Eq. (4-35a) reduces to Eq. (4-35b), which shows $I_E$ to be independent of $\beta_F$ and $V_{BE}$:

$$I_E \cong \frac{V_{EE}}{R_E} \qquad (4\text{-}35\text{b})$$

Collector current $I_C$ is essentially equal to $I_E$ but may be expressed exactly from Eq. (4-34) as

$$I_C = \beta_F I_B = \beta_F \frac{V_{EE} - V_{BE}}{R_B + (\beta_F + 1)R_E} \qquad (4\text{-}36)$$

The effect of $I_{CBO}$ is analyzed by reference to Fig. 4-18. Current $kI_{CBO}$ crosses the emitter junction to cause $\beta_F k I_{CBO}$. Both currents traverse $R_E$ so that $R_E$ is apparently multiplied by $(\beta_F + 1)R_E$ and $k$ is identical to single-battery biasing with an emitter resistor, or

Sec. 4-8　　　　　　　　　　　　　　　　　Other Basic Biasing Circuits　　131

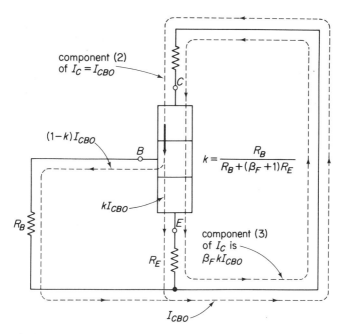

**Fig. 4-18.** Current-source model to account for leakage currents when a germanium transistor is used in Fig. 4-17.

$$k = \frac{R_B}{R_B + (\beta_F + 1)R_E} \quad (4\text{-}37)$$

The collector-current expression may now be written directly by adding to Eq. (4-36) leakage-current component (2) of $I_{CBO}$, plus component (3) of $k\beta_F I_{CBO}$. The resulting equation is expressed by Eq. (4-17) with $V_B = 0$. Since $R_E$ can be made large in the emitter-bias circuit the factor $k$ becomes small, showing excellent stabilizing properties.

Another biasing arrangement shown in Fig. 4-19 illustrates bootstrapping, which is a technique to raise the ac input resistance. Consider leakage current to be negligible. Analysis proceeds by applying Thévenin's theorem to points $B$ and $B'$ to develop an equivalent dc input-circuit model as shown in Fig. 4-19(b). From the resulting equivalent input model write the input loop and solve for $I_B$ to obtain

$$I_B = \frac{V_B - V_{BE}}{R_B + (\beta_F + 1)R_E} \quad (4\text{-}38)$$

where

$$R_B = R_3 + \frac{R_1 R_2}{R_1 + R_2} \quad (4\text{-}39)$$

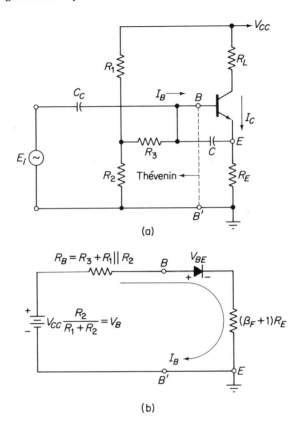

**Fig. 4-19.** A bootstrapped common-emitter circuit in (a) is modeled by its simplified input-circuit model in (b).

$$V_B = \frac{R_2}{R_1 + R_2} V_{CC} \tag{4-40}$$

$I_C$ is expressed by $\beta_F$ times $I_B$ in Eq. (4-38). The leakage-current split is the same as that shown in Fig. 4-18, except that $R_B$ is expressed by Eq. (4-39).

**Example 4-12.** In the circuit of Fig. 4-19(a), $R_1 = R_2 = R_3 = 500 \text{ k}\Omega$, $R_E = 5 \text{ k}\Omega$, $R_L = 10 \text{ k}\Omega$, $V_{CC} = 30$ V, and $\beta_F = 49$. Consider $I_{CBO}$ to be negligible and solve for $I_C$ and $V_{CE}$.

**Solution.** Values for the equivalent input circuit are found: From (4-39),

$$R_B = R_3 + (R_1 \| R_2) = 500 + (500 \| 500) = 750 \text{ k}\Omega$$

From (4-40),

$$V_B = \frac{R_2}{R_1 + R_2} V_{CC} = \left(\frac{500}{500 + 500}\right) 30 = 15 \text{ V}$$

Solve for $I_B$:

$$I_B = \frac{V_B - V_{BE}}{R_B + (\beta_F + 1)R_E} = \frac{15 - 0.6}{(750 + 250)(10^3)} = 14.4 \; \mu A$$

$$I_C = \beta_F I_B = 49(14.4 \times 10^{-6}) = 0.7 \; mA$$

$$V_{RL} = I_C R_L = 0.7 \times 10^{-3} \times 10^4 = 7 \; V$$

$$V_{RE} \cong I_C R_E = (0.7 \times 10^{-3})(5 \times 10^3) = 3.5 \; V$$

$$V_{CE} = V_{CC} - V_{RE} - V_{RL} = 19.5 \; V$$

*Example 4-13.* If a leakage-current component of $I_{CBO} = 10 \; \mu A$ were present in Example 4-12, what changes would result in the base and emitter currents?

*Solution.* $I_{CBO}$ would see $R_B = 750 \; k\Omega$ in the base leg and $(\beta_F + 1)R_E = 250 \; k\Omega$ in the emitter leg. Therefore the base current would be decreased by

$$\frac{(\beta_F + 1)R_E}{R_B + (\beta_F + 1)R_E} I_{CBO} = \frac{250}{750 + 250}(10) = 2.5 \; \mu A$$

Collector current would be increased by $I_{CBO} = 10 \; \mu A$ plus

$$\beta_F \frac{R_B}{R_B + (\beta_F + 1)R_E} I_{CBO} = \frac{750}{750 + 250}(10)(49) = 367 \; \mu A$$

for a total increase of 377 $\mu A$.

## 4-9 Biasing with Diodes for Temperature Compensation

When the bleeder current through $R_2$ in Fig. 4-2 cannot be tolerated because of the drain on $V_{CC}$, a nonlinear biasing technique may be adopted. A nonlinear element such as a thermistor or a diode replaces $R_2$ in Fig. 4-20 and automatically adjusts the collector current so that it remains constant.

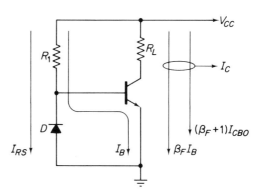

Fig. 4-20. Nonlinear biasing technique to offset $I_C$ variations due to temperature dependence of $I_{cbo}$.

The forward voltage drop across the base-emitter junction of the transistor insures that diode $D$ is reverse biased so that it conducts only the small value of reverse saturation current $I_{RS}$. From the input loop equation,

$$I_B = \frac{V_{CC} - V_{BE}}{R_1} - I_{RS}$$

Collector current is found by substituting for $I_B$ in the expression

$$I_C = \beta_F I_B + (\beta_F + 1) I_{CBO}$$
$$= \beta_F \frac{V_{CC} - V_{BE}}{R_1} - \beta_F I_{RS} + (\beta_F + 1) I_{CBO} \cong \beta_F \frac{V_{CC} - V_{BE}}{R_1} \quad (4\text{-}41)$$

If $I_{RS}$ is approximately equal to $I_{CBO}$, and if diode and transistor are exposed to the same thermal environment, then $I_{RS}$ will equal $I_{CBO}$ at any temperature and will not affect $I_C$ as shown by the approximation in Eq. (4-41). The two leakage currents cancel one another. This method of leakage-current compensation is effective with a germanium transistor (and diode) where the most significant temperature-dependent parameter is $I_{CBO}$. It is not particularly effective with silicon transistors.

A *thermistor* is a block of fairly intrinsic semiconductor material, with two ohmic contacts, whose resistance depends primarily on temperature through the mechanism of thermal rupture of covalent bonds. Thus its resistance will decrease with increasing temperature. A thermistor may therefore be substituted for diode $D$ in a bias circuit for silicon transistors.

By exposing the transistor and thermistor to the same thermal environment, collector current can be held constant within a restricted range of temperature variation. For example, an increasing temperature will increase $\beta_F$. But the thermistor's resistance decreases, lowering both $R_B$ and $V_B$ and consequently lowering base current. The product of a higher $\beta_F$ and lower $I_B$ results in a constant value of $I_C$.

As shown in Fig. 4-21, two more or diodes illustrate a technique

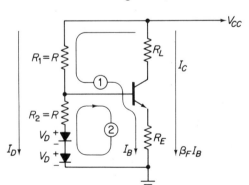

Fig. 4-21. An integrated-circuit biasing technique which minimizes operating-point dependence on temperature variations.

employed in integrated circuits (IC) to stabilize collector current. Using silicon devices we can ignore $I_{CBO}$ and write equations for loop (1) and loop 2:

Loop 1 $\quad V_{CC} = I_D R_1 + I_B[R_1 + (\beta_F + 1)R_E] + V_{BE}$

Loop 2 $\quad I_D R_2 + 2V_D = V_{BE} + I_B(\beta_F + 1)R_E$

Solving both equations for $I_B$ and multiplying the result by $\beta_F$ gives an exact expression for collector current:

$$I_C = \frac{\left[\left(\dfrac{R_2}{R_1 + R_2}\right)V_{CC} + \left(\dfrac{R_1}{R_1 + R_2}\right)2V_D - V_{BE}\right]\beta_F}{R_1 \| R_2 + (\beta_F + 1)R_E} \qquad (4\text{-}42)$$

If we make $R_1 = R_2 = R$ and $V_{BE} = V_D$, then the temperature-dependent diode voltage $V_D$ cancels the transistor-dependent voltage $V_{BE}$. (Since diode and transistor are fabricated on the same IC chip, $V_D$ will track $V_{BE}$.) Also, if we divide numerator and denominator by $\beta_F$ and assume $\beta_F \cong (\beta_F + 1)$, Eq. (4-42) reduces to

$$I_C = \frac{V_{CC}}{\dfrac{R}{\beta_F} + 2R_E}, \qquad R_1 = R_2 = R \qquad (4\text{-}43)$$

Usually $(R/\beta_F) \ll 2R_E$ so that Eq. (4-43) reduces to Eq. (4-44), showing that this bias technique renders an operating point insensitive to variations in $\beta_F$:

$$I_C \cong \frac{V_{CC}}{2R_E} \qquad (4\text{-}44)$$

Actually, $I_C$ is so stable it is used primarily as a constant-current source to bias other transistors.

### 4-10 Transistor-Stabilized Biasing

Another biasing technique popular in integrated-circuit technology is shown in Fig. 4-22. Characteristics of both transistors are identical because they are fabricated on the same chip (or matched, in the case of discrete components). Since $V_{BE}$ and $R$ are identical in both base circuits and both are connected to the same voltage at point 0, the base currents are equal. Furthermore, with identical $\beta_F$ for both $Q_1$ and $Q_2$, the collector currents must be equal. Writing the loop 1 equation and solving for $I_{B1}$:

$$I_{B1} = \frac{V_{CC} - V_{BE}}{R + (\beta_F + 2)R_1} \qquad (4\text{-}45a)$$

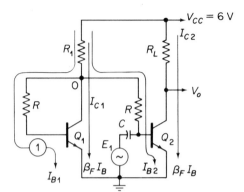

Fig. 4-22. $Q_1$ established equal collector bias currents in both transistors, which are stabilized against temperature variations.

Stipulating that $(\beta_F + 2) \cong \beta_F$ and $\beta_F R_1 \gg R$ will reduce Eq. (4-45) to

$$I_{B1} \cong \frac{V_{CC} - V_{BE}}{\beta_F R_1} \qquad (4\text{-}45b)$$

Assume $V_{BE} \leq 0.1 V_{CC}$ and multiply both sides of Eq. (4-45) by $\beta_F$ to obtain a simple expression for the equal collector currents:

$$I_C = I_{C1} = I_{C2} \cong \frac{V_{CC}}{R_1} \qquad (4\text{-}46)$$

Thus $V_{CC}$ and $R_1$ establish both collector currents equal and independent of temperaure variation. Point 0, the collector of $Q_1$, will always be at a low potential. But often we want the collector of $Q_2$ to be set at $V_{CC}/2$ or halfway up the load line so that $V_{CC}/2 = I_{C2} R_L$. Substituting in Eq. (4-46) for $I_{C2}$ gives

$$R_1 = 2R_L \qquad \text{for } V_o = \frac{V_{CC}}{2} \qquad (4\text{-}47)$$

*Example 4-14.* Given $R_L = 3\,\text{k}\Omega$ in the circuit of Fig. 4-22. Bias $Q_2$ for $V_o = V_{CC}/2$ by approximate methods.

*Solution.* Since $V_o = 3$ V, $I_{C2} = 1$ mA, $R_1 = 2R_L = 6\,\text{k}\Omega$, and $R \leq 6\,\text{k}\Omega$.

*Example 4-15.* Given $\beta_F = 100$ and $V_{BE} = 0.5$ V for each transistor, solve for the exact value of $V_o$, using the resistance values in Example 4-14.

*Solution.* From Eq. (4-45a)

$$I_{B1} = \frac{6 - 0.5}{6000 + (102)6000} = 8.9\,\mu\text{A}$$

Therefore $I_C = \beta_F I_B = (100)(8.9\,\mu\text{A}) = 0.89$ mA. The voltage drop across $R_L$ is 2.7 V and $V_o$ is $6 - 2.7 = 3.3$ V.

## 4-11 Demonstration of Operating-Point Stability

A visual demonstration and summary of operating-point stability may be presented through the circuit of Fig. 4-23. (1) At room temperature,

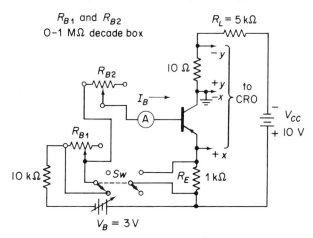

Fig. 4-23. Demonstration of operating-point stability. Sensitivities: $y$ amplifier at 10 mV/div for $I_C = 1$ mA/div, $x$ amplifier at 1 V/div for $V_{CE}$. CRO zeroed at bottom left of screen.

and with switch $S\omega$ in the down or stabilized position, adjust $R_{B1}$ to 0 Ω. (2) Then adjust $R_{B2}$ for a reference value of base current $I_{BR}$, so that the CRO displays an operating point at the load line's center. (See Fig. 3-12 and associated discussion.) (3) Throw $S\omega$ up to the unstabilized position so that $R_E$ is short circuited. $R_{B1}$ is increased to obtain the reference value of $I_{BR}$. This demonstrates clearly the resistance-transformation properties of the transistor since $R_{B1}$ must be increased to a value of $(\beta_F + 1)R_E$. In addition *we have a new method of measuring $\beta_F$* at the operating point since both $R_{B1}$ and $R_E$ are measured values in Eq. (4-48):

$$\text{Decade reading of } R_{B1} = (\beta_F + 1)R_E \qquad (4\text{-}48)$$

(4) Apply considerable heat directly to the transistor with a match or electrically isolated soldering iron (25 W). Observe the operating point move up the dc load line. (5) Throw $S\omega$ to the stabilized position and observe that the operating point jumps back toward its original room-temperature location.

## PROBLEMS

**4-1** In Example 4-1, what is the maximum possible value of $I_C$? (Assume $V_{CE(\text{sat})} = 0$.) Find $I_B$ and $R_B$ required to cause this value of $I_C$.

**4-2** If a transistor with $\beta_F = 50$ is substituted into the circuit resulting from the solution of Example 4-1, what are the new values of $I_B$, $I_C$, and $V_{CE}$? Repeat for a transistor with $\beta_F = 200$, and $\beta_F = 400$. Compare results with Problem 4-1 and resolve any apparent conflict.

**4-3** When the load resistor in Example 4-2 is increased to 5 k$\Omega$, what is the resulting maximum possible output signal swing and operating point?

**4-4** Derive Eqs. (4-5) and (4-6) from Eqs. (4-3) and (4-4).

**4-5** Choose $R_B = 100$ k$\Omega$ in Example 4-3 and solve for $V_B$, $R_1$, and $R_2$.

**4-6** If $\beta_F = 100$ in Example 4-4, what are the new values of $R_1$ and $R_2$? Repeat for $\beta_F = 200$.

**4-7** Compare the changes in Example 4-5 caused by substituting a transistor with $\beta_F = 100$.

**4-8** Given $V_{\text{collector}} = 9$ V in Fig. 4-5(a). Assume $V_{EE} = 0.6$ V and $\beta_F$ is unknown. What will be the value of (a) $V_{\text{base}}$, (b) $V_{CE}$, (c) $I_2$, (d) $I_1$, (e) $I_B$, and (f) $\beta_F$? (Hint: assume $I_C \cong I_E$)

**4-9** In the circuit of Fig. 4-1(a), $V_{CC} = 12$ V. Allowing 1 V to guard against saturation, refer to Fig. 4-6 and for a 2-V p/p output voltage what is (a) the best $V_{CE}$ compromise location for the operating point (b) maximum and minimum $V_{CE}$ location?

**4-10** Let $V_{EE} = 0$ in Fig. 4-10(a). (a) Assuming a silicon transistor, write the base-current and collector-current expressions from an inspection of the circuit. (b) Now include the leakage-current components in the equations of (a) by drawing the model of Fig. 4-11 and evaluating $k$.

**4-11** What are the new values of $I_B$ and $I_C$ in Example 4-7(b) if $I_{CBO} = 40$ μA?

**4-12** What are the new values of $I_B$ and $I_C$ in Example 4-8(b) if $I_{CBO} = 40$ μA?

**4-13** Given $R_B = 10$ k$\Omega$, $\beta_F = 50$, and $R_E = 500$ $\Omega$ in Fig. 4-12. (a) Evaluate $k$. What is the new $k$ if (b) only $\beta_F$ is doubled, (c) only $R_B$ is doubled and (d) only $R_E$ is doubled?

**4-14** Write the base- and collector-current expressions with the superposition method by inspection of Fig. 4-13 for (a) a silicon transistor, and (b) a germanium transistor.

**4-15** Evaluate $I_B$ in Example 4-9(b). Does $R_F$ adjust $I_B$ to reduce changes in $I_C$ due to changes in $\beta_F$?

**4-16** What changes result in Example 4-10 if $\beta_F$ is given as 100?

**4-17** Calculate $I_C$ for $\beta_F = 75$ in Example 4-9 and plot $I_C$ versus $\beta_F$. Note that the slope $I_C/\beta_F$ is the stability factor for $\beta_F$.

**4-18** Write the expression for the base and collector-current expressions by inspection of Fig. 4-15 for (a) a silicon transistor, (b) a germanium transistor.

**4-19** In Example 4-11(a), calculate $I_B$ for $\beta_F = 50$, and $\beta_F = 150$. Plot $I_B$ and $I_C$ versus $\beta_F$ for $50 < \beta_F < 199$.

**4-20** In Example 4-11(c), calculate $I_B$ and $I_C$ for $I_{CBO} = 10$ μA, and 15 μA. Plot $I_B$ and $I_C$ versus $I_{CBO}$ from $0 < I_{CBO} < 15$ μA.

**4-21** What are the new values of $I_{B1}$, $I_C$, and $V_{CE}$ in Example 4-12 if $\beta_F = 99$?

**4-22** In Fig. 4-22, $R_L = 5$ kΩ and $V_{CC} = 10$ V. Select approximate values for $R_1$ and $R$ for $V_{CE2} = V_{CC}/2$.

**4-23** In Fig. 4-22, $R_L = 6$ kΩ, $V_{CC} = 12$ V, $R_1 = 12$ kΩ, and $\beta_F = 50$ for both silicon transistors. Solve for the exact value of $R$ to establish $V_{CE2} = 6$ V, assuming $V_{BE} = 0.6$ V. Repeat the solution for $\beta_F = 100$ and compare the values of $R_1$.

# Chapter 5

**5-0** INTRODUCTION ................................. 141
**5-1** BASIC LOW-FREQUENCY HYBRID-PI MODEL .......... 142
**5-2** COMMON-EMITTER AMPLIFIER ..................... 145
**5-3** COMMON-EMITTER AMPLIFIER WITH EMITTER
       RESISTANCE .................................. 153
**5-4** COMMON-COLLECTOR AMPLIFIER ................. 159
**5-5** OUTPUT RESISTANCE AND RESISTANCE
       TRANSFORMATIONS ............................. 161
**5-6** COMMON-BASE AMPLIFIER ....................... 166
**5-7** MEASUREMENT OF OUTPUT AND INPUT RESISTANCE .. 170
**5-8** GAIN IN DECIBELS ............................ 171
       PROBLEMS .................................... 173

# Small-Signal Low-Frequency Amplifiers

## 5-0 Introduction

An amplifier is a device that accepts signals at its input terminals and delivers an amplified reproduction to a load at its output terminals. *Small signal* is generally defined to mean an alternating or incremental voltage that is small enough for a linear relationship between output and input signals Specifically, 5 mV is the maximum signal voltage allowed across the base-emitter junction for small-signal operation.

When a BJT is the amplifying element we require a *small-signal model* consisting of *linear circuit elements* that can replace the transistor. We can then use basic circuit theory to analyze its performance in a circuit. Linear circuit elements are resistors, capacitors, and dependent current or voltage generators. Dependent generators are signal sources, usually in the output of the transistor, which depend linearly on an externally applied, real signal generator.

There are many types of models in the literature and most are based on measurements. The measurement-dependent models require a huge amount of supporting data, including correction curves and data tables. One model is based on the physics of BJTs and consequently demands a minimum of supporting data. A much simpler model results and evaluation of its basic circuit elements is both simple and swift. Often it is desirable to gain facility with one model and go deeper into circuit analysis rather than study several models, at the expense of practice with circuit analysis. For these reasons, this text will be fully committed to the physics-derived *hybrid-pi* model.

Finally, in this chapter we shall be concerned strictly with low-frequency amplifiers and will define low frequency as the range of frequencies where circuit coupling and bypass capacitors act as short circuits to signal currents and internal transistor capacitances act as open circuits. This frequency range extends approximately from 100–10,000 Hz.

## 5-1  Basic Low-Frequency Hybrid-Pi Model

Basic elements of the low-frequency hybrid-pi model are shown in Fig. 5-1, where $V_{be}$ is an ideal ac voltage source that divides between resistors

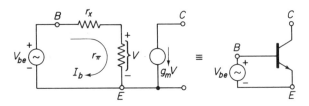

Figure 5-1

$r_x$ and $r_\pi$. $V$ is the ac voltage drop across $r_\pi$ and completes the input circuit model. Dependent-current generator $g_m V$ is in the collector circuit and depends on voltage $V$ by the small-signal transconductance $g_m$.

Physics of the model's elements are explained in volumes 1–4 of the SEEC series (Semiconductor Electronics Education Committee), as noted in the bibliography. For our purposes the physical signficance of each element will describe its role in the model.

RESISTANCE $r_\pi$.  Because $V_{be}$ increases forward bias by an increment (in Fig. 5-1), an increased quantity of carriers is injected across the emitter junction into the active base region where they diffuse through to the collector. Recombination increases because of the increase in carriers in the base. The base terminal must feed these recombinations by $I_b$. The relationship between $I_b$ and the base-emitter junction voltage $V$ is $r_\pi$.

BASE-SPREADING RESISTANCE $r_x$.  Signal voltage $V_{be}$ causes a signal current $I_b$ to flow in Fig. 5-1. Recombination carriers, fed by the base terminal as $I_b$, must cross a bulk resistance of the base on their way to the emitter junction. This incremental bulk resistance is modeled by $r_x$.

TRANSCONDUCTANCE $g_m$.  The increment of carriers due to $V_{be}$ which reach the collector terminal constitutes a collector signal current increment $\Delta I_C$. This increment is determined by the incremental voltage across the emitter junction $V$ through $g_m$ so the collector signal current is modeled by the dependent-current generator $g_m V$. Since $V_{be}$ increases forward bias, $g_m V$ represents a collector current increase and must have the direction shown.

Elements of the hybrid-pi model are evaluated quickly and simply. Transconductance $g_m$ is directly proportional to the dc collector bias current $I_C$ and is evaluated from

$$g_m = \frac{I_C}{\frac{kT}{q}} \qquad (5\text{-}1)$$

Sec. 5-1  Basic Low-Frequency Hybrid-Pi Model  143

where $I_C$ = magnitude of the dc collector–bias current,
$k$ = Boltzmann's constant = $1.38 \times 10^{-23}$ J/°K,
$q$ = charge on an electron = $1.60 \times 10^{-19}$ C, and
$T$ = temperature (°K).

The rms signal-collector current $I_c$ is proportional to the rms base-emitter junction voltage $V$ times the transconductance $g_m$, or

$$I_c = g_m V \tag{5-2}$$

Thus collector-signal current is modeled by the dependent-current generator in Fig. 5-1 and acts between emitter and collector. $V$ results from $I_b$ flowing through $r_\pi$, or

$$V = I_b r_\pi \tag{5-3}$$

Substituting Eq. (5-3) into Eq. (5-2) yields

$$I_c = g_m r_\pi I_b \tag{5-4}$$

Express Eq. (5-4) in terms of incremental currents and compare it with Eq. (2-15):

$$\begin{matrix} (5\text{-}4) & (2\text{-}15) \\ g_m r_\pi = \dfrac{\Delta I_C}{\Delta I_B} & \dfrac{\Delta I_C}{\Delta I_B} = \beta_0 \end{matrix}$$

We conclude that small-signal current gain $\beta_0$ is related to $g_m$ by

$$\beta_0 = g_m r_\pi \tag{5-5}$$

Measure or obtain $\beta_0$ from data sheets by the methods given in Section 2-6 and with knowledge of $g_m$ from Eq. (5-1) it is easy to evaluate $r_\pi$ from Eq. (5-5) The element $r_x$ is often negligible at low frequencies, but input resistance $h_{ie}$, which was discussed in Section 3-4 and Fig. 2-20, is equal to the sum of $r_x$ and $r_\pi$, or

$$h_{ie} = r_x + r_\pi \quad \text{and} \quad r_x = h_{ie} - r_\pi \tag{5-6}$$

SUMMARY OF THE PROCEDURE TO CALCULATE HYBRID-PI ELEMENTS

1. Begin by finding the dc collector bias current $I_C$ and evaluate $g_m$ from Eq. (5-1). Coefficient $kT/q$ will equal 25 mV at room temperature.
2. Obtain $\beta_0$ from (a) typical values of $\beta_0(h_{fe})$ on a data sheet and correct to the actual operating point, or (b) measure $\beta_0$ from collector characteristics of your transistor at your operating point.

3. Calculate $r_\pi$ from Eq. (5-5).
4. Obtain $h_{ie}$ from (a) the slope of the input characteristic at the operating point, or (b) typical values on a data sheet, also corrected to the actual operating point. Find $r_x$ from Eq. (5-6).

Almost always, $h_{ie}$ is needed at a point on the $I_B$ versus $V_{BE}$ characteristic which is not amenable to accurate graphical analysis. Also, the corrected value of $h_{ie}$, from published data, is for a typical transistor and not your transistor. Values of $h_{ie}$ will be approximately equal to the value of $r_\pi$ in (3), so $r_x$ is the difference between two nearly equal quantities. Sometimes $h_{ie}$ is calculated to be smaller than $r_\pi$. In this event, assume $r_x = 0$ and $h_{ie}$ equals the magnitude of $r_\pi$, obtained in step (3).

**Example 5-1.** Plot $kT/q$ versus temperature over the range of $T = -50°C$ to $50°C$ for future use in Eq. (5-1).

**Solution.** The fraction $k/q$ is a constant with a value of

$$\frac{k}{q} = \frac{1.38 \times 10^{-23} \text{ J/°K}}{1.60 \times 10^{-19} \text{ C}} = 8.62 \times 10^{-5} \text{ V/°K}$$

$T$ must be converted to degrees Kelvin by adding 273 to the value of degrees Celsius. Three solution points will be $-50°C$ and 19.2 mV, $17°C$ and 25.0 mV, $50°C$ and 27.8 mV; the results are plotted in Fig. 5-2.

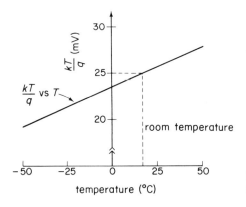

Fig. 5-2. Solution to Example 5-1.

In Example 5-2 we analyze how $g_m$ varies with dc collector current.

**Example 5-2.** Plot $g_m$ versus $I_C$ for values of $I_C$ between 0.1 and 10 mA at a temperature of $17°C$.

**Solution.** From Example 5-1, $kT/q = 0.025$ and we plot the curve $g_m = I_C/0.025$ in Fig. 5-3.

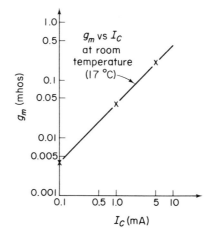

**Fig. 5-3.** Variation of $g_m$ with collector bias current in Example 5-2.

*Example 5-3.* A transistor is to be operated in a common-emitter circuit biased at $I_C = 1$ mA, and at 17 °C room temperature. Assume $\beta_0$ remains constant at 100 and $r_x$ is negligible. (a) Find the hybrid-pi parameters for the low-frequency model. (b) Repeat (a) for $I_C = 100$ μA and compare the effect on input resistance. (c) Repeat (a) and (b) for a temperature rise of 50°C.

*Solution.* (a) From Eq. (5-1), $g_m = 1$ mA/25 mV $= 0.04$ mho. From Eq. (5-5), $r_\pi = \beta_0/g_m = 100/0.04 = 2.5$ kΩ.
(b) $g_m = 0.10$ mA/25 mV $= 0.004$ mho, $r_\pi = 100/0.004 = 25$ kΩ.
(c) From Example 5-1, $kT/q = 8.62 \times 10^{-5}(67 + 273) = 29.3$ mV. $g_m = 1$ mA/29.3 mV $= 0.034$ mho, $r_\pi = 100/0.034 = 2.93$ kΩ; $g_m = 0.1$ mA/29.3 mV $= 0.0034$, $r_\pi = 100/0.0034 = 29.3$ kΩ.

It is concluded from Example 5-3 that the input resistance of the transistor increased with an increase in temperature and a decrease in collector current.

## 5-2 Common-Emitter Amplifier

The most basic common-emitter (CE) amplifier, shown in Fig. 5-4(a), is driven by an ideal ac voltage source $E_i$ that has no internal resistance. We draw the circuit model in Fig. 5-4(b), replacing capacitors and the dc supply voltage by short circuits since their internal impedance to signal-current flow is assumed to be negligible. Bias resistor $R_B$ has purposely been drawn to the left of $E_i$ to show that it does not affect *input resistance of the transistor* $R_i$. The dc bias currents are shown in Fig. 5-4(a) and are to be compared with the ac signal currents in Fig. 5-4(b). $E_i$ is shown at the instant when the input base terminal is driven positive with respect to the common

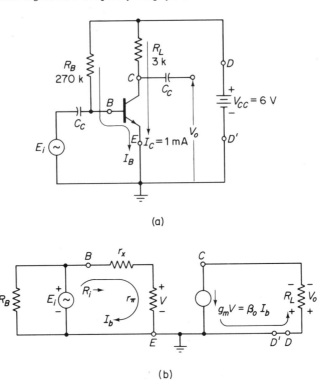

**Fig. 5-4.** The common-emitter amplifier circuit in (a) is modeled for analysis in (b).

terminal. The collector output terminal is driven negative by $g_m V$ and $V_o$ will be 180° out of phase with $E_i$. Voltage gain $A_V$ is the relationship between $E_i$ and $V_o$ or $V_o = -A_V E_i$. The minus sign accounts for the phase reversal. Voltage gain is derived in terms of circuit and transistor parameters by observing the voltage division of $E_i$ in Fig. 5-4(b):

$$V = \frac{r_\pi}{(r_x + r_\pi)} E_i \qquad (5\text{-}7)$$

$V_o$ equals $g_m V$ times $R_L$ and, substituting from Eq. (5-7),

$$V_o = -g_m V R_L = -g_m \left(\frac{r_\pi}{r_x + r_\pi}\right) E_i R_L \qquad (5\text{-}8)$$

Employ Eq. (5-5) and write Eq. (5-8) in the form of voltage gain:

$$A_V = -\beta_0 \frac{R_L}{(r_x + r_\pi)} = \frac{V_o}{E_i} \qquad (5\text{-}9)$$

Sec. 5-2  Common-Emitter Amplifier  147

An extremely useful method of interpreting Eq. (5-9) and all other voltage-gain equations is to look for the format:

$$\text{Voltage gain} = \text{Current gain} \times \frac{\text{Resistance presented to output voltage}}{\text{Resistance presented to input voltage}}$$

(5-10)

$R_L$ is the resistance presented to output voltage $V_o$, and $r_x + r_\pi$ is the input resistance presented to $E_i$.

Current gain $A_i$ is output collector current $g_m V$ divided by input base current $I_b$ or (in magnitude):

$$A_i = \frac{g_m V}{I_b} = g_m \frac{(I_b r_\pi)}{I_b} = g_m r_\pi = \beta_0 \qquad (5\text{-}11)$$

Thus the voltage-gain equation contains just about all the information we need to know about the low-frequency operation of the circuit. It pinpoints the input resistance of the transistor, the current gain, and the load resistance across which the output voltage is developed. We shall try to manipulate all our voltage-gain expressions into this form *to gain an ability to write them by inspection.*

Equations (5-8) and (5-9) show that voltage gain depends directly on $R_L$. The dependence of $A_v$ on operating-point current is illustrated by neglecting $r_x$ in Eq. (5-9) to write

$$\frac{V_o}{E_i} = A_v \cong -\frac{\beta_0 R_L}{r_\pi} = -g_m R_L, \quad \text{for } r_x \ll r_\pi \qquad (5\text{-}12)$$

Substituting for $g_m$ from Eq. (5-1),

$$\frac{V_o}{E_i} = A_v \cong -\frac{I_C R_L}{\frac{kT}{q}} \qquad (5\text{-}13)$$

Thus voltage gain increases with increasing operating-point current and decreasing temperature.

*Example 5-4.* (a) What is the voltage gain from $E_i$ to $V_o$ in Fig. 5-4 if $\beta_0 = 50$? (b) What ac resistance value is presented by the transistor to $E_i$? Assume $r_x = 0$. (c) What is the small-signal current gain of the transistor?

*Solution.* (a) Evaluating $g_m$ and $r_\pi$ from Eqs. (5-1) and (5-5),

$$g_m = \frac{I_C}{\frac{kT}{q}} = \frac{1}{25} = .0040 \qquad r_\pi = \frac{\beta_0}{g_m} = \frac{50}{0.040} = 1250 \ \Omega$$

From Eq. (5-9),

$$A_v = \frac{V_o}{E_i} = -\beta_0 \frac{R_L}{(r_x + r_\pi)} = -50\left(\frac{3000}{1250}\right) = 120$$

(b) From (a), $h_{ie} = r_\pi = 1250 \, \Omega = R_i$.
(c) $A_i = \beta_0 = 50$.

It is necessary to be acutely aware of the limitations of the hybrid-pi model as applied in the solution of Example 5-4. One limitation is that the frequency of $E_i$ should be between 100 and 10,000 Hz. A second is that the incremental signal voltage between base and emitter, $V_{be}$, must not exceed 5 mV and in the circuit of Fig. 5-4, this stipulation limits $E_i$ to a peak value of 5 mV. However, *if $\beta_0$ is reasonably constant, the validity of the voltage-gain equation often may be extended considerably for larger values of* $V_{be}$. A third limitation is imposed by the circuit, and is identified as the maximum permissible p/p variation in $V_o$ and depends directly on $V_{CE}$. Example 5-5 investigates these limitations.

*Example 5-5.* (a) For the circuit in Fig. 5-4(a), what is the maximum permissible p/p value of $V_o$ and $I_b$ to meet the restrictions imposed by the hybrid-pi model? (b) Assuming $\beta_0 = 50$ and does not vary with $I_C$ or $V_{CE}$, and also assuming that the hybrid-pi model is valid, what is the maximum permissible peak value of $E_i$ before $V_o$ distorts? (c) What are the peak signal currents entering the base terminal for the conditions of (a) and (b)? Use the hybrid-pi model for this calculation. (d) What are the corresponding peak collector currents in (c)?

*Solution.* (a) Since $E_i$ is limited to 5 mV peak and $A_v$ is $-120$ from Example (5-4), the peak value of $V_o$ is $120 \times 5 \times 10^{-3} = 0.6$ V.
(b) Since $V_{CE} - V_{RL} = 3$ V in Fig. 5-4(a), $V_{op} = 3$ V and the maximum peak value of $E_i$ is $E_{i(\max)} = \frac{3}{120} = 25$ mV.
(c) The peak signal base currents are

$$I_{bp} = \frac{E_{ip}}{r_\pi} = \frac{5 \text{ mV}}{1250} = 4 \, \mu A, \quad I_{bp(\max)} = \frac{E_{i(\max)}}{r_\pi} = \frac{25 \text{ mV}}{1250} = 20 \, \mu A$$

(d) The peak collector current swing for the hybrid-pi model's limitation is $\beta_0$ times the value found in (c), or

$$I_{cp} = \beta_0 I_{bp} = 50(4) \, \mu A = 0.2 \text{ mA}, \quad E_{ip} = 5 \text{ mV}$$

With $E_i$ at its maximum value of 25 mV,

$$I_{cp(\max)} = \beta_0 I_{bp(\max)} = 50(20) \, \mu A = 1 \text{ mA}$$

It is informative to portray that region of the collector characteristics, where the hybrid-pi model is unconditionally valid. This region is a window-shaped area and will be called the *hybrid-pi window*. Its dimensions are found from the signal current and voltage peaks which exist when $V_{be} = 5$ mV. In Example 5-5 the values were found to be as follows: At $E_{i_p} = 5$ mV, $I_{b_p} = 4$ μA, $I_{c_p} = 0.2$ mA, $V_{cep} = V_{op} = 0.6$ V. In Fig. 5-5, points $h$ and $h'$ locate

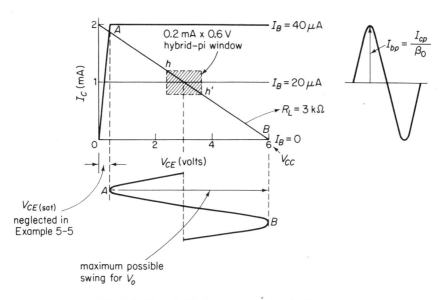

Fig. 5-5. The hybrid-pi window graphically illustrates the unconditional limits of the hybrid-pi model.

the maximum operating-point excursion to meet the model's limitations. The crosshatched area is the window which shows the p/p excursions of collector current and collector voltage, corresponding to $E_{i_p} = 5$ mV peak.

Voltage gain is reduced by generator resistance $R_g$ in the CE circuit of Fig. 5-6(a). From the model in Fig. 5-6(b), $E_g$ divides between $R_g$ and $R_{in}$ according to

$$E_i = \frac{R_{in}}{R_{in} + R_g} E_g \qquad (5\text{-}14)$$

where

$$R_{in} = R_B \| (r_x + r_\pi) = \frac{R_B(r_x + r_\pi)}{R_B + r_x + r_\pi} \qquad (5\text{-}15)$$

Since $V_o$ is already known in terms of $E_i$ from Eq. (5-9) we substitute for $E_i$ from Eq. (5-14) to obtain

$$A_v = \frac{V_o}{E_g} = -\beta_0 \left(\frac{R_L}{r_x + r_\pi}\right)\left(\frac{R_{in}}{R_{in} + R_g}\right) \qquad (5\text{-}16)$$

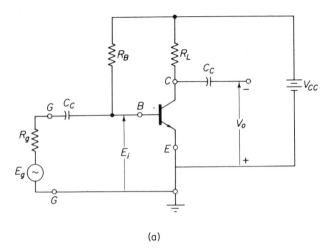

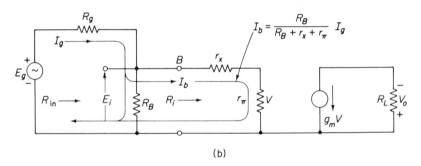

**Fig. 5-6.** In (a) we are interested in voltage gain between $E_g$ and $V_o$, where $V_o/E_g$ is derived from the model in (b).

Equation (5-16) is a gain expression and illustrates clearly the effect of generator resistance on voltage gain. $R_g$ should be low or much of the gain $V_o/E_i$ will be lost right at the input. Alternatively, $R_{in}$ should be high to avoid a loss in voltage gain. Physically this means that $R_B$ should be large so that it will not contribute to a small $R_{in}$. If $R_B$ is large with respect to $r_\pi$, then the ac input resistance of the transistor $R_i$ and $R_g$ determine how much gain is lost at the input.

Current gain in the circuit is $A_i$, where

$$A_i = \frac{I_o}{I_{in}} = \frac{g_m(I_b I_\pi)}{I_g} = \frac{\beta_0 I_b}{I_g} \tag{5-17}$$

Since $I_g$ divides between $R_B$ and $r_x + r_\pi$ we can write $I_b$ in terms of the input current divider,

## Sec. 5-2     Common-Emitter Amplifier

$$I_b = \frac{R_B}{(R_B + r_x + r_\pi)} I_g \tag{5-18}$$

and substitute for $I_b$ in Eq. (5-17) to give

$$A_i = \frac{\beta_0 I_b}{I_g} = \frac{R_B}{(R_B + r_x + r_\pi)} \beta_0 \tag{5-19}$$

Equation (5-16) can be manipulated into our standard format but we shall write the expression *by inspection* of Fig. 5-6. $V_o$ sees $R_L$ and $E_g$ sees $R_g + r_{in}$. Current gain of the transistor is $\beta_0$ but it is reduced by the input divider as shown in Eq. (5-19). We can now write directly

$$A_v = \frac{V_o}{E_g} = -\beta_0 \left(\frac{R_B}{R_B + r_x + r_\pi}\right) \left(\frac{R_L}{R_g + R_{in}}\right) \tag{5-20}$$

The student should verify that Eqs. (5-20) and (5-16) are identical.

**Example 5-6.** (a) Assuming $I_{CBO}$ and $r_x$ are negligible in Fig. 5-6(a), and given $\beta_F = \beta_0 = 50$, $V_{BE} = 0.6$ V, $V_{CC} = 6$ V, and $R_L = 5$ kΩ, choose $R_B$ to establish $I_C$ at 0.5 mA. (b) Plot voltage gain $V_o/E_g$ against $R_g$ between $R_g = 100$ Ω and 100 kΩ.

**Solution.** (a) Solve for $I_B = I_C/\beta_F = 0.5$ mA/50 = 10 μA, and

$$R_B = \frac{V_{CC} - V_{BE}}{I_B} = \frac{6 - 0.6}{10\ \mu A} = 540\ k\Omega$$

The operating point is at $I_B = 10$ μA, $I_C = 0.5$ mA, $V_{CE} = 3.5$ V.
(b) The value of $r_\pi$ is found by

$$g_m = \frac{I_C}{0.025} = \frac{0.5}{0.025} = 0.02\ \text{mho} \qquad r_\pi = \frac{\beta_0}{g_m} = \frac{50}{0.02} = 2500\ \Omega$$

Calculate

$$\frac{V_o}{E_i} = \beta_0 \left(\frac{R_L}{r_x + r_\pi}\right) = -50 \left(\frac{5000}{2500}\right) = -100$$

Now calculate the magnitude of voltage divider $R_{in}/(R_{in} + R_g)$ as a function of $R_g$. Multiplying the result by 100 will give $V_o/E_g$:

$$R_{in} = R_B \| (r_x + r_\pi) = \frac{(540\ k\Omega)(2.5\ k\Omega)}{542\ k\Omega} \cong 2.5\ k\Omega$$

Evidently $R_B$ does not seriously change the magnitude of $R_{in}$ so that $R_{in} \cong r_\pi$. The calculations are tabulated in Table 5-1 and the results are plotted in Fig. 5-7.

## Table 5-1

| $R_g$ | $R_{in} + R_g$ | $R_{in}/(R_g + R_{in}) \times$ | $V_o/E_i$ | $V_o/E_g$ |
|---|---|---|---|---|
| 0 | 2,500 | 1.0 $\times$ | 100 | 100 |
| 1,000 | 3,500 | 0.715 | 100 | 71.5 |
| 2,500 | 5,000 | 0.500 | 100 | 50 |
| 5,000 | 7,500 | 0.334 | 100 | 33.4 |
| 25,000 | 27,500 | 0.091 | 100 | 9.1 |

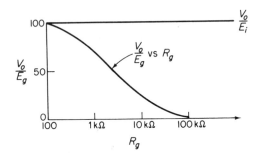

Fig. 5-7. Solution to Example 5-6.

In Example 5-7 we investigate how $R_B$ can affect voltage gain and results should be compared with Example 5-6.

**Example 5-7.** (a) Find the dc operating point of the circuit in Fig. 5-8(a). (b) Calculate the incremental resistance presented to $E_g$ if $R_g = 0\,\Omega$ and $2500\,\Omega$. (c) Find $V_o/E_g$ for values of $R_g$ in (b).

**Solution.** (a) Find the dc collector current from

$$R_B = 12.5\text{ k}\Omega \,\|\, 49.6\text{ k}\Omega = 10\text{ k}\Omega$$

$$V_B = \frac{12.5 \times 10^3}{(12.5 + 49.6) \times 10^3} \times 6 = 1.21 \text{ V}$$

$$I_B = \frac{V_B - V_{BE}}{R_B + (\beta_F + 1)R_E} = \frac{1.21 - 0.6}{(10 + 51) \times 10^3} = \frac{0.61}{61 \times 10^3} = 10\,\mu\text{A}$$

$$I_C = \beta_F I_B = 50(10)\,\mu\text{A} = 0.5\text{ mA} \qquad V_{CE} = 3.5\text{ V}$$

(b) The small-signal model is shown in Fig. 5-8(b). Note that $R_E$ and $V_{CC}$ are short circuited in the small-signal model and $R_1$ is in parallel with $R_2$. The transistor parameters are found from

$$g_m = \frac{I_C}{25\text{ mV}} = \frac{0.5}{25} = 0.020 \qquad r_\pi = \frac{\beta_0}{g_m} = \frac{50}{0.020} = 2.5\text{ k}\Omega$$

$$R_{in} = R_B \,\|\, r_\pi = 10\text{ k}\Omega \,\|\, 2.5\text{ k}\Omega = 2.0\text{ k}\Omega$$

The resistance presented to $E_g$ is $2000\,\Omega$ for $R_g = 0$, and $4500\,\Omega$ for $R_g = 2500\,\Omega$.

Sec. 5-3  Common-Emitter Amplifier with Emitter Resistance  153

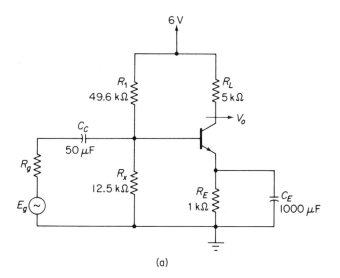

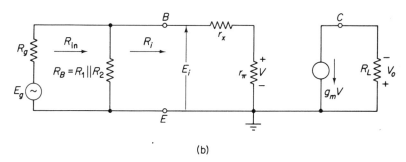

**Fig. 5-8.** Circuit in (a) and model in (b) for Example 5-7. Given $\beta_o = \beta_F = 50$, $I_{cbo} = 0$, $V_{BE} = 0.6$ V, $r_x = 0$.

(c) $V_o/E_i$ is calculated from Eq. (5-9) and is equal to $V_o/E_g$ when $R_g = 0$:

$$\frac{V_o}{E_i} = -\beta_o \left(\frac{R_L}{r_x + r_\pi}\right) = -50\left(\frac{5000}{2500}\right) = -100$$

When $R_g = 2500$ Ω, from Eq. (5-14), $E_i/E_g = \frac{2000}{4500} = 0.42$ and $V_o/E_g = 0.424(100) = 42.4$. Compare this value with 50 in Example 5-6.

## 5-3 Common-Emitter Amplifier with Emitter Resistance

An unbypassed emitter resistor $R_E$, in the CE circuit of Fig. 5-9(a), will increase input resistance and decrease voltage gain. Resistance-transformation properties of the transistor, introduced in Section 4-2, will be employed to simplify circuit analysis.

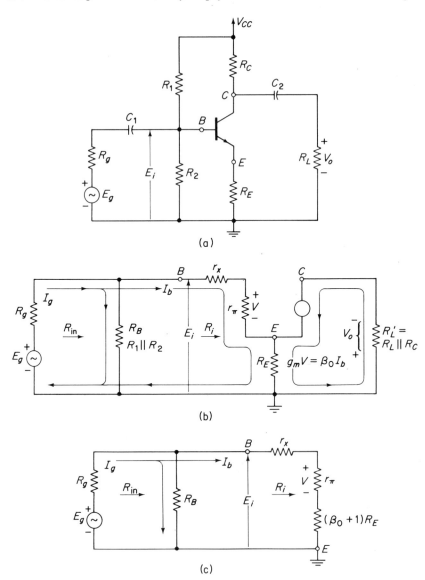

**Fig. 5-9.** The CE circuit in (a) is modeled in (b). An equivalent input resistance model in (c) allows $V$ to be written in terms of $E_i$ or $E_g$, by inspection.

The small-signal model is drawn in Fig. 5-9(b) to find general expressions for $V_o/E_g$, $V_o/E_i$, and $R_i$. The signal voltage drop across $R_E$ can be written in several forms as

$$V_{RE} = (I_b + g_m V)R_E = (I_b + \beta_0 I_b)R_E = I_b(\beta_0 + 1)R_E = I_e R_E \qquad (5\text{-}21)$$

We interpret the term $I_b(\beta_0 + 1)R_E$ as saying that the voltage drop across $R_E$ is the product of base signal current $I_b$ and a *new and larger resistance* whose magnitude is $(\beta_0 + 1)R_E$. This means that we can look into the base terminal and see the emitter resistance $R_E$, as if it were *transformed into* the base leg as a resistance $(\beta_0 + 1)R_E$. An equivalent input circuit is constructed in Fig. 5-9(c) to illustrate the transformation.

The equivalent input circuit simplifies the task of finding $A_v$ in terms of $V_o$ and $E_i$. $E_i$ divides between $r_x$, $r_\pi$, and $(\beta_0 + 1)R_E$ so that $V$ is, by inspection of Fig. 5-9(c),

$$V = \frac{r_\pi}{[r_x + r_\pi + (\beta_0 + 1)R_E]} E_i \qquad (5\text{-}22)$$

Output voltage $V_o$ is the product of current and resistance:

$$V_o = -g_m V R'_L$$

where

$$R'_L = R_L \parallel R_C \qquad (5\text{-}23)$$

Substituting for $V$ and dividing by $E_i$:

$$A_v = \frac{V_o}{E_i} = -\frac{g_m r_\pi R'_L}{[r_x + r_\pi + (\beta_0 + 1)R_E]} = -\frac{\beta_0 R'_L}{R_i} \qquad (5\text{-}24)$$

where

$$R_i = r_x + r_\pi + (\beta_0 + 1)R_E \qquad (5\text{-}25)$$

Equation (5-24) is in our standard format with current gain of the transistor times $R_L$ as seen by $V_o$, divided by $R_i$ as seen by $E_i$. Input resistance as measured from the base terminal to ground is given by $R_i$.

Voltage gain, in terms of $E_g$, is a simple matter if we refer to the equivalent input circuit model in Fig. 7-9(c). $E_g$ divides between $R_g$ and $R_{in} = R_B \parallel R_i$ so that $E_i$ is found in terms of $E_g$ by

$$E_i = \left(\frac{R_B \parallel R_i}{R_g + R_B \parallel R_i}\right) E_g = \left(\frac{R_{in}}{R_g + R_{in}}\right) E_g \qquad (5\text{-}26)$$

Substituting Eq. (5-24) for $E_i$ gives

$$A_v = \frac{V_o}{E_g} = -\left(\frac{R_B \parallel R_i}{R_g + R_B \parallel R_i}\right) \frac{\beta_0 R'_L}{R_i} \qquad (5\text{-}27)$$

Equation (5-27) is maneuvered into standard format by writing out the numerator term of $R_B \| R_i$ and cancelling $R_i$ to obtain

$$A_v = \frac{V_o}{E_g} = -\left(\frac{R_B}{R_B + R_i}\right)\beta_0\left(\frac{R'_L}{R_g + R_B \| R_i}\right) \qquad (5\text{-}28)$$

The first term is current gain $\beta_0$ times the input current divider of $R_B/(R_B + R_i)$. The resistance presented to $E_g$ is $R_{ig}$, where

$$R_{ig} = R_g + R_B \| R_i \qquad (5\text{-}29)$$

The general method of analysis for a common-emitter circuit with an unbypassed emitter resistor is summarized as follows: (1) Employ the resistance-transformation property of the transistor to draw the equivalent input circuit model directly from the circuit diagram so that we can skip part (b) of Fig. 5-9. (2) Find voltage gain in terms of the simpler relationship $V_o/E_i$. (3) Use the equivalent input circuit to get $E_i$ in terms of $E_g$ and write the more complex expression for $V_o/E_g$ (4) After facility is gained in steps (1) through (3) the voltage gain may be written by inspection directly from the circuit diagram by identifying each term in the standard format. Resistance levels at any point in the input circuit may also be written by inspection when $R_E$ is transformed by the factor of $(\beta_0 + 1)$ into the base circuit.

It is possible to express $V_o/E_i$ by a simple approximation if Eq. (5-24) is shown by

$$A_v = \frac{V_o}{E_i} = -\frac{\beta_0 R'_L}{r_x + r_\pi + (\beta_0 + 1)R_E} \qquad (5\text{-}24)$$

We can neglect $r_x + r_\pi$ in the denominator if it is small with respect to $(\beta_0 + 1)R_E$ and since $(\beta_0 + 1) \cong \beta_0$, Eq. (5-24) simplifies to

$$A_v = \frac{V_o}{E_i} \cong -\frac{R'_L}{R_E} \qquad (5\text{-}30)$$

*Example 5-8.* Values are given for the circuit of Fig. 5-9 as follows: $V_{CC} = 10.6$ V, $R_C = R_L = 5$ k$\Omega$, $I_C = 1$ mA, $\beta_0 = 50$, $r_x = 0$, $R_2 = \infty$, $V_{BE} = 0.6$ V. Assume $R_1$ can be adjusted to set $I_C$ at 1 mA for each of the conditions. (a) Find $V_o/E_i$ and the resistance $R_i$ for $R_E = 0$. (b) Repeat (a) for $R_E = 500$ $\Omega$, but compute $V_o/E_i$ with both exact and approximate equations.

*Solution.* The value of $g_m$ is 1 mA/25 mV = 0.040 mho and $r_\pi = \beta_0/g_m = 50/0.040 = 1.25$ k$\Omega$. $R_L$ is equal to $R_C \| R_L = 5$ k$\Omega \| 5$ k$\Omega = 2.5$ k$\Omega$.
(a) When $R_E = 0$, $R_i = r_x + r_\pi = 1.25$ k$\Omega$ and $A_v$ is, from Eq. (5-9),

$$A_v = \frac{V_o}{E_i} = -\frac{\beta_0 R'_L}{R_i} = -\frac{50 \times 2.5 \times 10^3}{1.25 \times 10^3} = -100$$

(b) When $R_E = 500 \, \Omega$, from Eq. (5-25),

$$R_i = r_x + r_\pi + (\beta_0 + 1)R_E = (1.25 + 51 \times 0.5) \times 10^3 = 26.7 \, \text{k}\Omega$$

From Eq. (5-24)

$$A_v = -\beta_0 \left(\frac{R_L'}{R_i}\right) = -50\left(\frac{2.5 \times 10^3}{26.7 \times 10^3}\right) = 4.68$$

From Eq. (5-30)

$$A_v = -\frac{R_L'}{R_E} = \frac{2.5 \times 10^3}{0.5 \times 10^3} = 5$$

Comparison of voltage gains calculated from the approximate Eq. (5-30) and actual Eq. (5-24) are shown in Fig. 5-10 as a function of $R_E$ for

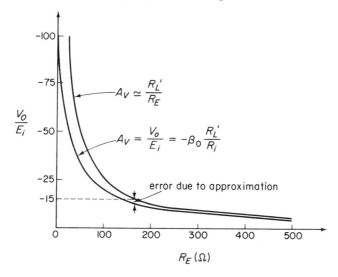

Fig. 5-10. Voltage gain $V_o/E_i$ as a function of $R_E$ for Example 5-8.

Example 5-8. It is seen that the approximate relationship is valid for gains below 15. Serious error is introduced by using the approximation for gains which are much above 15.

In Example 5-9 it is necessary to consider the input resistance of the circuit in order to bias the transistor. It will be seen that an arbitrary choice for $R_B$ can seriously reduce the voltage gain.

*Example 5-9.* In the circuit of Fig. 5-11, it is seen that $V_o/E_i$ is roughly 30, from $R_L/R_E$. The voltage loss from $E_g$ to $E_i$ will be about $\frac{1}{3}$ if $R_{\text{in}}$ is specified to be twice $R_g$. Assume we desire a voltage gain greater than 20 between $E_g$ and $V_o$. Observe how $R_B$ affects the outcome in the following:

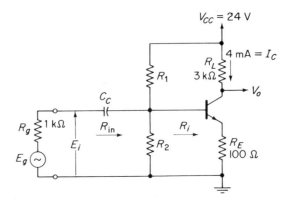

**Fig. 5-11.** Circuit for Example 5-9. $\beta_O = \beta_F = 50$, $r_x = 0$, $V_{BE} = 0.6$ V, and $I_{cbo} = 0$.

(a) Bias the transistor to establish $I_C$ at 4 mA. Pick $R_B$ as low as possible for stability but the resistance presented to $E_i$ must be a minimum of 2.7 kΩ.
(b) Find voltage gains $V_o/E_i$ and $V_o/E_g$.

*Solution.* (a) Evaluate $r_\pi$ from (1) $g_m = I_C/25$ mV $= 4/25$ mV $= 0.16$ mho. (2) $r_\pi = \beta_0/g_m = 50/0.16 = 313$ Ω. $R_i$ is evaluated from Eq. (5-25), at $R_i = 313 + 51 \times 100 = 5.4$ kΩ. Since $R_{in}$ must be greater than 2.7 kΩ and

$$R_{in} = 2.7 \text{ k}\Omega = R_B \| R_i = R_B \| 5.4 \text{ k}\Omega$$

we see that $R_B$ must be equal to at least 5.4 kΩ to meet the input-resistance specification. Now employ the dc bias equation to find $V_B$, with $I_B = I_C/\beta_F = 80$ μA:

$$V_B = I_B[R_B + (\beta_F + 1)R_E] + V_{BE}$$
$$= 80 \times 10^{-6} \times 10.5 \times 10^3 + 0.6 = 1.44 \text{ V}$$

$$R_1 = \frac{V_{CC}}{V_B} R_B = \frac{24}{1.44} \times 5.4 \times 10^3 = 90 \text{ k}\Omega$$

$$R_2 = \frac{V_{CC}}{V_{CC} - V_B} R_B = \frac{24}{22.56} \times 5.4 \times 10^3 = 5.75 \text{ k}\Omega$$

(b) From Eq. (5-24),

$$\frac{V_o}{E_i} = -\frac{\beta_0 R_L}{R_i} = -\frac{50(3000)}{5400} = -28.3$$

From Eq. (5-26),

$$\frac{E_i}{E_g} = \frac{R_{in}}{(R_g + R_{in})} = \frac{2700}{3700} = 0.73$$

Finally,

$$A_v = \frac{V_o}{E_g} = \frac{V_o}{E_i} \frac{E_i}{E_g} = 28.3(0.73) = 20.6$$

## 5-4 Common-Collector Amplifier

When an input signal is connected to the base and an output signal is extracted from the emitter, the configuration is designated as a common collector (CC) and is shown in Fig. 5-12(a). Biasing methods employed for

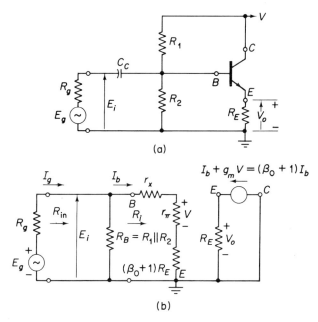

**Fig. 5-12.** CC circuit in (a) has the equivalent input circuit of (b). $R_E$ is transformed into the base leg by $(\beta_o + 1)$.

the common-collector circuit are indentical to those employed for the common-emitter circuit. At the instant shown in Fig. 5-12(b), $E_i$ puts an increment of forward bias on the *npn* transistor, causing an incremental increase in $I_C$, which is equal to $g_m V$. The increment $g_m V$ plus $I_b$ causes an incremental voltage increase across $R_E$ that is equal to the incremental output voltage $V_o$. We conclude that when the input base goes positive, the output emitter also goes positive. Therefore there is no phase shift. Observe that the collector terminal is at ac ground, through $V_{CC}$, so that it is common to the negative terminal of both $E_i$ and $V_o$.

We shall explore the resistance-transformation properties of the transistor to sketch the *equivalent input* circuit in Fig. 5-12(b). An input resistance $R_i$ is then defined and determined by inspection to be

$$R_i = \frac{E_i}{I_b} = r_x + r_\pi + (\beta_0 + 1)R_E \tag{5-31}$$

Output voltage $V_o$ is the product of $R_E$ times the sum of both currents $I_b$ and $\beta_0 I_b$, or

$$V_o = R_E(\beta_0 + 1)I_b \qquad (5\text{-}32)$$

Substituting for $I_b$ from Eq. (5-31),

$$\frac{V_o}{E_i} = \frac{(\beta_0 + 1)R_E}{R_i} = \frac{(\beta_0 + 1)R_E}{r_x + r_\pi + (\beta_0 + 1)R_E} \qquad (5\text{-}33)$$

Stipulating that $(r_x + r_\pi) \ll (\beta_0 + 1)R_E$, we can drop $r_x + r_\pi$ in Eq. (5-33) and it will be approximated by

$$A_v = \frac{V_o}{E_i} \cong 1, \qquad (r_x + r_\pi) \ll (\beta_0 + 1)R_E \qquad (5\text{-}34)$$

Note that Eq. (5-33) is in our standard format. The input current is $I_b$ and the output current through $R_E$ is $(\beta_0 + 1)I_b$. Current gain is the ratio of the two, or

$$A_i = \frac{I_o}{I_{in}} = \frac{(\beta_0 + 1)I_b}{I_b} = \beta_0 + 1 \qquad (5\text{-}35)$$

$V_o$ sees resistance $R_E$ and $E_i$ sees resistance $R_i$. We should now be able to construct Eq. (5-33) directly from the standard format and the circuit of Fig. 5-12(a).

Voltage gain of the common collector must also be expressed in terms of $V_o/E_g$. We see from Fig. 5-12(b) that $E_g$ divides between $R_g$ and $R_{in}$, where $R_{in} = R_B \| R_i$. $E_i$ is written in terms of $E_g$ from the voltage-divider relationship of Eq. (5-14) and substituted for $E_i$ in Eq. (5-35) to obtain

$$A_v = \frac{V_o}{E_g} = \frac{(\beta_0 + 1)R_E}{R_i}\left(\frac{R_{in}}{R_g + R_{in}}\right) \qquad (5\text{-}36)$$

Write $R_{in}$ in the numerator as $R_B R_i/(R_B + R_i)$ to cancel the $R_i$ terms and obtain Eq. (5-36) in our standard format:

$$A_v = \frac{V_o}{E_g} = \left(\frac{R_B}{R_B + R_i}\right)(\beta_0 + 1)\left(\frac{R_E}{R_g + R_{in}}\right) \qquad (5\text{-}37)$$

The contribution of each term in the standard format of Eq. (5-37) is described as follows:

1. $\beta_0 + 1$ is the *current gain of the transistor* from base to collector.
2. $R_B/(R_B + R_i)$ is the input current divider and tells how much of $I_b$ enters the base terminal to be multiplied by $\beta_0 + 1$. Therefore the current gain of the circuit, referred to $E_g$, is $(\beta_0 + 1)R_B/(R_B + R_i)$
3. $R_E$ is the resistance presented to $V_o$.
4. $R_g + R_{in}$ is the resistance presented to $E_g$.

## 5-5 Output Resistance and Resistance Transformations

The concept of *output resistance* will be introduced from the CC circuit. We shall *define output resistance* as the *resistance seen looking into the output terminal of the device, with the load disconnected.*

It has already been estabilished in Section 5-3 that looking into the base terminal we see any resistance-carrying base current unchanged and any resistance-carrying emitter current apparently multiplied by $\beta_0 + 1$. Looking into a collector terminal we see the very high resistance of current generator $g_m V$ and can consider it as infinite. Since the output terminal of a CE circuit is the collector terminal we would expect its output resistance to be high.

A standard procedure for deriving the output resistance, or any other resistance, is to apply a test voltage source and measure current drawn from this source. This is one of those procedures that may be difficult actually to accomplish in a laboratory. So assume the transistor circuit modeled in Fig. 5-12(b) is properly biased and we are able to connect an ac test voltage $E$ as shown in Fig. 5-13. Voltage source $E_g$ is replaced by its internal resistance

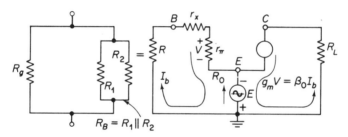

**Fig. 5-13.** Circuit to determine output resistance of the emitter in Fig. 5-12.

and $R_g$ and $R_B$ are combined for simplicity into an equivalent resistance $R$. $R_E$ is removed to simplify the circuit analysis. Test voltage $E$ causes loop current $I_b$ to traverse $r_\pi$, exciting the $g_m V$ generator and causing both $\beta_0 I_b$ and $I_b$ to flow through $E$. Output resistance $R_o$ is the resistance seen by $E$ and determines the current drawn from $E$, or

$$R_o = \frac{E}{(\beta_0 + 1)I_b} \qquad (5\text{-}38)$$

From the input loop $I_b = E/(r_x + r_\pi + R)$. Substituting into Eq. (5-38) for $I_b$,

$$R_o = \frac{r_x + r_\pi + R}{\beta_0 + 1} \qquad (5\text{-}39)$$

From Eq. (5-39) we conclude that *all resistance traversed by $I_b$ is divided by $\beta_0 + 1$ when we look into the emitter terminal.*

Should the occasion arise where it is necessary to obtain the resistance from the emitter to ground, with $R_E$ connected, we merely calculate the value of $R_o \| R_E$. Note that $R_E$ could have been shown across $E$ in Fig. 5-13 but would *not* have changed $I_b$, $g_m V$, and consequently $R_o$.

The basic characteristics of the CC configuration have been established and may be summarized as follows: (1) Input resistance is high, Eq. (5-31). (2) Output resistance is low, Eq. (5-39). (3) Voltage gain is approximately unity, Eq. (5-34), (4). Current gain is high at $\beta_0 + 1$, Eq. (5-35).

Since voltage gain is about 1, the input voltage approximately equals the output voltage. Therefore the output at the emitter terminal follows the base signal. This explains why the *common collector* is often referred to as an *emitter follower*.

***Example 5-10.*** Values for the circuit in Fig. 5-12 are given as follows: $V_{CC} = 9$ V, $V_{BE} = 0.6$ V, $I_C = 1$ mA, $\beta_0 = \beta_F = 50$, $I_{CBO} = 0$, $R_E = 3$ k$\Omega$, $r_x = 0$. (a) Calculate values of $R_1$ and $R_2$ so that their total shunting resistance will be at least equal to $R_i$. (b) Find the voltage gain $V_o/E_i$. (c) What is the maximum value of $E_i$ before the input voltage limitation on the hybrid-pi model is reached?

***Solution.*** (a) To find the input resistance $R_i$ find $r_\pi$ from (1) $g_m = I_C/25 = 0.040$, (2) $r_\pi = \beta_0/g_m = 50/0.040 = 1.25$ k$\Omega$. From Eq. (5-31), $R_i = r_x + r_\pi + (\beta_0 + 1)R_E = (1.25 + 51 \times 3) \times 10^3 = 154$ k$\Omega$. $R_B = R_1 \| R_2$ should be equal or greater than 154 k$\Omega$. Pick 154 k$\Omega$ and solve for $V_B$ from the dc input circuit with the dc base current of $I_C/\beta_F = 1$ mA/50 = 20 $\mu$A.

$$V_B = I_B[R_B + (\beta_F + 1)R_E] + V_{BE}$$
$$= 20 \times 10^{-6}(154 + 153) \times 10^3 + 0.6 = 6.74 \text{ V}$$

$$R_1 = \frac{V_{CC}}{V_B} R_B = \frac{9}{6.74}(154) \text{ k}\Omega = 205 \text{ k}\Omega$$

$$R_2 = \frac{V_{CC} R_B}{V_{CC} - V_B} = \frac{9 \times 154 \times 10^3}{9 - 6.74} = 614 \text{ k}\Omega$$

(b) From Eq. (5-33),

$$A_v = \frac{(\beta_0 + 1)R_E}{R_i} = \frac{51 \times 3 \times 10^3}{154 \times 10^3} = 0.994$$

(c) The signal voltage between base and emitter should not exceed 5 mV if our model is to be valid. The peak value of $I_b$ will be

$$I_{bp} = \frac{V_{be}}{r_\pi} = \frac{5 \times 10^{-3}}{1.25 \times 10^3} = 4 \ \mu\text{A}$$

The corresponding maximum value of $E_i$ will occur when the peak base current flows through $R_i$ or

$$E_{ip} = I_{bp}R_i = (4 \times 10^{-6})(154 \times 10^3) = 0.616 \text{ V}$$

It is concluded from part (c) of Example 5-10 that the common collector can accept relatively large input signal swings and still remain within the limitations of the hybrid-pi model. This is because most of $E_i$ is developed across $R_E$ and not across $r_\pi$.

**Example 5-11.** Extend Example 5-10 by assuming $R_g = 10 \text{ k}\Omega$ and $E_{gp} = 0.5$ V peak. (a) Find $I_{gp}$, (b) $I_{bp}$, (c) signal current through $R_E$, (d) $E_{ip}$, (e) $V_o/E_g$, (f) $V_o$.

**Solution.** (a) From Example 5-10, $R_{in} = R_i \| R_B = 154 \| 154 = 77 \text{ k}\Omega$. From Fig. 5-12(b),

$$I_{gp} = \frac{E_g}{R_g + R_{in}} = \frac{0.5}{(10 + 77)10^3} = 5.74 \text{ }\mu\text{A}$$

(b) From Fig. 5-12(b),

$$I_{bp} = \left(\frac{R_B}{R_B + R_i}\right)I_g = \left(\frac{154}{154 + 154}\right)5.74 = 2.87 \text{ }\mu\text{A}$$

(c) Load $R_E$ carries the current $(\beta_0 + 1)I_b = 51(2.87) \text{ }\mu\text{A} = 147 \text{ }\mu\text{A}$
(d) From Fig. 5-12(b),

$$E_i = \left(\frac{R_{in}}{R_g + R_{in}}\right)E_g = \left(\frac{77}{10 + 77}\right)0.50 = 0.442 \text{ V}$$

(e) From Eq. (5-37),

$$A_v = \left(\frac{154}{154 + 154}\right)(51)\left(\frac{3}{10 + 77}\right) = 0.88$$

(f)
$$V_o = A_v E_g = 0.88(0.5) = 0.44 \text{ V}$$

Using the data from Examples 5-11 and 5-10 we can illustrate a technique of finding output voltage by means of the output resistance. Take a Thévenin equivalent of Fig. 5-12(a), looking to the left from the base terminal as in Fig. 5-14(a). Transform the resulting base-leg resistance into the emitter as in Fig. 5-14(b). Solve for $V_{op}$ from the voltage division of $E_{Th}$ to get

$$V_{op} = \left(\frac{R_E}{R_E + R_o}\right)E_{Th} = \left(\frac{3000}{3208}\right)0.47 = 0.44 \text{ V}$$

An early appreciation for the possibilities inherent in looking from the load to the source as well as from the source to the load in the preceding examples should illuminate the reason why concepts of input and output

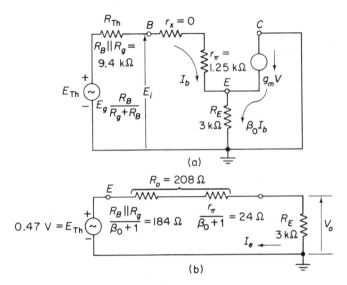

**Fig. 5-14.** A Thévenin equivalent of the input circuit in Fig. 5-12(a) is modeled in (a). Base-leg resistance is transformed into the emitter leg in (b). $V_o$ may then be found from inspection of the voltage divider.

resistance are always studied. The next example is an exercise, designed to give experience in performing resistance transformations by inspection. It is important to gain facility in estimating resistance levels, directly from the circuit schematic, and save the time normally spent in drawing the model and developing an equivalent model.

*Example 5-12.* Two basic rules are employed to apply the resistance transformation action of a transistor. (1) When looking into the emitter terminal, divide all resistances carrying base current by $\beta_0 + 1$. (2) When looking into the base terminal, multiply all resistances carrying emitter current by $\beta_0 + 1$. Note that (a) $r_x + r_\pi$ carries base current and (b) the collector terminal presents the large impedance of $g_m V$. Evaluate the ac resistance seen between ground and the specified terminal in Fig. 5-15. In each circuit $\beta_0 = 49$ and $r_x + r_\pi = 1 \text{ k}\Omega$. Assume the transistors are operating in the active mode even though the circuits are incomplete.

**Solution.**

|  | Terminal no. |
|---|---:|
| (a) $R = (r_x + r_\pi) \| R_B = 500 \, \Omega$ | ( 1) |
| $R = 2 \text{ k}\Omega$ | ( 2) |
| (b) $R = r_x + r_\pi + (\beta_0 + 1)R_E = 1 \text{ k}\Omega + 50 \text{ k}\Omega = 51 \text{ k}\Omega$ | ( 3) |
| (c) This resistance is the same as (3) | ( 4) |
| (d) $R = 2 \text{ k}\Omega + 1 \text{ k}\Omega + (50)(1 \text{ k}\Omega) = 53 \text{ k}\Omega$ | ( 5) |
| (e) $R = 2 \text{ k}\Omega + 1 \text{ k}\Omega + (50)(1 \text{ k}\Omega) = 53 \text{ k}\Omega$ | ( 6) |

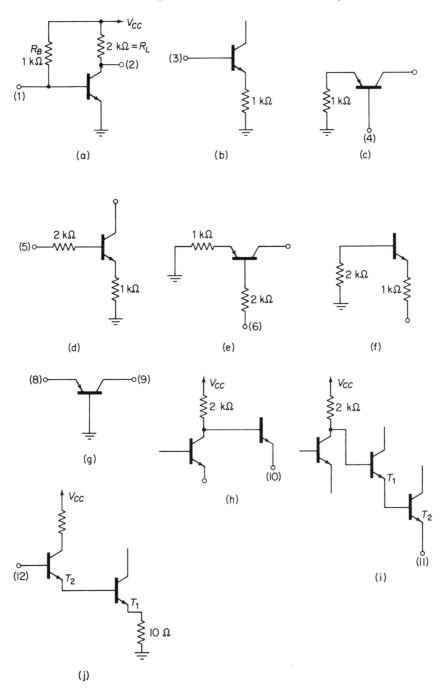

**Fig. 5-15.** Resistance-transforming exercise for Example 5-12.

(f) $R = 1 \text{ k}\Omega + 1 \text{ k}\Omega/50 + 2 \text{ k}\Omega/50 = 1.06 \text{ k}\Omega$ (7)
(g) $R = 1 \text{ k}\Omega/50 = 20 \text{ }\Omega$ (8)
$R = \infty$ (9)
(h) $R = 1 \text{ k}\Omega/50 + 2 \text{ k}\Omega/50 = 60 \text{ }\Omega$ (10)
(i) Analyze this circuit in two steps. Look into the emitter of $T_1$ and see $r_x + r_\pi$ of $T_1$ divided by $\beta_0 + 1$ or 20 $\Omega$, in series with $2 \text{ k}\Omega/(\beta_0 + 1) = 40 \text{ }\Omega$. Then look into the emitter of $T_2$ to see $r_x + r_\pi$ of $T_2$ as 20 $\Omega$ in series with the equivalent resistance of $T_1$ equal to $60/(\beta_0 + 1) = 1.2 \text{ }\Omega$. The answer is 21.2 $\Omega$. (11)
(j) Looking into the base of $T_1$, we see $r_x + r_\pi = 1 \text{ k}\Omega$ of $T_1$ plus $(50)(10) = 500$ for a total of 1.5 k$\Omega$. Looking into the base of $T_2$, we see the 1.5-k$\Omega$ equivalent resistance of $T_1$, multiplied by $\beta_0 + 1$ or 75 k$\Omega$, plus $r_x + r_\pi$ of $T_1 = 1 \text{ k}\Omega$. The answer is 76 k$\Omega$. (12)

## 5-6 Common-Base Amplifier

Early applications of transistors featured the common-base (CB) configuration almost to the exclusion of all others because of its inherently superior operating stability and high-frequency performance. (Recall that collector bias current will change with $I_{CBO}$ in the common-base circuit and $I_{CBO}$ cannot cross the emitter junction to be multiplied by $\beta_F$.) When the technology of fabricating silicon transistors had been developed, $I_{CBO}$ was no longer a major problem. As will be shown, the CB requires some sort of impedance transformer at its input or much of the available voltage gain of the transistor is lost by voltage division at the input. Also, when we cover the frequency dependence of voltage gain, it will be seen that the CB can provide usable voltage gain at frequencies considerably higher than the other configurations.

The CB circuit in Fig. 5-16(a) appears to be remarkably similar to a CE circuit with an unbypassed emitter resistor. Both circuits are identical with regard to the dc biasing model. Differences exist strictly in the ac or small-signal model. For example, a large value of $C_b$ will make its reactance negligible at 1000 Hz and consequently the base will be at ac ground. $C_E$ also has negligible reactance, so $E_i$ really exists between the emitter and base terminal. Because $V_{CC}$ has negligible reactance, the top terminal of $R_L$ is at ac ground and $V_o$ is developed between collector and base. This relationship defines the configuration whereby the base terminal is common to the input- and output-signal voltages. The circuit model is drawn in Fig. 5-16(b). $R_E$ must be present to avoid grounding the emitter-input terminal. $C_B$ masks the presence of $R_B$. The output resistance, considered in Fig. 5-14, is now the input resistance of the common-base circuit. But we really no longer care about definition of the configuration as long as we can employ the resistance-transformation properties of the transistor.

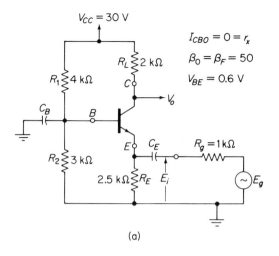

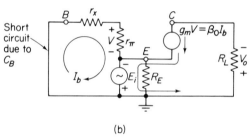

**Fig. 5-16.** Common-base circuit in (a) and model in (b).

Assume $E_i$ is the source voltage in the model in Fig. 5-16(b). $R_E$ acts as a current drain on $E_i$ but cannot affect $I_b$ and consequently voltage gain. Hence ignoring $R_E$, $E_i$ will carry $I_b$ and $\beta_0 I_b$ and we define input resistance *of the transistor* as $R_i$:

$$R_i = \frac{E_i}{(\beta_0 + 1)I_b} \tag{5-40}$$

But from the input loop $E_i/I_b = r_x + r_\pi$. Substituting for $E_i/I_b$ gives

$$R_i = \frac{r_x + r_\pi}{\beta_0 + 1} \tag{5-41}$$

Since $V_o = \beta_0 I_b R_L$ and $I_b = E_i/(r_x + r_\pi)$,

$$A_v = \frac{V_o}{E_i} = \beta_0 \frac{R_L}{(r_x + r_\pi)} \tag{5-42}$$

Voltage gains are identical for the CB and CE circuits except that there is no phase reversal in the CB.

Equation (5-42) fails to show any fundamental difference between the two configurations. To find the difference, construct a voltage-gain equation from the standard format. Referring to Fig. 5-16(b) define current gain of the transistor as $A_i = I_o/I_{in}$, where $I_{in}$ is the total current drawn from $E_i$ by the transistor of $I_b$ plus $\beta_0 I_b$ or $(\beta_0 + 1)I_b$. $I_o$ is collector current $\beta_0 I_b$. $A_i$ may be expressed as

$$A_i = \frac{I_o}{I_{in}} = \frac{\beta_0 I_b}{(\beta_0 + 1)I_b} = \frac{\beta_0}{\beta_0 + 1} = \alpha_0 \quad (5\text{-}43)$$

where $\alpha_0$ is the forward current gain of a common base ($h_{fb}$ on data sheets). Resistance seen by $E_i$ is given in Eq. (5-41), and $V_o$ sees $R_L$. Voltage gain is therefore

$$A_v = \frac{V_o}{E_i} = \frac{\beta_0}{(\beta_0 + 1)} \left( \frac{R_L}{\frac{r_x + r_\pi}{\beta_0 + 1}} \right) \quad (5\text{-}44)$$

Cancelling the $\beta_0 + 1$ terms simplifies Eq. (5-44) to (5-43) but the latter relationship is more indicative of the actual circuit performance.

Output resistance $R_o$ is determined by looking into the collector terminal to see the high resistance of $g_m V$. Resistance to ground from the collector terminal is simply, $R_L$.

Voltage gain, $A_v = V_o/E_g$, is found by reference to the equivalent input circuit in Fig. 5-17. From the voltage divider, $E_i = E_g R_{in}/(R_g + R_{in})$,

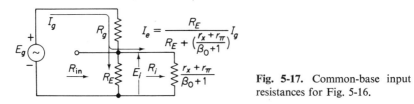

Fig. 5-17. Common-base input resistances for Fig. 5-16.

$$A_v = \frac{V_o}{E_g} = \frac{V_o}{E_i} \frac{E_i}{E_g} = \frac{\beta_0}{(\beta_0 + 1)} \left( \frac{R_L}{\frac{r_x + r_\pi}{\beta_0 + 1}} \right) \frac{R_{in}}{(R_g + R_{in})} \quad (5\text{-}45)$$

Manipulate Eq. (5-45) into the standard format by substituting for $R_{in}$ and cancelling the $(r_x + r_\pi)/(\beta_0 + 1)$ terms to obtain

$$A_v = \frac{V_o}{E_g} = \frac{R_E}{R_E + \left(\frac{r_x + r_\pi}{\beta_0 + 1}\right)} \left(\frac{\beta_0}{\beta_0 + 1}\right) \left(\frac{R_L}{R_g + R_{in}}\right) \quad (5\text{-}46)$$

where

$$R_{in} = R_E \| R_i \quad (5\text{-}47)$$

In Eq. (5-46) the current gain of the circuit is reduced by the division of $I_g$ between $R_E$ and $R_i$ to give

$$A_i = \frac{I_L}{I_g} = \left(\frac{R_E}{R_E + R_i}\right)\left(\frac{\beta_0}{\beta_0 + 1}\right) \tag{5-48}$$

*Example 5-13.* Analyze Fig. 5-16 to find (a) the dc operating point, (b) $V_o/E_i$, (c) $R_{in}$, (d) $V_o/E_g$.

*Solution.* (a) $R_B = (3\|4) \times 10^3 = 1.72 \text{ k}\Omega$, $V_B = 30(\frac{3}{7}) = 12.8$ V

$$I_B = \frac{V_B - V_{BE}}{R_B + (\beta_F + 1)R_E} = \frac{12.2}{[1.72 + 51(2.5)] \times 10^3} = 94.2 \text{ } \mu\text{A}$$

$$I_C = \beta_F I_B = 50(94.2) \text{ } \mu\text{A} = 4.71 \text{ mA}$$

$$V_{CE} = V_{CC} - (V_{RE} + V_{RL}) = 30 - (12.0 + 9.4) = 8.6 \text{ V}$$

where

$$V_{RE} = I_E R_E = (4.71 \text{ mA} + 0.94 \text{ } \mu\text{A})2500 = 12.0 \text{ V}$$

$$V_{RL} = I_C R_L = 4.71 \text{ mA} \times 2000 = 9.4 \text{ V}$$

(b) $g_m = \dfrac{I_C}{25 \text{ mV}} = \dfrac{4.71}{25} = 0.187 \text{ mho}$, $\quad r_\pi = \dfrac{\beta_0}{g_m} = \dfrac{50}{0.187} = 267 \text{ }\Omega$

From Eq. (5-42),

$$A_v = \frac{V_o}{E_i} = \beta_0 \frac{R_L}{(r_x + r_\pi)} = 50\left(\frac{2000}{267}\right) = 374$$

(c) From Eqs. (5-41) and (5-47),

$$R_i = \frac{r_x + r_\pi}{\beta_0 + 1} = \frac{267}{51} = 5.2 \text{ }\Omega \qquad R_{in} = R_E \| R_i = \frac{2500 \times 5.2}{2500 + 5.2} = 5.2 \text{ }\Omega$$

(d) $\quad E_i = \dfrac{R_{in}}{R_g + R_{in}}$, $\quad E_g = \dfrac{5.1}{1000 + 5.1} = E_g = 0.005$

$$A_v = \frac{V_o}{E_g} = \frac{V_o}{E_i}\frac{E_i}{E_g} = 374(0.005) = 1.87$$

$R_i$ is the single most important element, responsible for the tremendous loss in voltage gain.

Since the input resistance of a CB amplifier is extremely low it must be driven by a low-impedance source. For the amplifier of Fig. 5-16(a), a signal generator with a low internal resistance of only 50 $\Omega$ would still result in a voltage loss of over 90% right at the transistor's input terminals. This problem will be studied further in Chapter 9.

## 5-7 Measurement of Output and Input Resistance

In this text $E_i$ has represented a perfect voltage source. In practice, a signal generator with a 50-$\Omega$ impedance is about the nearest that is available. Many signal generators have internal or output impedances of 600 $\Omega$ or more In the laboratory a generator is connected to a load (for example a transistor input), and a signal voltage level is monitored at the generator's terminals. We may hold this load voltage constant by the generator's volume control while we vary some circuit element in the load. If this is the case, we have created a perfect voltage source $E_i$ and the gain equations in terms of $E_i$ are used.

Should we remove the load from the generator, its output voltage will increase to $E_g$. That is, $E_g$ *can be measured only when the generator is open circuited.* If we measure the open-terminal voltage of the signal generator,

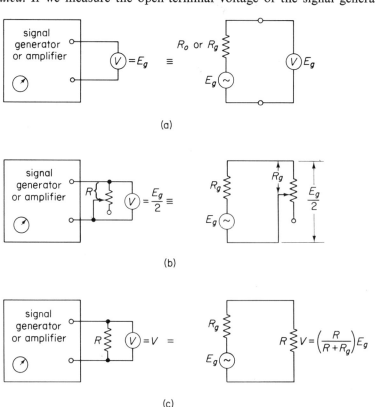

Fig. 5-18. Measure open-terminal voltage $E_g$ in (a). In (b) adjust $R$ to measure $E_g/2$, measure $R$ to find $R_g$. If too much current is drawn in (b), place a known resistor $R$ across the generator in (c) to measure $V$. Calculate $R_g$ from known values of $V$, $R$, and $E_g$.

connect it to a circuit and then make tests or circuit variations *without touching the generator's volume control*, the gain equations in terms of $E_g$ are used. Of course it is important to know the value of the generator's internal resistance. Two simple methods of measuring $R_g$ are given in Fig. 5-18. Output impedance of any device or circuit can also be measured by one of these methods.

The symbol $R_i$ refers to the input resistance of the transistor without accounting for $R_B$. $R_{\text{in}}$ will be reserved for the input resistance of the circuit, and $R_i$ is in parallel with the biasing resistors. We must be able to measure input resistance, wherever it might be defined, in order to verify our calculations. Two simple methods of doing so are shown in Fig. 5-19.

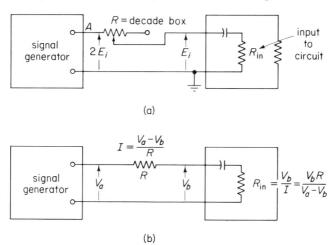

**Fig. 5-19.** In (a) vary $R$ for equal voltage across $R$ and $R_{\text{in}}$, where $R = R_{\text{in}}$. In (b) measure $V_a$ and $V_b$ with a known $R$ to calculate $R_{\text{in}}$.

## 5-8 Gain in Decibels

The relationship between an output and an input variable, such as signal power, is often expressed in a dimensionless unit called the *decibel* (dB) and given by

$$N = 10 \log \frac{P_o}{P_i} \tag{5-49}$$

where $N$ = number of decibels,
$P_o$ = the larger power (to avoid working with negative logarithms), and
$P_i$ = the smaller power (to avoid working with negative logarithms).

**172**  Small-Signal Low-Frequency Amplifiers  Chap. 5

Voltage gain is commonly expressed in decibels and may be expressed by substituting for power in terms of voltage and resistance by

$$N = 10 \log \frac{\frac{V_o^2}{R_o}}{\frac{V_i^2}{R_i}} = 20 \log \frac{V_o}{V_i} + 10 \log \frac{R_i}{R_o} \qquad (5\text{-}50)$$

Equation (5-50) is correct. However, it has become standard practice to make the normally erroneous assumption that $R_i = R_o$ so that the term $\log R_i/R_o$ goes to zero and Eq. (5-50) simplifies to

$$\text{Voltage gain} = N = 20 \log \frac{V_o}{V_i}, \qquad R_i = R_o \qquad (5\text{-}51)$$

Alternatively, the power ratio may be expressed in terms of current by

$$N = 10 \log \frac{I_o^2 R_o}{I_i^2 R_i} = 20 \log \frac{I_o}{I_i} + 10 \log \frac{R_o}{R_i} \qquad (5\text{-}52)$$

Again erroneously assuming $R_o = R_i$, Eq. (5-52) reduces to

$$\text{Current gain} = N = 20 \log \frac{I_o}{I_i}, \qquad R_i = R_o \qquad (5\text{-}53)$$

*Example 5-14.* An input signal of 0.2 V to an amplifier's input resistance of 2.0 kΩ yields 5 W into a 10-Ω load. Find, in decibels, (a) power gain, (b) voltage and current gain assuming $R_o = R_i$, (c) power gain with input and output voltages.

*Solution.* (a) The input current $I_i$ is $0.2/2000 = 0.1$ mA and the input power is $P_i = (0.2)^2/2000 = 20\ \mu\text{W}$. Output voltage is $V_o = (P_o R_o)^{1/2} = 7.07$ V and output current is $I_o = (P_o/R_o)^{1/2} = 0.707$ A. Power gain is found from Eq. (5-49):

$$N = 10 \log \frac{P_o}{P_i} = 10 \log \frac{5}{20 \times 10^{-6}} = 10 \log 250000$$

Evaluate by the following procedure of extracting the powers of 10:

$$N = 10 \log (10^5 \times 2.50) = 10(\log 10^5 + \log 2.50)$$
$$= 10(5 + 0.398) = 54 \text{ dB}$$

(b) Assuming $R_i = R_o$, the decibel voltage gain is calculated from Eq. (5-51):

$$N = 20 \log \frac{V_o}{V_i} = 20 \log \frac{7.07}{0.2} = 20 \log 35.3 = 31 \text{ dB}$$

From Eq. (5-53), current gain is

$$N = 20 \log \frac{I_o}{I_i} = 20 \log \frac{0.707}{0.1 \times 10^{-3}} = 20 \log 7070 = 77$$

(c) From Eq. (5-50),

$$N = 20 \log \frac{V_o}{V_i} + 10 \log \frac{R_i}{R_o} = 20 \log \frac{7.07}{0.2} + 10 \log \frac{2000}{10}$$
$$= 30.9 + 10 \log 200 = 30.9 + 23.1 = 54 \text{ dB}$$

Observe that the values of voltage, current, and power gain must be identical *when the resistance levels are included* because they derive from the power relationship. Even when the resistance levels are *not* included, the average of the sum of voltage and current gains will give the actual power gains. To prove this point, add Eqs. (5-50) and (5-52) and group like terms to obtain

$$2N = 20 \log \frac{V_o}{V_i} \frac{I_o}{I_i} + 10 \log \frac{R_i}{R_o} \frac{R_o}{R_i}$$

The right-hand term reduces to $10 \log 1 = 0$ and we have

$$N = 10 \log \frac{V_o I_o}{V_i I_i} = 10 \log \frac{P_o}{P_i}.$$

As a test, average the answers in Example 5-14 to get $(31 + 77)/2 = 54$ dB, which is the power gain.

## PROBLEMS

**5-1**  Evaluate $kT/q$ at 0°C.

**5-2**  Plot $g_m$ versus $I_C$ for values of $I_C$ between 1 $\mu$A and 100 $\mu$A at 17°C.

**5-3**  Sketch a low-frequency hybrid-pi model of a transistor and name the transistor parameters.

**5-4**  Given $\beta_0 = 50$. Plot $r_\pi$ versus $I_C$ for operating-point currents between 0.1 and 10 mA. Repeat for $\beta_0 = 100$ on the same graph. Assume $kT/q = 25$ mV.

**5-5**  In the circuit of Fig. 5-4(a), plot $V_o/E_i$ for collector currents between 0.1 and 1.6 mA. Assume $r_x$ is negligible and $kT/q = 25$ mV.

**5-6**  Compare the changes in Example 5-4 if $\beta_0 = 100$.

5-7 What changes occur in Example 5-5 if a transistor with $\beta_0 = 100$ is substituted?

5-8 Repeat Example 5-6 for $\beta_0 = 100$ and compare results.

5-9 If $E_g = 10$ mV in Example 5-6, evaluate signal currents $I_g$, $I_b$, and $I_c$ for (a) $R_g = 0$; (b) $R_g = 2500\ \Omega$.

5-10 If $E_g = 10$ mV in Example 5-7, evaluate signal currents $I_g$, $I_b$, and $I_c$ for (a) $R_g = 0$; (b) $R_g = 2500\ \Omega$.

5-11 Will the values of $R_L$ or $R_E$ affect the answers in Problems 5-9 and 5-10 as long as the transistor operates in the active region? Explain.

5-12 (a) Draw the hybrid-pi model for Fig. 5-9(a). (b) By inspection write the resistance expressions for $R_i$, $R_{\text{in}}$, and $R_{ig}$. (c) By inspection write the expression for $V_o/E_i$, $E_i/E_g$, and $V_o/E_g$.

5-13 $R_L$ is removed in Example 5-8 and $R_C = 5$ k$\Omega$ is the new load. What are the new results in (a) and (b) of Example 5-8.

5-14 If $R_B$ is reduced to 2.7 k$\Omega$ in Example 5-9, but $V_B$ is adjusted to keep the same operating point what is the resulting resistance presented to $E_i$ and $V_o/E_g$?

5-15 Double $R_B$ to 10.8 k$\Omega$ in Example 5-9; retain the same operating point and evaluate the change in $V_o/E_g$.

5-16 Write the expression for $V_o/E_i$ from inspection of the common-collector circuit in Fig. 5-12(a) or (b). Write the expression for $E_i/E_g$ by inspection of the voltage divider in Fig. 5-12(b).

5-17 In the circuit of Fig. 5-12, $V_{CC} = 9$ V, $R_E = 3$ k$\Omega$, $I_C = 1.5$ mA, $\beta_0 = \beta_F = 100$, $V_{BE} = 0.6$ V, $R_B = 10 R_E = 30$ k$\Omega$. Find (a) base-to-emitter signal-current gain, (b) $R_1$ and $R_2$, (c) $V_o/E_i$, (d) $R_i$, (e) $R_{\text{in}}$.

5-18 Find (a) $V_o$ in Problem 5-17 if $R_g = 10$ k$\Omega$ and $E_g = 0.5$ V; (b) signal currents $I_g$, $I_b$, and load current through $R_E$.

5-19 What is the maximum peak output voltage in Problems 5-17 and 5-18, assuming $V_{CE} = 0$ when the transistor is on the edge of saturation? Find the corresponding values of $E_i$ and $E_g$ assuming the hybrid-pi model is valid.

5-20 Evaluate all the ac resistance seen between ground and the specified terminals in Fig. 5-15 if $\beta_0 = 99$ and $r_x + r_\pi = 1$ k$\Omega$.

5-21 Repeat Problem 5-20 for $\beta_0 = 49$ and $r_x + r_\pi = 2$ k$\Omega$. Compare results.

5-22 A phonograph cartridge generates an open-circuit voltage of 0.5 V rms and has an internal resistance of 50 k$\Omega$. Which voltage gain formulas are applicable, those with $E_i$ or those with $E_g$?

5-23 In Fig. 5-16, $V_o$ is connected through a coupling capacitor to the input of an identical common-base stage (without another $R_g$) whose output voltage is $V'_o$. Derive an approximate expression for $V_o/E_i$ and show that it is less than 1.

5-24 In Fig. 5-19(b), $V_a = 11V_b$ and $R = 1000\ \Omega$. What is $R_{in}$?

5-25 In Example 5-14(c), calculate power gain using input and output currents and resistances.

5-26 If you were writing the amplifier's gain specifications for sales or publicity from Example 5-14 which one should you choose?

5-27 If voltage across a load resistor is (a) doubled, what is the power increase in decibels? (b) halved, what is the power decrease in decibels?

5-28 Design an amplifier stage based on Fig. 5-11 to (a) operate at $I_C = 1$ mA, $V_{CE} = 12$ V; (b) yield a voltage gain of $V_o/E_i = 25$; (c) with a minimum $R_{in}$ of 2500. (Hint: Find the sum of $R_L + R_E$ for $V_{CE} = 12$ V. Let (sum $- R_E$) $= R_L$ in Eq. (5-24), substituting $R_L$ for $R'_L$ and solve for $R_E$, then for $R_L$. Make $R_B$ as small as possible but make $R_{in}$ at least 2500 $\Omega$.)

# Chapter 6

6-0  INTRODUCTION .............................. 177
6-1  IDENTIFICATION OF CUTOFF FREQUENCY ........... 177
6-2  LOW-FREQUENCY CUTOFF BY THE COUPLING
     CAPACITOR .................................. 179
6-3  LOW-FREQUENCY CUTOFF BY THE BYPASS
     CAPACITOR .................................. 182
6-4  CUTOFF FREQUENCY WITH BOTH COUPLING AND
     EMITTER BYPASS CAPACITORS ................... 184
6-5  HIGH-FREQUENCY MODEL OF A BJT .............. 187
6-6  DEPENDENCE OF $\beta$ ON FREQUENCY ................ 189
6-7  COMMON-EMITTER HIGH-FREQUENCY CUTOFF ...... 191
6-8  DERIVATION OF MILLER EFFECT ................. 195
6-9  EMITTER RESISTANCE AND HIGH-FREQUENCY CUTOFF
6-10 COMMON-BASE AND COMMON-COLLECTOR UPPER
     CUTOFF FREQUENCY .......................... 197
6-11 MEASUREMENT OF THE BASIC HYBRID-PI
     PARAMETERS ................................ 199
     PROBLEMS................................... 204

# Frequency Limitations of Voltage Gain

## 6-0 Introduction

Our objectives in this chapter are to learn how to predict the lower and upper useful frequency limits of an amplifier circuit, quickly and with reasonable accuracy. Our approach will be directed toward identifying the specific elements responsible for a particular frequency limitation to see if there is anything we can do about it, rather than studying circuit behavior beyond the frequency limits.

## 6-1 Identification of Cutoff Frequency

*Cutoff frequency* is defined as the lower frequency $\omega_L$ or upper frequency $\omega_H$ where output power is decreased by one-half of its midfrequency value. Thus $\omega_L$ and $\omega_H$ specify the frequency limits or bandwidth where the circuit or device is usable. When an output-input relationship such as voltage gain is plotted against frequency the result is a *frequency-response* curve as shown in Fig. 6-1, which illustrates pictorially the frequency limits.

The cutoff frequency may also be called *break, 3-db, half-power,* or *0.707 frequency*. Since it is easier to measure voltage levels than power levels we focus attention on the 0.707 frequency by reference to the frequency-dependent circuit of Fig. 6-2, where

$$V_o = \frac{E_i R}{R - jX_C} \qquad (6\text{-}1)$$

Output power delivered to load $R$ is expressed generally as

$$P = \frac{V_o^2}{R} = \frac{E_i^2 R}{(R - jX_C)^2} \qquad (6\text{-}2)$$

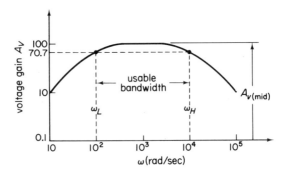

**Fig. 6-1.** Typical voltage gain versus frequency for a singe-stage amplifier.

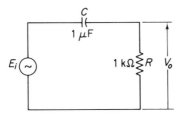

**Fig. 6-2.** $V_o$ depends on the frequency of $E_i$.

At high frequencies $C$ acts as a short circuit so $X_C = 1/\omega C = 0$ and Eq. (6-2) simplifies to

$$P = \frac{E_i^2}{R}, \quad \omega = \text{high}$$

At a frequency $\omega_L$, where the magnitude of the $j$ term in Eq. (6-2) equals the real term, substitute $R$ for $X_C$ to obtain a denominator magnitude of $R - jR = \sqrt{2}\,R$. and output power is then

$$P = \frac{E_i^2}{2R}, \quad \omega = \omega_L \qquad (6\text{-}3)$$

Comparing Eqs. (6-3) and (6-2) we see that output power is reduced by one-half at $\omega_L$ and the ratio of maximum power to power at $\omega_L$ is 3 dB from

$$N = 10 \log 2 = 3 \text{ dB}$$

Now, substituting for $X_C = 1/(\omega_L C) = R$ into Eq. (6-1) yields

$$\frac{V_o}{E_i} = \frac{R}{R - \dfrac{j}{\omega_L C}} = \frac{R}{R - jR} = \frac{R}{\sqrt{2}\,R\underline{/-45°}} = 0.707\underline{/45°} \qquad (6\text{-}4)$$

From Eq. (6-4), $V_o/E_i = 1$ at high frequencies and is down to 0.707 times high-frequency value at the cutoff frequency. Thus we can measure $\omega_L$ simply by decreasing the frequency of $E_i$ in Fig. 6-2 until $V_o$ drops to 0.707 $E_i$ and reading $\omega_L$ from the oscillator's dial.

There is a tremendous asset hidden in Eq. (6-4), namely, a method of determining $\omega_L$ *by inspection of the circuit*. Since $\omega_L$ was defined as the frequency where magnitudes of real and $j$ terms are equal,

$$R = \frac{1}{\omega_L C} \quad \text{or} \quad \omega_L = \frac{1}{RC} \qquad (6\text{-}5)$$

But $RC$ is a time constant that can be identified in the circuit without recourse to the formulas. To find the cutoff frequency, we therefore (a) look for the resistance $R$, present it to the capacitor (2) evaluate the time constant $\Upsilon$, and take its reciprocal to find $\omega_L$. In equation form:

$$\omega_L = \frac{1}{RC} = \frac{1}{\Upsilon} \qquad (6\text{-}6)$$

*Example 6-1.* What is the cutoff frequency of Fig. 6-2?
*Solution.* $\Upsilon = RC = 10^3 \times 10^{-6} = 10^{-3}$, $\omega_L = 1/10^{-3} = 1000$ rad/sec. In hertz, cutoff occurs at $f_L = \omega_L/(2\pi) = 159$ Hz.

From Eq. (6-4) we can see that at high frequencies the $j$ term is negligible and $V_o$ is in phase with $E_i$. At a frequency of $\omega_L$, $V_o$ leads $E_i$ by 45° and we shall usually associate a phase shift of 45° with the cutoff frequency. In the rest of this chapter we first shall look for the time constant to write the cutoff frequency by inspection of the circuit and then develop the proof.

## 6-2 Low-Frequency Cutoff by the Coupling Capacitor

The CE circuit in Fig. 6-3(a) is modeled in Fig. 6-3(b), where coupling capacitor $C_C$ is retained because we are interested in how $C_C$ affects voltage gain. We choose $E_i$ as our reference, for simplicity, in the model of Fig. 6-3(b). We maintain $E_i$ constant in amplitude and monitor $V_o$ to read voltage gain at the midfrequency range where $C_C$ acts as a short circuit. As the frequency of $E_i$ is reduced, the reactance of $C_C$ increases to reduce $V$ and consequently output voltage $V_o$. The resistance presented to $C_C$ is $R_B \| (r_x + r_\pi)$ and forms the time constants that allow us to express the low cutoff frequency as

$$\omega_L = \frac{1}{C_C [R_B \| (r_x + r_\pi)]} = \frac{1}{\Upsilon} \qquad (6\text{-}7)$$

We can conclude that voltage gain will decrease to 0.707 times its

**Frequency Limitations of Voltage Gain**    Chap. 6

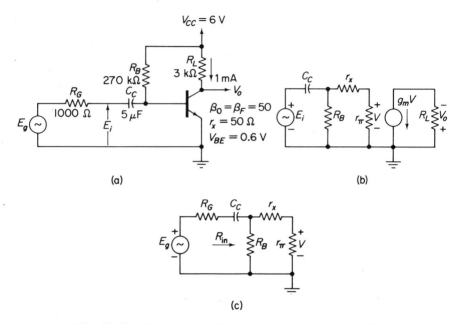

**Fig. 6-3.** The time constant for the coupling capacitor in (a) is shown for the $E_i$ reference in (b). In (c) resistance $R_g + R_B \| (r_x + r_\pi)$ forms the time constant with $C_c$ for the $E_g$ reference.

midfrequency value at $\omega_L$. Proof of this conclusion is accomplished by writing the expression for $V$ in Fig. 6-3(b).

$$V = \frac{r_\pi}{r_x + r_\pi}\left[\frac{R_{in}}{R_{in} + \frac{1}{j\omega C_c}}\right]E_i, \text{ where } R_{in} = R_B \| (r_x + r_\pi)$$

Then $|V_o| = g_m V R_L$, or

$$|V_o| = \frac{g_m r_\pi R_L}{r_x + r_\pi}\left[\frac{R_{in}}{R_{in} + \frac{1}{j\omega C_c}}\right]E_i = \frac{\beta_0 R_L E_i}{r_x + r_\pi}\left[\frac{R_{in}}{R_{in} + \frac{1}{j\omega C_c}}\right]$$

Divide numerator and denominator by $R_{in}$ and substituting Eq. (5-9):

$$\frac{V_o}{E_i} = \frac{\frac{\beta_0 R_L}{r_x + r_\pi}}{1 + \frac{1}{j\omega C_c R_{in}}} = A_{v(mid)}\frac{1}{1 + \frac{\omega_L}{j\omega}} \qquad (6\text{-}8)$$

where

$$A_{v(mid)} = -\frac{\beta_0 R_L}{r_x + r_\pi} \text{ and } \omega_L = \frac{1}{C_c R_{in}}$$

Sec. 6-2        Low-Frequency Cutoff by the Coupling Capacitor        181

The phase angle will be 45° at $\omega = \omega_L$ and $\theta$ is given by

$$\theta = \arctan \frac{\omega_L}{\omega} \tag{6-9}$$

In Eq. (6-8), when $\omega = \omega_L$ the $j$ term is 1 and $A_{v(\text{mid})}$ is divided by $1 + j1 = \sqrt{2}$ or multiplied by $1/\sqrt{2} = 0.707$. Therefore $\omega_L$ locates the low cutoff frequency.

When $E_g$ is the reference voltage, the lower cutoff frequency is formed by the time constant of Fig. 6-3(c). Capacitor $C_C$ sees $R_g$ in series with a parallel combination of $R_B$ and $r_x + r_\pi$. The midfrequency voltage expression is given by Eq. (5-16) and the low cutoff frequency by

$$\omega_L = \frac{1}{\Upsilon} = \frac{1}{C_C(R_g + R_{\text{in}})} \tag{6-10}$$

where $R_{\text{in}} = R_B \| (r_x + r_\pi)$.

Addition of resistance in the emitter leg will increase the time constant because $(\beta_0 + 1)R_E$ is added to the term $r_x + r_\pi$ in Eqs. (6-7) and (6-10). Of course, increasing the time constant will decrease the lower 0.707 frequency and improve low-frequency response, but also decrease voltage gain.

*Example 6-2.* If $E_i$ is held at a constant amplitude, what is the low-frequency cutoff for the circuit in Fig. 6-3?

*Solution.* Evalute $g_m = 1$ mA/25 mV $= 0.040$, $r_\pi = \beta_0/g_m = 1250\ \Omega$.

$$R_B \| (r_x + r_\pi) = (270 \| 1.3)\ \text{k}\Omega \cong 1300\ \Omega$$

From Eq. (6-7),

$$\omega_L = \frac{1}{(5 \times 10^{-3})(1300)} = 154\ \text{rad/sec} \cong 25\ \text{Hz}$$

*Example 6-3.* If $E_g$ is held at constant amplitude, what is the low-frequency cutoff in Example 6-2?

*Solution.* Evaluate $R_{\text{in}} = (270 \| 1.3) \times 10^3 \cong 1.3$ k$\Omega$. From Eq. (6-10),

$$\omega_L = \frac{1}{(5 \times 10^{-6})(10 + 1.3) \times 10^3} = 17.7\ \text{rad/sec} \cong 3\ \text{Hz}$$

*Example 6-4.* If an emitter resistor of $R_E = 100\ \Omega$ were added to the circuit in Example 6-2, what would be the new low-frequency cutoff, assuming $I_C$ was held constant? Neglect $R_B$.

*Solution.* In Fig. 6-3(b) we would add $(\beta_0 + 1)R_E = 5100\ \Omega$ in series with $r_\pi$ and modify Eq. (6-7) to

$$\omega_L = \frac{1}{C_C[r_x + r_\pi + (\beta_0 + 1)R_E]} \cong \frac{1}{(5 \times 10^{-6})(6400)}$$
$$= 31.2\ \text{rad/sec} \cong 5\ \text{Hz}.$$

## 6-3 Low-Frequency Cutoff by the Bypass Capacitor

In the circuit of Fig. 6-4(a), $C_E$ short circuits $R_E$ at midfrequencies. As the frequency of $E_i$ is lowered the reactance of $C_E$ increases to such a large value that $R_E$ becomes unbypassed and lowers voltage gain to approximately $R_L/R_E$. Thus voltage gain at midfrequencies is given by Eq. (5-9) and at very low frequencies by Eq.s (5-24) or (5-30). At some intermediate value of low frequency, $\omega_L$, the cutoff frequency, is reached where gain is reduced to 0.707 times its midfrequency value. To evaluate $\omega_L$ we look for the resistance presented to $C_E$ in Fig. 6-4(a). $C_E$ sees $R_E$ in parallel with the incremental resistance seen looking into the emitter. We replace voltage source $E_i$ by its zero internal impedance (which shorts out $R_B$) to draw the model in Fig. 6-4(c). From the resulting time constant we can express $\omega_L$ by

$$\omega_L = \frac{1}{T} = \frac{1}{C_E(R_E \| R)} \qquad (6\text{-}11)$$

where $R = (r_x + r_\pi)/(\beta_0 + 1)$.

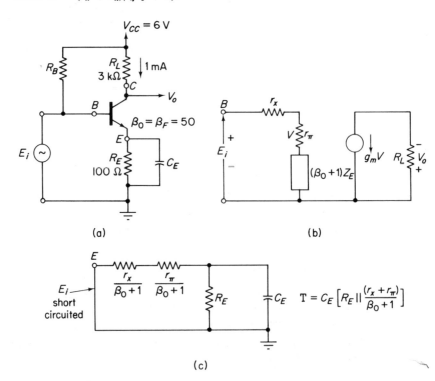

**Fig. 6-4.** The circuit in (a) is modeled by its equivalent input impedance in (b). Resistance presented to $C_E$ is shown in (c).

Sec. 6-3   Low-Frequency Cutoff by the Bypass Capacitor   183

*Example 6-5.* Assuming $r_x = 0$ in Fig. 6-4(a), find a value for $C_E$ to establish a low-frequency cutoff at 500 rad/sec.

*Solution.* Since $I_C = 1$ mA, $g_m = 0.040$ and $r_\pi = 50/0.040 = 1.25$ kΩ. Evaluate $R$ from Eq. (6-11) as $R = 1250/51 = 24.5$ Ω, and $R_E \| R = 100 \| 24.5 \cong 20$ Ω. Substituting into Eq. (6-11),

$$C_E = \frac{10^6}{(500)(20)} = 100 \ \mu\text{F}$$

It is evident in Example 6-5 that the low impedance presented by the emitter terminal is the most significant factor responsible for the large value of $C_E$.

In order to verify Eq. (6-11), we transform $Z_E = R_E \| X_{CE}$ into the base leg in Fig. 6-4(b) and express $V$ in terms of $E_i$, after considerable manipulation, in the form shown below:

$$V = \left(\frac{r_\pi}{r_x + r_\pi}\right) \left[\frac{1 + \dfrac{1}{j\omega C_E R_E}}{1 + \dfrac{1}{j\omega C_E (R_E \| R)}}\right] E_i \tag{6-12}$$

where $R = (r_x + r_\pi)/(\beta_0 + 1)$. There are two time constants in Eq. (6-12) and two corresponding cutoff frequencies:

$$\omega_1 = \frac{1}{T_1} = \frac{1}{C_E R_E} \tag{6-13}$$

$$\omega_2 = \frac{1}{T_2} = \frac{1}{C_E(R_E \| R)} \tag{6-14}$$

Since $R_E \| R$ is always considerably less than $R_E$, the denominator cutoff frequency $\omega_2$ will affect $E_i$ at a higher frequency and be the dominant effect on $V$ in Eq. (6-12). That is, the denominator's $j$ term will become significant and lower $V$ at frequencies where the numerator's $j$ term is still negligible. Thus $\omega_2$ agrees with our intuitively obtained expression in Eq. (6-11).

To complete our analysis substitute Eq. (6-12) for $V$ in $V_o = -g_m V R_L$ and substitute Eqs. (6-13) and (6-14) to obtain

$$A_V = \frac{V_o}{E_i} = -\left(\frac{g_m r_\pi R_L}{r_x + r_\pi}\right) \frac{\left(1 + \dfrac{\omega_1}{j\omega}\right)}{\left(1 + \dfrac{\omega_2}{j\omega}\right)} = A_{v(\text{mid})} \frac{\left(1 + \dfrac{\omega_1}{j\omega}\right)}{\left(1 + \dfrac{\omega_2}{j\omega}\right)} \tag{6-15}$$

In Eq. (6-15), when $\omega = \omega_2 = \omega_L$, the term $\omega_1/\omega$ is negligible and $\omega_2/\omega_L = 1$, so Eq. (6-15) reduces to

$$A_V = \frac{V_o}{E_i} = A_{v(\text{mid})}\left(\frac{1}{1 + j1}\right) = 0.707 \, A_{v(\text{mid})} \quad \text{at } \omega = \omega_L$$

## 6-4 Cutoff Frequency with Both Coupling and Emitter Bypass Capacitors

Our final concern with low-frequency response is to find the low cutoff frequency when both coupling and emitter bypass capacitors are present in a CE circuit. Our approach will be aimed at finding only the cutoff frequency which identifies that frequency where midfrequency gain is reduced by a factor of 0.707. We will make a separate analysis for the effects of each capacitor. When two energy-storage elements, such as capacitors, are present in a circuit they can interact in a complicated manner to give more than two cutoff frequencies. However, only two of the cutoff frequencies are important for our purposes, and are found from the circuit's *short-circuit time constants*.

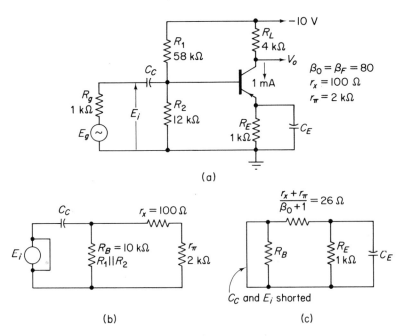

**Fig. 6-5.** Effects of $C_c$ and $C_e$ in (a) are shown independently by short circuiting $C_c$ in (b) and $C_E$ in (c).

Refer to Fig. 6-5(a), where both capacitors $C_C$ and $C_E$ are present. We find the short-circuit time constant for $C_C$ by assuming that $C_E$ is an ac short circuit. Physically this means that we assume $C_E$ is so large that $C_C$ alone will introduce the cutoff frequency $\omega_C$. $C_C$ will then start reducing $E_i$ at least a decade before $C_E$ starts connecting $R_E$ into the circuit. Using $E_i$ as our reference, the short-circuit time constant $\Upsilon_C$ and $\omega_C$ are shown in Fig. 6-5(b). A resistance of $R_B \| (r_x + r_\pi)$ is presented to $C_C$ and forms the time constant:

Sec. 6-4    Cutoff Frequency Coupling and Emitter Bypass Capacitors    185

$$\omega_C = \frac{1}{\tau_C} = \frac{1}{C_C[R_B \| (r_x + r_\pi)]} \quad (6\text{-}16)$$

To find the *short-circuit time constant of* $C_E$ we short circuit $C_C$ and see what resistance is presented to $C_E$ in Fig. 6-5(c). Physically this means that we assume $C_C$ is so large that it remains an ac short circuit during the frequency range where the reactance of $C_E$ becomes large, and $C_E$ alone causes cutoff at $\omega_E$. Employing $E_i$ as our reference, the short-circuit time constant in Fig. 6-5(c) shows $\omega_E$ to be

$$\omega_E = \frac{1}{\tau_E} = \frac{1}{C_E(R_E \| R)} \quad \text{where} \quad R = \frac{r_x + r_\pi}{\beta_0 + 1} \quad (6\text{-}17)$$

By comparing Eqs. (6-16) and (6-17), it is seen that factor $(r_x + r_\pi)/(\beta_0 + 1)$ is an extremely low value and will tend to make $\omega_E$ very high unless $C_E$ is made very large. We now must choose either $\omega_E$ or $\omega_C$ to cause the dominant cutoff frequency. Our choice will be based on economics and once our choice is made we shall increase the other capacitor's size by a factor of 10 to validate our use of the short-circuit time constant assumption.

*Example 6-6.* In the circuit of Fig. 6-5, assume $E_i$ is the reference and (a) calculate $C_C$ to cutoff the amplifier at $\omega_C = 100$ rad/sec. (b) calculate $C_E$ for cutoff at the same frequency, $\omega_E$. (c) Select one capacitor to cut off the amplifier and change the other for cutoff a decade lower at 10 rad/sec. (d) calculate $V_o/E_i$ at midfrequency, and at the cutoff frequency.

*Solution.* (a) From Eq. (6-16),

$$C_C = \frac{1}{100(10^4 \| 2100)} \cong 5.8 \ \mu\text{F}$$

where $R_B = (12 \| 58) \ \text{k}\Omega = 10 \ \text{k}\Omega$.
(b) From Eq. (6-17),

$$C_E = \frac{1}{100(1000 \| 26)} \cong 400 \ \mu\text{F}$$

(c) If we selected $C_C$ to cutoff the amplifier at 100 rad/sec then $C_E$ must be increased to 4000 $\mu$F to cutoff at 10 rad/sec. It is more economical to retain $C_E$ at 400 $\mu$F and increase $C_C$ to 58 $\mu$F. In practice we would select the nearest commercially available capacitors. Voltage ratings can be quite low because even with a shorted resistor, only 10 V is present in the circuit. (A note of caution is inserted concerning the equivalent series resistance of an electrolytic capacitor, which is obtained from a reading of the dissipation factor or $Q$ when capacitance is measured. This resistance can be as high as 15 $\Omega$ and in part (b) would increase the net resistance presented to $C_E$ from 25 to 40 $\Omega$ and decrease the required value of $C_E$ to 250 $\mu$F.)
(d) From Eq. (5-9), $V_o/E_i = 80(4000)/2100 = 152$ at midfrequency and $0.707(152) = 108$ at $\omega_E$.

The actual cutoff frequency $\omega_L$ in Example 6-6 is cased by the sum of the time constants

$$\omega_L \cong \frac{1}{\tau_C} + \frac{1}{\tau_E} = \omega_E + \omega_C \qquad (6\text{-}18)$$

but since we have standardized on making $\omega_C$ a decade lower than $\omega_E$, for reasons of economy, we can approximate $\omega_L$ by $\omega_L \cong \omega_E$.

There is an even shorter method for evaluating $C_C$ once $C_E$ is known. Assume that (1) $\omega_E = \omega_C$, (2) $R_B \gg (r_x + r_\pi)$ in Eq. (6-16), (3) $R_E \gg (r_x + r_\pi)/(\beta_0 + 1)$ in Eq. (6-17), and (4) equate the result as follows:

$$\text{Eq. (6-16)} \qquad \text{Eq. (6-17)}$$

$$\omega_C \cong \frac{1}{C_C(r_x + r_\pi)} = \omega_E \cong \frac{1}{C_E\left(\frac{r_x + r_\pi}{\beta_0 + 1}\right)}$$

Solve for $C_C$ to find

$$C_C \cong \frac{C_E}{\beta_0 + 1} \quad \text{at } \omega_C = \omega_E \qquad (6\text{-}19)$$

Thus in Example (6-6) we could calculate $C_E$ from (b) and then immediately evaluate $C_C$ from Eq. (6-19) $\cong 5\ \mu F$ and multiply the result by 10.

**Example 6-7.** With $E_g$ as reference, sketch the short-circuit time constants for Fig. 6-5. Calculate the break frequency for each, if $C_C = 50\ \mu F$ and $C_E = 500\ \mu F$.

**Solution.** The circuits are drawn in Fig. 6-6. $\omega_C$ and $\omega_E$ are evaluated from Figs. 6-6(a) and 6-6(b).

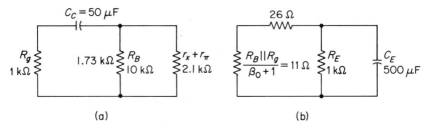

**Fig. 6-6.** Short-circuit time constants for $C_C$ in (a) and $C_E$ in (b) for Example 6-7.

$$\omega_C = \frac{10^6}{50(2.73 \times 10^3)} \cong 7\ \text{rad/sec} \qquad \omega_E = \frac{10^6}{500(1000\ \|\ 37)} = 56\ \text{rad/sec}$$

## 6-5  High-Frequency Model of a BJT

A varying signal voltage will cause a variation in voltage across both collector and emitter junctions of a BJT and must vary the amount of charge carriers traversing the space-charge regions as well as the base region. A signal that momentarily increases forward bias of the emitter junction increases the charges temporarily in these regions, and when the signal is removed these excess charges are removed and equilibrium is restored at the bias condition. A capacitor also stores charges and it follows that we can model charge storage in the junctions by *junction capacitances* and charges stored in the base by a *diffusion capacitance* to represent the excess charges stored as they diffuse across the base region. Base diffusion capacitance and emitter-junction capacitance are essentially in parallel so that they are modeled by $C_\pi$ in Fig. 6-7. Collector-junction capacitance is modeled by $C_\mu$.

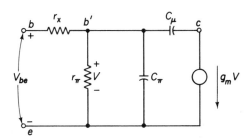

**Fig. 6-7.** High-frequency model of a BJT.

EVALUATION OF $C_\mu$. Values for $C_\mu$ may be given in some transistor data sheets under the symbol $C'_{bc}$.

An approximate value for $C_\mu$ is available from other data sheets as $C_{obo}$, the measured output capacitance of a common base with the emitter *incrementally* open circuited. This measurement is made between base and collector terminals and includes header capacitance plus capacitance between the base and collector leads so that $C_{obo}$ is actually slightly higher than $C_\mu$. $C_{obo}$ decreases with increasing collector voltage and their relationship is often given on a curve as in Fig. 6-8.

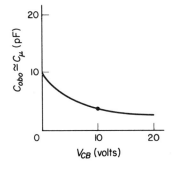

**Fig. 6-8.** Variation of $C_\mu$ with $V_{CB}$.

188  Frequency Limitations of Voltage Gain    Chap. 6

EVALUATION OF $C_\pi$. $C_\pi$ may be specified by some typical value as $C'_{be}$ on some data sheets. However, on many other data sheets $C_\pi$ must be calculated from one of three possible presentations of data.

1. The manufacturer may give a value for $\omega_\beta$ or $f_\beta$ and we evaluate $C_\pi$ from

$$C_\pi = \frac{1}{r_\pi \omega_\beta} - C_\mu \qquad (6\text{-}20)$$

where $\omega_\beta = 2\pi f_\beta$.

2. The manufacturer can give a value for $\omega_T$ or $f_T$ and we evaluate

$$C_\pi = \frac{g_m}{\omega_T} - C_\mu \qquad (6\text{-}21)$$

where $\omega_T = 2\pi f_T$.

3. The manufacturer can measure the magnitude of $\beta = \beta_M$ at some high frequency $\omega_M$ (where $\omega_M > 3\omega_\beta$) and

$$C_\pi = \frac{g_m}{\beta_M \omega_M} - C_\mu \qquad (6\text{-}22)$$

Names and equivalent symbols appearing in the literature for $f_\beta$ and $f_T$ are $f_\beta = f_{hfe}$, $f$-beta, and beta cutoff frequency; $f_T = f_\alpha = f_{hfb}$, alpha cutoff or current gain-bandwidth product.

The magnitude of $f_T$ or $f_\beta$ varies in a complex manner with both operating-point collector current and collector voltage. The manufacturer locates operating points where $f_T$ is a particular value and connects them together to form a curve of constant gain-bandwidth product. This process is repeated for other values of $f_T$ and results in a family of curves like those in Fig. 6-9.

**Example 6-8.** At its intended operating point of $I_C = 1$ mA and a frequency of 10 Mrad, $C_{obo} = 15$ pF, and $\beta = 10$. Find $C_\pi$ and $C_\mu$.

**Solution.** $C_{obo} = C_\mu = 15$ pF, $g_m = 1/25 = 0.040$. From Eq. (6-22),

$$C_\pi = \frac{g_m}{\beta_M \omega_M} - C_\mu = \frac{0.040}{10 \times 10^7} - 15 \times 10^{-12} = 400 - 15 = 385 \text{ pF}$$

**Example 6-9.** A BJT with characteristics given in Figs. 6-8 and 6-9 is operated at $I_C = 1$ mA, $V_{CE} = 10$ V. Find $C_\pi$ and $C_\mu$.

**Solution.** From Fig. 6-8, $C_\mu \cong 3$ pF at $V_{CE} = 10$ V. From Fig. 6-9, $f_T = 150$ MHz at the operating point. From Eq. (6-21),

$$C_\pi = \frac{0.040}{2\pi(150)10^6} - 3 = 42 - 3 = 39 \text{ pF}$$

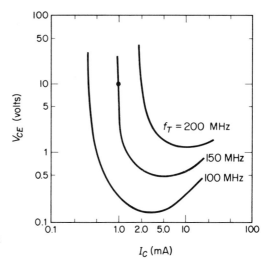

**Fig. 6-9.** Variation of $f_T$ with operating point.

## 6-6 Dependence of $\beta$ on Frequency

We derive Eqs. (6-20)–(6-22) by placing an ac short circuit across the output of a CE circuit in Fig. 6-10(a) and draw the high-frequency model in Fig. 6-10(b). $C_\pi$ is in parallel with $C_\mu$ to form an equivalent capacitance of

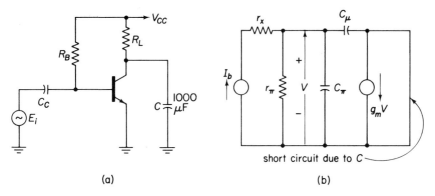

**Fig. 6-10.** Test circuit in (a) is modeled in (b) for use at high frequencies.

$C_\pi + C_\mu$. Write $V$ in terms of $I_b$ and the circuit elements as

$$V = \frac{I_b}{\dfrac{1}{r_\pi} + j\omega(C_\pi + C_\mu)}$$

From the definition of $\beta = I_c/I_b = g_m V/I_b$, substitute for $I_b$ to obtain

190    Frequency Limitations of Voltage Gain                Chap. 6

$$\beta = \frac{g_m}{\frac{1}{r_\pi} + j\omega(C_\pi + C_\mu)} \quad (6\text{-}23)$$

Equation (6-23) shows $\beta$ to be dependent on frequency and we analyze this equation in three steps. First, at low frequencies the $j$ term is negligible and Eq. (6-23) is

$$\beta = g_m r_\pi = \beta_0, \quad \omega = \text{low} \quad (6\text{-}24)$$

Thus our low frequency $\beta_0$ comes out of Eq. (6-23). Second, $\omega_\beta$ is defined as the frequency where real and $j$ terms are equal in Eq. (6-23) and the magnitude of $\beta$ equals $0.707\,\beta_0$, or

$$\frac{1}{r_\pi} = \omega_\beta(C_\pi + C_\mu), \quad \omega = \omega_\beta \quad (6\text{-}25)$$

Equations (6-25) and (6-20) are identical. Third, at frequencies higher than $3\omega_\beta$ the real term is negligible and Eq. (6-23) reduces to

$$\beta \cong \frac{g_m}{\omega(C_\pi + C_\mu)}, \quad \omega \geq 3\omega_\beta \quad (6\text{-}26)$$

We can measure the value of $\beta = \beta_M$ at a measured frequency $\omega_M$, which is above $3\omega_\beta$, and substitute into Eq. (6-26) to obtain Eq. (6-22). We can also raise the measuring frequency until the *value of $\beta$ falls to unity* and *define this frequency as* $\omega_T$. Substituting $\omega = \omega_T$ and $\beta = 1$ into Eq. (6-23) yields Eq. (6-21). A plot of Eq. (6-23) is given in Fig. 6-11 for a BJT with $\beta_0 = 100$ and $\omega_T = 10$ Mrad/sec.

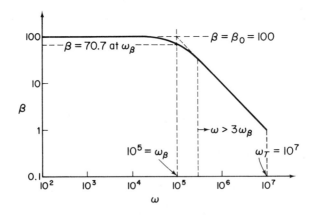

**Fig. 6-11.** From values of $\omega_\beta$ or $\omega_T$ we can find values of $C_\pi$ and $C_\mu$.

**Example 6-10.** Assume $C_\mu = 20$ pF, $g_m = 0.006$ mho, and evaluate $C_\pi$ by three different methods from the data in Fig. 6-11.

*Solution.* (a) From Eq. (6-21),

$$C_\pi = \frac{g_m}{\omega_T} - C_\mu = \frac{0.006}{10^7} - 20 \times 10^{-12} = 600 - 20 = 580 \text{ pF}$$

(b) From $\omega_\beta = 10^5$ rad/sec, $r_\pi = \beta_0/g_m = 100/0.006 = 16.6$ k$\Omega$ and from Eq. (6-20)

$$C_\pi = \frac{1}{r_\pi \omega_\beta} - C_\mu = \frac{1}{16.6 \times 10^3 \times 10^5} - 20 \text{ pF}$$
$$= 600 - 20 = 580 \text{ pF}$$

(c) Choosing the point $\beta = 10$ when $\omega = 10^6$ rad/sec and using Eq. (6-22),

$$C_\pi = \frac{g_m}{\beta_M \omega_M} - C_\mu = \frac{0.006}{10 \times 10^6} - C_\mu = 600 - 20 = 580 \text{ pF}$$

## 6-7 Common-Emitter High-Frequency Cutoff

To find the upper cutoff frequency of the CE amplifier in Fig. 6-12(a) we would begin by drawing the model in Fig. 6-12(b), which is complicated by the coupling of $C_\mu$ between output and input circuit. As will be shown we can derive the simpler equivalent model in Fig. 6-12(c). Observe that $C_\mu$ appears to be increased by the voltage gain $g_m R_L$ by a mechanism known as the *Miller effect*. This effect will be explained shortly but we employ the simplified model to define the equivalent input capacitance $C$ from Fig. 6-12(c) as

$$C = C_\pi + C_\mu(1 + g_m R_L) \tag{6-27}$$

$C$ and $r_x \| r_\pi$ determine the high-frequency time constant with respect to $E_i$ and the time constant's reciprocal gives us the upper cutoff frequency $\omega_H$, where

$$\omega_H = \frac{1}{\Upsilon} = \frac{1}{C(r_x \| r_\pi)} \cong \frac{1}{Cr_x} \tag{6-28}$$

**Example 6-11.** In Fig. 6-12 the BJT is operated at $I_C = 1$ mA, with $\beta_0 = \beta_F = 100$, $r_x = 50$ $\Omega$, $r_\pi = 2.5$ k$\Omega$, $\omega_\beta = 800$ krad/sec, $g_m = 0.040$, $C_\pi = 490$ pF, $C_\mu = 10$ pF. (a) Find low-frequency voltage gain $V_0/E_i$; (b) upper cutoff frequency $\omega_H$.

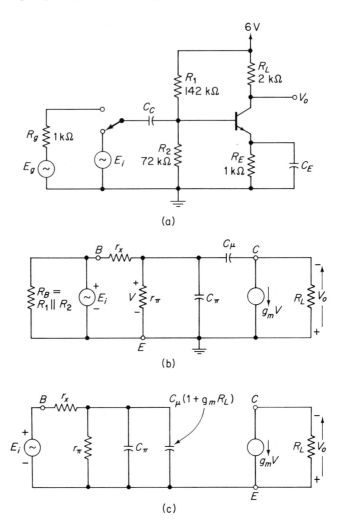

**Fig. 6-12.** The CE circuit in (a) is represented by its high-frequency model in (b) and simplified version in (c).

**Solution.** (a)

$$\frac{V_o}{E_i} = -\frac{\beta_0 R_L}{r_x + r_\pi} = -\frac{(100)(2000)}{2550} = -78.4$$

(b) From Eq. (6-27), $C = 490 + 10(1 + 0.040 \times 2000) = 1300$ pF. From Eq. (6-28),

$$\omega_H = \frac{10^{12}}{(1300)(50 \parallel 2500)} = 1.54 \times 10^7 \text{ rad/sec}$$

## Sec. 6-7 Common-Emitter High-Frequency Cutoff 193

**Example 6-12.** Using the data in Example 6-11, throw the switch in Fig. 6-12 to connect generator $E_g$. (a) Calculate the voltage gain $V_o/E_g$. (b) Find the new upper cutoff frequency.

**Solution.** (a) Evaluate $R_B = R_1 \| R_2 = 50 \text{ k}\Omega$, and find $E_i/E_g$ from the input voltage divider $R_g$ and $R_\text{in}$, where $R_\text{in} = R_B \| (r_x + r_\pi) = (50 \| 2.55) \times 10^3 = 2.43 \text{ k}\Omega$.

$$E_i = \frac{R_\text{in}}{R_g + R_\text{in}} E_g = \frac{2430}{3430} = 0.70$$

From Example 6-11,

$$\frac{V_o}{E_g} = \frac{E_i}{E_g} \frac{V_o}{E_i} = 0.70(78.4) = 55.5$$

Since $E_g$ is the reference, capacitor $C$ sees the new resistance shown in Fig. 6-13 of $R$, where

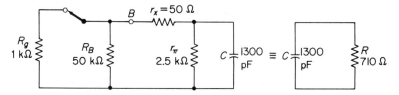

**Fig. 6-13.** High-frequency time constant for Example 6-12.

$$R = r_\pi \| (r_x + R_B \| R_g) = 2500 \| [50 + (50 \| 1)10^3] = 710 \ \Omega$$

Cutoff frequency is

$$\omega_H = \frac{1}{RC} = \frac{10^{12}}{(710)(1300)} = 1.08 \times 10^6 \text{ rad/sec}$$

There are several important conclusions to be drawn from comparing Examples 6-11 and 6-12. First, voltage gain is approximately constant for signals between very low frequencies and $\omega_H$. At $\omega_H$ the gain is down by 0.707, and less than half power is delivered to the load for all frequencies above $\omega_H$. Consequently, *the band of frequencies up to and including $\omega_H$ approximates the bandwidth* of the amplifier. Second, it is vital that we define how the bandwidth was measured. For example, a measurement of $\omega_H$ with respect to $E_i$ is much larger than a measurement of $\omega_H$ with respect to $E_g$. Third, generator resistance has a significant effect on bandwidth, decreasing bandwidth with increasing $R_g$. To illustrate this point, frequency-response curves for Examples 6-11 and 6-12, together with a curve for $R_g = 10 \text{ k}\Omega$, are plotted in Fig. 6-14.

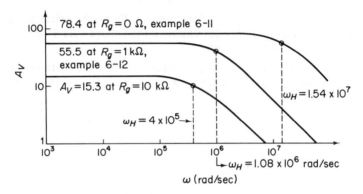

**Fig. 6-14.** Variation of gain and bandwidth with generator resistance for Fig. 6-12(a).

Our final conclusion is concerned with the maximum bandwidth at maximum gain, or $\omega_H$ when $A_V$ is defined as $V_o/E_i$. Assuming $r_x \ll r_\pi$ we combine the approximation in Eq. (6-28) with Eq. (6-27) and express maximum gain as $g_m R_L$ to create an expression for *maximum gain-bandwidth product:*

$$A_V \omega_{H(\max)} \simeq \frac{g_m R_L}{r_x[C_\pi + C_\mu(1 + g_m R_L)]} \simeq \frac{g_m R_L}{r_x(C_\pi + g_m R_L C_\mu)}$$

Divide numerator and denominator by $g_m R_L$ to simplify:

$$A_V \omega_{H(\max)} = \frac{1}{r_x\left(\dfrac{C_\pi}{g_m R_L} + C_\mu\right)} \qquad (6\text{-}29)$$

Equation (6-29) and the examples prove that a low-impedance source is necessary to obtain maximum available bandwidth of a transistor. If this condition is met, then Eq. (6-29) shows that a low value of $R_L$ is required to realize this available bandwidth. Of course, low $R_L$ leads to low gain and we are introduced to the reality that one pays for increased bandwidth with lost gain or vice versa in a tradeoff.

To investigate the tradeoff between gain and bandwidth further, we let $R_L$ vary between 500 Ω and 2500 Ω in the circuit of Fig. 6-12(a) and Example 6-11. The cutoff frequencies and voltage gain were calculated for 500-Ω increments and the resultant plots of $A_V$ versus frequency are shown in Fig. 6-15. These curves were made with the stipulation that (1) $I_C$ was held constant to hold $g_m$ at 0.040 and (2) voltage gain was specified with respect to $E_i$ so we could concentrate strictly on the effect of $R_L$ on bandwidth. We see that if gain is increased by increasing $R_L$, then bandwidth is decreased although not in direct proportion. Finally, it is evident from Eq. (6-29)

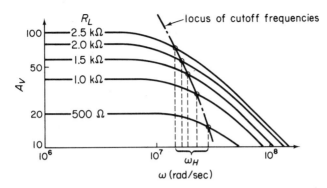

**Fig. 6-15.** Varying $R_L$ in Fig. 6-12(a) changes both gain and bandwidth.

that a transistor must have small $r_x$, $C_\pi$, and $C_\mu$ plus high $g_m$ (large $I_C$) for good high-frequency response.

## 6-8 Derivation of Miller Effect

In order to simplify the proof that we can approximate input capacitance by Fig. 6-12(c), refer to the simplified model in Fig. 6-16(a). Summing currents at the input node and expressing them in terms of $V$ and $V_o$,

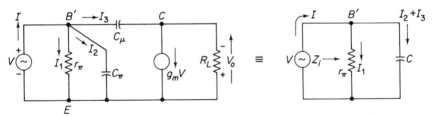

**Fig. 6-16.** Simplified model to derive the Miller-effect relationship.

$$I = I_1 + I_2 + I_3 = \frac{V}{r_\pi} + j\omega C_\pi V + j\omega C_\mu (V + V_o)$$

$I_3$ is small with respect to $g_m V$ in the output circuit, so $V_o = g_m V R_L$ and substituting for $V_o$ yields

$$\frac{V}{I} = \frac{r_\pi}{1 + j\omega r_\pi[C_\mu(1 + g_m R_L) + C_\pi]}$$

Now multiply numerator and denominator by $1/j\omega[C_\pi + C_\mu(1 + g_m R_L)]$ to obtain

$$\frac{V}{I} = Z_i = \frac{r_\pi \left(\frac{1}{j\omega C}\right)}{r_\pi + \frac{1}{j\omega C}} = r_\pi \left\|\frac{1}{j\omega C}\right. \qquad (6\text{-}30)$$

where $C = C_\pi + C_\mu(1 + g_m R_L)$

Equation (6-30) says that $V$ sees $r_\pi$ in parallel not only with $C_\pi$ but also with a much larger version of $C_\mu$. The apparent increase in $C_\mu$ is called *Miller effect*, and is modeled in Fig. 6-16(b).

## 6-9 Emitter Resistance and High-Frequency Cutoff

The presence of $R_E$ in Fig. 6-17 lowers voltage gain and raises the upper cutoff frequency. An exact high-frequency analysis is lengthy and should be left for computer handling but we can employ the *equivalent* high-frequency circuit model which is given without proof in Fig. 6-17.

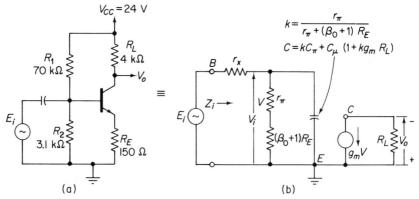

Fig. 6-17. Equivalent high-frequency model of a CE circuit with unbypassed emitter resistance.

In Fig. 6-17 the input impedance is given by $Z_i$,

$$Z_i = r_x + R \left\|\frac{1}{j\omega C}\right. \qquad (6\text{-}31)$$

where $R = r_\pi + (\beta_0 + 1)R_E$,
$C = kC_\pi + C_\mu(1 + kg_m R_L)$, and
$k = r_\pi/[r_\pi + (\beta_0 + 1)R_E]$,

and we see that $R_E$ acts to reduce input capacitance by approximately the constant $k$. The time constant and upper break frequency are expressed by

$$\omega_H = \frac{1}{\tau_H} = \frac{1}{C(R\|r_x)} \qquad (6\text{-}32)$$

*Example 6-13.* (a) Calculate voltage gain and upper cutoff frequency for the circuit of Fig. 6-17, assuming $R_E$ is bypassed by a 100-$\mu$F capacitor. (b) Repeat for $R_E$ unbypassed. (c) Calculate the gain-bandwidth product for (a) and (b). Given: $I_C = 2$ mA, $g_m = 0.080$, $r_x = 100$ Ω, $r_\pi = 625$ Ω, $\beta_0 = \beta_F = 50$, $C_\pi = 580$ pF, and $C_\mu = 10$ pF.

*Solution.* (a) With $R_E$ bypassed, voltage gain is

$$|A_v| = \frac{V_o}{E_i} = \frac{\beta_0 R_L}{r_x + r_\pi} = \frac{50(4000)}{725} = 276$$

From Eq. (6-27), $C = 580 + 10[1 + 0.080(4000)] = 3790$ pF. For Eq. (6-28),

$$R = r_x \| r_\pi = 100 \| 625 = 86.4 \ \Omega$$

$$\omega_H = \frac{1}{RC} = \frac{10^{12}}{86.4(3790)} = 3.0 \ \text{Mrad/sec}$$

(b) With $R_E$ unbypassed,

$$|A_v| = \frac{V_o}{R_i} = \frac{\beta_0 R_L}{r_x + r_\pi + (\beta_0 + 1)R_E} = \frac{50(4000)}{725 + (51)(150)} = 23.9$$

From Eq. (6-31),

$$k = \frac{625}{625 + 7650} = 0.075$$

$$C_T = (0.075)(580) + 10[1 + 0.075(320)] = 295 \text{ pF}$$

$$r_x \| R = 100 \| (625 + 7650) = 99 \ \Omega$$

$$\omega_H = \frac{10^{12}}{99(295)} = 34.2 \ \text{Mrad/sec}$$

(c) Gain-bandwidth products are (a) 276(3.0) Mrad/sec = $8.3 \times 10^8$ rad/sec, and (b) 23.9(34.2) Mrad/sec = $8.2 \times 10^8$ rad/sec.

We conclude from this example that $R_E$ does not significantly effect the gain-bandwidth product and there is a direct tradeoff between gain and bandwidth.

## 6-10 Common-Base and Common-Collector Upper Cutoff Frequency

The high-frequency model of a CC circuit is shown in Fig. 6-18, where $C_\mu$ is connected to ground through the grounded collector terminal. No signal voltage is developed at the collector terminal and no feedback can be coupled through $C_\mu$. To develop an expression for cutoff frequency, write the input loop equation from $E_i$ and substitute $I_b Z_\pi$ for $V$:

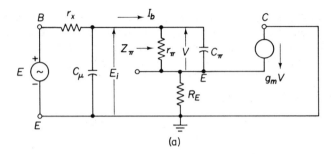

**Fig. 6-18.** High-frequency model of the CC circuit in Fig. 5-12(a).

$$E_i = I_b[Z_\pi + (I_b + g_m V)R_E] = I_b[Z_\pi + (1 + g_m Z_\pi)R_E]$$

Substitute for $Z_\pi$ and $V/Z_\pi$ for $I_b$ to obtain

$$\frac{V}{E_i} = \left[\frac{r_\pi}{r_\pi + (\beta_0 + 1)R_E}\right]\left[\frac{1}{1 + j\omega\dfrac{C_\pi r_\pi R_E}{r_\pi + (\beta_0 + 1)R_E}}\right] \quad (6\text{-}33)$$

By setting the $j$ term equal to 1 at $\omega = \omega_H$ in Eq. (6-32) we can force out a time constant formed by a capacitance $C_\pi/(\beta_0 + 1)$ and a resistance $r_\pi \| (\beta_0 + 1)R_E$. Neglecting $C_\mu$, which is negligible, $\omega_H$ is expressed by

$$\omega_H = \frac{1}{\Upsilon_H} = \frac{1}{\left(\dfrac{C_\pi}{\beta_0 + 1}\right)[(\beta_0 + 1)R_L \| r_\pi]} \quad (6\text{-}34)$$

Since $(\beta_0 + 1)R_L \gg r_\pi$ and $\beta_0 + 1 \cong \beta_0$, Eq. (6-34) can be approximated by

$$\omega_H \cong \frac{\beta_0}{C_\pi r_\pi} = \frac{g_m}{C_\pi} \cong \omega_T \quad (6\text{-}35)$$

In Fig. 6-19, output voltage $V_o$ does couple a signal back through $C_\mu$

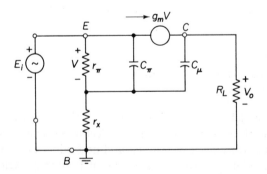

**Fig. 6-19.** High-frequency model of a CB circuit.

but it is bypassed through $r_x$ to ground at the base terminal and the high-frequency cutoff for the common-base approaches $\omega_T$.

Derivations leading to Eq. (6-35) are not accurate since the hybrid-pi model, without extrinsic lead and header capacitance, is not useful much above $10\,\omega_\beta$. However, we may safely conclude that the absence of any significant coupling between output and input allows the common-base and common-collector configurations to exhibit a high cutoff frequency.

## 6-11 Measurement of the Basic Hybrid-Pi Parameters

A simple test circuit is shown in Fig. 6-20 that allows measurement of $\beta_0$, $r_\pi$, and $g_m$ with reasonable accuracy. The dc operating point of the circuit in which the transistor is to be used should be known. This same operating point should be established in the test circuit by the following procedure:

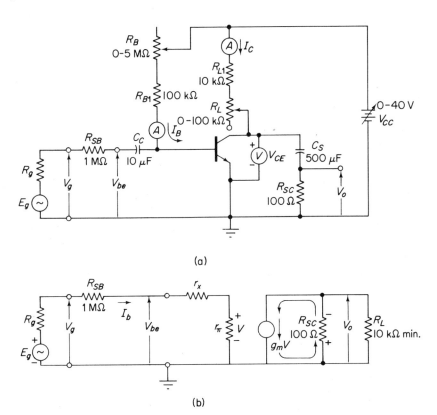

**Fig. 6-20.** Test circuit in (a) measures $\beta_0$ and $h_{ie}$. A small-signal model is given in (b).

1. Set $V_{CC}$ at 30 V and adjust $R_B$ so that $I_B$ is equal to the desired $I_B$.
2. With a high-impedance dc voltmeter (VTVM) across the emitter and collector, adjust $R_L$ to establish $V_{CE}$ equal to the intended operating-point voltage. $I_C$ should be at the predicted value because it is set by $I_B$ and should not vary appreciably with $V_{CE}$. It is convenient to open one terminal of $C_s$ during this step because the time constant is large. When $C_s$ is reconnected, it will take a few seconds for $V_{CE}$ to stabilize. Retrim $R_L$, $R_B$, and $V_{CC}$ as required.
3. Record $I_C$ and $I_B$ to calculate $\beta_F$. The magnitude of $\beta_0$ is measured in the next sequence and should be reasonably close to $\beta_F$.

AC MEASUREMENTS—THEORY OF OPERATION

Resistor $R_{SB}$ is the signal base-current sampling resistor. Its value should be accurately known because the voltage drop across it determines $I_b$ from measurement of $V_g$ and $V_{be}$. A value for $I_b$ is calculated from

$$I_b = \frac{V_g - V_{be}}{R_{SB}} \qquad (6\text{-}36)$$

Capacitor $C_C$ is selected to prevent $R_g$ and $R_{SB}$ from affecting the dc operating point. The reactance of $C_C$ at 1000 Hz is about 40 Ω and should develop a negligible signal voltage drop. $V_{be}$ is measured with an ac VTVM with a minimum sensitivity of 10 mV, full-scale deflection. This measurement indicates the small-signal voltage, which will be called $E_i$ in our derivations.

The fixed resistor $R_{B1}$ insures that a negligible amount of $I_b$ will be shunted away from the transistor's base, for values of $r_\pi$ up to 10% of $R_{B1}$. This resistance must be present in the event $R_B$ is maladjusted to a low value.

$I_b$ excites the $g_m V$ generator and $R_{L1}$ ensures that there is at least 100 times more resistance in the collector supply lead than there is in the shunt path through $C_s$. Despite the setting of $R_L$, at least 99% of $g_m V$ is directed through $R_{SC}$ to cause the signal voltage drop $V_o$, as shown in Fig. 6-20(b).

SMALL-SIGNAL MEASUREMENTS—PROCEDURE

Parameters $\beta_0$ and $h_{ie}$ are measured directly.

1. After establishing the dc operating point adjust $E_g$ so that $V_{be}$ is not greater than 3.5 mV rms. This will ensure that we do not exceed one limit of the hybrid-pi model. Measure $V_{be}$ and $V_g$ to calculate $I_b$ from the measured values by Eq. (6-36).
2. Measure $V_o$ and calculate $I_c$ from $I_c = g_m V = V_o/R_{SC}$.

### Sec. 6-11  Measurement of the Basic Hybrid-Pi Parameters

3. Find $\beta_0$ from $\beta_0 = I_c/I_b$.
4. Find $h_{ie} = r_x + r_\pi$ from $h_{ie} = V_{be}/I_b$.
5. Calculate $g_m$ from $I_c/(kT/q)$.
6. Compute $r_\pi$ from $\beta_0/g_m$.
7. Find $r_x$ by subtracting $r_\pi$ (step 6) from $h_{ie}$ (step 4).
8. As a matter of interest, an approximate value of $g_m$ is obtained from $I_c/V_{be}$ and should be compared with step 5. The difference in values is due to the fact that $V_{be}$ is not equal to $V$, because of $r_x$.

**Example 6-14.** Measured data on a 40406 *npn* transistor was obtained at 25°C as follows: $I_C = 1$ mA, $V_{CE} = 6.1$ V, $I_B = 8$ μA. Calculate $\beta_F = 1.0 \times 10^{-3}/8 \times 10^{-6} = 125$.

**Solution.** Ac measurements: $V_g = 300$ mV, $V_{be} = 1.0$ mV, $V_o = 3.7$ mV. Calculations:

1. $I_b = \dfrac{V_g - V_{be}}{R_s} = \dfrac{(300 - 1.0)}{1.0 \times 10^6} \times 10^{-3} = 0.299$ μA

2. $I_c = \dfrac{V_o}{R_L} = \dfrac{3.7 \times 10^{-3}}{100} = 37$ μA

3. $\beta_0 = \dfrac{I_c}{I_b} = \dfrac{3.7 \times 10^{-3}}{0.299 \times 10^{-6}} = 124$

4. $h_{ie} = \dfrac{V_{be}}{I_b} = \dfrac{1.0 \times 10^{-3}}{0.299 \times 10^{-6}} = 3.35$ kΩ

5. Parameter $g_m$ is found from $kT/q$ at 25°C equals 25.7 mV.

$$g_m = \dfrac{I_C}{\dfrac{kT}{q}} = \dfrac{1.0}{25.7} = 0.039 \text{ mho}$$

6. $r_\pi = \dfrac{124}{0.039} = 3.18$ kΩ

7. Subtracting this from $h_{ie}$ gives an estimated but inaccurate $r_x$ of

$$r_x = h_{ie} - r_\pi = 3.35 - 3.18 = 0.17 \text{ kΩ}$$

8. The *approxinate* value of $g_m$ is found from the measurements

$$g_m = \dfrac{I_c}{V_{be}} = \dfrac{37 \times 10^{-6}}{1.0 \times 10^{-3}} = 0.037 \text{ mho}$$

**Example 6-15.** Measured data on a 40408 *pnp* transistor was obtained as follows: $I_C = 1.0$ mA, $V_{CE} = 0.5$ V, $I_B = 3.18$ μA.

**Solution.** Ac measurements: $V_g = 2.12$ V, $V_{be} = 2.0$ mV, $V_o = 7.4$ mV.

$$\beta_F = \frac{I_C}{I_B} = \frac{1.00 \times 10^{-3}}{31.8 \times 10^{-6}} = 31.4$$

1. $I_b = \dfrac{V_g - V_{be}}{R_s} = \dfrac{2.12 - 0.002}{1 \times 10^6} = 2.12 \ \mu A$

2. $I_c = \dfrac{V_o}{R_c} = \dfrac{7.40 \times 10^{-3}}{100} = 74.0 \ \mu A$

3. $\beta_0 = \dfrac{I_c}{I_b} = \dfrac{74.0}{2.12} = 34.8$

4. $h_{ie} = \dfrac{V_{be}}{I_b} = \dfrac{2 \times 10^3}{2.12 \times 10^{-6}} = 0.94 \ k\Omega$

5. $g_m = 0.039$ mho (from Example 6-14)

6. $r_\pi = \dfrac{34.8}{0.039} = 0.89 \ k\Omega$

7. $r_x = h_{ie} - r_\pi = 0.94 - 0.89 \cong 50$

8. $g_m = \dfrac{I_c}{V_{be}} = \dfrac{74 \times 10^{-6}}{2.0 \times 10^{-3}} = 0.037$ mho

MEASUREMENT OF $C_\pi$ AND $C_\mu$. A slight modification in the test circuit of Fig. 6-20 is made in Fig. 6-21(a) to allow indirect measurement of $C_\pi$ and $C_\mu$. We want a long time constant or low value of $\omega_H$ to allow use of signal generators below radio frequencies. Choose $R_s = 10 \ k\Omega$ for this reason and also to swamp out the unreliable $r_x$. Voltage $E_i$ is held constant in amplitude by varying the volume control on $E_g$ as required. $E_i$ thus becomes a simulated perfect voltage source, with zero internal resistance.

We measure $C_\pi$ plus $C_\mu$ indirectly by measuring $\omega_H$. With switch $Sw$ on point 1, set the value of $V_i$ to 2 mV at a frequency of 1 kHz. Measure the value of $E_i$ at this frequency. Hold $E_i$ constant in amplitude and increase the frequency until $V_i$ drops to 0.707(2) mV. Record this value of frequency as the cutoff frequency $\omega_H$. Since $r_x \ll R_s \| R_B$ and $R_B \gg R_s$, the cutoff frequency is approximated from Fig. 6-21(b).

$$\omega_H \cong \frac{1}{(C_\pi + C_\mu)(r_\pi \| R_s)} \quad \text{and} \quad C_\pi + C_\mu = \frac{1}{\omega_H(r_\pi \| R_s)} \quad (6\text{-}37)$$

$\omega_H$, $r_\pi$, and $R_s$ are measured values and the magnitude of $C_\pi + C_\mu$ may be calculated from Eq. (6-37).

Since there are two unknowns, a second measurement is necessary. Throw switch $Sw$ to point 2, connecting the 100-$\Omega$ load resistor $R_L$. $C_\mu$ will now be increased by $g_m R_L$ or $0.040(100) = 4$ as shown in Fig. 6-21(c). Repeat the procedure to measure $\omega_{H2}$, which will occur at a lower frequency. Assuming again that $r_x$ and $R_B$ do not significantly affect cutoff frequency, $\omega_{H2}$ is expressed by

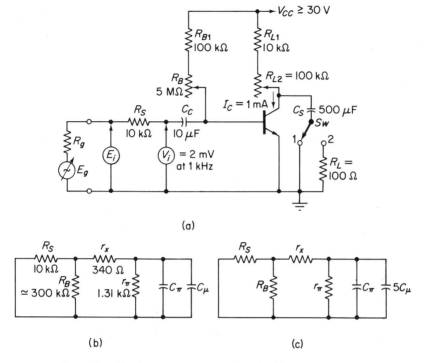

**Fig. 6-21.** Circuit to measure $C_\pi$ and $C_\mu$ in (a). Equivalent input circuit models are shown in (b) and (c) for switch $Sw$ on points 1 and 2, respectively.

$$\omega_{H2} = \frac{1}{(C_\pi + 5C_\mu)(r_\pi \| R_s)} \quad \text{and} \quad C_\pi + 5C_\mu = \frac{1}{\omega_{H2}(r_\pi \| R_s)} \quad (6\text{-}38)$$

The values of $\omega_{H2}$, $r_\pi$, and $R_s$ are known in Eq. (6-38) so the magnitude of $C_\pi + 5C_\mu$ may be calculated. Subtracting Eq. (6-37) from (6-38),

$$\begin{aligned} C_\pi + 5C_\mu &= \text{known value in pF} \\ -(C_\pi + C_\mu &= \text{known value in pF}) \\ \hline 4C_\mu &= \text{difference in pF} \end{aligned}$$

$$C_\mu = \frac{\text{difference in pF}}{4} \quad (6\text{-}39)$$

Finally, the value of $C_\mu$ is substituted in Eq. (6-37) to obtain $C_\pi$.

**Example 6-16.** A low-power *pnp* transistor is connected into the circuit of Fig. 6-21(a), with $I_C = 1$ mA and $R_B = 10.3$ kΩ. $E_i$ is held constant at

14 mV and $V_i$ measures 2 mV at 1 kHz. The frequency of $E_i$ is increased to $\omega_H = 240$ kHz, where $V_i$ drops to 1.41 mV. Switch $Sw$ is thrown to point 2 and the new breakpoint frequency $\omega_{H2}$ is measured at 228 kHz. Prior tests from the circuit of Fig. 6-20 yielded $r_\pi = 1.31$ k$\Omega$, $r_x = 340$ $\Omega$, $\beta_0 = 52.4$. Calculate $C_\pi$ and $C_\mu$.

*Solution.* From Eq. (6-37),

$$C_\pi + C_\mu = \frac{1}{(2\pi \times 240 \times 10^3)(1.31 \parallel 10.3) \times 10^3} = 572 \text{ pF} \qquad (1)$$

From Eq. (6-38),

$$C_\pi + 5C_\mu = \frac{1}{(2\pi \times 228 \times 10^3)(1.31 \parallel 10.3) \times 10^3} = 602 \text{ pF} \qquad (2)$$

Subtracting (1) from (2), $4C_\mu = 30$ pF and $C_\mu = 7.5$ pF. Substituting for $C_\mu$ in (1), $C_\pi = 572 - 7.5 = 565$ pF.

## PROBLEMS

**6-1** Interchange $R$ and $C$ in Fig. 6-2 such that $V_o$ is taken across $C$. Develop formulas to correspond with Eqs. (6-1) to (6-6) with $\omega_L$ replaced by $\omega_H$.

**6-2** If operating-point collector current is halved to 0.5 mA in Examples 6-2 and 6-3, what are the new values for $\omega_L$?

**6-3** In Example 6-4, $R_E$ is doubled to 200 $\Omega$. What is the effect on $\omega_L$?

**6-4** If $C_E$ is doubled in Example 6-5, what is the resultant $\omega_L$?

**6-5** Reducing $I_C$ by one-half to 0.5 mA in Example 6-5 has what affect on $\omega_L$, where $C_E = 100$ $\mu$F?

**6-6** For the circuit of Fig. 6-6, evaluate $\omega_C$, $\omega_E$, and $\omega_L$ with (a) $C_C = 2$ $\mu$F and $C_E = 100$ $\mu$F; (b) $C_C = 100$ $\mu$F and $C_E = 2$ $\mu$F.

**6-7** A low-resistance source with $R_g = 50$ $\Omega$ is employed in the circuit of Fig. 6-5. Compare the values of $\omega_C$ and $\omega_E$ with those of Example 6-7.

**6-8** With $C_\mu = 5$ pF, $\beta_F = 628$ kHz, and $\beta_0 = 100$ at $I_C = 1$ mA, find $C_\pi$.

**6-9** A specification sheet shows $f_T = 62.8$ MHz and $C_\mu = 3$ pF at $I_C = 1$ mA. Find $C_\pi$.

**6-10** $\beta_M$ is measured as 5 at 20 Mrad/sec and $C_{obo} = 10$ pF. Find $C_\pi$ if $I_C = 1$ mA.

6–11 If $\beta_0 = 50$ for a particular transistor, what will be the value of $\beta$ at $\omega_\beta$ and $\omega_T$?

6–12 Calculate $\omega_H$ and $A_v = V_o/E_g$ for (a) $R_g = 10$ k$\Omega$ in Example 6-12. (b) $R_g = 500$ $\Omega$. (c) Plot these cutoff points in Fig. 6-14 and connect the cutoff points with a smooth curve.

6–13 Assuming $R_g$ approaches $\infty$ in Prob. 6-12, $A_v = V_o/E_g$ will approach zero since $E_i/E_g$ approaches zero. Employ Example 6-12 to show that $\omega_H$ approaches a value determined by $C$ and $r_\pi$, assuming $R_B \gg r_\pi$.

6–14 In Example 6-11, calculate $V_o/E_i$ and $\omega_H$ for (a) $R_L = 500$ $\Omega$, (b) $R_L = 1.0$ k$\Omega$, (c) $R_L = 1.5$ k$\Omega$, (d) $R_L = 2.5$ k$\Omega$. Verify your results against Fig. 6-15.

6–15 For Problem 6-14, use Eqs. (6-27) and (6-28) to show that $\omega_H$ approaches the *transverse cutoff frequency* $\omega_b = 1/[(C_\pi + C)(r_x \| r_\pi)]$ when $R_L$ approaches zero.

6–16 If $R_E$ is halved to 75 $\Omega$ in Example 6-13b and all other data is unchanged, what is the effect on $V_o/E_i$, $\omega_H$, and the gain-bandwidth product?

6–17 Sketch modifications required in Fig. 6-20 when testing a *pnp* transistor.

6–18 An *npn* transistor is tested at the same operating point as in Example 6-14, yielding test data $I_B = 10$ $\mu$A, $V_{be} = 0.79$ mV, $V_o = 3.0$ mV, and $V_g = 300$ mV. Evaluate $\beta_0$, $r_\pi$, and $h_{ie}$.

6–19 Measurements of $r_\pi = 1.31$ k$\Omega$, $f_{H1} = 300$ kHz, and $f_{H2} = 285$ kHz are found for a transistor in the circuit of Fig. 6-21. Evaluate $C_\pi$ and $C_\mu$.

## Chapter 7

**7-0** INTRODUCTION .................................207
**7-1** BOOTSTRAPPING THE EMITTER FOLLOWER ...........207
**7-2** BOOTSTRAPPING THE COMMON EMITTER ............210
**7-3** SINGLE-STAGE COLLECTOR FEEDBACK .............211
**7-4** CONTROL OF UPPER CUTOFF FREQUENCY BY
       COLLECTOR FEEDBACK ..........................216
**7-5** SINGLE-STAGE COLLECTOR AND EMITTER FEEDBACK ..219
**7-6** THE DARLINGTON PAIR .........................221
**7-7** THE INVERTED DARLINGTON AMPLIFIER ...........225
**7-8** COMMON EMITTER TO COMMON-BASE CASCODE........227
**7-9** LINEAR MIXING ...............................231
       PROBLEMS ....................................232

# Selected Applications for Analysis and Design

## 7-0 Introduction

Each application of a basic circuit configuration usually introduces some constraint that is not considered during an introductory study. External elements may be added to improve one aspect of the circuit's performance but always at some cost, usually gain. In this chapter we shall investigate the effects of connecting external elements to a BJT, including other BJTs, to modify the behavior of the basic circuits.

## 7-1 Bootstrapping the Emitter Follower

A technique for reducing shunting of signal current by the bias resistors is known as *bootstrapping*. In Fig. 7-1(a), resistor $R_3$ increases input resistance together with bootstrap capacitor $C_B$, which acts as a short circuit in the model of Fig. 7-1(b). (Biasing was covered in Section 4-8.) Since $R_1$, $R_2$, and $R_E$ are in parallel they are replaced, for simplicity, by their series equivalent resistance $R$ in Fig. 7-1(c). We derive an equivalent simplified input circuit model by noting that

$$I_b = \frac{R_3}{r_\pi + R_3} I = CI \qquad (7\text{-}1)$$

where $C$ is the current divider,

$$C = \frac{R_3}{r_\pi + R_3} \qquad (7\text{-}2)$$

Write the input loop equation

$$E_i = I(R_3 \| r_\pi) + (\beta_0 I_b + I)R \qquad (7\text{-}3)$$

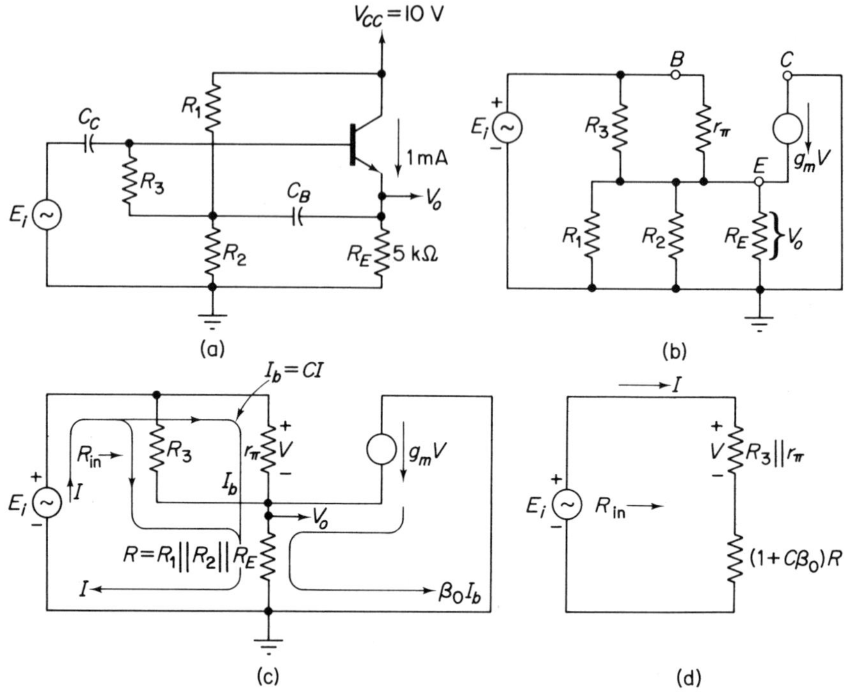

**Fig. 7-1.** The bootstrapped emitter follower in (a) is modeled in (b) and (c) to obtain the equivalent input model in (d).

Substitute for $I_b = CI$ to obtain

$$\frac{E_i}{I} = (R_3 \| r_\pi) + (1 + C\beta_0)R = R_{in} \qquad (7\text{-}4)$$

Current divider $C$ tells us what fraction of $I$ enters the base to be multiplied by $\beta_0$ and we should be able to construct Eq. (7-4) directly from the model in Fig. 7-1(c).

An expression for voltage gain is developed from

$$V_o = (I + \beta_0 I_b)R = I(1 + C\beta_0)R$$

Substituting for $I$ from Eq. (7-4) gives, in standard format,

$$\frac{V_o}{E_i} = (1 + C\beta_0)\frac{R}{R_{in}} \qquad (7\text{-}5)$$

where current gain is

$$\frac{I_o}{I_{in}} = \frac{I + \beta_0 I_b}{I} = 1 + C\beta_0 \qquad (7\text{-}6)$$

## Sec. 7-1  Bootstrapping the Emitter Follower

We analyze the effect of bootstrapping on resistance seen looking into the emitter by applying test voltage $E$ in Fig. 7-2(a). Defining $R_o$ and

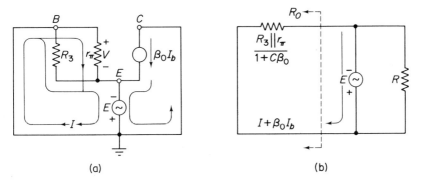

**Fig. 7-2.** The circuit to test output resistance in (a) results in the model of (b).

substituting from Eq. (7-1) for $I_b$, with $E = I(R_3 \| r_\pi)$,

$$R_o = \frac{E}{I + \beta_0 I_b} = \frac{E}{I(1 + C\beta_0)} = \frac{R_3 \| r_\pi}{1 + C\beta_0} \tag{7-7}$$

The modified resistance-transformation properties are shown in Fig. 7-2(b). If any generator resistance were present in the circuit, $R_g$ would also be divided by $(1 + C\beta_0)$ and the result would be added to $R_o$ in Eq. (7-7). We compare the bootstrapped and conventional circuits in Examples 7-1 and 7-2.

**Example 7-1.** Assume $R_3 = 0$ and $C_B$ is removed in Fig. 7-1(a) to construct a conventional emitter follower. Assume $R_B = R_1 \| R_2 = 7.5 \text{ k}\Omega$, $\beta_0 = \beta_F = 100$, $I_C = 1$ mA, and $r_\pi = 2.5 \text{ k}\Omega$. Calculate (a) $R_1$ and $R_2$, (b) $R_{\text{in}}$, (c) voltage gain.

**Solution.** (a) Required $I_B$ is $I_C/\beta_F = 10$ μA, and $V_B$ is

$$V_B = I_B[R_B + (\beta_F + 1)R_E] + V_{BE}$$
$$= 10 \times 10^{-6}[7.5 + (101)5] \times 10^3 + 0.6 = 5.72 \text{ V}$$

$$R_1 = \frac{V_{CC}}{V_B}R_B = 13.1 \text{ k}\Omega \qquad R_2 = \frac{V_{CC}}{V_{CC} - V_B}R_B = 17.5 \text{ k}\Omega$$

(b) $\quad R_{\text{in}} = R_B \| [r_\pi + (\beta_0 + 1)R_E] = (7.5 \| 507) \text{ k}\Omega = 7.4 \text{ k}\Omega$

(c) $\quad \dfrac{V_o}{E_i} = \dfrac{(\beta_0 + 1)R_E}{r_\pi + (\beta_0 + 1)R_E} = \dfrac{(101)(5000)}{2500 + (101)(5000)} = 0.99$

**Example 7-2.** Use the data given in Example 7-1 for $R_B$, $\beta_0$, $\beta_F$, and $r_\pi$. But let $R_3 = 3r_\pi = 7500 \text{ }\Omega$ and install $C_B$ in Fig. 7-1(a). Calculate (a) the new values of $R_1$ and $R_2$ for $I_C = 1$ mA, (b) $R_{\text{in}}$, (c) voltage gain. Compare results.

**Solution.** (a) Refer to Section 4-8 to calculate the slight change in $V_B$:

$$V_B = I_B[R_B + R_3 + (\beta_F + 1)R_E] + V_{BE} = 5.8 \text{ V}$$

$$R_1 = \frac{10}{5.8}(7500) = 12.9 \text{ k}\Omega \qquad R_2 = \frac{10}{4.2}(7500) = 17.8 \text{ k}\Omega$$

(b) $\qquad C = \dfrac{R_3}{(R_3 + r_\pi)} = 0.75 \qquad R = R_1 \| R_2 \| R_E \cong 3.2 \text{ k}\Omega$

From Eq. (7-4),

$$R_{in} = (7.5 \| 2.5) \text{ k}\Omega + (1 + 75)3.2 \text{ k}\Omega \cong 245 \text{ k}\Omega$$

(c) From Eq. (7-5),

$$\frac{V_o}{E_i} = \frac{(1 + 75)3 \times 10^3}{245 \times 10^3} = 0.99$$

We conclude from Examples 7-1 and 7-2 that there is no significant change in $R_1$ and $R_2$, dc operating point, or voltage gain by the addition of bootstrapping. However, input resistance is increased by a factor of 31.

## 7-2 Bootstrapping the Common Emitter

Input resistance relationships derived for the emitter follower are identical to the common emitter. The first difference encountered is in the high output resistance of current source $g_m V$. An expression for voltage gain is developed by adding load resistance $R_L$ to the circuit of Fig. 7-1 to give the simplified common-emitter model in Fig. 7-3.

Voltage $V$ is expressed by the voltage divider in Fig. 7-3(a) as

$$V = \frac{\dfrac{R_3 r_\pi}{R_3 + r_\pi}}{R_{in}} E_i = \frac{C r_\pi}{R_{in}} E_i \qquad (7\text{-}8)$$

Substituting for $V$ from Eq. (7-8) and $\beta_0 = g_m r_\pi$ into $V_o = -g_m V R_L$, we obtain voltage gain in the standard format

$$\frac{V_o}{E_i} = -C\beta_0 \frac{R_L}{R_{in}} \qquad (7\text{-}9)$$

Equation (7-9) contains the current gain expressed by $A_i = g_m V/I = C\beta_0$.

*Example 7-3.* The resistance levels and voltage gains for the circuit of Fig. 7-3 are analyzed in Problem 7-4. In Problem 7-5 we repeat the analysis for the same operating point but with bootstrapping eliminated ($R_3 = 0$, $C_B$ removed). The results are shown in Table 7-1 for comparison.

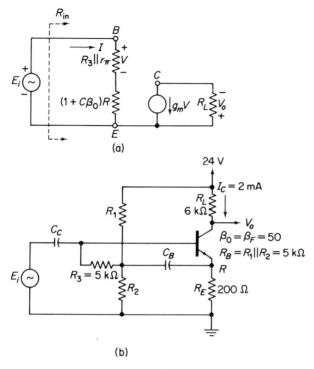

**Fig. 7-3.** The equivalent circuit in (a) models input resistance for the bootstrapped CE circuit in (b).

Table 7-1

|              | $I_C$ (mA) | $R_1$ (kΩ) | $R_2$ (kΩ) | $R_{in}$ (kΩ) | $V_o/E_i$ |
|---|---|---|---|---|---|
| Conventional | 2 | 99.4 | 5.3 | 3.4 | 27.8 |
| Bootstrapped | 2 | 85   | 5.3 | 9.1 | 29.4 |

## 7-3 Single-Stage Collector Feedback

The common-emitter amplifier with external feedback resistance $R_F$ between base and collector is of major importance because it serves to introduce (1) the basic operating principles of operational amplifiers, (2) standard methods of controlling high-frequency amplifier response, and (3) feedback principals. The circuit of Fig. 7-4(a) is modeled in Fig. 7-4(b), where $r_x$ is eliminated for simplicity. To determine the affect of $R_F$ on input resistance we write the outside loop equation

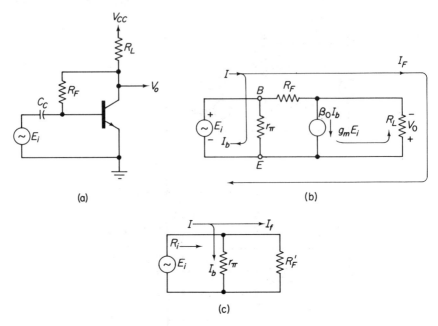

**Fig. 7-4.** Common emitter with collector feedback in (a) is modeled in (b) to develop the equivalent input model in (c).

$$E_i = I_f(R_F + R_L) - g_m E_i R_L$$

and solve for $E_i/I_f$ to find the resistance presented to $E_i$ by the $R_F$ branch, or

$$\frac{E_i}{I_f} = \frac{R_F + R_L}{1 + g_m R_L} = R'_F \qquad (7\text{-}10)$$

$R'_F$ is the equivalent resistance seen by $E_i$, looking into the base terminal of the transistor, and represents completely how the output affects the input. Collector-base resistor $R_F$ *lowered* the transistor's input resistance because it placed $R'_F$ in parallel with $r_\pi$. Input resistance $R_i$, as presented to $E_i$, is modeled in Fig. 7-4(c), and is lowered by the addition of $R_F$ with respect to the regular CE input.

Voltage gain is simplified if we discard the small component $I_f$ in Fig. 7-4 to express $V_o$ as $g_m E_i R_L$ and

$$\frac{V_o}{E_i} = -g_m R_L \qquad (7\text{-}11)$$

If we did not discard $I_f$ the gain would be $g_m R_L \| R_F$. Current gain is defined as $\beta_0 I_b / I$ by neglecting $I_f$ in the output circuit and from the input current divider of Fig. 7-4, where $I_b = I R'_F / (r_\pi + R'_F)$,

Sec. 7-3                    Single-Stage Collector Feedback    213

$$A_i = \frac{\beta_0 I_b}{I} = \frac{R'_F}{r_\pi + R'_F}\beta_0 \qquad (7\text{-}12)$$

Substitute for $R'_F$ from Eq. (7-10) into Eq. (7-12); assume $r_\pi$ is small and $\beta_0 \cong \beta_0 + 1$:

$$A_i = \frac{\beta_0(R_L + R_F)}{r_\pi + R_F + (\beta_0 + 1)R_L} \cong \frac{\beta_0(R_L + R_F)}{R_F + \beta_0 R_L} \qquad (7\text{-}13)$$

Divide numerator and denominator of Eq. (7-12) by $\beta_0$ to show that current gain is stabilized against changes in $\beta_0$ when $R_F \ll \beta_0 R_L$:

$$A_i = \frac{R_L + R_F}{\frac{R_F}{\beta_0} + R_L} \cong \frac{R_L + R_F}{R_L} \qquad (7\text{-}14)$$

To test our standard format on Eq. (7-11), $V_o$ sees $R_L$ and $E_i$ sees $R_i$, and with Eq. (7-12),

$$-\frac{V_o}{E_i} = A_i\frac{R_L}{R_i} = \left(\frac{R'_F}{r_\pi + R'_F}\beta_0\right)\left(\frac{R_L}{\frac{r_\pi R'_F}{r_\pi + R'_F}}\right) = \frac{\beta_0 R_L}{r_\pi} = g_m R_L$$

We conclude that current gain is stabilized, input resistance is lowered, and voltage gain is not affected by $R_F$ as long *as the amplifier is driven with a voltage source.*

CURRENT-SOURCE DRIVE    When $E_i$ is replaced by a generator with internal resistance, or when series resistance $R_g$ is deliberately introduced, we have a current-source drive as in Fig. 7-5. Current gain is still given by Eq. (7-12) and input resistance $R_{in}$ is increased directly by $R_g$ to

$$R_{in} = R_g + r_\pi \| R'_F \qquad (7\text{-}15)$$

Voltage gain is constructed by applying our standard format to Fig. 7-5 and using the current-gain approximation in Eq. (7-14):

$$-\frac{V_o}{E_g} \cong \frac{R_L + R_F}{R_L}\left(\frac{R_L}{R_g + R_i}\right) \cong \frac{R_F}{R_g} \qquad (7\text{-}16)$$

The approximation in Eq. (7-16) is valid under normal conditions, where $R_L \ll R_F$ and $R_i \ll R_g$ and by "large" we mean a ratio of 10 to 1.

Collector feedback is normally operated with current-source drive so that most of $I_f$ will not be shunted from the transistor's input by the signal generator to yield the stabilization of voltage gain indicated in Eq. (7-16). To realize this physically, $R'_F$ should be less than $\frac{1}{10}$ of $r_\pi$ and $\frac{1}{10}$ of $R_g$ or

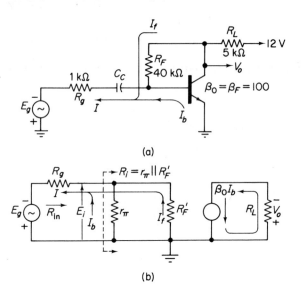

**Fig. 7-5.** Current-source drive in (a) is modeled by the equivalent input model in (b). Incremental currents are shown.

$$R'_F = \frac{R_F + R_L}{1 + g_m R_L} \lesssim \frac{r_\pi}{10} \quad \text{or} \quad R_F + R_L \lesssim \frac{\beta_0 R_L}{10} \qquad (7\text{-}17)$$

To realize our simplification in Eq. (7-14),

$$\frac{R_F}{\beta_0} \le \frac{R_L}{10} \quad \text{or} \quad R_F \le \frac{\beta_0 R_L}{10} \qquad (7\text{-}18)$$

For transistors with $\beta_0 > 100$, both stipulations in Eqs. (7-17) and (7-18) are satisfied when $R_F \le 10 R_L$. $R_i$ is typically low at 200 Ω and $R_g$ is typically greater than $10 R_i$ or 2.0 kΩ.

**Example 7-4.** In Fig. 7-5, find (a) $I_C$, $g_m$, and $r_\pi$; (b) $R'_F$, $R_i$, and $R_{\text{in}}$; (c) $V_o/E_i$; (d) $V_o/E_g$.

**Solution.** (a) From Eq. (4-22), $I_B = 21$ μA and $I_C = \beta_F I_B = 2.1$ mA, $g_m = 2.1/25 = 0.084$, $r_\pi = 100/0.084 = 1190$ Ω.
(b) From Eq. (7-10),

$$R'_F = \frac{(40 + 5)10^3}{1 + (0.084)(5000)} = 107 \text{ Ω}$$

Note $R'_F \lesssim r_\pi/10$.

$$R_i = r_\pi \| R'_F = 1190 \| 107 = 98 \text{ Ω}$$
$$R_{\text{in}} = R_g + R_i = 1000 + 98 = 1.09 \text{ kΩ}$$

Sec. 7-3          Single-Stage Collector Feedback     215

(c) From Eq. (7-11), $V_o/E_i = (0.084)(5000) = 420$.
(d) Calculate $V_o/E_g$ by two methods:
   (1) From Fig. 7-5(b),

$$\frac{V_o}{E_g} = \left(\frac{R_i}{R_g + r_i}\right)\left(\frac{V_o}{E_i}\right) = \left(\frac{98}{1000 + 98}\right)(420) \cong 38$$

(2) From Eq. (7-16),

$$\frac{V_o}{E_g} = \frac{R_F}{R_g} = \frac{40}{1} = 40$$

From Example 7-4 we conclude that $I_f$ approximately equals generator current $I$, and since $R_i$ is very low the transistor's base is close to ground potential. We can therefore, portray the circuit's condition simply by Fig. 7-6. Since $E_g \cong IR_g$ and $V_o \cong IR_F$, $V_o/E_g \cong R_F/R_g$.

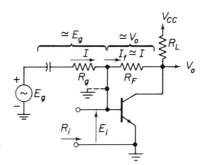

**Fig. 7-6.** Simplification of signal voltages in Fig. 7-5.

The lower cutoff frequency $\omega_L$ of Fig. 7-5 is determined by inspection of the equivalent input model in Fig. 7-7, where

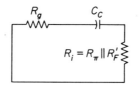

**Fig. 7-7.** Lower cutoff frequency model for Fig. 7-5.

$$\omega_L = \frac{1}{C_C R_{in}}, \qquad R_{in} = R_g + R_i \qquad (7\text{-}19)$$

Output resistance $R_o$ is found by (1) replacing $E_g$ in Fig. 7-5 with its internal resistance, (2) applying test voltage $E$, and (3) eliminating $R_L$ for simplicity in the circuit of Fig. 7-8. $R_o$ is the ratio of $E$ to current drawn from $E$, or $I_f + \beta_o I_b$. But $I_b$ is related to $I_f$ by the simple current division of $I_f$ between $R_g$ and $r_\pi$. Thus we write

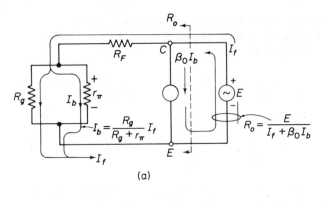

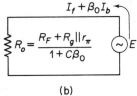

**Fig. 7-8.** The test circuit in (a) measures $R_o$ and develops the model in (b).

$$R_o = \frac{E}{I_f + \beta_0 I_b} = \frac{E}{I_f + C\beta_0 I_f} \quad \text{where} \quad C = \frac{R_g}{R_g + r_\pi} \quad (7\text{-}20)$$

Write the outside loop of $E = I_f(R_F + R_g \| r_\pi)$ and substitute for $E/I_f$ in Eq. (7-20) to obtain

$$R_o = \frac{R_F + R_g \| r_\pi}{1 + C\beta_0} \quad (7\text{-}21)$$

We conclude that output resistance is lowered by $R_F$ and, where $r_\pi \ll R_g$, $C \cong 1$ and $R_o$ is approximately $R_F/(\beta_0 + 1)$.

## 7-4 Control of Upper Cutoff Frequency by Collector Feedback

Transducers such as crystal phonograph cartridges may be characterized by a high-impedance source whose output voltage increases with frequency. The frequency dependence does not have to come from the transducer alone but may come from the element it is reading, as is the case with a phonograph record. An RC or RL network has an impedance which changes at 6 dB/octave. This convenient feature is employed both to increase high-frequency signals as they are recorded and to decrease them on playback.

Sec. 7-4    Feedback Control of Upper Cutoff Frequency    217

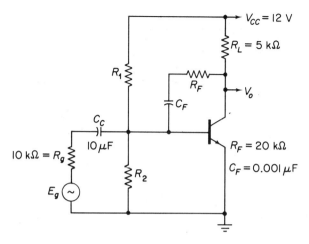

**Fig. 7-9.** $C_F$ and $R_F$ control the upper cutoff frequency.

The circuit of Fig. 7-9 can be designed to decrease gain of the amplifier at a rate of 6 dB/octave for all frequencies larger than some desired middle frequency $\omega_M$ that is smaller than $\omega_H$. Should the transducer output increase with frequency at the same rate as the amplifier gain decreases, then the resultant output is a flat frequency-response curve.

At low signal frequencies $C_F$ acts as an open circuit and from the model of Fig. 7-10 we develop voltage gain and lower break frequency $\omega_L$:

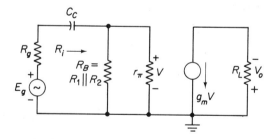

**Fig. 7-10.** Low-frequency model of Fig. 7-9.

$$\omega_L = \frac{1}{(R_g + R_i)C_C} \quad \text{where} \quad R_i = R_B \| r_\pi \quad (7\text{-}22)$$

$$\frac{V_o}{E_g} = -\frac{R_i}{R_i + R_g}(g_m R_L), \quad \omega_L < \omega < \omega_M \quad (7\text{-}23)$$

As frequency is increased to $\omega_M$, $C_F$ begins to connect $R_F$ into the circuit to reduce voltage gain. By substituting $Z_F = R_F + 1/j\omega C_F$ for $R_F$ in Eq. (7-10) we obtain an equivalent imput impedance, shown in Fig. 7-11, of

$$Z_{\text{in}} = \frac{R_F + R_L}{1 + g_m R_L} + \frac{1}{j\omega C_F(1 + g_m R_L)} = R'_F + \frac{1}{j\omega C'_F} \quad (7\text{-}24)$$

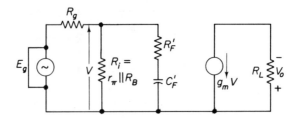

**Fig. 7-11.** Model of Fig. 7-9 in the frequency range near $\omega_M$.

where $C'_F = C_F(1 + g_m R_L)$

$C_C$ and $E_g$ are short circuited to find the time constant presented to $C'_F$ in Fig. 7-11 and $\omega_M$ is

$$\omega_M = \frac{1}{C'_F R} \qquad (7\text{-}25)$$

where $R = R'_F + (R_g \| R_i)$. Observe that $C_F$ has been increased by the Miller effect just as $C_\mu$ was and a decade above $\omega_M$ the circuit reduces to the model in Fig. 7-5(b).

**Example 7-5.** In the circuit of Fig. 7-11, $\beta_0 = \beta_F = 100$, $I_C = 2$ mA, $R_1 = 15$ k$\Omega$, $R_2 = 10.8$ k$\Omega$, and $R_B = 10$ k$\Omega$, so $g_m = 0.08$, $r_\pi = 1.25$ k$\Omega$, and $R_i \cong 1.1$ k$\Omega$. Find (a) $\omega_L$, (b) $\omega_M$, (c) $V_o/E_g$ between $\omega_L$ and $\omega_H$. (d) $V_o/E_g$ at $\omega$ greater than 10 $\omega_H$. (e) $V_o/E_g$ at $\omega_H$.

**Solution.** (a) From Eq. (7-22),

$$\omega_L = \frac{1}{(10 + 1.1)(10^3)(10)(10^{-6})} = 9 \text{ rad/sec}$$

(b) For Eq. (7-25),

$$R'_F = \frac{(20 + 5)10^3}{1 + (0.08)(5000)} = 62.5 \text{ }\Omega$$

$$C'_F = (0.001)(401) = 0.4 \text{ }\mu\text{F}$$

$$R_g \| R_i = (10 \| 1.1)10^3 = 1 \text{ k}\Omega$$

$$R = R'_F + (R_g \| R_i) = 1062 \text{ }\Omega$$

$$\omega_M = \frac{10^6}{(0.4)(1062)} = 2350 \text{ rad/sec}$$

(c) From Eq. (7-23),

$$\frac{V_o}{E_g} = -\frac{1100}{11100}(0.08)(5000) \cong 40$$

(d) From Eq. (7-16)

$$\frac{V_o}{E_g} = -\frac{20 \times 10^3}{10 \times 10^3} = 2$$

(e) $\frac{V_o}{E_g} = -0.707(40) = -28.3$

In amplifiers with more than one stage, there may be feedback paths at high frequencies that cause oscillation because some portion of output voltage can drive the input. The principles developed in this section show that connecting a small capacitor between collector and base of one stage will introduce a rolloff in voltage gain at some predictable frequency. If the voltage gain of the amplifier at the frequency of oscillation has been reduced below unity, then it is impossible for enough energy to be fed back to the input to cause oscillation, and the amplifier is stabilized.

## 7-5 Single-Stage Collector and Emitter Feedback

In Fig. 7-12(a), emitter resistor $R_E$ will increase input resistance and stabilize voltage gain. From the model in Fig. 7-12(b) we write the outside loop equation and the $I_b$ loop,

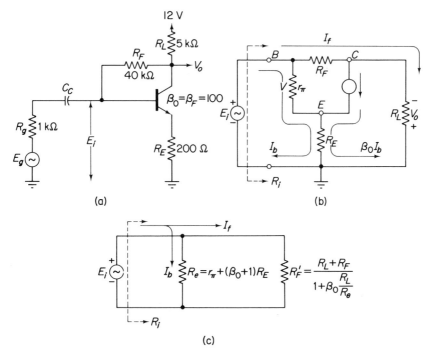

**Fig. 7-12.** Collector and emitter feedback resistors in (a) are modeled in (b) to derive the simplified input model in (c).

## 220 Applications for Analysis and Design    Chap. 7

$$E_i = I_f(R_F + R_L) - \beta_0 I_b R_L \tag{7-26}$$

$$I_b = \frac{E_i}{r_\pi + (\beta_0 + 1)R_E} = \frac{E_i}{R_e} \tag{7-27}$$

where $R_e = r_\pi + (\beta_0 + 1)R_E$. Substitute for $I_b$ in Eq. (7-26) to get

$$\frac{E_i}{I_f} = R'_F = \frac{R_L + R_F}{1 + \beta_0 \dfrac{R_L}{R_e}} \tag{7-28}$$

By assuming $\beta_0 \cong \beta_0 + 1$ and $1/g_m \ll R_E$, the term $\beta_0/R_e$ is approximated by $1/R_E$ and Eq. (7-28) simplifies to

$$R'_F \cong \frac{R_L + R_F}{R_L} R_E \tag{7-29}$$

The equivalent input presented to $E_i$ is shown in Fig. 7-12(c) as $R_i$, where

$$R_i = R_e \| R'_F \tag{7-30}$$

*Example 7-6.* Calculate (a) operating-point current and (b) $R_i$ for Fig. 7-12.
*Solution.* (a) From Eq. (4-28),

$$I_C = 100 \frac{(12 - 0.6)}{[40 + 101(5.2)]10^3} = 2 \text{ mA}$$

$g_m = 2/25 = 0.08$ and $r_\pi = 100/0.08 = 1250 \; \Omega$.
(b) From Eq. (7-27), $R_e = 1250 + 101(200) = 21.4 \text{ k}\Omega$. From Eq. (7-28),

$$R'_F = \frac{45 \times 10^3}{1 + 100\left(\dfrac{5000}{21{,}400}\right)} = 1.85 \text{ k}\Omega$$

Check from Eq. (7-29):

$$R'_F = \frac{45(0.2)}{5} = 1.8 \text{ k}\Omega$$

From Eq. (7-30),

$$R_i = (21.4 \| 1.8) = 1.66 \text{ k}\Omega$$

To express voltage gain we see that $E_i + V_o = I_f R_F$ in Fig. 7-12(b); substituting for $I_f$ from Eq. (7-28) and $R'_F$ from Eq. (7-29) yields

$$-\frac{V_o}{E_i} = \left(\frac{R_F}{R_F + R_L} - 1\right) + \frac{\beta_0}{R_E}(R_L \| R_F) \tag{7-31}$$

Since $R_F/(R_F + R_L) \cong 1$, and $\beta_0/R_e \cong 1/R_E$, Eq. (7-31) simplifies to

$$-\frac{V_o}{E_i} \cong \frac{R_L \| R_F}{R_E} \tag{7-32}$$

*Example 7-7.* Calculate (a) $V_o/E_i$ and (b) $V_o/E_g$ for Fig. 7-12.

*Solution.* (a) From Eq. (7-32),

$$-\frac{V_o}{E_i} = \frac{(5\|40)10^3}{200} = 22.2$$

(b) From Fig. 7-12(a), $E_i/E_g = R_i/(R_g + R_i) = 1.66/2.66 = 0.625$, and

$$-\frac{V_o}{E_g} = \frac{E_i}{E_g}\left(\frac{V_o}{E_i}\right) = 0.625(22.2) = 13.9$$

There is a fundamental limitation imposed by $R_F$ on the allowable output voltage swing. The peak-to-peak output voltage is limited to a maximum value equal to the dc voltage drop $V_{CE}$. But $V_{CE}$ is equal to $V_{BE}$ plus the drop across $R_F$ or $V_{BC}$. $R_F$ carries $I_B$ and $R_L$ carries $(\beta_0 + 1)I_B$. For a large output swing $R_F I_B \cong V_{CE}$ should equal $R_L(\beta_F + 1)I_B$, or $R_F \cong (\beta_F + 1)R_L$. This large value of $R_F$ would give no stabilizing action since it has been shown that $R_F$ should be about $\beta_0 R_L/10$. Thus $V_{CC}$ will split into roughly 5–20 parts for $R_L$ and 1 part for $R_F$ or $V_{CE}$, depending on $\beta_F$.

## 7-6 The Darlington Pair

A compound connection of an emitter follower driving a common emitter is referred to as a *Darlington pair*. As will be shown from the circuit of Fig. 7-13(a), both transistors act as a single unit with large $\beta$. Darlington units are available commercially, in one case with leads brought out from the equivalent base, collector, and emitter.

Biasing calculations are made from the dc model of Fig. 7-13(b), where $R_E$ is transformed into the base of $Q_1$ in two steps: (1) Transform $R_E$ into the base of $Q_2$ as $(\beta_{F2} + 1)R_E$. (2) Transform this resistance into the base leg of $Q_1$ by multiplying by $\beta_{F1} + 1$ to obtain the effective resistance seen by $I_{B1}$. Then find $I_{C1}$ from $\beta_{F1} I_{B1}$ and $I_{C2}$ from $\beta_{F2} I_{C1}$.

The Darlington's small-signal model, in Fig. 7-14, represents an equivalent input resistance developed from the input loop equation:

$$R_i = \frac{E_i}{I_{b1}} = r_{x1} + r_{\pi 1} + (\beta_{01} + 1)(r_{x2} + r_{\pi 2}) \tag{7-33}$$

Voltage gain is developed by tracing signal currents in Fig. 7-14. $I_{b1}$ traverses

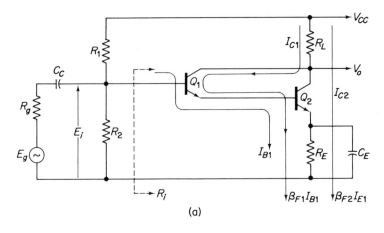

(a)

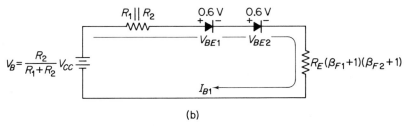

(b)

**Fig. 7-13.** DC currents in the Darlington circuit of (a) are found from the equivalent input bias circuit of (b).

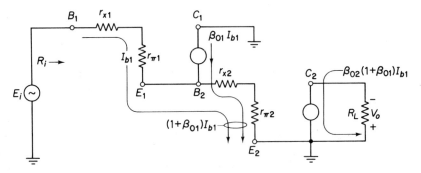

**Fig. 7-14.** Small-signal model of Fig. 7-13.

$r_{\pi 1}$ causing $\beta_{01} I_{b1}$. Both currents traverse $r_{\pi 2}$ to cause an output current of $\beta_{02}(1 + \beta_{01}) I_{b1}$ and a current gain of

$$A_i = \frac{I_o}{I_{in}} = \beta_{02}(1 + \beta_{01}) \cong \beta^2 \quad \text{for} \quad \beta_{01} = \beta_{02} \quad (7\text{-}34)$$

Since $V_o = I_o R_L$ and from Eq. (7-33) $E_i = R_i I_{in}$,

$$-\frac{V_o}{E_i} = A_i \frac{R_L}{R_i} \qquad (7\text{-}35)$$

Silicon transistors are employed in the Darlington configuration because of their low leakage current. Any leakage current exiting from the emitter of $Q_1$ will enter the base of $Q_2$ and be multiplied by $\beta_{F2} + 1$ to appear in the load resistor. Connecting an additional resistor from the base of $Q_2$ to ground will shunt some of the leakage (as well as signal) current and mitigate operating-point shift.

The Darlington's high current gain makes a versatile impedance-transforming device when operated as an emitter follower. In Fig. 7-15 the

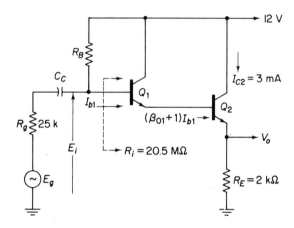

**Fig. 7-15.** Darlington emitter follower.

load $R_E$ is driven through $Q_1$ and $Q_2$ by high-resistance source $E_g$. By inspection of the circuit it is evident that signal-current gain from base of $Q_1$ to $R_E$ is

$$A_i = (\beta_{01} + 1)(\beta_{02} + 1) \cong \beta_0^2 \quad \text{for} \quad \beta_{01} = \beta_{02} \qquad (7\text{-}36)$$

The incremental resistance seen looking into the base of $Q_2$ is $r_{x2} + r_{\pi 2} + (\beta_{02} + 1)R_E = R_{i2}$. Then, looking into the base of $Q_1$ we see $r_{x1} + r_{\pi 1}$ plus $R_{i2}$ multiplied by $\beta_{02} + 1$, so

$$R_i = r_{x1} + r_{\pi 1} + (\beta_{01} + 1)[r_{x2} + r_{\pi 2} + (\beta_{02} + 1)R_E] \qquad (7\text{-}37)$$

The $r_x$ and $r_\pi$ terms are normally small with respect to the other terms so Eq. (7-37) may be approximated by

$$R_i \cong \beta_{01}\beta_{02}R_E \cong \beta_0^2 R_E \qquad (7\text{-}38)$$

$R_i$ is extremely large since the $\beta$ product can multiply $R_E$ by a factor of 2500 to 40,000 or more. Voltage gain is expressed by

$$\frac{V_o}{E_i} = A_i \frac{R_E}{R_i} \cong 1 \tag{7-39}$$

It is useful to examine the Darlington's resistance-transformation properties by viewing the amplifier as seen by $E_g$ in Fig. 7-16(a). Values shown are developed in Example 7-8. To see how the Darlington's output looks to $R_L$ we first look into the emitter of $Q_1$ to see $r_{x1} + r_{\pi 1}$ in series with $R_B \| R_G$, both divided by $\beta_{01} + 1$, and express the result by $R_{01}$ in Fig. 7-16(b)

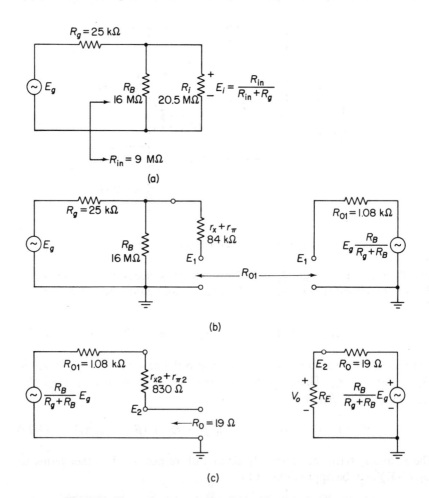

**Fig. 7-16.** The resistance seen by $E_g$ in Fig. 7-15 is given in (a) The resistance seen by load $R_E$ is modeled by two steps in (b) and (c).

Then look into the emitter of $Q_2$ to see $r_{x2} + r_{\pi 2}$ and $R_{o1}$, both divided by $\beta_{02} + 1$, to obtain $R_o$ in Fig. 7-16(c), where

$$R_o = \frac{R_{01} + r_{x2} + r_{\pi 2}}{\beta_{02} + 1} \tag{7-40}$$

where

$$R_{o1} = \frac{r_{x1} + r_{\pi 1} + R_g \| R_B}{\beta_{01} + 1}$$

*Example 7-8.* Given $\beta_0 = \beta_F = 100$ in Fig. 7-15. Find (a) $R_B$ to establish $I_{E2} = 3$ mA, (b) $R_i$, (c) $V_o/E_g$, (d) $C_c$ for $\omega_L = 100$ rad/sec.

*Solution.* (a) The emitter of $Q_2$ must be at 6 V, placing the base of $Q_2$ at 7.2 V because of the 0.6 V drops across $V_{BE}$ of $Q_1$ and $Q_2$. $I_{B2} = I_{E2}/(\beta_{F2} + 1) = 30\ \mu A = I_{E1}$ and $I_{B1} = I_{E1}/(\beta_{F1} + 1) = 0.3\ \mu A$. Voltage across $R_B$ is $12 - 7.2 = 4.8$ V and it carries $0.3\ \mu A$ to give $R_B = 4.8/3\ \mu A = 16$ MΩ. (b) Neglecting $r_x$ and assuming $I_{C1} \cong I_{E1}$, $I_{C2} \cong I_{E2}$,

$$g_{m1} = \frac{30 \times 10^{-3}}{25} = 1.2 \text{ mmho} \qquad g_{m2} = \frac{3}{25} = 0.12 \text{ mho}$$

$$r_{\pi 1} = \frac{100}{1.2 \times 10^{-3}} = 83 \text{ k}\Omega \qquad r_{\pi 2} = \frac{100}{0.12} = 0.83 \text{ k}\Omega$$

From Eq. (7-37), $R_i = \{83 + 101[0.83 + 101(2)]\}10^3 = 20.5$ MΩ.
(c) From Eq. (7-39),

$$\frac{V_o}{E_i} = \frac{(101)^2(2000)}{20.5 \times 10^6} = 0.99$$

From Fig. 7-16(b),

$$R_{in} = 20.5 \| 16 = 9 \text{ M}\Omega \qquad \text{and} \qquad \frac{E_i}{E_g} = \frac{9}{9 + .025} = 0.99$$

$$\frac{V_o}{E_g} = \frac{V_o}{E_i}\left(\frac{E_i}{E_g}\right) = 0.99(0.99) = 0.99$$

(d) From Fig. 7-16(b), $C$ would form a low-frequency time constant with $R_g$ and $R_{in}$, so

$$C_C = \frac{1}{\omega_L(R_g + R_{in})} = \frac{1}{100(9.025)10^6} = 1100 \text{ pF}$$

## 7-7 The Inverted Darlington Amplifier

The inverted Darlington is quite similar in performance to the Darlington pair. Transistor $Q_2$ has been changed to a *pnp* in Fig. 7-17(a), where the bias current directions are shown. An $R_1 - R_2$ combination may be

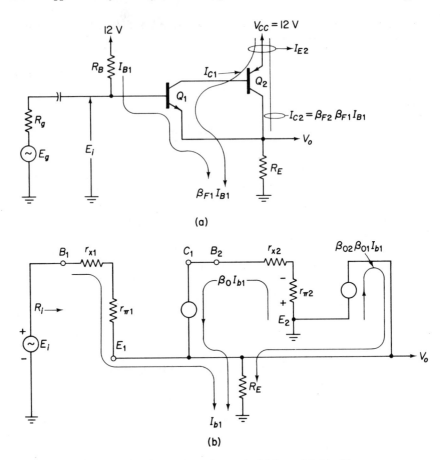

**Fig. 7-17.** The inverted Darlington in (a) is modeled in (b).

substituted for $R_B$ to improve operating-point stability at the sacrifice of input resistance. The dc input loop equation is written for evaluation of $I_{B1}$ and all other bias currents:

$$V_{CC} = I_{B1}\{R_B + R_E[1 + \beta_{F1}(1 + \beta_{F2})]\} + V_{BE1} \tag{7-41}$$

Small-signal analysis proceeds from the model of Fig. 7-17(b), where current gain is

$$A_i = \frac{I_{b1} + \beta_{01}I_{b1} + \beta_{01}\beta_{02}I_{b1}}{I_{b1}} = 1 + \beta_{01}(1 + \beta_{02}) \tag{7-42a}$$

Dropping the small unity terms gives the simple approximation,

$$A_i \cong \beta_{01}\beta_{02} \tag{7-42b}$$

Input resistance $R_i$ is derived from the input loop equation of

$$\frac{E_i}{I_b} = R_i = r_{x1} + r_{\pi 1} + [1 + \beta_{01}(1 + \beta_{02})]R_E \qquad (7\text{-}43)$$

Voltage gain is expressed simply by

$$\frac{V_o}{E_i} = A_i \frac{R_E}{R_i} \cong 1 \qquad (7\text{-}44)$$

We can derive output resistance by short circuiting $E_i$ in Fig. 7-17(b) and replacing $R_E$ with a test voltage $E$ so that the positive-going terminal of $E$ is grounded. All currents shown will flow through $E$ as $A_i I_{b1}$. Substituting for $I_{b1} = E/(r_x + r_\pi)$,

$$R_o = \frac{E}{A_i I_{b1}} = \frac{r_x + r_\pi}{A_i} \cong \frac{r_x + r_\pi}{\beta_{01}\beta_{02}} \qquad (7\text{-}45)$$

The inverted Darlington has two advantages over the Darlington pair. First, only $V_{BE}$ of $Q_1$ affects bias current $I_{B1}$ in the inverted version, while both $V_{BE}$ of $Q_1$ and $Q_2$ affect $I_{B1}$ in the regular pair. Hence, operating-point stability against temperature dependence of $V_{BE}$ is inherently superior in the inverted version. Second, if $Q_1$ and $Q_2$ are driven to saturation in Fig. 7-15 the maximum instantaneous voltage developed across $R_E$ is $V_{CC} - 2V_{BE} = 12 - 1.2 - 10.8$ V. But in the inverted Darlington of Fig. 7-17, assume the base of $Q_1$ can be driven to $V_{CC}$ and the maximum output voltage is limited by only one $V_{BE}$ to $12 - 0.6 = 11.4$ V.

## 7-8 Common-Emitter to Common-Base Cascode

The cascode is the basis for an important family of integrated circuits. Two advantages are obtained from the *cascode* arrangement of common-emitter to common-base stages in Fig. 7-18. First there is a minimum of internal feedback from output to input terminal. Voltage swings at the collector of $Q_2$ will couple energy back through $C_{\mu 2}$ to its base and be bypassed through $C_B$ to ground. Voltage swings at the collector of $Q_1$ are developed across the low-input resistance of $Q_2$. The $g_m R_L$ multiplier for $C_{\mu 1}$ will be low and superior high-frequency performance can be expected. This reduced internal feedback makes it possible to have a tuned cirucit for a load without incurring oscillations. Second, the output terminal can fluctuate between two extremes. When both $Q_1$ and $Q_2$ are near saturation, $V_o$ approaches the voltage across $C_E$ or $V_E$. When both $Q_1$ and $Q_2$ are near cutoff $V_o$ approaches $V_{CC}$. The total peak-to-peak output-voltage swing is $V_{CC} - V_E$.

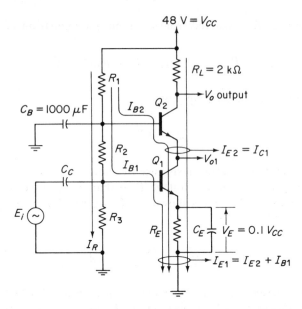

Fig. 7-18. Common-emitter to common-base cascode circuit.

Near cutoff both transistors are in series and share the division of $V_{CC} - V_E$ which can be larger than the breakdown voltage of either one. This circuit makes available larger output voltages then are possible with a single transistor.

As will be shown, the $Q_1 - Q_2$ combination acts as a single common-emitter stage with a load resistor of $R_L$. To bias this circuit, choose $R_E$ to develop a voltage drop $V_E$ equal to approximately 10% of $V_{CC}$. The remainder of $V_{CC}$ is equal to the maximum available peak-to-peak output swing and is to be divided equally between $R_L$ and the equivalent collector emitter formed by both collector emitters of $Q_2$ and $Q_1$. A bleeder current $I_R$ flows through bias resistor network $R_1$, $R_2$, and $R_3$ and should be 10 times greater than the base currents. $R_3$ is picked to establish $V_E + 0.6$ V, and the sum of $R_1$ and $R_2$ establishes $V_{CC} - V_E - 0.6$ V across both $R_L$ and the equivalent collector emitter.

$R_2$ establishes a bias voltage to feed $V_{CE1}$. $Q_1$ will develop a relatively small output-voltage swing across the small input resistance of $Q_2$ so that $V_{CE1}$ can be smaller than $V_{CE2}$. Of the remaining 90% of $V_{CC}$, allot 15% to $V_{CE1}$ and 75% to $V_{CE2}$. Finally, $I_{E2} = I_{C1}$ by identity and therefore $I_{C2} \cong I_{E1}$ despite any differences in $\beta_{F1}$ and $\beta_{F2}$.

An expression for $V_o/E_i$ is developed by showing the load on $Q_1$ as the input resistance to $Q_2$ in Fig. 7-19(a). Voltage gain of $Q_1$ is written by inspection, neglecting $r_x$ and assuming $\beta_{o2} + 1 \cong \beta_{o1}$:

Sec. 7-8        Common-Emitter to Common-Base Cascode        229

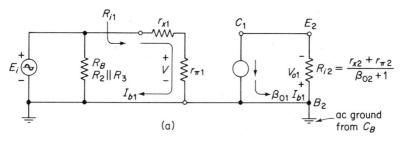

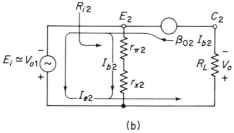

**Fig. 7-19.** The load seen by $Q_1$ of Fig. 7-18 is modeled in (a) and voltage gain for $Q_2$ is modeled in (b).

$$\frac{V_{o1}}{E_i} = \beta_{01}\left(\frac{\frac{r_{x2}+r_{\pi 2}}{\beta_{02}+1}}{r_{x1}+r_{\pi 1}}\right) \simeq \frac{\beta_{01}}{r_{\pi 1}}\left(\frac{r_{\pi 2}}{\beta_{02}}\right) = \frac{g_{m1}}{g_{m2}} \simeq 1 \qquad (7\text{-}46)$$

Since collector currents are approximately equal, the $g_m$'s of $Q_1$ and $Q_2$ are approximately equal, and voltage gain of $Q_1$ is unity. Thus $Q_2$ is driven by a signal equal to $E_i$. Voltage gain of $Q_2$ is found from the CB model in Fig. 7-19(b), where current gain is $\beta_{02}/(\beta_{02}+1) \simeq 1$.

$$\frac{V_o}{V_{o1}} = \frac{\beta_{02}R_L}{r_{x2}+r_{\pi 2}}, \qquad V_{o1} \simeq E_i \qquad (7\text{-}47)$$

Overall voltage gain of the cascode is provided by $Q_2$ and current gain by $Q_1$. Since there is a phase reversal between $E_i$ and $V_o$, it acts like a single CE circuit with minimum internal feedback and high voltage rating.

*Example 7-9.* An output p/p voltage swing of 40 V is required from the circuit of Fig. 7-18. This fixes the dc voltage drop across $R_L$ at 20 V with at least 20 V across $V_{CE1}$ and $V_{CE2}$. Assuming $\beta_{F1} = \beta_{F2} = 50$, $V_{CC} = 48$ V, and $V_{BE} = 0.6$ V, calculate (a) $I_{C2}$. (b) The remaining bias current. (c) Pick $I_R$ and calculate values for $R_3$ and $R_E$. (d) Find $R_1$ and $R_2$ and show the dc circuit voltages in Fig. 7-20.

*Solution.* (a) $I_{C2} = 20/2000 = 10$ mA, $V_{RL} = 20$ V.

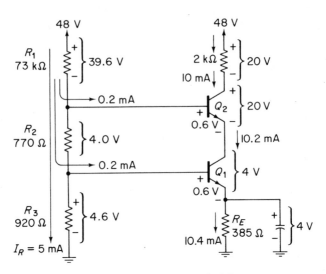

Fig. 7-20. Solution to Example 7-9.

$$V_E = 0.10(V_{CC}) = 4\text{ V} \qquad V_{CE1} + V_{CE2} = V_{CC} - V_{RL} - V_E = 24\text{ V}$$

Choose 4 V for $V_{CE1}$ and 20 V for $V_{CE2}$.

(b) $\quad I_{B2} = I_{C2}/\beta_F = 0.2\text{ mA}, \; I_{E2} = I_{C2} + I_{B2} = 10.2\text{ mA} = I_{C1}.$
$\quad I_{B1} = I_{C1}/\beta_F = 10.2/50 = 0.2\text{ mA}, \; I_{E1} = I_{B1} + I_{C1} = 10.4\text{ mA}.$

(c) Choose arbitrarily $I_R = 25 I_B = 5$ mA. $R_3$ carries 5 mA and must develop $V_{BE1} + V_E = 4.6$ V in Fig. 7-20 and $R_3 = 4.6/0.005 = 920\;\Omega$. $R_E$ develops 4 V with $I_{E1}$ and is $R_E = 4/(10.4 \times 10^{-3}) = 380\;\Omega$.
(d) $R_2$ carries 5.2 mA to set $V_{CE1}$ at 4 V, $R_2 = 4.0/(5.2 \times 10^{-3}) = 770\;\Omega$. $R_1$ carries 5.4 mA to feed 39.4 V to both $R_L$ and $V_{CB2}$, so $R_1 = 39.4/(5.4 \times 10^{-3}) = 7.3\text{ k}\Omega$. Voltage levels are shown in Fig. 7-20.

**Example 7-10.** Using the data in Example 7-9, find (a) input resistance and (b) $V_o/E_i$ of the cascode in Fig. 7-18.

**Solution.** (a) Assuming $r_x = 0$ and from $I_{C1} = 10.2$ mA, $g_{m1} = 0.41$, $R_i = r_{\pi 1} = \beta_{o1}/g_{m1} = 50/0.41 = 122\;\Omega$.

$$R_{in} = R_2 \| R_3 \| R_i = 770 \| 920 \| 122 = 94\;\Omega$$

(b) From Eq. (7-47),

$$\frac{V_o}{E_i} = \frac{50(2000)}{122} = 820$$

## 7-9 Linear Mixing

Linear mixing is the process of combining two or more signals from different sources into one composite signal. All frequencies contained in the original signals must not be changed in phase or amplitude with *respect to one another* and no additional frequencies are to be created in the mixing process. For example, it might be necessary to assign separate microphones for (1) lead guitar and vocalist, (2) rhythm section, and (3) electronic organ to reproduce their efforts with either fidelity or special effects.

Linear mixers should have three basic characteristics:

1. A separate volume control to adjust the signal level from each source prior to the mixing process. Often one source signal is much smaller than the largest source signal and they must be equalized so that, for example, the organ does not override the vocalist.
2. Signals from one source should not appear at any of the other sources. It is disconcerting to the singer when bass guitar beats come out of her microphone.
3. No additional frequencies or distortion should be added by the circuitry.

In the two-channel linear mixer of Fig. 7-21, volume is controlled individually for each channel by potentiometers $R_V$. Isolation between channels is provided by the collectors of $Q_1$ and $Q_2$; that is, a large output by $Q_1$

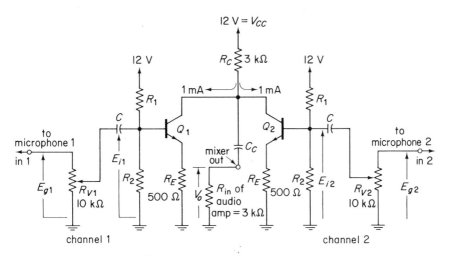

**Fig. 7-21.** Two-channel linear mixer.

cannot pass through the collector of $Q_2$ to drive the microphone at $E_{g2}$. Mixing of signal currents from both $Q_1$ and $Q_2$ is accomplished in coupling resistor $R_C$ and input resistance $R_i$ of an audio amplifier.

AC load resistance for each BJT in Fig. 7-21 is $R_C \| R_{in} = 1500\ \Omega$ and with $R_E = 500\ \Omega$, voltage gain $V_o/E_i \cong 1500/500 = 3$. Biasing and volume control considerations are shown in Examples 7-11 and 7-12.

**Example 7-11.** In the mixer of Fig. 7-21, design the biasing network to bias the collectors of $Q_1$ and $Q_2$ at $V_{CC}/2$, and choose $R_B = 20R_E$ as a compromise between operating-point stability and input resistance. $\beta_0 = \beta_F = 50$.

**Solution.** $R_C$ must carry 2 mA to drop 6 V so that $I_C$ of each BJT is to be 1 mA, and $I_B$ must be $I_C/\beta_F = 20\ \mu A$. From the input loop equation and $R_B = 10\ k\Omega$,

$$V_B = I_B[R_B + (\beta_F + 1)R_E] + V_{BE}$$
$$= 20[10 + 51(0.5)]10^{-3} + 0.6 = 1.31\ V$$

$$R_1 = \frac{V_{CC}}{V_B} R_B = 91.5\ k\Omega \qquad R_2 = \frac{V_{CC}}{V_{CC} - V_B} R_B = 11.2\ k\Omega$$

Voltage drop across $R_E$ is 0.5 V, leaving 5.5 V for both $V_{CE}$'s.

**Example 7-12.** Let the wiper contact of $R_{V1}$ be adjusted to its center point in Fig. 7-21. (a) Show that $E_i/E_{g1}$ does *not* equal $\frac{1}{2}$. (Assume $C$ acts as a short circuit, $R_B = 10\ k\Omega$, and $r_x + r_\pi = 1250\ \Omega$.) (b) What is $V_o/E_{g1}$?

**Solution.** (a) The bottom half (5 k$\Omega$) of $R_{V1}$ is in parallel with $R_B = 10\ k\Omega$ and $R_i$ of $Q_1$, where

$$R_i = r_x + r_\pi + (\beta_0 + 1)R_E = 1250 + 51(500) = 26.7\ k\Omega$$

Thus $E_g$ divides between the top half of $R_{V1}$ and $5 \| 10 \| 26.7 = 3\ k\Omega$:

$$\frac{E_i}{E_{g1}} = \frac{3}{3 + 5} = 0.375$$

(b) $\qquad V_o/E_{g1} = 0.375(3) = 1.1$

# PROBLEMS

**7-1** If $R_3$ is changed to 5 k$\Omega$ in Example 7-2, what is the new value of $R_{in}$ and $V_o/E_i$? Repeat for $R_3 = 10\ k\Omega$ and $R_3 = 2.5\ k\Omega$.

**7-2** Using the results of Examples 7-1, 7-2, and Problem 7-1, plot $C$, $R_{in}$, and $V_o/E_i$ versus $R_3$ for $R_3 = 0$ to $R_3 = 10\ k\Omega$ on the same graph.

**7-3** Repeat Example 7-2 with $R_2$ eliminated.

# Problems

**7-4** Calculate input resistance and $V_o/E_i$ for the circuit of Fig. 7-3.

**7-5** Calculate input resistance and $V_o/E_i$ for the nonbootstrapped version of Fig. 7-3 with $C_B$ and $R_3$ removed.

**7-6** A transistor with $\beta_F = 200$ is substituted in Example 7-4. What are the new values of $R_i$ and $V_o/E_g$?

**7-7** In Example 7-4, calculate $C_C$ to give a lower break frequency of 100 rad/sec for $V_o/E_g$.

**7-8** In Example 7-5, what is the amplifier's gain at $\omega_L$ and $\omega_M$?

**7-9** In Example 7-5, $C_C$ is reduced to 5 $\mu$F and $C_F$ is halved to 0.0005 $\mu$F. Find the new values of $\omega_L$ and $\omega_M$.

**7-10** Choose $C_F$ in Fig. 7-9 to begin reducing voltage gain $V_o/E_g$ at 5000 Hz.

**7-11** How is $R_i$ changed in Example 7-6 if the transistor has a $\beta_0 = \beta_F = 50$? How would this change effect Example 7-7?

**7-12** In Fig. 7-13, show that $I_E \cong (V_B - 2V_{BE})/R_E$, where the bias voltage $V_B = R_2 V_{CC}/(R_1 + R_2)$.

**7-13** Ideally, $V_o$ can be varied by a peak value of almost 5 V in Example 7-8. (a) What peak value of $E_g$ is required and (b) what peak signal currents are drawn from $E_g$ and delivered to $R_E$?

**7-14** In Example 7-8, how is input resistance and voltage gain affected if $\beta_0$ of each transistor is halved to 50?

**7-15** Evaluate $R_o$ in Example 7-8 if $\beta_{01} = \beta_{02} = 50$.

**7-16** In Fig. 7-17(a), $R_E = 2$ k$\Omega$, $I_{C2} = 2$ mA, $R_g = 25$ k$\Omega$, and $\beta_0 = \beta_F = 100$ for each transistor. Evaluate (a) $R_B$, (b) $R_i$, (c) $V_o/E_i$, (d) $V_o/E_g$, (e) $R_o$.

**7-17** Rework Example 7-9 assuming $\beta_{F1} = \beta_{F2} = 100$.

**7-18** Employ the results of Problem 7-17 to evaluate $R_{in}$ and $V_o/E_i$, when $\beta_{01} = \beta_{02} = 100$.

**7-19** The mixer of Fig. 7-21 drives an audio amplifier with an input resistance of 1.5 k$\Omega$. Find $V_o/E_{g1}$ when $R_{V1}$ is centered. Compare results with Example 7-12.

**7-20** Evaluate $C_C$ in Fig. 7-21 for a lower break frequency of 600 rad/sec.

# Chapter 8

8-0   Introduction .................................235
8-1   Current and Voltage Limitations ..............236
8-2   Thermal and Power Dissipation Limitations ....239
8-3   Thermal Resistance .........................240
8-4   Heat Sinking................................242
8-5   Review of Power Fundamentals ...............246
8-6   Power Supply to a Transistor Circuit .........247
8-7   Class A Direct-Coupled Amplifier. ............248
8-8   Class A Transformer-Coupled Power Amplifier 253
8-9   Collector-Emitter Voltage Considerations with Transformer Coupling .......................258
8-10 Class A Power Relationships under Nonsinusoidal Drive .........................260
8-11 Class B Amplifier............................261
      Problems....................................265

# Power Amplifiers

## 8-0 Introduction

A power amplifier delivers large amounts of power (more than 100 mW) from a power supply to a load. Power amplifiers are classified with respect to that percentage of a cycle where the amplifier operates in the linear region, as shown in Table 8-1.

Table 8-1

| Classification | Active region operation in per cent input signal cycle |
|---|---|
| A | 100 |
| AB | 50–100 |
| B | 50 |
| C | less than 50 |

Comparison of the three major classifications are shown in Fig. 8-1, where we see that for class A operation point A and input signal are such that output current flows for the entire period of the input signal. This type of operation is often defined as linear operation, even if the relationship between input and output is not precisely linear. When the B operating point is set so that current flows in the output for exactly 1/2 cycle, then operation of the amplifier is class *B* because for half the time it is cut off, and for the other half it operates in the active region, duplicating and amplifying the input signal. Classification AB identifies operation whereby the amplifier is active for more than 180°, yet less than 360°. A class *C* amplifier requires a reverse bias and a large input signal to drive the amplifier into the active region. The resultant output signal exists for less than 180° and is decidedly

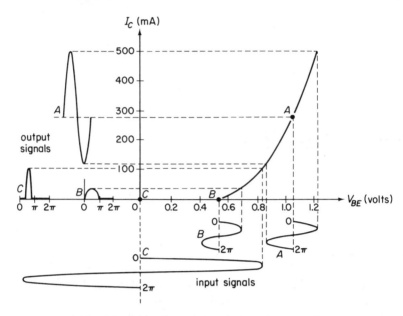

**Fig. 8-1.** Comparison of class A, B, and C operation.

nonlinear. The class C amplifier is useful for amplifying large amounts of power at a constant frequency where large collector current pulses are generally fed to an LC tank circuit tuned to the frequency of the input signal. The circulating current in the tank may be coupled inductively to a load to extract a sinusoidal signal from the oscillating tank circuit even though the pulses driving the tank are far from sinusoidal.

Power amplifiers operate with large current and voltage swings that may approach the limitations of the transistor. We must examine these limitations and undertake a brief review of power fundamentals in the first half of this chapter in order to understand the basic power relationships in the second half.

## 8-1 Current and Voltage Limitations

Dependence of BJT operation on collector-current magnitude is observed from the variation of $\beta_F$ in Fig. 8-2. When $I_C$ is less than 1 mA the base current furnishes a small, relatively fixed component of carriers to surface recombinations, imperfections, and emitter depletion-region recombinations. As injection level increases toward $I_C = 10$ mA, more of the base current feeds recombinations in the active base region and $\beta_F$ increases. Over 10 mA, the number of minority carriers traversing the base region becomes comparable to the fixed base-doping atoms, so that base doping

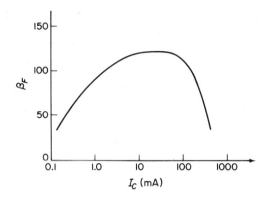

Fig. 8-2. Variation of $\beta_F$ with collector current level.

is effectively increased. Consequently base width is apparently increased, and together with the increased minority carrier density increases recombinations in the base, reducing $\beta_F$. As emitter injection is increased further, base majority carriers begin injection into the emitter and this carrier loss is furnished by the base, increasing base current with no corresponding increase in collector current, and resulting in a dramatic decrease in $\beta_F$ at high injection levels.

*Reach-through voltage* is a breakdown mechanism *occurring in alloy transistors.* When the collector-base reverse bias increases, the electric field across the collector space-charge region increases, and the space-charge region widens. If the base region is thin, and the collector is doped more heavily than the base, then the space-charge region expands further into the base region than into the collector, The upper limit of reverse bias occurs when the collector space-charge region stretches through the base to reach the emitter. The collector and emitter are then short circuited so that collector current increases sharply, limited only by external circuit resistance. This phenonenon is associated with alloy transistors and not generally with mesa, planar, or diffused transistors.

AVALANCHE BREAKDOWN

When reverse bias across the collector junction becomes large enough some of the minority carriers are accelerated and obtain enough energy to collide with the lattice and rupture a covalent bond. For every such ionizing collision an additional hole carrier and electron carrier is freed to cause further ionization. Freed electrons may enter the base region from the junction, supplying electrons to the base which are normally furnished by the base terminal (*pnp* transistor) and reducing the base current. The freed hole, together with the ionizing hole, are swept into the collector body. The collector current apparently increases for the same base current, because of the additional holes, and more collector current leaves the collector space-charge region than enters. The collector current is apparently multiplied by the *avalanche multiplication factor M*, the ratio of current leaving to current entering the collector space-charge region. Since the base current is reduced by avalanche multiplication and the collector current increases, $\beta_0$ and $\beta_F$ will increase rapidly, as shown by the characteristic curves in Fig. 8-3 which converge with increasing slope after the onset of avalanche.

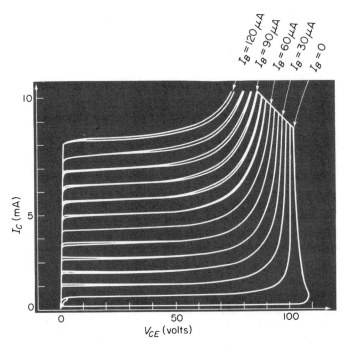

Fig. 8-3. Increase of $\beta_F$ near avalanche breakdown.

*Sustaining voltage* is the upper limit of reverse collector-junction bias imposed by avalanche increase of $\beta_F$. Two methods of specifying sustaining voltages are illustrated by their method of measurement in Fig. 8-4(a). With the switch up, breakdown voltage $BV_{CEO}$ is measured with open base by increasing $V_{CC}$ to 200 V either manually or with a sweep circuit to show the result in Fig. 8-4(b). Obviously the collector current must be specified for the voltage measurement. Throwing the switch down shorts base and emitter for a measurement or display of $BV_{CES}$. Since fewer carriers are avilable for avalanche, $BV_{CES}$ is higher than $BV_{CEO}$. A resistor $R$ may replace the short circuit in Fig. 8-4(a) and a breakdown voltage $BV_{CER}$ results, which is of an intermediate value.

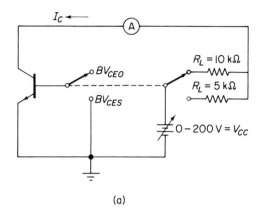

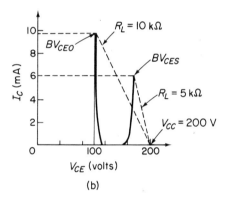

**Fig. 8-4.** Sustaining voltages $BV_{CEO}$ and $BV_{CES}$ may be measured by the circuit in (a) and plotted in (b).

An avalanche-multiplication breakdown voltage $BV_{EBO}$ is also specified for the emitter junction. Since the emitter and base may be heavily doped, any reverse biasing will add to a strong electric field already present. For this reason $BV_{EBO}$ is usually small, with values ranging down to 2 V.

## 8-2 Thermal and Power Dissipation Limitations

When a transistor is operated in the active region the collector junction is reverse biased and the emitter junction is forward biased. The collector junction is therefore a region of high resistance while the emitter junction is a region of low resistance. Collector current will cause most of the voltage drop in the transistor to exist across the high-resistance collector junction. Therefore most of the heat generated within the transistor will flow from this region, by conduction, to the header or case where it is passed to the surrounding medium, mostly by convection. The actual average junction temperature is determined by the average rate at which heat is generated minus the average rate at which heat is removed. The average rate at which heat is generated depends on $P_D$, the average power dissipated in the transistor, which is assumed to be developed entirely at the collector junction. One fundamental operating limitation is therefore the *maximum junction* temperature, $T_{J(\max)}$. Typical values for $T_{j(\max)}$ are 200 °C for silicon transistors and 85–100 °C for germanium transistors. There is also a maximum temperature limitation for storage and a maximum lead or pin temperature during soldering which represents the highest allowable temperature, above which the solder or other alloys may soften.

The manufacturing processes are conducted at elevated temperatures when contact metal is deposited on the base, emitter, and collector of the transistor wafer. Connector wires are bonded or soldered to the metalized deposits and brought out to the package leads. For good thermal conductivity, the collector metalized area is bonded to a metal header which may be the transistor enclosure. During soldering operations a heat sink, such as a long-nose pliers or metal clip, should contact the transistor lead between the case and soldering iron to remove heat that might damage the connector wire or contact metal.

## 8-3 Thermal Resistance

Heat flows from the collector junction to the surrounding medium which is at some measurable average or ambient temperature $T_A$. Within the small temperature differences and distances normally encountered, an average amount of power, $P_D$, will generate heat at the junction that will be removed at a rate directly proportional to the difference between the junction temperature $T_J$ and the ambient temperature $T_A$, or

$$P_D \theta_{JA} = T_J - T_A \qquad (8\text{-}1)$$

*Thermal resistance* $\theta_{JA}$ relates $P_D$ to the temperature difference and has units of degrees centigrade per watt. The path between the junction and ambient is interrupted by the case and thus $\theta_{JA}$ has two components which, together with Eq. (8-1), may be represented by the model of Fig. 8-5. $P_D$ is analogous to current flow and as power or heat issues from the collector junction it flows through thermal resistance, causing a temperature drop across the resistance so that temperature is analogous to voltage in the model. From the model, we can add the temperature rises across each thermal resistance (product of $\theta P$) to the reference temperature $T_A$ in order to find $T_J$:

## Sec. 8-3  Thermal Resistance  241

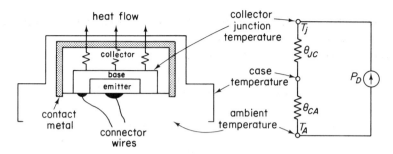

**Fig. 8-5.** Heat-flow model of a transistor.

$$T_A + (\theta_{JC} + \theta_{CA})P_D = T_J \tag{8-2}$$

and

$$\theta_{JA} = \theta_{JC} + \theta_{CA}$$

where $\theta_{JC}$ = thermal resistance, junction to case, and
$\theta_{CA}$ = thermal resistance, case to ambient.

The transistor manufacturer in many cases gives values for both $\theta_{JA}$ and $\theta_{JC}$. The methods of stating this information are fairly uniform but there are differences between manufacturers. Some give the information contained in Eq. (8-2) as a *dissipation derating curve*. For example, a silicon transistor encapsuled in a TO-5 package may have limitations specified as "Maximum transistor power dissipation is $P_{D(\max)} = 1$ W, up to 25 °C at free-air temperatures." The operating junction temperature range is −65 °C to +200 °C.

The first interpretation to be placed on this type of information is that the collector junction will be at 200 °C ($T_{J(\max)} = 200$ °C) if $P_D$ is at its maximum rating of 1 W and the transistor is in free air of temperature 25 °C or below. The transistor is right at the edge of its heat dissipating capability. If the ambient, free-air temperature is raised, the power dissipation must be decreased so that the junction temperature will not rise above 200 °C If the ambient temperature is raised to equal the maximum junction temperature then no heat can be dissipated from the junction without raising its temperature above the $T_{J(\max)}$ rating. This data is shown as the dissipation derating curve of Fig. 8-6.

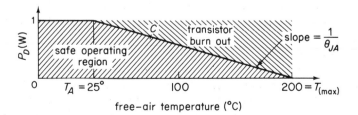

**Fig. 8-6.** Dissapation derating curve for a TO-5 case.

**Example 8-1.** (a) Find $\theta_{JA}$ for the BJT in Fig. 8-6. (b) What power is dissipated if the BJT is operating at point $C$?

**Solution.** (a) Modify Eq. (8-1) for maximum values

$$\theta_{JA} \cong \frac{T_{J(\max)} - T_A}{P_{D(\max)}} = \frac{200 - 25}{1} = 175 \text{ °C/W}$$

(b) At point $C$, $T_A = 75°$ C, $T_J = 200°$ C. From Eq. (8-1),

$$P_{D(\max)} = \frac{200 - 75}{175} = 0.7 \text{ W}$$

**Example 8-2.** Consider the BJT in Example 8-1. (a) $P_D$ is maintained at 1 W in an ambient air of 75° C. What is $T_J$? (b) If $P_D$ must be 2 W in an ambient of 25° C, what new value of $\theta_{JA}$ is required?

**Solution.** (a) From Eq. (8-1), $T_J = 175(1) + 75 = 250°$ C.

(b) $\qquad \theta_{JA} = (200 - 25)/2 = 87.5 \text{ °C/W}.$

## 8-4 Heat Sinking

Operation of the BJT in Example 8-2(b) is made possible by increasing the transistor's surface area by addition of a radiator or *heat sink* as shown in Fig. 8-7(a). The ability of a heat sink to transfer heat to the ambient depends on its material, volume, area, shape, and contact surface condition, plus the movement of the ambient fluid (generally air). Finned aluminum heat sinks yield the best heat transfer per unit cost, copper per unit volume, and magnesium per unit weight. A thermal resistance exists between case and sink, $\theta_{CS}$, and between sink and ambient, $\theta_{SA}$. Together, they shunt $\theta_{CA}$ of the transistor in the thermal model of Fig. 8-7(b). Normally $\theta_{CA}$ is large with respect to $\theta_{CS} + \theta_{SA}$ and is neglected in the model. For example, $\theta_{JC} \cong 35$ °C/W and $\theta_{CA} = 140$ °C/W for a TO-5 package. ($\theta_{CA}$ can be estimated roughly by measuring the case area in contact with the ambient and approximating heat transfer by 100–125 °C/W in.²).

Since the collector is usually connected electrically to the case, mica insulating separators and nylon shoulder washers insulate the transistor from heat sink and mounting screws. Silicon grease is used to fill minute dead air spaces in the mating surfaces between case, separator, and sink. Surfaces are forced together with a pressure established by torque values specified for the fastenings by the manufacturer. In the absence of manufacturer's data, assume $\theta_{CS}$ is 0.05–0.5 °C/W for a silicon-greased separator. Values for $\theta_{SA}$ are obtained from manufacturer's test data similar to the convection curves in Fig. 8-8. Point $x$ of curve $A$ means that when 5 W of power are applied to the heat sink the surface next to the semiconductor will experience a tempera-

Sec. 8-4  Heat Sinking  243

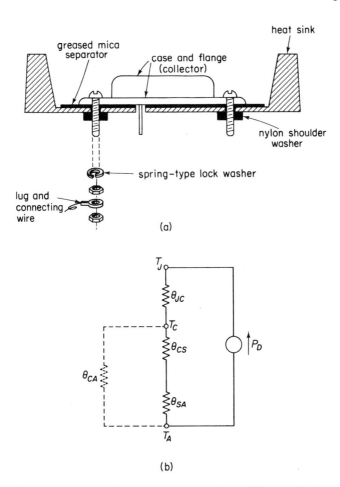

**Fig. 8-7.** Thermal resistance of the heat sink in (a) lowers $\theta_{CA}$ in the model of (b).

ture rise of 55 °C above the ambient temperature. Conversely, if the surface temperature rises by 55 °C above the ambient the sink will dissipate 5 W to the ambient. Thus if the transistor must dissipate 5 W the type-$A$ sink will offer a mating-surface temperature rise of 55 °C and the type-$D$ sink will offer a mating-surface temperature rise of about 10 °C. The respective average thermal resistances, $\theta_{SA}$, are located from points $x$ and $y$ of Fig. 8-8 as $55/5 = 11$ °C/W and $10/5 = 2$ °C/W. Heat exchange occurs predominantly by convection. However, the decreased slope of the curves at elevated temperatures indicates additional heat exchange by radiation.

A transistor with the thermal characteristics of Fig. 8-6 has a measured case temperature of $T_C = 95$ °C in an ambient of 25 °C. $P_D$ is calculated

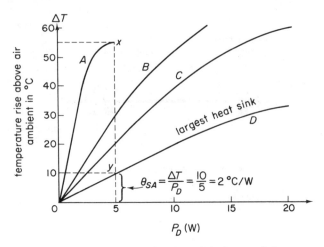

**Fig. 8-8.** Natural convection heat sink characteristics.

from a measurement of $I_C$ and $V_{CE}$ to be $P_D = 0.5$ W. From Fig. 8-5 or Fig. 8-7,

$$\theta_{CA} = \frac{T_C - T_A}{P_D} = \frac{95 - 25}{0.5} = 140 \,°C/W$$

Thus from measured values of $I_C$, $V_{CE}$, $T_C$, and $T_A$ we can measure $\theta_{CA}$.

**Example 8-3.** What is $T_C$ for the BJT of Fig. 8-6 if $P_D = 1$ W in an ambient of 25° C? Given $\theta_{CA} = 140 \,°C/W$.

**Solution.** From Fig. 8-6, $\theta_{JA} = 175 \,°C/W$ and $\theta_{JC}$ must be $175 - 140 = 35 \,°C/W$. Also, $P_D$ is at its maximum value, so $T_J$ must be 200° C. Manipulate Eq. (8-2) to express case temperature with respect to both junction and ambient temperature:

$$T_A + P_D\theta_{CA} = T_C = T_J - P_D\theta_{JC}$$
$$25 + 1(140) = 165° \text{ C} = 200 - 1(35)$$

**Example 8-4.** A data sheet specifies $\theta_{JC} = 1.5 \,°C/W$, $P_{D(max)} = 115$ W, and $T_{J(max)} = 200°$ C at $T_C = 25°$ C, for a general-purpose power transistor in a TO-3 package. In a circuit the BJT'S operating point will be at $I_C = 1.25$ A, $V_{CE} = 12.5$ V. (a) What is $P_D$ and what value will $T_J$ try to reach? (b) Select a heat sink to hold $T_J$ at or below $T_{J(max)}$.

**Solution.** (a) $P_D = 1.25(12.5) = 16$ W. This BJT can dissipate 115 W, with $T_J$ at 200° C when the case is immersed in a circulating fluid of 25° C that carries excess heat away as fast as it reaches the case. Our application has the BJT immersed in air and we must approximate $\theta_{CA}$ (usually not given) by measuring the top surface area of 1.5 in.$^2$ and estimating $\theta_{CA}$ from

$$\theta_{CA} = \frac{125 \,°C/W \text{ in.}^2}{1.5 \text{ in.}^2} = 83 \,°C/W$$

From Eq. (8-2), $T_J = 25 + 16(1.5 + 83) = 1300°$ C. Clearly a heat sink is required. The minimum thermal resistance is found from Eq. (8-2) and Fig. 8-7(b) to be (assuming $\theta_{CS} = 0.5\ °C/W$)

$$T_J = T_A + P_D(\theta_{JC} + \theta_{CS} + \theta_{SA})$$
$$200 = 25 + 16(1.5 + 0.5 + \theta_{SA})$$
$$\theta_{SA} = 8.9\ °C/W$$

Heat sink $B$ of Fig. 8-9 would be satisfactory and should be mounted to the

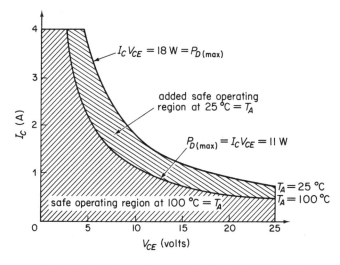

**Fig. 8-9.** Variation of $P_{D(max)}$ with $T_A$ for Example 8-5.

transistor not only for operation in the circuit but also while testing the transistor.

In Example 8-5 we see how maximum power dissipation depends on $T_A$ and plot $P_{D(max)}$ on a graph of $I_C$ versus $V_{CE}$. Then we can tell at a glance what operating points are allowed without exceeding thermal limitations.

*Example 8-5.* Assume the BJT of Example 8-4 is heat sinked for $\theta_{JA} = 9\ °C/W$, ($\theta_{SA} = 7\ °C/W$). Find $P_{D(max)}$ for air-ambient temperatures of 25 and 100° C.
*Solution.* From Eq. (8-1),

$$P_{D(max)} = \frac{T_J - T_A}{\theta_{JA}} = \frac{200 - 25}{9} = 18\ W = I_C V_{CE}, \quad T_A = 25°\ C$$
$$= \frac{200 - 100}{9} = 11\ W = I_C V_{CE}, \quad T_A = 100°\ C$$

Assume values for $I_C$ to calculate corresponding values of $V_{CE}$ and plot the result as in Fig. 8-9.

## 8-5 Review of Power Fundamentals

It is mandatory that we review basic power concepts as applied to a resistor or dc battery, carrying current with both dc and ac components.

### RESISTOR POWER DISSIPATION

The load resistor $R$ in a collector circuit usually conducts a dc current $I_C$, together with alternating components which we shall represent simply with the signal current sinewave ($I_{cp} \sin t$). The total instantaneous power $p$ is the product of the square of the total instantaneous current and the resistance:

$$p = (I_C + I_{cp} \sin t)^2 R \qquad (8\text{-}3)$$

Expanding Eq. (8-3) yields three terms:

$$p = I_C^2 R + 2I_C I_{cp} R \sin t + (I_{cp} \sin t)^2 R \qquad (8\text{-}4)$$

Our interest is in the *average power P* of each component in Eq. (8-4), so we assume $R = 1 \text{ k}\Omega$, $I_C = 5$ mA, $I_{cp} = 3$ mA and plot each component in Fig. 8-10.

The first term in Eq. (8-4) has an average value $P_{DR}$ due to the dc component $I_C^2 R$, as shown in Fig. 8-10(a). The second term, shown in Fig.

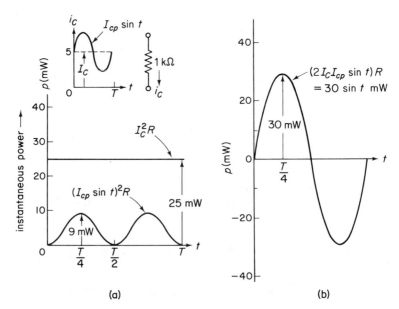

**Fig. 8-10.** Instantaneous power for components 1 and 3 in (a) and 2 in (b) for Eq. (8-4).

8-10(b), has no average value because every positive value is canceled by an equal negative value and sum to zero over one cycle. The third term is a double frequency wave whose average value $P_{RA}$ is found by measuring the total area under the curve of $9(\sin t)^2$ mW in Fig. 8-10(a) and dividing by the period of $2\pi$, or

$$P_{RA} = \frac{I_{cp}^2 R}{2} = I_c^2 R = 4.5 \text{ mW} \qquad (8\text{-}5)$$

where $I_c = I_{cp}/\sqrt{2}$ = rms value of the ac component. $I_c$ is the effective or rms value of current and therefore Eq. (8-5) represents heating effect on the resistor due to the alternating component of the current wave. Total average power dissipated in the resistor is therefore

$$P_R = P_{DR} + P_{RA} \qquad (8\text{-}6\text{a})$$
$$= I_C^2 R + I_c^2 R = 25 + 4.5 = 29.5 \text{ mW} \qquad (8\text{-}6\text{b})$$

Usually the alternating power component of Fig. 8-10(b) is ignored but it should be pointed out that a peak value of *instantaneous power* occurs at $T/4$ and equals 64 mW while a minimum value occurs at $3T/4$ of 4 mW. The source must be capable of supplying the peak value of 64 mW even though the average value is 29.5 mW.

## 8-6 Power Supply to a Transistor Circuit

$V_{CC}$ supplies power to the transistor circuit of Fig. 8-11(a), and with *no signal* the total power $P_S$ equals $V_{CC}(I_C + I_B)$. Since $I_B \ll I_C$ we neglect $I_B$ and express $P_S$ by

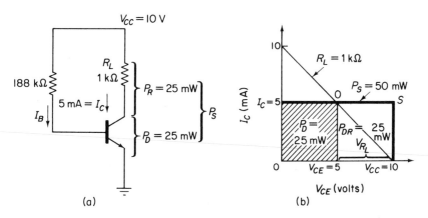

**Fig. 8-11.** Power-supply distribution for the circuit in (a) is shown pictorially in (b).

$$P_S = I_C V_{CC} \tag{8-7}$$

Assuming an operating point $O$ of $I_C = 5$ mA, $V_{CE} = 5$ V on the load line of Fig. 8-11(b), we can show transistor power $P_D$, by the crosshatched area under the operating point. By plotting point $S$ at $I_C = 5$ mA, $V_{CC} = 10$ V we outline the supply power rectangle to show supply power $P_S$ of 50 mW. The remaining area is the resistor dissipation $P_{DR}$.

When the collector bias current $I_C$ is varied sinusoidally by $I_{cp} \sin t$, the instantaneous power furnished by the supply is

$$P_S = V_{CC} I_C + V_{CC} I_{cp} \sin t \tag{8-8}$$

The second power term in Eq. (8-8) has an average value of zero so that average supply power is expressed by Eq. (8-7) regardless of the presence of an alternating component. We cannot ignore $I_{cp}$ even though a dc ammeter in a power supply will read only $I_C$, because the supply must be capable of furnishing a maximum peak current of $I_C + I_{cp}$. Thus any current-limiting feature on the power supply must be set with this point in mind or the power supply will introduce distortion.

The above discussion applies only to current excursions between cutoff and saturation. Assume the BJT in Fig. 8-11(a) is biased to cut off for class $B$ operation at $I_C = 0$, $V_{CE} = V_{CC}$ in Fig. 8-11(b). Then assume the base is driven as in Fig. 8-1 to yield a half-sine collector current of $I_{cp} \sin t\,]_0^\pi$. The instantaneous supply power is

$$p_s = V_{CC} I_{cp} \sin t \,\Big]_0^\pi \tag{8-9}$$

The average value of $P_S$ over a complete period of $2\pi$ is

$$P_S = \frac{V_{CC} I_{cp}}{2\pi} = 0.318\, V_{CC} I_{cp} \tag{8-10}$$

Thus the power supply furnishes power only when signal current flows and power is not wasted under no-signal conditions. In Eq. (8-10), $0.318 I_{cp}$ is the value read on a dc milliammeter and corresponds to the average value of a half-rectified sine wave.

## 8-7 Class A Direct-Coupled Amplifier

In Fig. 8-12 the circuit, output characteristics, and load line are shown for a linear class A amplifier. For maximum possible undistorted output power operating point $Q$ must be located at the midpoint of the load line. The $Q$-point collector current is then $I_{CQ} = V_{CC}/(2R_L)$ and the $Q$-point collector

Sec. 8-7    Class A Direct-Coupled Amplifier    249

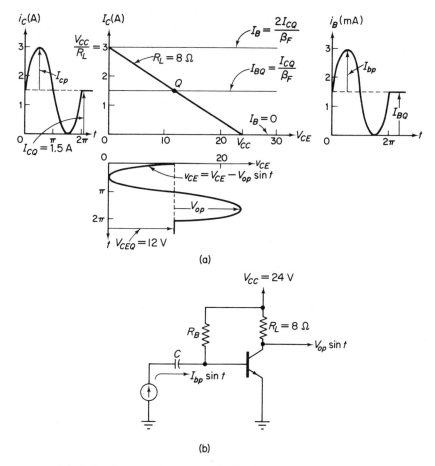

**Fig. 8-12.** Current-voltage waves in (a) develop maximum output power in (b).

voltage must be $V_{CEQ} = V_{CC}/2$. For linear operation the peak signal output voltage $V_{op}$ must not exceed $V_{CC}/2$ to prevent cutoff and the peak value of collector current $I_{cp}$ must not exceed $V_{CC}/(2R_L)$ to avoid saturation. Presumably the signal generator can supply any required values of driving current or voltage.

SUPPLY POWER. Supply power $P_S$ is independent of signal-current level and is expressed either in terms of $I_{CQ}$ or in terms of $V_{CC}$ and $R_L$ from $I_{CQ} = V_{CC}/2R_L$:

$$P_S = V_{CC}I_{CQ} = \frac{V_{CC}^2}{2R_L} \tag{8-11}$$

LOAD RESISTOR POWER. Load resistor power $P_R$ consists of $P_{RD}$, due to the dc component, plus the useful output power $P_{RA}$ or $P_o$ due to the rms signal current. Each component is expressed generally by

$$P_{RD} = I_{CQ}^2 R_L = \frac{V_{CC}^2}{4R_L} \tag{8-12}$$

$$P_o = \left(\frac{V_{op}}{\sqrt{2}}\right)^2 \frac{1}{R_L} = \frac{V_{op}^2}{2R_L} \tag{8-13}$$

At maximum output $V_{op} = V_{CC}/2$ so that Eq. (8-13) reduces to

$$P_{o(\max)} = \frac{V_{CC}^2}{8R_L} \tag{8-14}$$

Total power delivered to $R_L$ at maximum output is found from Eqs. (8-12) and (8-14) in terms of $V_{CC}$ and $R_L$ as

$$P_R = P_{RD} + P_{o(\max)} = \frac{V_{CC}^2}{4R_L} + \frac{V_{CC}^2}{8R_L} \tag{8-15a}$$

Substituting from Eq. (8-11), express $P_R$ in terms of $P_S$ as

$$P_R = P_{RD} + P_{o(\max)} = \frac{P_S}{2} + \frac{P_S}{4} \tag{8-15b}$$

Obviously, applying maximum signal increases power dissipation in the load $R_L$.

TRANSISTOR DISSIPATION. The transistor's dissipation depends on the direct component of collector current $P_{DD}$ and on the alternating component of collector current $P_{DA}$ and is expressed by

$$P_D = P_{DD} - P_{DA} \tag{8-16}$$

where

$$P_{DD} = V_{CEQ} I_{CQ} \tag{8-17}$$

and

$$P_{DA} = -\frac{V_{op}}{\sqrt{2}} \frac{I_{cp}}{\sqrt{2}} \tag{8-18}$$

The minus sign in the expression for $P_{DA}$ will be considered in detail because of its physical significance. We cannot derive an expression for average transistor power in terms of $I^2 R$ or $V^2/R$ because there is no readily available value of $R$ corresponding to the operating resistance between collector and emitter. We therefore express instantaneous transistor power $P_d$ and find its

Sec. 8-7  Class A Direct-Coupled Amplifier  251

average value. [Note that the minus sign in the expression for the instantaneous voltage across the collector and emitter results from Fig. 8-12(a).]

$$P_d = (I_C + I_{cp} \sin t)(V_{CEQ} - V_{op} \sin t)$$

Expanding:

$$P_d = I_C V_{CEQ} - I_C V_{op} \sin t + I_{cp} V_{CEQ} \sin t - I_{cp} V_{op} \sin^2 t$$

The second and third terms of this expression have an average value of zero. The first term is given by Eq. (8-17). The average value of the last term is found by reference to the power review; it is due to the ac components and is expressed by Eq. (8-18).

The minus sign means that the transistor gives up or transfers power to the resistor as output power. The transistor therefore runs cooler when a signal is applied. It is at maximum dissipation with no signal. Alternatively, we can reason that if current through a device is *increasing* when voltage across it is *increasing*, the device must be absorbing power. However, if current through a device is *increasing* when voltage across it is *decreasing*, the device must be giving up power.

Transistor dissipation at maximum output is expressed in terms of $R_L$ and $V_{CC}$ from Eqs. (8-16) to (8-18) and $I_{cp} = I_{CQ} = V_{CC}/2R_L$, $V_{op} = V_{CC}/2$:

$$P_{DD} = \frac{V_{CC}^2}{4R_L} = \frac{P_S}{2} \tag{8-19}$$

$$P_{DA} \text{ at maximum output} = \frac{V_{CC}^2}{8R_L} = \frac{P_S}{4} \tag{8-20}$$

$$P_D \text{ at maximum output} = P_{DD} - P_{DA} = \frac{P_S}{4} \tag{8-21}$$

It is evident from Eqs. (8-20) and (8-14) that $P_o = P_{DA}$ and $P_{DA}$ represents output power transferred from the transistor to the load.

*Example 8-6.* For the circuit of Fig. 8-12 find the power levels at no signal and maximum output for (a) power supply, (b) load, and (c) transistor.

*Solution.* (a) From Eq. (8-11), $P_S = 24(1.5) = 36$ W.

(b) At no signal, from Eq. (8-12),

$$P_{RD} = \frac{24^2}{4(8)} = 18 \text{ W} = P_R$$

From Eq. (8-14), $P_{o(\max)} = 9$ W, so maximum output is found from Eq. (8-15b) to be $18 + 9 = 27$ W.

(c) At no signal, from Eq. (8-17), $P_{DD} = 18 \text{ W} = P_D$. At maximum output, from Eqs. (8-19) and (8-20),

$$P_D = 18 - 9 = 9 \text{ W}$$

We conclude from Example 8-6 that the power amplifier transforms power available from the power supply into a useful output component $P_o$ plus other wasted components $P_D$ and $P_R$. If the load resistor in Example 8-6 represented a loud speaker and the maximum output signal was delivered, only the power component $P_o$ would drive the speaker cone back and forth at the signal frequency. The power component $P_D$ would appear as heat, emanating from the collector to emitter junction of the transistor. The power component $P_{RD}$ would severely and uselessly heat the speaker winding.

EFFICIENCY. The efficiency $\eta$ of a class $A$ power amplifier is the ratio of useful power output $P_o$ to power supplied by the power source and is expressed in per cent:

$$\eta = \frac{P_o}{P_S} 100 \tag{8-22a}$$

As already shown, $P_S$ is constant in the linear amplifier. *However*, $P_o$ *is dependent* on the signal level $V_{op}$, being zero at no signal and a maximum of $P_S/4$ at maximum output. Therefore the efficiency $\eta_{\max}$ at maximum output signal is determined by dividing Eq. (8-20) by Eq. (8-11) to obtain $\eta_{\max} = 25\%$.

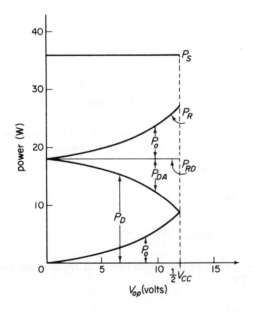

**Fig. 8-13.** Power levels versus $V_{op}$ for class A direct-coupled amplifier (Example 8-6).

The ratio of maximum output power $P_{o(\text{max})}$ to total power dissipated in the transistor, $P_D$, is useful when selecting a transistor of adequate power dissipation capability. Dividing Eq. (8-14) by Eq. (8-19) gives the *maximum output to dissipation ratio*

$$\frac{P_{o(\text{max})}}{P_{DD(\text{max})}} = \frac{1}{2} \tag{8-22b}$$

Thus a transistor which is rated to dissipate 5 W can supply a maximum of only 2.5 W of output power when operated class *A* from a supply power of 10 W. The variation of each power level as a function of output voltage for Example 8-6 is shown in Fig. 8-13.

## 8-8 Class A Transformer-Coupled Power Amplifier

Load $R_L$ is coupled by a 1 : 1 transformer into a transistor's collector circuit in Fig. 8-14(b). DC resistance of the primary winding is negligible so the dc load line rises vertically from $V_{CC}$ in the output characteristics of Fig. 8-14(a).

Operating point $Q$ is chosen for maximum power output and minimum transistor distortion. Due to the 1 : 1 turns ratio, the transistor sees an ac load of 8 Ω. We could draw ac load lines through any arbitrary point on the dc load line. Points below $Q$ would limit $V_{op(\text{max})}$ by cutoff to values less than $V_{CC}$. Points above $Q$ would allow $V_{op(\text{max})}$ of $V_{CC}$ but would waste power needlessly. From the geometry of Fig. 8-14(a) the operating point $Q$ is selected at

$$I_{CQ} = \frac{V_{CC}}{R_L} = I_{cp(\text{max})} \tag{8-23}$$

$$V_{CEQ} = V_{CC} = V_{op(\text{max})} \tag{8-24}$$

SUPPLY POWER. Whether collector current contains a signal component or not, supply power is constant (linear operation) at

$$P_S = I_{CEQ}V_{CEQ} = \frac{V_{CC}^2}{R_L} \tag{8-25}$$

LOAD RESISTOR POWER. Load resistor power $P_R$ consists only of the usable signal component since the transformer will not transform dc collector bias current $I_{CQ}$. It must be realized, however, that if the signal current in the secondary is distorted (not a pure sine wave), one half-cycle will differ in area from the other half-cycle and the load resistor will experience this average difference as a dc component. Neglecting distortion, $P_o$ is expressed by

## 254 Power Amplifiers

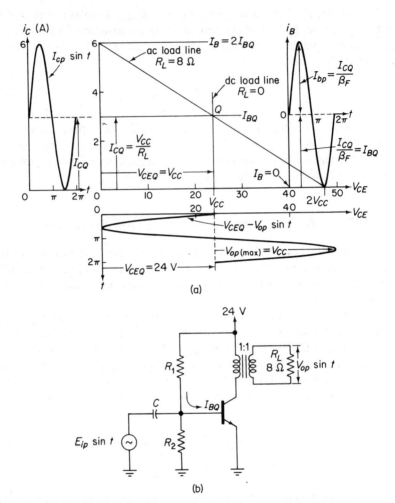

**Fig. 8-14.** Current and voltage waves in (a) develop maximum power output for the circuit in (b).

$$P_o = P_R = \left(\frac{V_{op}}{\sqrt{2}}\right)^2 \frac{1}{R_L} = \frac{V_{op}^2}{2R_L} \qquad (8\text{-}26)$$

Since $V_{op(\text{max})} = V_{CC}$, maximum output power is, from Eqs. (8-26) and (8-25),

$$P_{o(\text{max})} = \frac{V_{CC}^2}{2R_L} = \frac{P_s}{2} \qquad (8\text{-}27)$$

Transistor dissipation $P_D$ still consists of a dc component $P_{DD}$ and an ac component $P_{DA}$. Since $I_{cp} = V_{op}/R_L$ the relationships are

## Sec. 8-8  Class A Transformer-Coupled Power Amplifier

$$P_{DD} = I_{CQ}V_{CEQ} = \frac{V_{CC}^2}{R_L} = P_s \qquad (8\text{-}28a)$$

$$P_{DA} = \left(\frac{V_{op}}{\sqrt{2}}\right)\left(\frac{I_{cp}}{\sqrt{2}}\right) = -\frac{V_{op}^2}{2R_L} = -P_o \qquad (8\text{-}28b)$$

$$P_D = P_{DD} - P_{DA} \qquad (8\text{-}28c)$$

At maximum output $V_{op} = V_{CC}$ and

$$P_{DD} \text{ at maximum output} = P_s \qquad (8\text{-}29a)$$

$$P_{DA} \text{ at maximum output} = \frac{V_{CC}^2}{2R_L} = \frac{P_s}{2} \qquad (8\text{-}29b)$$

$$P_D \text{ at maximum output} = P_s - \frac{P_s}{2} = \frac{P_s}{2} \qquad (8\text{-}29c)$$

We conclude that maximum transistor dissipation occurs under no-signal conditions from Eq. (8-28a) and the transistor runs cooler at maximum output when it transfers half its no-signal power to the load as $P_{o(\max)}$ in Eq. (8-27).

Maximum efficiency $\eta_{\max}$ occurs at full signal and is found through dividing Eq. (8-27) by (8-25) to be 50%. Ratio of maximum output power to maximum transistor dissipation results from dividing Eq. (8-27) by (8-28a):

$$\text{Maximum output to dissipation ratio} = \frac{1}{2} \qquad (8\text{-}30)$$

Thus a transistor rated as $P_{D(\max)} = 5$ W can supply 2.5 W from a 5-W source. Compared with the direct-coupled amplifier, there is a saving of 50% in supply power for the same output due to higher efficiency of transformer coupling.

*Example 8-7.* For the circuit in Fig. 8-14 find the power levels at no-signal and maximum output for (a) supply, (b) load, and (c) transistor.

*Solution.* (a) The $Q$ point is located from Eqs. (8-23) and (8-24) to be $I_{CQ} = 3$ A, $V_{CEQ} = 24$ V. From Eq. (8-25), $P_S = 24^2/8 = 72$ W.
(b) At no signal, from Eq. (8-28a), $P_D = 72$ W. At maximum output, from Eqs. (8-29), $P_{DA} = 36$ W and $P_D = 36$ W.
(c) At no signal, $P_R = P_o = 0$. At maximum output, from Eq. (8-27), $P_{o(\max)} = 36$ W.
Power levels will vary with the amount of drive and are plotted in Fig. 8-15 as a function of $V_{op}$.

*Design Example 8-8.* Using Fig. 8-16(a) as a guide, design a transformer-coupled power output stage that will deliver a maximum output of 4 W to a 4-$\Omega$ speaker. Transformers $T_1$ and $T_2$ are 80% efficient. Characteristic curves for the transistor are given in Fig. 8-16(b). The transistor is heat

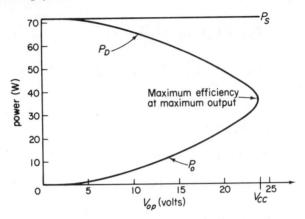

**Fig. 8-15.** Power levels versus $V_{op}$ for class A transformer-coupled amplifier (Example 8-7).

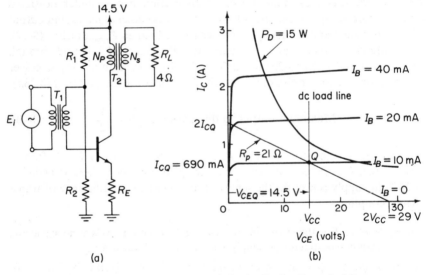

**Fig. 8-16.** Circuit and characteristics for Design Example 8-8.

sinked for maximum dissipation of 15 W. $R_E$ is included for stability and is chosen to drop 5% $V_{CC}$ at no signal.

**Solution Procedure.** (a) Calculate the transistor's ac load resistance $R_p$ to develop the required $P_{o(max)}$, together with operating point $Q$. (Assume signal power lost in $R_E$ is negligible.) (b) Select $T_2$. (c) Calculate values for bias resistors and $R_E$. (d) Find the power required by $T_1$ for maximum output.

**Solution.** (a) The primary winding of $T_2$ must receive 5 W of signal power from the transistor to deliver $5(0.80) = 4$ W to the secondary speaker load. Assuming $R_E$ is small with respect to $R_p$ then $P_{o(max)} = 5$ W and at maximum output $V_{op} = V_{CC}$. We discount the small saturation voltage as negligible with respect to the peak swing of $V_{op}$, although in actual practice we should allow about 1 V to be deducted from the maximum allowable peak-to-peak swing. From Eq. (8-26) or (8-27),

$$R_p = \frac{V_{CC}^2}{2P_o} = \frac{(14.5)^2}{10} = 21\ \Omega$$

The $Q$-point voltage $V_{CEQ} = V_{CC} = 14.5$ V. The peak current swing and consequently the $Q$-point collector current must be

$$I_{CQ} = \frac{V_{CC}}{R_p} = \frac{14.5}{21} = 690\ \text{mA}$$

Maximum transistor dissipation $P_{D(\text{max})}$ occurs at no signal and from Eq. (8-28a), $P_{D(\text{max})} = 690(10^{-3})(14.5) = 10$ W. This does not exceed the 15-W maximum plotted as the hyperbola in Fig. 8-16(b). To locate operating point $Q$, assume initially that $R_E$ is negligible and plot the dc load line straight up from $V_{CC}$ in Fig. 8-16(b). Draw the ac load line with slope of $1/R_p$ through $Q$. $P_{o(\text{max})}$ is the area under the triangle formed by the axes and the ac load line.

(b) Turns ratio for $T_2$ is found from $R_p N_s^2 = R_L N_p^2$, $R_p = 21\ \Omega$, $R_L = 4\ \Omega$:

$$\frac{N_p}{N_s} = 2.29$$

(c) $V_{RE} = 0.05 V_{CC} = 0.725$ V. Since $I_{EQ} \cong I_{CQ}$,

$$R_E = \frac{V_{RE}}{I_{CQ}} = \frac{0.725}{0.690} \cong 1\ \Omega$$

To find the bias resistors, we read $I_{BQ} = 10$ mA from Fig. 8-16(b) and calculate $\beta_F = I_{CQ}/I_{BQ} = 69$. For stability let $R_B = 15R_E = 15\ \Omega$. Assuming the secondary resistance of $T_2$ is negligible and $V_{BEQ} = 0.6$ V, find $V_B$ from the equivalent dc input model of Fig. 8-17(a).

$$V_B = I_B[R_B + (\beta_F + 1)R_E] + V_{BE} = 1.45\ \text{V}$$

$$R_1 = \frac{V_{CC}}{V_B} R_B = 150\ \Omega \qquad R_2 = \frac{V_{CC}}{V_{CC} - V_B} R_B = 16.6\ \Omega$$

Select the nearest standard value of resistors.

(d) At the high collector bias current, input resistance $h_{IE}$ of the transistor is small and usually negligible with respect to any other circuit resistance. $h_{IE}$ may be approximated by $r_\pi \cong 25\beta_0/I_{CQ} = 25(69/690) = 2.5\ \Omega$. From the ac model of Fig. 8-17(b) the peak secondary voltage of $T_1$ must be

$$E_{sp} = I_{bp}[R_B + r_\pi + (\beta_F + 1)R_E] = 10(10^{-3})(87.5) = 0.875\ \text{V}$$

Secondary signal power of $T_1$ is $E_{sp}I_{bp}/2 \cong 4.4$ mW. Since $T_1$ is 80% efficient, input power required is $4.4/0.8 = 5.5$ mW.

**258** Power Amplifiers Chap. 8

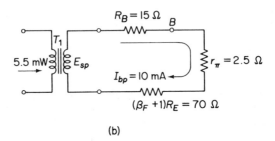

**Fig. 8-17.** Input circuit models for dc and ac in (a) and (b), respectively, for Design Example 8-9.

## 8-9 Collector-Emitter Voltage Considerations with Transformer Coupling

The fact that voltage across the transistor can exceed supply voltage with a transformer-coupled load is evident from the graphical analysis. An intuitive understanding, however, is not so obvious unless we look for the presence of an energy-storage element in the inductance of the transformer's primary winding. An audio transformer is wound on an iron core so that most of the flux generated in the primary links the secondary. Therefore the primary ampere-turns equals the secondary ampere-turns and the voltage per turn of the primary equals the voltage per turn of the secondary. These relationships depend on the mutual flux linkages caused by an alternating current, and are given by

$$\frac{V_p}{N_p} = \frac{V_s}{N_s} \qquad (8\text{-}31)$$

$$I_p N_p = I_s N_s \qquad (8\text{-}32)$$

$$\frac{R_p}{N_p^2} = \frac{R_s}{N_s^2} \qquad (8\text{-}33)$$

where $R_s$ = secondary load resistance,
$N_p$ = number of primary turns,
$N_s$ = number of secondary turns,
$R_p$ = resistance coupled into the primary by $R_s$, and
$\eta$ = turns ratio of the transformer = $N_p/N_s$.

Sec. 8-9            Collector-Emitter Voltage Considerations    259

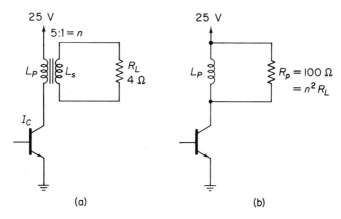

Fig. 8-18. The circuit in (a) may be modeled by (b) only when an ac signal is present.

Equation (8-33) shows that the effect of the secondary circuit on the primary may be modeled by a resistor $R_p = n^2 R_L$ as shown in Fig. 8-18.

If the reactance of $L_p$ in Fig. 8-18(b) is large with respect to $R_p$, $L_p$ will be an open circuit to incremental currents. But $L_p$ has low dc re-

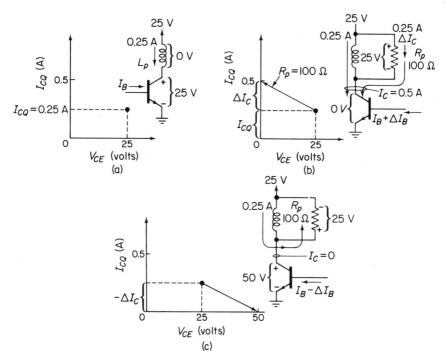

Fig. 8-19. $L_p$ acts as a dc current source in (a) that is passed or blocked by the BJT in (b) and (c), respectively.

sistance and shunts the dc collector current around $R_p$. Once collector bias current has been established through $L_p$, its large inductance will tend to maintain $I_{CQ}$ constant and $L_p$ can be shown as a constant-currrent source in Fig. 8-19(a).

Assume a positive-going increment of base current $\Delta I_B$ causes an incremental increase of collector current $\Delta I_C$ along the ac load line, into saturation as shown in Fig. 8-19(b). The incremental change in collector current for 0.25 to 0.50 A cannot pass through $L_p$ since it presents a large reactance to a current change and is so large that the dc current through it cannot change. Therefore, the collector-current increment must take the direction shown and develop the polarity and magnitude of the voltage drop shown across $R_p$. The voltage across the transistor is $V_{CE} = V_{CC} - V_{RP} = 0 \text{ V}$.

The transistor is cut off in Fig. 8-19(c). Since current in $L_p$ cannot change, 0.25 A is diverted up through $R_p$ so that $V_{CE}$ goes to 50 V, or twice the supply voltage.

It is informative at this point to consider the problem encountered when buying an output transformer. To purchase a transformer for the circuit of Fig. 8-18 we might see three transformers listed in a catalog with the same turns ratio but with different impedance ratings, as seen in Table 8-2.

Table 8-2

| Transformer | Turns ratio | Primary impedance (Ω) | Secondary load impedance (Ω) |
|---|---|---|---|
| A | 5:1 | 100 | 4 |
| B | 5:1 | 1000 | 40 |
| C | 5:1 | 50 | 2 |

Recalling our stipulation that $L_p$ must be large we would surmise that transformer C would be unsatisfactory since the primary impedance ratio of 50 Ω indicates a large $L_P$ if $R_L$ is 2 Ω but not if $R_L$ is greater than 2 Ω. Transformer A would be the most economical selection. Transformer B would perform well but would be more expensive because of the greater number of winding turns indicated by the primary impedance of 1 kΩ.

## 8-10 Class A Power Relationships Under Nonsinusoidal Drive

It must be stressed that all the power relationships developed were based on sinusoidal currents and voltages. Any other wave shape will cause markedly different expressions. For example, if the circuit of Fig. 8-12 is driven to maximum output with a square wave the resultant wave shapes are as shown in Fig. 8-20.

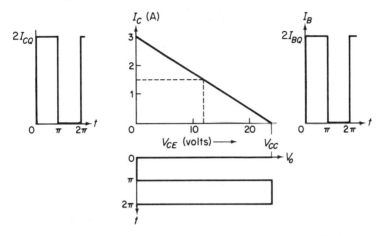

Fig. 8-20. Signal shapes for Fig. 8-12 under square-wave drive.

$P_{o(\max)}$ is determined from $V_{CC}$ existing across $R_L$ for half a cycle and $P_S$ equals $I_{CQ}V_{CC}$, where $I_{CQ} = V_{CC}/2R_L$:

$$P_{o(\max)} = \frac{V_{CC}^2}{R_L}\frac{1}{2} = P_S = \frac{V_{CC}^2}{2R_L} \qquad (8\text{-}34)$$

The ratio of $P_o$ to $P_S$ is 1 or an efficiency of 100%. During the first half-cycle, a maximum current flows with zero voltage across the transistor; during the second half-cycle, no current flows when there is maximum voltage across the transistor. Thus there is no power dissipation in the transistor. This example represents an ideal case, which is never found in real transistors. Actually, during the first half-cycle a small voltage $V_{CE(\text{sat})}$ is measured across the transistor, and in the second half-cycle the leakage current $I_{CEO}$ flows in the transistor. However, these components will normally be inconsequential.

## 8-11 Class B Amplifier

To operate under the $B$ classification, $R_1$ and $R_2$ of Fig. 8-21(a) must be selected to bias the transistor exactly at cutoff or at $I_B = 0$. To be more exact, the transistor must be on the verge of conduction. If $R_1$ and $R_2$ were not present it is true that base current would not flow, but $E_1$ would have to increase to a positive value of about 0.6 V (for a silicon transistor) before base current would begin to drive the transistor into the active region. It is

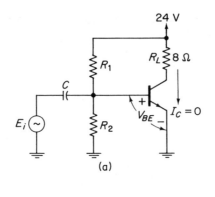

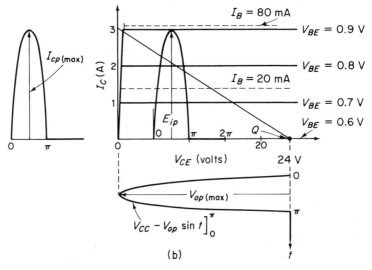

**Fig. 8-21.** Class B circuit with load line and wave shapes at maximum output.

seen from Fig. 8-21(b) that the $Q$ point must be established at $V_{BE} \cong 0.6$ V, $I_{CQ} = 0$, $V_{CEQ} = V_{CC}$. Since no base current is taken at the $Q$ point, the bias resistors are found from the voltage-divider relationship

$$V_{BE} = 0.6 \text{ V} = \frac{R_2}{(R_1 + R_2)} V_{CC}$$

$R_2$ should be large with respect to the ac input resistance $R_i$. For maximum allowable input signal the transistor is driven from cutoff to the verge of saturation. We shall again discount the loss in signal swing due to the small saturation voltage.

The average ac, large-signal input resistance, for maximum allowable drive, can be estimated from Fig. 8-21(b) as the ratio of $\Delta E_{i_p}$ to $\Delta I_{b_p}$ or $R_i$

= 0.3/(80 × 10⁻³) ≅ 4 Ω. The restrictions on the bias resistors are not severe and the ac input resistance is low, as would be expected for large-signal operation.

SUPPLY POWER.  Under no-signal conditions the transistor passes zero current so that the supply furnishes no idling power and results in a substantial increase in efficiency. When a signal voltage is applied to the base, the resultant collector current is a half-sinewave whose average value is $I_{cp}/\pi$. Since $I_{cp} = V_{op}/R_L$, supply power depends on $V_{op}$ by

$$P_S = \frac{V_{CC}I_{cp}}{\pi} = \frac{V_{CC}V_{op}}{\pi R_L} \tag{8-35}$$

At maximum output, $V_{op} = V_{CC}$ and $P_S$ is maximum at

$$P_{S(\max)} = \frac{V_{CC}^2}{\pi R_L} \tag{8-36}$$

TRANSISTOR POWER.  Instantaneous transistor power is the product of instantaneous collector voltage and current, where

$$p_d = \left(I_{cp} \sin t \Big]_0^\pi\right)\left(V_{CC} - V_{op} \sin t \Big]_0^\pi\right) \tag{8-37}$$

The average value of Eq. (8-37) can be simplified to

$$P_D = \frac{V_{op}V_{CC}}{\pi R_L} - \frac{V_{op}^2}{4R_L} \tag{3-38}$$

and, by comparison with Eq. (8-35),

$$P_D = \text{supply power} - \text{output power} \tag{8-39}$$

In Eq. (8-38), supply power increases directly as $V_{op}$ is increased but the output power increases as the square of $V_{op}$. We must differentate Eq. (8-38) with respect to $V_{op}$ and set the result equal to zero to find $V_{op}$ at *maximum dissipation*

$$V_{op} = \frac{2V_{CC}}{\pi} = 0.636 V_{CC} \tag{8-40}$$

Maximum transistor dissipation is found by substituting Eq. (8-40) into (8-38):

$$P_{D(\max)} = \frac{1}{\pi^2}\frac{V_{CC}^2}{R_L} = 0.1\frac{V_{CC}^2}{R_L} \tag{8-41}$$

LOAD RESISTOR POWER.  Load $R_L$ is driven by a half-sine wave $V_{op} \sin t$ from 0 to $\pi$ and the average power is

$$P_o = \frac{V_{op}^2}{4R_L} \tag{8-42}$$

Maximum output occurs when $V_{op} = V_{CC}$ and is

$$P_{o(\text{max})} = \frac{V_{CC}^2}{4R_L} \tag{8-43}$$

EFFICIENCY. Efficiency is the ratio of Eq. (8-42) to (8-35), or

$$\eta = \frac{P_o}{P_s} = \frac{\pi V_{op}}{4V_{CC}}(100\%) \tag{8-44}$$

At maximum output power $V_{op} = V_{CC}$ and the *maximum* efficiency is found from Eq. (8-44) to be $\eta_{(\text{max})} = 100\pi/4 = 78\%$. Finally, the ratio of Eq. (8-43) to Eq. (8-41) gives the maximum power output available from a transistor with known maximum dissipation rating $P_D$ as

$$\frac{P_{o(\text{max})}}{P_{D(\text{max})}} = \frac{\pi^2}{4} = 2.5$$

Thus a 5-W transistor could give 12.5 W output under class $B$.

*Example 8-9.* For the class B circuit in Fig. 8-21, plot output, supply, and transistor power versus peak output voltage for $0 < V_{op} < V_{CC}$.

*Solution.* From Fig. 8-21, Eq. (8-35) is expressed by

$$P_S = \frac{24}{8\pi} V_{op} = \frac{3}{\pi} V_{op}$$

This supply power curve is a straight line with slope of $3/\pi$, from the origin at $P_s = 0$, $V_{op} = 0$ to a maximum value at $V_{op} = 24$ V, $P_s = 22.9$ W. From Eq. (8-42),

$$P_o = \frac{V_{op}^2}{32}$$

This output power curve is parabolic, beginning at the origin and terminating at $V_{op} = V_{CC} = 24$ V, $P_{o(\text{max})} = 18$ W. From Eq. (8-39), we subtract the $P_o$ curve from the $P_s$ curve to obtain $P_D$ versus $V_{op}$. The dissipation curve begins at the origin and goes to a maximum value at $V_{op} = 0.636 V_{CC} = 15.3$ V, with a magnitude of $0.1 V_{CC}^2/R_L = 7.2$ W. At maximum output the value of $P_D$ is found from Eq. (8-38):

$$P_D = \frac{V_{CC}^2}{\pi R_L} - \frac{V_{CC}^2}{4R_L} = 0.068 \frac{V_{CC}^2}{R_L} = 4.9 \text{ W} \quad \text{at } V_{op} = V_{CC}$$

Power curves are shown in Fig. 8-22.

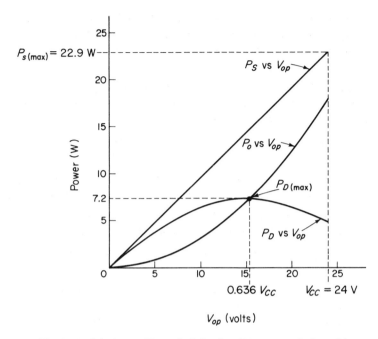

**Fig. 8-22.** Solution to Example 8-9: class B power variation with peak output voltage.

## PROBLEMS

8-1  Manufacturer's data is given for a TO-18 case as $T_{j(max)} = 200\,°C$, $P_{D(max)} = 1.2$ W at 25 °C case temperature. $P_{D(max)} = 0.36$ W at 25 °C ambient temperature (a) Sketch the dissipation derating curves for both ambient and case temperatures. (b) Find $\theta_{JC}$, $\theta_{JA}$, and $\theta_{CA}$.

8-2  Thermal power ratings for a junction diode are $P_{d(max)} = 25$ W. $T_{j(max)} = 150\,°C$, $\theta_{JC} = 1.0\,°C/W$, and $\theta_{CS} = 0.6\,°C/W$. If $T_A = 50\,°C$, find the maximum $\theta_{SA}$ requirement for the heat sink.

8-3  Given $P_{d(max)} = 8$ W at $T_A = 20\,°C$ with $\theta_{JA} = 10\,°C/W$. If $T_{j(max)} = 100\,°C$, what is the transistor's maximum power dissipation when operated in an ambient of 75 °C?

8-4  Given $P_{d(max)} = 15$ W at $T_C = 25\,°C$ and $T_{j(max)} = 100\,°C$. (a) Find $\theta_{JC}$. (b) If a heat sink is added to yield $\theta_{JA} = 15\,°C/W$, what is the maximum power that can be dissipated by the transistor in an ambient of 25 °C?

8-5  In Example 8-5, plot $P_{d(max)}$ for $T_A = 50\,°C$.

8–6 A sinusoidal current with peak value of 1 A is superimposed on a dc component of 1 A through a 10-$\Omega$ resistor. Find (a) average power dissipation, (b) maximum and minimum instantaneous power dissipation.

8–7 In Fig. 8-11, an operating point is located at $V_{CE} = 7.5$ V, $I_C = 2.5$ mA. Find the average power in the transistor, load, and supply. Repeat for $V_{CE} = 2.5$ V, $I_C = 7.5$ mA.

8–8 $V_{CC}$ is halved to 12 V in Fig. 8-12. Calculate supply, load, and transistor power at no signal and at maximum output.

8–9 Show the ratio of $V_{op}$ to $V_{CC}$ is 0.353 in a class $A$ power amplifier when the amplifier delivers one-half of the maximum power output.

8–10 If an 8-$\Omega$ speaker is coupled capacitively to the transistor's collector in Fig. 8-12, show that the maximum efficiency is 6.25%.

8–11 Verify by calculations each point on the curves of Fig. 8-13 located at $V_{op} = 9$ V.

8–12 A transistor can dissipate 10-W maximum. What is the maximum signal power output obtainable and supply power required for class $A$ operation (a) directly coupled; (b) transformer-coupled?

8–13 Verify by calculations each point on the curves of Fig. 8-15 located at $V_{op} = 16$ V.

8–14 What changes are required in the solution of Design Example 8-8 if it is required to deliver 4 W to an 8-$\Omega$ speaker?

8–15 Revise Design Example 8-8 for $V_{CC} = 12$ V.

8–16 What is (a) the maximum output power available from the circuit of Fig. 8-16; (b) the supply power; (c) the operating-point location? Assume ideal transformer $T_2$ with 2 : 1 turns ratio.

8–17 In Example 8-9, calculate power levels at $V_{op} = V_{CC}/\pi$ and verify your results from Fig. 8-22.

8–18 Calculate power levels and efficiency in Example 8-9 for $V_{op} = V_{CC}/2$.

8–19 A 12-V battery is rated at 6 W/hr and supplies a portable radio that should deliver a maximum of 500 mW to a speaker. Neglect current drain except at the output stage and assume that the battery delivers full voltage during its lifetime. Show that the battery will last for 3 hr at full volume with direct-coupling class $A$ and for 6 hr with transformer-coupling class $A$.

# Chapter 9

- **9-0** INTRODUCTION ...............................269
- **9-1** THE BASIC COMPLEMENTARY AMPLIFIER ............269
- **9-2** BIASING THE COMPLEMENTARY AMPLIFIER ..........270
- **9-3** QUALITATIVE ANALYSIS OF THE COMPLEMENTARY AMPLIFIER ..................................273
- **9-4** POWER RELATIONSHIPS ........................274
- **9-5** QUANTITATIVE ANALYSIS OF THE COMPLEMENTARY OUTPUT STAGE .............................276
- **9-6** BOOTSTRAPPING THE COMPLEMENTARY OUTPUT STAGE ....................................279
- **9-7** THE DRIVER STAGE ..........................282
- **9-8** DESIGN EXAMPLE FOR A MULTISTAGE AUDIO AMPLIFIER ..................................285
- **9-9** OPERATION OF A QUASICOMPLEMENTARY AMPLIFIER ..290
- **9-10** A PRACTICAL QUASICOMPLEMENTARY MULTISTAGE AMPLIFIER ..................................293
- PROBLEMS ..................................296

# Complementary and Multistage Operation

## 9-0 Introduction

One prime advantage of BJTs over vacuum tubes is the semiconductor's ability to conduct either holes or electrons. This advantage is exploited by connecting an *npn* and a *pnp* power transistor in a class *B complementary* output arrangement to drive loads that require high current and low voltage. A speaker and power output stage will be connected to illustrate the principles involved.

Signal sources inherently develop very small output voltages. We shall encounter several problems when trying to amplify the source signal to a magnitude sufficient to drive a speaker load. From their solutions it will be seen that we merely provide necessary amplification with one stage and use the transistor's impedance-transforming properties to couple amplification and output stages together.

## 9-1 The Basic Complementary Power Amplifier

In Fig. 9-1, a rudimentary complementary output stage illustrates how input signal $E_i$ turns on $Q_3$ and turns off $Q_4$ for the polarities shown. When $E_i$ reverses in polarity, $Q_3$ turns off and $Q_4$ conducts. $E_i$ must rise to 0.6 V before $Q_3$ conducts and since $V_o$ does not change during this interval, *crossover* distortion is introduced. Likewise $E_i$ must rise to $-0.2$ V before $Q_4$ conducts so that crossover distortion results during both half-cycles.

From Fig. 9-1 we see that when $E_i = 0$, capacitor $C$ is charged to $V_{cc}/2$. Thus when $Q_3$ conducts, its collector-emitter sees a load of $R_L$ and a supply voltage of $V_{cc} - V_{cc}/2 = V_{cc}/2$. When $Q_3$ is off, $Q_4$ conducts and sees the capacitor's voltage as a supply voltage with a load of $R_L$. Thus $C$ should be large (typically 500 $\mu$F) to maintain a reasonably steady supply voltage.

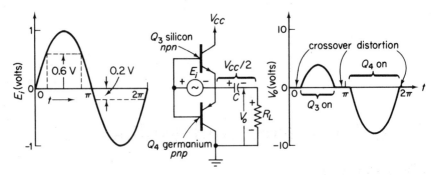

Fig. 9-1. Rudimentary complementary stage with crossover distortion.

To eliminate crossover distortion, two biasing batteries are added in Fig. 9-2 to place $Q_3$ and $Q_4$ on the verge of conduction. When $E_i$ goes positive as shown, $Q_3$ immediately conducts and $Q_4$ becomes reverse biased.

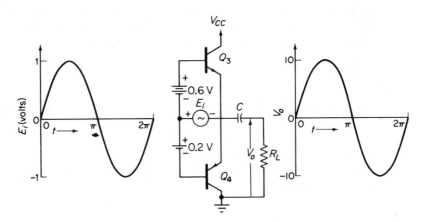

Fig. 9-2. Biasing batteries eliminate crossover distortion in the complementary amplifier.

## 9-2 Biasing the Complementary Amplifier

Biasing with batteries is undesirable and we employ the voltage-divider arrangement in Fig. 9-3 to introduce the fundamental concepts of a practical biasing arrangement. First, resistor $R_2$ develops a voltage drop to forward bias $Q_3$ and $Q_4$ so that a small *idle* current flows though the collectors as insurance against crossover distortion.

Second, since emitter currents of $Q_3$ and $Q_4$ are equal, their dc collector currents must be nearly equal despite differences in $\beta_F$. We reason that $I_{E3} = I_{E4}$, so

Sec. 9-2        Biasing the Complementary Amplifier        271

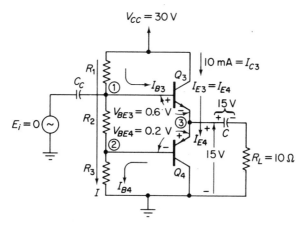

Fig. 9-3. Biasing circuit for the complementary amplifier.

$$I_{C3} + I_{B3} = I_{C4} + I_{B4} \quad \text{or} \quad I_{C3} = I_{C4} + I_{B4} - I_{B3}$$

and collector currents can differ only by the small difference in base current. Third, point 3 must be biased to $V_{CC}/2 = 15$ V to allow a maximum output signal swing. Finally, bleeder current $I$ is chosen to be at least 10 times larger than $I_{B3}$ or $I_{B4}$ so that we can neglect the base current in calculations for bias resistance.

The student might be tempted to try a graphical approach to find $V_B$, such that idling current $I_{CQ3} = I_{CQ4}$ equals some small value of 5–25 mA. There are two practical objections to this approach. First, one cannot read accurately $I_{CQ} = 5$ mA on the usual published data curve of $I_C$ versus $V_{BE}$. A value of 5 mA for a power transistor is often smaller than the thickness of one of the grid lines. Second, we can photograph characteristics on a curve plotter. However, we must display $I_C$ versus $V_{BE}$ over a large current swing region, heating the transistor considerably so that the curves are plotted at elevated temperatures. But in class B or AB operation, the transistors usually run cooler than they do on the curve tracer. $V_{BE}$ will be higher in a cooler transistor than it will be for a hot transistor by 2 mV/°C for the same collector current. After considering these factors, we concluded that we are on firm ground if we select $V_{BE} = 0.6$ V for a silicon transistor and 0.2 V for a germanium transistor, then make minor adjustments in the circuit to our final selection for $R_2$ to give the desired idle current.

An alternate biasing method is to substitute one or more forward-biased diodes for $R_2$, or a combination of diode(s) and a resistor to generate $V_B$. Placing the diodes in the same thermal environment as $Q_3$ and $Q_4$ ensures that the variation of $V_{BE}$ with temperature for both diodes and transistors will track one another and hold a constant idle current. For example, a temperature rise would cause $V_{BE}$ of each transistor to decrease, causing a rise in base bias current. But if the diode's $V_{BE}$ also decreases it lowers the base bias current back to its original value.

*Example 9-1.* Given $\beta_{F3} = \beta_{F4} = 100$, $V_{BE3} = 0.6$ V, and $V_{BE2} = 0.2$ V in Fig. 9-3. (a) What bias potentials are required with respect to ground at points 3, 1, and 2? (b) Calculate the minimum bleeder current to give a collector idle current of 10 mA.

*Solution.* (a) Point 3 must be at $V_{CC}/2 = 15$ V. Point 1 must be higher by $V_{BE3}$ at 15.6 V, and point 2 must be lower by 0.2 V at 14.8 V. (b) Both base currents will be equal at 10 mA/100 = 0.1 mA and bleeder current $I$ should be equal to or greater than 10(0.1) = 1 mA.

Voltage across $R_2$ must be made equal to $V_{BE3} + V_{BE4} = 0.8$ V. Thus from Example 9-1 we can pick any value for $I$ greater than 1 mA, and solve for $R_2$ from

$$R_2 = \frac{V_{BE3} + V_{BE4}}{I} \tag{9-1}$$

Since $I \gg I_{B3} = I_{B4}$, $R_3$ and $R_1$ are calculated from

$$R_3 = \frac{\frac{V_{CC}}{2} - V_{BE4}}{I} \tag{9-2}$$

$$R_1 = \frac{\frac{V_{CC}}{2} - V_{BE3}}{I} \tag{9-3}$$

As will be seen in Section 9-3, the smallest possible value of $R_2$ is desired. But as $R_2$ is made small, bleeder current must increase. So we must pay for the smaller value of $R_2$ in terms of power loss. To have a record of

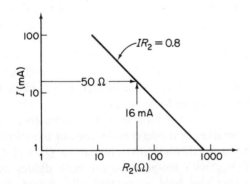

Fig. 9-4. Bleeder current versus choice of $R_2$.

Sec. 9-3   Qualitative Analysis of the Complementary Amplifier   273

the tradeoff between $I$ and $R_2$ we plot Eq. (9-1) in Fig. 9-4 for selections of $I$ between 1 mA and 100 mA.

**Example 9-2.** Arbitrarily choose $R_2 = 50\ \Omega$ and solve for $I$, $R_1$, and $R_3$. $R_2$ can then be trimmed to give the idle current of Fig. 9-1.

**Solution.** From Eq. (9-1), $I = 0.8/50 = 16$ mA. From Eq. (9-2), $R_3 = 14.8/0.016 = 925\ \Omega$. From Eq. (9-3), $R_1 = 14.4/0.016 = 900\ \Omega$.

SUMMARY AND RESTATEMENT. The dc voltage drop across $R_1$ determines collector dc idle current. $V_{CC}$ and the sum of $R_1$, $R_2$, and $R_3$ (practically $R_1 + R_3$) determine bleeder current $I$. DC voltage to ground at point 3 is established by the voltage to ground at point 1. This last point is now clear but may become obscure when we replace $R_1$ with a driver transistor.

## 9-3 Qualitative Analysis of the Complementary Amplifier

Dc bias voltages are enclosed in rectangles in Fig. 9-5(a) to agree with Examples 9-1 and 9-2. Assume $E_i$ goes positive by an increment of $\Delta E_i = 10$ V as shown. Point 1 will go positive by 10 V to 25.6 V. The 10-V increment divides between $R_2$ and $R_3$ with 0.5 V and 9.5 V, respectively, to increase the drop across $R_2$ to 1.3 V. Base-emitter *incremental resistance* of $Q_3$ and $Q_4$ is negligible so the entire 10-V increment is developed across $R_L$. Observe that both $Q_3$ and $Q_4$ are heavily forward biased. $Q_4$ must be shut off during positive excursions of $E_i$ or it will shunt current from $Q_3$ and $V_{CC}$ away from $R_L$.

Introducing a small resistance of 0.47–1 $\Omega$ in series with each emitter will force the voltage drop between points 1 and 3 to increase faster than the voltage drop across $R_2$. The emitter resistance also tends to prevent thermal runaway and may take the form of a fuse.

In Fig. 9-5(b) the same 10-V increment of $E_i$ causes the same voltage changes at points 1 and 2. $Q_3$ must conduct because its base goes positive and $\Delta E_i$ divides between $R_{E3}$ and $R_L$ according to their values as

$$\Delta V_{RE3} = \frac{1}{1+10}(10) = 0.9\ \text{V}, \qquad \Delta V_o = \frac{10}{1+10}(1) = 9.1\ \text{V}$$

Since base-emitter resistance of $Q_3$ is negligible $V_{BE3}$ does not increase appreciably. The dc drop across $R_{E3}$ with no signal is due to the idle current of 10 mA or 0.01 V and does not change the bias conditons. However $Q_4$ is now reverse biased by 0.2 V at the peak of $E_i$ and $Q_4$ is definitely off during the entire positive input signal. A similar analysis will show that $Q_3$ will turn off during a negative input signal because of $R_{E4}$.

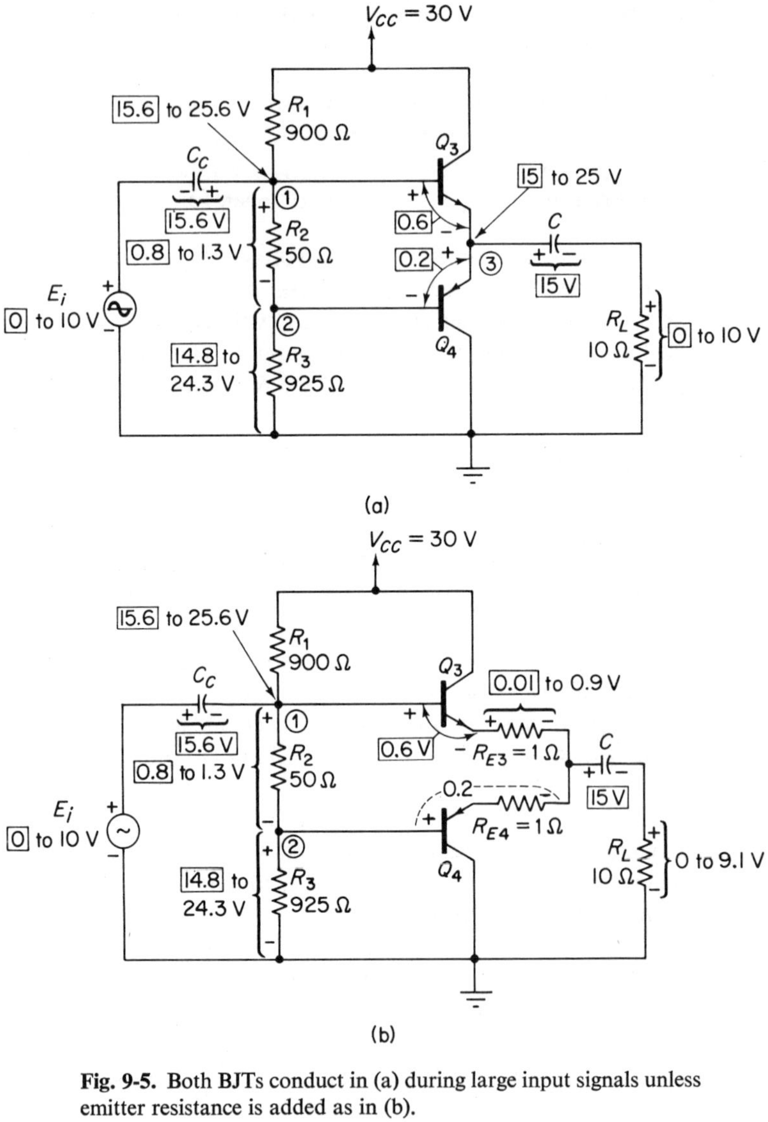

**Fig. 9-5.** Both BJTs conduct in (a) during large input signals unless emitter resistance is added as in (b).

## 9-4 Power Relationships

Power output can be calculated from Fig. 9-5(a). $R_1$ will be replaced eventually by a transistor so that point 1 can be driven from $V_{CC}$ to ground. Point 3 can then be driven to a maximum voltage of $V_{CC} - V_{BE3}$ and a minimum voltage of $0 + V_{BE4}$ for a maximum peak-to-peak swing $V_{p/p}$ of $V_{CC} -$

$V_{BE3} - V_{BE4}$. If we include $R_E$ the peak-to-peak swing across $R_L$ will be reduced to $V_{p/p}R_L/(R_E + R_L)$ so that maximum rms output power is

$$P_{o(\max)} = \left[\frac{V_{p/p}}{2\sqrt{2}}\frac{R_L}{(R_E + R_L)}\right]^2 \frac{1}{R_L} \tag{9-4}$$

where $V_{p/p} = V_{CC} - V_{BE3} - V_{BE4}$. Neglecting $V_{BE}$ and $R_E$ simplifies Eq. (9-4) to

$$P_{o(\max)} \simeq \frac{V_{CC}^2}{8R_L} \tag{9-5}$$

The class B power relationships in Section 8-11 may be modified to apply here if we note that (1) twice as much power is delivered to the load because both half-cycles are used and (2) peak output voltage $V_{op}$ is approximately equal to $V_{CC}/2$. From Eqs. (8-35) and (8-36),

$$P_S = \frac{V_{CC}V_{op}}{\pi R_L} \tag{9-6a}$$

$$P_{S(\max)} = \frac{V_{CC}^2}{2\pi R_L} \tag{9-6b}$$

From Eqs. (8-42) and (8-43),

$$P_o = \frac{V_{op}^2}{2R_L} \tag{9-7a}$$

$$P_{o(\max)} = \frac{V_{CC}^2}{8R_L} \tag{9-7b}$$

From Eq. (8-38),

$$P_D = \frac{V_{op}V_{CC}}{2\pi R_L} - \frac{V_{op}^2}{4R_L} \tag{9-8a}$$

From Eq. (8-40), maximum transistor dissipation occurs at $V_{op} = V_{CC}/\pi$ and from Eqs. (8-41) and (9-8a) is

$$P_{D(\max)} = \frac{V_{CC}^2}{4\pi^2 R_L} \tag{9-8b}$$

Efficiency is the ratio of Eq. (9-7a) to (9-6a):

$$\eta = \frac{\pi V_{op}}{2V_{CC}} \tag{9-9}$$

Efficiency is greatest at maximum output where $V_{op} = V_{CC}/2$ and $\eta_{(\max)} = 78\%$. Finally, the ratio of maximum output to maximum dissipation is found from Eqs. (9-7b) and (9-8b) to be

$$\frac{P_{o(\max)}}{P_{D(\max)}} = \frac{\pi^2}{2} = 5 \qquad (9\text{-}10)$$

*Example 9-3.* Neglecting $R_{E3}$ and $V_{BE}$, what is (a) the maximum output power available from the circuit of Fig. 9-5; (b) Supply power at maximum output; (c) maximum transistor dissipation?

*Solution.*

(a) From Eq. (9-7b), $\quad P_{o(\max)} = \dfrac{900}{80} = 11.5 \text{ W}$

(b) From Eq. (9-6b),

$$P_{s(\max)} = \frac{900}{20\pi} = 14.3 \text{ W}$$

(c) From Eq. (9-8b),

$$P_{D(\max)} = \frac{900}{40\pi^2} = 2.25 \text{ W}$$

## 9-5 Quantitative Analysis of the Complementary Output Stage

Collector characteristics of a power transistor are given in Fig. 9-6. To use this transistor for $Q_3$ in the examples thus far, we do not have to know $\beta_F$ at operating point $Q$. As already stated, we shall trim $R_2$ or a diode-resistor combination to give a collector idle current of $I_{CQ} = 10$ mA. From Fig. 9-5 we see that $V_{CEQ}$ will be approximately equal to 15 V and plot the

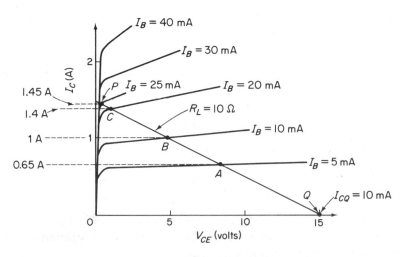

**Fig. 9-6.** Calculation of $\beta_F$ for large-signal operation.

$Q$ point accordingly. Furthermore, due to the ampere current scale, the $Q$ point's location will not change much even if $I_{CQ}$ is raised to 25 or 50 mA.

Neglecting $R_E$ for simplicity we plot the ac load line for $R = 10\,\Omega$ through $Q$. If the transistor is driven by a peak base current signal of $I_{bp} = 5$ mA, the operating point will rise to point $A$, where $I_{cp} = 0.65$ A and $\beta_F = 650/5 = 130$. However, if $I_{bp}$ drives $Q_3$ to the verge of saturation at point $P$ then $\beta_F$ is reduced to $\beta_F = 1450/25 = 58$. It is therefore apparent that we should select that value of $\beta_F$ corresponding to maximum expected output. This principle is stated succinctly as "taking data where its at." Since we normally are most concerned with what happens at full output we would work with $\beta_F = 58$. On manufacturer's data sheets, look for typical values of $h_{FE}$ or $\beta_F$ (dc or large-signal forward current gain) at a low value of $V_{CE}$ and high value of $I_C$.

Signal gain and resistance levels are analyzed in two phases: (1) $Q_3$ on and $Q_4$ off for positive input signals; and (2) $Q_3$ off and $Q_4$ on for negative input signals. $R_1$ of Fig. 9-5 will be replaced by a driver transistor so we neglect it in the large-signal model for positive inputs in Fig. 9-7. Values correspond

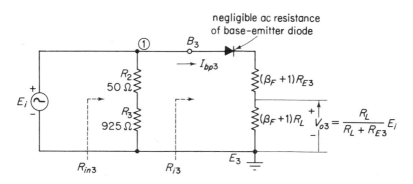

Fig. 9-7. Large-signal model of Fig. 9-5(b), $Q_3$ and $Q_4$ off.

to those developed in our examples. $R_E$ and $R_L$ are transformed into the base leg so that transistor input resistance is

$$R_{i3} = (\beta_F + 1)(R_{E3} + R_L) \tag{9-11}$$

Resistance presented to $E_i$ is $R_{in3}$ or

$$R_{in3} = (R_2 + R_3) \| R_{i3} \tag{9-12}$$

$V_{o3}$ equals $I_{bp3}$ times $(\beta_F + 1)R_L$ and $E_i$ equals $I_{bp3}$ times $R_{i3}$ so voltage gain $A_{V3}$ is

$$A_{V3} = \frac{V_{o3}}{E_i} = \frac{R_L}{R_L + R_{E3}} \tag{9-13}$$

For negative-going inputs, $Q_4$ is on and $Q_3$ is off so we employ the large-signal model in Fig. 9-8 to see that $R_{i3} = R_{i4}$, if $\beta_F$ of the transistors

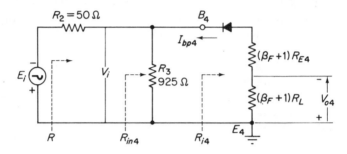

**Fig. 9-8.** Large-signal model of Fig. 9-5(b), $Q_3$ and $Q_4$ on.

are equal. Just as in Fig. 9-7 and Eq. (9-13) we see that because $R_{E3} = R_{E4}$

$$\frac{V_{o4}}{V_i} = \frac{R_L}{R_L + R_{E4}} = A_{V3} \tag{9-14}$$

$E_i$ divides between $R_2$ and $R_{in4}$ so that

$$\frac{V_i}{E_i} = \frac{R_{in4}}{R_2 + R_{in4}} \tag{9-15}$$

where $R_{in4} = R_3 \| R_{i4}$. Multiplying Eqs. (9-14) and (9-15) gives voltage gain

$$\frac{V_{o4}}{E_i} = \frac{V_{o4}}{V_i}\frac{V_i}{E_i} = \frac{R_{in4}}{R_2 + R_{in4}} A_{V3} \tag{9-16}$$

*Example 9-4.* Assume $E_{ip} = 10$ V in the circuit of Fig. 9-5. For both positive and negative half-cycles, find (a) input resistance presented to $E_i$ and (b) $V_{op}$. Assume $\beta_F = 60$ for both transistors.

*Solution.* (a) *Positive-going $E_i$:* From Eq. (9-11), $R_{i3} = (61)(11) = 671 \,\Omega$. From Eq. (9-12), $R_{in3} = 975 \| 671 = 398 \,\Omega$.
*Negative-going $E_i$:* From Eq. (9-15) and Fig. 9-8,

$$R_{in4} = 925 \| 671 = 388 \,\Omega \qquad R = R_2 + R_{in4} = 50 + 388 = 438 \,\Omega$$

(b) *Positive-going $E_i$:* From Eq. (9-13),

$$V_{o3p} = \frac{10(10)}{11} = 9.1 \text{ V}$$

*Negative-going $E_i$:* From Eq. (9-16),

$$V_{o4p} = \frac{388}{438}\left(\frac{10}{11}\right)(10) = 8.0 \text{ V}$$

Sec. 9-6   Bootstrapping the Complementary Output Stage   279

From Example 9-4 we see that $R_2$ must be minimized to prevent serious differences between positive- and negative-going output voltages, with the resulting distortion. Replacing $R_2$ with a diode, wherever possible, is desirable not only for operating point stability but also because the low incremental resistance of the diode minimizes differences between $V_{o3}$ and $V_{o4}$.

*Example 9-5.* In Example 9-4 and Fig. 9-7, find the (a) peak signal base and load current, and (b) peak shunt current $I_{sp}$ through $R_2$ and $R_3$.

*Solution.*

(a) $I_{bp} = \dfrac{E_{ip}}{(\beta_F + 1)(R_E + R_L)} = \dfrac{10}{(61)(11)} = 14.9$ mA

Peak load current is $(\beta_F + 1)I_{bp} = (61)(14.9) = 910$ mA.

(b) $\qquad I_{sp} = \dfrac{E_{ip}}{R_2 + R_3} = \dfrac{10}{975} = 10.2$ mA

Signal-current loss through the input shunt of $R_2 + R_3$ in Example 9-5 is intolerable and will be reduced by bootstrapping in Section 9-6. A loss of almost the same magnitude occurs during the negative-going input signal.

## 9-6  Bootstrapping the Complementary Output Stage

The basic complementary amplifier of Fig. 9-5(b) is bootstrapped by transferring the bottom connection of $R_3$ to the junction of $C$ and $R_L$ as in Fig. 9-9. Decreasing $R_3$ by $R_L = 10\,\Omega$ ensures that bias conditions are unchanged because of the bootstrap connection. Load $R_L$ does carry the small bleeder current but ordinarily this is not a disadvantage.

For positive inputs we derive input resistance beginning with the model in Fig. 9-10(a), where $R_1$ is eliminated and $r_{\pi 3}$ is negligible. $R_E$ is transformed

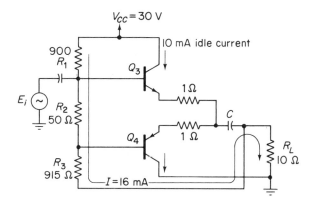

Fig. 9-9. Bootstrapping the complementary amplifier.

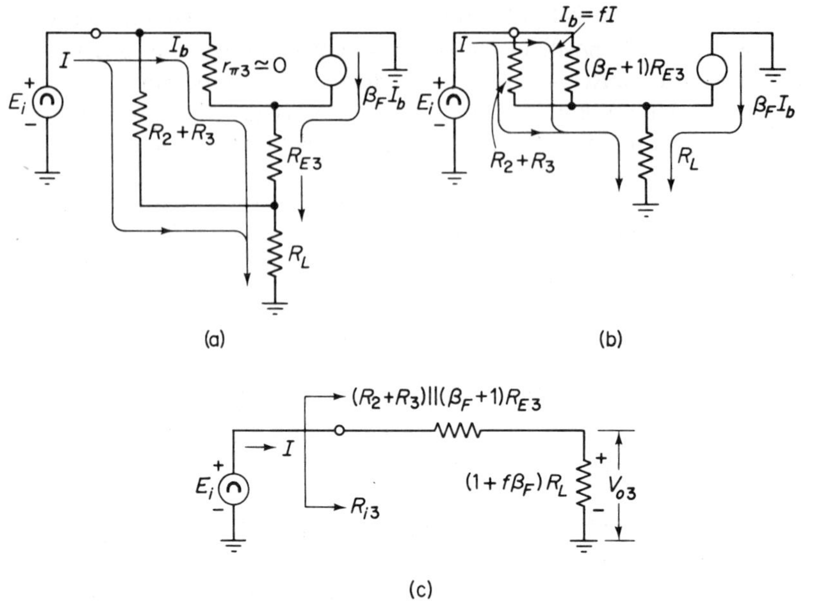

280     *Complementary and Multistage Operation*     Chap. 9

**Fig. 9-10.** Positive input model of Fig. 9-9 in (a) for $Q_3$ on and $Q_4$ off. $R_{E3}$ is transformed into the base leg in (b) to develop an equivalent model in (c).

into the base leg in Fig. 9-10(b) to show the current divider $f$ that expresses $I_b$ in terms of $I$, where

$$I_b = fI \tag{9-17}$$

and

$$f = \frac{R_2 + R_3}{R_2 + R_3 + (\beta_F + 1)R_{E3}}$$

Writing the input loop equation

$$E_i = I[(R_2 + R_3) \| (\beta_F + 1)R_{E3}] + \beta_F I_b R_L + IR_L$$

and substituting for $I_b$ from Eq. (9-17) yields input resistance

$$R_{i3} = \frac{E_i}{I} = (R_2 + R_3) \| (\beta_E + 1)R_{E3} + R_L(1 + f\beta_F) \tag{9-18}$$

Since $V_{o3} = (I + \beta_F I_b)R_L = I(1 + f\beta_F)R_L$, substitute for $I$ from Eq. (9-18) to get voltage gain as shown in Fig. 9-10(c):

$$\frac{V_{o3}}{E_i} = \frac{(1 + f\beta_F)R_L}{R_{i3}} \cong 1 \tag{9-19}$$

## Sec. 9-6    Bootstrapping the Complementary Output Stage    281

For negative inputs we transform $R_{E4}$ into the base leg as in Fig. 9-11(a) to develop the current division of $I_b$ and express input resistance $R_{i4}$ as shown in Fig. 9-11(b) by

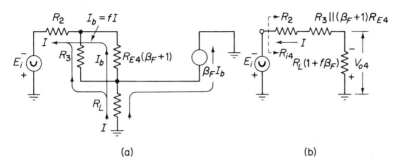

**Fig. 9-11.** Negative input model of Fig. 9-9 in (a) simplified equivalent model in (b).

$$R_{i4} = R_2 + R_3 \| (\beta_F + 1)R_{E4} + R_L(1 + f\beta_F) \qquad (9\text{-}20)$$

where

$$f = \frac{R_3}{R_3 + (\beta_F + 1)R_{E4}}$$

Voltage gain is given by

$$\frac{V_{o4}}{E_i} = \frac{(1 + f\beta_F)R_L}{R_{i4}} \simeq 1 \qquad (9\text{-}21)$$

*Example 9-6.* Calculate the resistance presented to $E_i$ in Fig. 9-9 for both positive and negative inputs. Compare your results with Example 9-4 to see the improvement by bootstrapping. Assume $\beta_{F3} = \beta_{F4} = 60$.

*Solution. Positive inputs:* From Eq. (9-17),

$$f = \frac{965}{965 + (61)1} = 0.95$$

and from Eq. (9-18),

$$R_{i3} = 965 \| 61 + 10[1 + (0.95)60] = 57 + 580 = 637 \ \Omega$$

*Negative inputs:* From Eq. (9-20),

$$f = \frac{915}{915 + 61} = 0.95$$

and

$$R_{i4} = 50 + 915 \| 61 + 10[1 + (0.95)60] = 50 + 57 + 580 = 687 \ \Omega$$

See Table 9-1 for a comparison.

Table 9-1

| Input Resistance | Nonbootstrapped ($\Omega$) | Bootstrapped ($\Omega$) |
|---|---|---|
| Positive inputs | 398 | 637 |
| Negative inputs | 439 | 687 |

We conclude from Example 9-6 that bootstrapping has almost doubled input resistance and that we can approximate bootstrapped input resistance by $(\beta_F + 1)R_L = 610 \; \Omega$.

**Example 9-7.** Assuming $E_{ip} = 10$ V in Example 9-6(a), calculate peak values for $V_{o3}$ and $V_{o4}$ and compare results with Example 9-4(b). (b) Calculate peak current drawn from $E_i$ at its positive peak and compare with Example 9-5. Assume $\beta_{F3} = \beta_{F4} = 60$.

*Solution.* (a) From Eq. (9-19) and Example 9-6,

$$V_{o3p} = \frac{[1 + 0.95(60)]10}{637} = 8.9 \text{ V}$$

From Eq. (9-21),

$$V_{o4p} = \frac{[1 + 0.95(60)]10}{687} = 8.3 \text{ V}$$

Bootstrapping had no significant effect on output voltage.
(b) From Eq. (9-18),

$$I_p = \frac{E_{ip}}{R_{i3}} = \frac{10}{637} = 15.7 \text{ mA}$$

From Example 9-5,

$$I_p = I_{bp} + I_{sp} = 14.9 + 10.2 = 25.1 \text{ mA}$$

Example 9-7 shows that bootstrapping not only had no significant effect on output voltage but reduced the signal current drawn from $E_i$.

## 9-7 The Driver Stage

Even though bootstrapping doubled input resistance of the complementary amplifier, this resistance is much too low to be coupled directly or through a capacitor to a voltage amplifier. Shunting the voltage amplifier's collector resistor with $R_{i3}$ or $R_{i4} \cong 600 \; \Omega$ would eliminate any gain in the voltage amplifier. We therefore replace $R_1$ with a driver transistor in Fig.

Sec. 9-7  The Driver Stage  283

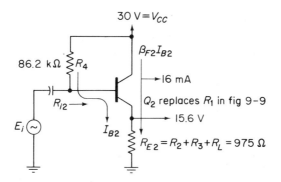

**Fig. 9-12.** $Q_2$ replaces $R_1$ in Fig. 9-9 and the complementary stage is modeled by $R_{E2}$ for dc calculations.

9-12 to function as a common collector and transform the input resistance of the complementary stage to a higher value.

If $Q_2$ is biased for a collector current of 16 mA, its emitter will go to 15.6 V and bias the complementary stage exactly as before. For dc purposes only, we can replace the complementary stage with $R_{E2} = 975$ Ω. The base of $Q_2$ requires a bias current (assuming $\beta_{F2} = 100$) of $I_{B2} = 16$ mA/$100 = 160$ μA. The base of $Q_2$ is 0.6 V above 15.6 V at 16.2 V so that $R_4$ must be

$$R_4 = \frac{30 - 16.2}{160 \times 10^{-6}} = 86.2 \text{ k}\Omega$$

Input resistance of $Q_2$ is shown in Fig. 9-13 by looking into the base

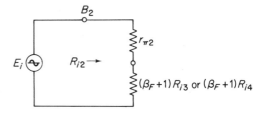

**Fig. 9-13.** AC or incremental model of Fig. 9-12.

terminal. We can neglect $R_4$ because it will be replaced later by the collector resistor of another stage. Validity of the hybrid-pi model is stretched somewhat in this application but $r_{\pi 2}$ will be negligible due to the large value of $R_{E2}$.

**Example 9-3.** Calculate $R_{i2}$ for Fig. 9-12, assuming $\beta_0 = \beta_F = 100$.

**Solution.** Since the emitter of $Q_2$ will be driven positive or negative by signal voltages exceeding 10 V peak, $r_{\pi_2}$ (or, more correctly, $h_{iE}$) will be small in comparison with $R_{i2} \cong (\beta_F + 1)R_{i3} = (101)(637) = 64.4$ kΩ for positive input signals and $R_{i2} = (\beta_F + 1)R_{i4} = (101)(687) = 69.4$ kΩ for negative input signals.

A dc load line for the driver transistor of Fig. 9-12 is drawn in Fig. 9-14. $I_E - V_{CE}$ and $I_C - V_{CE}$ characteristics are almost identical so we use the commonly available $I_C - V_{CE}$ axes. Locate voltage and current axis intercepts and $Q$ point as shown. From Example 9-6 assume an average input resistance of $R_{i34} = 650\ \Omega$ for the bootstrapped complementary stage. This forms the ac load for $Q_2$ and we plot the ac load line through operating point $Q$. The peak value of collector current is limited by cutoff to $I_{CQ} = I_{cp(\max)} = 16$ mA. Thus peak output voltage will be limited to

$$V_{op(\max)} = I_{cp(\max)} R_{i34} = 16(10^{-3})(650) = 10.4\ \text{V}$$

We concluded from this section that the ac load of $Q_2$ should be close in value to the dc load in order to avoid limiting the maximum signal swing. In our examples thus far we have not considered this factor, but we did obtain a reasonable maximum output. A guideline can be established for the selection

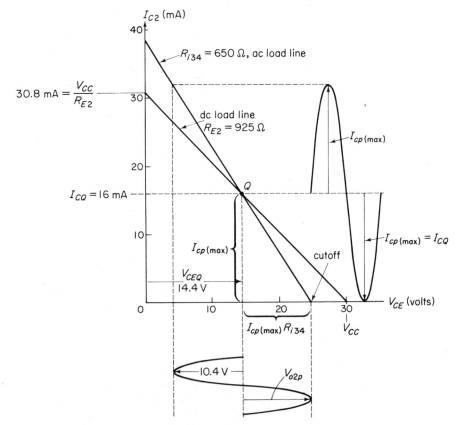

Fig. 9-14. Any difference in ac or dc loads for $Q_2$ reduces maximum output voltage.

Sec. 9-8  Multistage Audio Amplifier  285

of $R_3$ from Fig. 9-9. Ignoring $R_1$, the dc load resistance is $R_2 + R_3 + R_L$. From Fig. 9-11(b) the ac load resistance is given for $R_{i4}$. Set the two load resistances equal to see what $R_3$ must be to make them equal.

dc load = ac load

$$R_2 + R_3 + R_L = R_2 + R_3 \| [R_E(\beta_F + 1)] + R_L + f\beta_F R_L \quad (9\text{-}22)$$

$R_2$ and $R_L$ subtract out. The constant $f$ is close to unity and $R_3 \| [R_E(\beta_F + 1)]$ is small with respect to $f\beta_F R_L$ so that Eq. (9-22) simplifies to

$$R_3 \cong (\beta_F + 1)R_L, \quad \text{where } \beta_F = \beta_{F3} = \beta_{F4} \quad (9\text{-}23)$$

## 9-8 Design Example for a Multistage Audio Amplifier

Design a multistage amplifier that will deliver 5 W to an 8-Ω speaker with an input signal of 0.2 V. Input resistance is to exceed 2 kΩ and the lower break frequency is to be 30 Hz. Assume $\beta_0 = \beta_F = 100$ for all transistors and $V_{CC} = 30$ V.

GENERAL CONSIDERATIONS

1. *Voltage gain.* In order to develop 5 W in $R_L = 8\,\Omega$ we need an rms output voltage of $V_o^2 = PR = 5(8)$ or $V_o = 6.4$ V. Overall voltage gain must be $V_o/E_i = 6.4/0.2 = 32$.
2. *Maximum output voltage swing.* Peak output voltage will be 1.4(6.4) $\cong 9$ V or a peak-to-peak swing of 18 V. The supply voltage of 30 V is more than adequate and will allow the amplifier to be overdriven considerably.
3. *Circuit makeup.* One voltage amplifier consisting of a common emitter with emitter resistance will be directly coupled through a resistance-transforming driver transistor to the complementary output stage. All BJTs will be of silicon with $V_{BE} = 0.6$ V.

COMPLEMENTARY OUTPUT STAGE DESIGN

The junction of $R_{E3}$ and $R_{E4}$ should be biased at $V_{CC}/2 = 15$ V, and assuming $Q_3$ and $Q_4$ are silicon BJTs, their bases should be at the dc potentials shown in Fig. 9-15. From Eq. (9-23), select $R_3 = 101 R_L \cong 800\,\Omega$. Bleeder current $I$ is found from the voltage drop across $R_3$ and $R_L$ to be

$$I = \frac{14.4}{808} \cong 18 \text{ mA}$$

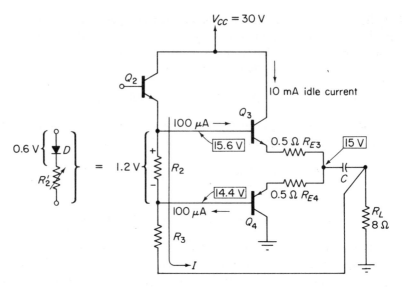

**Fig. 9-15.** Biasing the output stage.

Since $R_2$ must drop 1.2 V, it should be $1.2/0.018 = 66 \, \Omega$. To lower this value we replace $R_2$ with silicon diode D and $R'_2$ as shown. $R'_2$ will be $0.6/0.018 = 33 \, \Omega$ and assuming a diode incremental resistance of $7 \, \Omega$ gives an effective $R_2$ of $40 \, \Omega$. When the circuit is constructed $R'_2$ will be adjusted for the 10 mA idle current.

Input resistance is calculated from Eq. (9-20):

$$f = \frac{800}{800 + (101)0.5} = 0.94, \qquad R_{i4} = 40 + 800 \,||\, 50.5 + 8(95) = 847 \, \Omega$$

Assume $R_{i3} \cong R_{i4} \cong 800 \, \Omega$. Thus ac and dc load lines are approximately equal and bleeder current is large with respect to base current. Voltage gain is found from Eq. (9-21) to be

$$\frac{V_o}{E_i} = \frac{(1+94)8}{800} = 0.95$$

### DRIVER TRANSISTOR DESIGN

In Fig. 9-16(a), the dc biasing requirements for $Q_2$ are established. The base of $Q_2$ must be placed 16.2 V above ground by the preceeding stage and will draw 0.18 mA. Looking into the base of $Q_2$ we see that input resistance of the complementary stage has been transformed up to 80 kΩ. Since $I_{C2}$ is large, at 18 mA, $r_{\pi 2}$ is negligible.

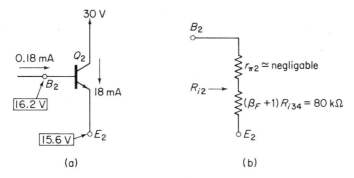

Fig. 9-16. Models of $Q_2$: dc in (a) and ac in (b).

VOLTAGE AMPLIFIER DESIGN

Coupling resistor $R_C$ is selected together with $I_{C_1}$ to put the base of $Q_2$ at 16.2 V and deliver 0.18 mA to $Q_2$. In Fig. 9-17(a) we select $I_{C_1} = 1.80$ mA

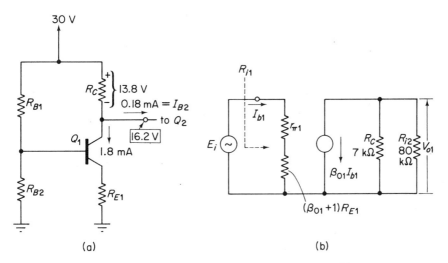

Fig. 9-17. Voltage amplifier in (a) and ac model in (b).

so that current through $R_C$ is $1.80 + 0.18 = 1.98$ mA and $R_C$ is

$$R_C = \frac{30 - 16.2}{(1.8 + 0.18)10^{-3}} = 7 \text{ k}\Omega$$

The ac load on $Q_1$ is $R_L$ and may be evaluated from Fig. 9-17(b), where

$$R_L = R_C \| R_{i2} = 7 \| 80 = 6.4 \text{ k}\Omega$$

## 288 Complementary and Multistage Operation  Chap. 9

Overall voltage gain of 32 is the product of the gain of each stage. Assuming the gain of $Q_2$ is unity,

$$32 = A_{V1}(1)(0.95)$$
$$A_{V1} \cong 34$$

We now select $R_{E1}$ to give the desired $A_{V1}$. Evaluate $g_{m1} = I_{C1}/25 = 1.8/25 = 0.072$ mho, to find $r_{\pi 1} = 100/0.072 = 1390 \, \Omega$.
Solve for $R_{E1}$ from

$$\frac{V_{o1}}{E_i} = \frac{\beta_{01} R_L}{r_{\pi 1} + (\beta_0 + 1)R_{E1}} = 34 = \frac{100(6400)}{1390 + 101 R_{E1}}$$

$$R_{E1} = 172 \, \Omega$$

Find input resistance $R_{i1}$ from Fig. 9-17(b) to be

$$R_{i1} = 1390 + 101(172) = 18.8 \text{ k}\Omega$$

We select $R_B$ to be as low as possible for operating-point stability but high enough to give an *amplifier* input resistance $R_{in}$ of 2 k$\Omega$, where

$$2 \text{ k}\Omega \leq R_{in} = R_B \| R_{i1} = R_B \| 18.8 \text{ k}\Omega$$
$$2.25 \text{ k}\Omega \leq R_B$$

For margin select $R_B \cong 14 R_{E1} = 2500 \, \Omega$ and since $I_{B1}$ must be $I_{C1}/\beta_{F1} = 18 \, \mu$A, we can solve for $V_B$ from the dc input loop of Fig. 9-17(a), or

$$V_B = I_B[R_B + (\beta_F + 1)R_{E1}] + V_{BE1}$$
$$= 18(10^{-6})[2500 + 101(172)] + 0.6 = 0.96$$

Solve for the bias resistors:

$$R_{B1} = \frac{V_{CC}}{V_B} R_B = \frac{30}{0.96}(2500) = 78.1 \text{ k}\Omega, \qquad R_{B2} = \frac{V_{CC}}{V_{CC} - V_B} R_B = 2.58 \text{ k}\Omega$$

The basic design is shown in Fig. 9-18.

TRANSISTOR SELECTION

$Q_1$ and $Q_2$ will exhibit maximum dissipation at no load and should be selected to have power ratings greater than

$$P_{D1} = I_{C1} V_{CE1} = (1.8)(16.2)(10^{-3}) = 30 \text{ mW}$$
$$P_{D2} = I_{C2} V_{CE2} = (18)(14.4)(10^{-3}) = 260 \text{ mW}$$

Sec. 9-8                                         Multistage Audio Amplifier    289

For $Q_3$ and $Q_4$ maximum power dissipation is found from Eq. (8-40) to occur at $V_{op} = 30/\pi = 9.5$ V or near full output. Maximum power dissipation is found from Eq. (9-8b) to be (neglecting $R_{E3}$ and $R_{E4}$)

$$P_{D(\max)} = \frac{30(30)}{4\pi 28} = 2.8 \text{ W}$$

CAPACITOR SELECTION

Output capacitor $C$ in Fig. 9-18 sees $R_L$ to ground on one side and $R_o$

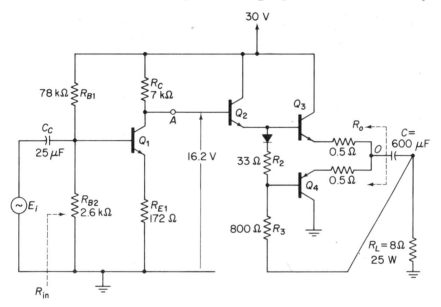

Fig. 9-18. Schematic of completed amplifier.

to an ac ground at $V_{CC}$ on the other. Bootstrapping has little effect on the resistance-transforming properties of $Q_3$ or $Q_4$ in this circuit. Assume $Q_3$ is conducting while $Q_4$ is off and look into point 0 from capacitor $C$. We see (1) 0.5 Ω plus (2) $r_{\pi 3}/(\beta_{F3} + 1)$ which is negligible plus (3) $r_{\pi 2}$ divided by $(\beta_{F2} + 1)(\beta_{F3} + 1)$ which is also negligible, plus (4) $R_C$ divided by $(\beta_{F2} + 1)(\beta_{F3} + 1)$ so that

$$R_o = R_{E3} + \frac{R_C}{(\beta_{F3} + 1)(\beta_{F2} + 1)} = 0.5 + \frac{7000}{(101)(101)} = 1.2 \text{ Ω}$$

Choose $C$ to give a lower break frequency of 30 Hz from

$$2\pi(30) = \omega_L = \frac{1}{C(R_L + R_o)} = \frac{1}{C(8 + 1.2)}, \quad C = 575 \text{ μF}$$

$C_C$ sees $R_{in} = R_B \| R_{i1} = (2.5 \| 18.8)$ k$\Omega$ = 2.2 k$\Omega$ so we choose $C_C$ to cut off a decade lower than $C$ at 3 Hz, or

$$C_C = \frac{1}{2\pi(3)2200} = 24 \ \mu\text{F}$$

PRACTICAL CONSIDERATIONS

When the amplifier is built we must use the nearest commercially available resistors and capacitors. The single most important dc voltage measurement is at point $A$ in Fig. 9-18. All dc potentials to the right of point $A$ depend solely on $A$. We can control potential at point $A$ *only by trimming* $R_{B1}$ or $R_{B2}$. In this design maximum possible peak output voltage is limited by the dc drop across $R_C$ of 13.8 V and may be overdriven with a peak input signal of 13.8/34 = 0.4 V to give a peak output power, before clipping, of approximately

$$P_{o(\max)} = \left(\frac{13.8}{\sqrt{2}}\right)^2 \frac{1}{8} \cong 12 \ \text{W}$$

Finally, emitter resistors $R_{E3}$ and $R_{E4}$ may be eliminated because bootstrapping prevents the increase in voltage across $R_2$ that tended to hold the off-transistor on.

## 9-9 Operation of a Quasicomplementary Amplifier

The quasicomplementary amplifier of Fig. 9-19 was developed originally to circumvent practical difficulties in manufacturing cheap, high-power silicon *pnp* transistors. One low-power driver transistor is employed for each output transistor. $Q_1$ is an *npn* to drive $Q_3$ and $Q_2$ is a *pnp* to drive $Q_4$. Typical maximum power ratings are 1 W and more than 10 W to compare the relative capabilities of driver and output transistors, respectively. Both driver and output transistors are biased for class B operation with minimum idle current to eliminate crossover distortion. Input and output terminals are at ground potential, with no signal, due to symmetry, since bias resistors $R_2 = R'_2$ and $R_1 = R'_1$.

POSITIVE INPUT SIGNALS

To understand operation with large positive input signals we see from Fig. 9-19 that $Q_1 - Q_3$ will conduct and $Q_2 - Q_4$ will turn off so that we can employ the model of Fig. 9-20. Assuming $R_2$ is negligible, $I_{b1}$ causes

Sec. 9-9                Operation of a Quasicomplementary Amplifier    291

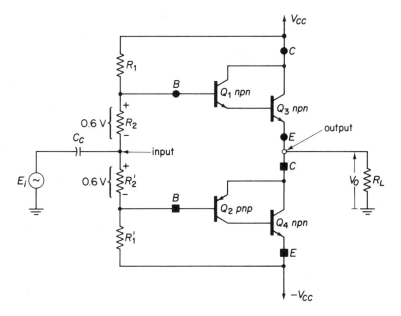

**Fig. 9-19.** Simplified quasicomplementary amplifier.

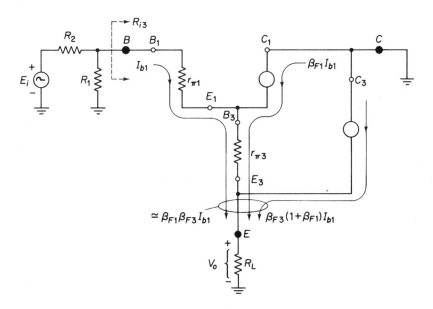

**Fig. 9-20.** Model of Fig. 9-19 for positive inputs.

$\beta_{F1}I_{b1}$ and both these currents flow through $r_{\pi 3}$ to be multiplied by $\beta_{F3}$. Current gain is

$$A_{i3} = \frac{I_o}{I_{in}} = I_b\left[\frac{1 + \beta_{F1} + \beta_{F3}(1 + \beta_{F1})}{I_b}\right] \cong \beta_{F1}\beta_{F3} \quad (9\text{-}24)$$

Input resistance $R_{i3}$ is obtained from Fig. 9-20:

$$R_{i3} = r_{\pi 1} + (\beta_{F1} + 1)[r_{\pi 3} + (\beta_{F3} + 1)R_L] \cong \beta_{F1}\beta_{F3}R_L \quad (9\text{-}25)$$

Voltage gain is found to be unity from

$$A_{V3} = \frac{V_o}{E_i} = \frac{I_o}{I_{in}}\frac{R_L}{R_{i3}} \cong (\beta_{F1}\beta_{F3})\left(\frac{R_L}{\beta_{F1}\beta_{F3}R_L}\right) \cong 1 \quad (9\text{-}26)$$

### SUMMARY

For positive input signals $Q_1 - Q_3$ acts as a composite *npn* transistor, connected as a common collector, with extremely high input resistance and current gain. Terminals of the composite *npn* are identified with solid circles in Figs. 9-20 and 9-19.

### NEGATIVE INPUT SIGNALS

If $E_i$ goes negative in Fig. 9-19, $Q_2 - Q_4$ conducts and the composite *npn* turns off. Biasing resistors are deleted for simplicity in the model of Fig. 9-21, where $I_{b2}$ causes $\beta_{F2}I_{b2}$ which, in turn, causes $\beta_{F4}\beta_{F2}I_{b2}$. By inspection, current gain and input resistance is

$$A_{i4} = \frac{I_{b2}(1 + \beta_{F2} + \beta_{F4}\beta_{F2})}{I_{b2}} \cong \beta_{F2}\beta_{F4} \quad (9\text{-}27)$$

$$R_{i4} = r_{\pi 2} + [\beta_{F2}(1 + \beta_{F4})]R_L \cong \beta_{F2}\beta_{F4}R_L \quad (9\text{-}28)$$

Voltage gain is approximately unity from (neglecting $R'_2$)

$$A_{V4} = \frac{V_o}{E_i} = A_{i4}\frac{R_L}{R_{i4}} \cong 1 \quad (9\text{-}29)$$

From the model of Fig. 9-21, $Q_2$ operates as a common collector. $Q_4$ receives an input signal at its base and output is taken from its collector so that $Q_4$ operates as a common emitter. However, since $V_o$ is in series opposition to $E_i$ there is almost 100% feedback and consequently unity voltage gain.

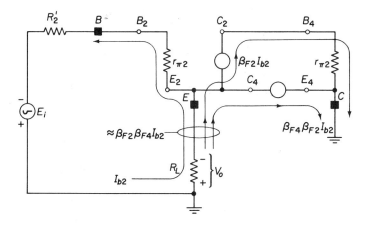

**Fig. 9-21.** Model of Fig. 9-19 for negative inputs.

We conclude that $Q_2 - Q_4$ forms a composite *pnp* transistor, operating as a common collector with extremely high input resistance and current gain. Note that a negative going $E_i$ increases the composite collector current and this is characteristic of a *pnp* transistor. Terminals of the composite *pnp* are identified as solid squares in Figs. 9-19 and 9-21.

Operation of the quasicomplementary amplifier in Fig. 9-19 can now be summarized quite briefly. If $E_i$ goes positive sinusoidally to a peak of 10 V, composite *npn* transistor $Q_1 - Q_3$ develops this voltage swing across $R_L$, less the negligible incremented drop across its composite base-emitter junction. If $E_i$ goes negative sinusoidally to $-10$ V, composite *pnp* transistor $Q_2 - Q_4$ develops this voltage swing across $R_L$.

## 9-10 A Practical Quasicomplementary Multistage Amplifier

A more practical complementary amplifier is shown in Fig. 9-22. Output capacitor $C'$ is added because output terminal 3 is above ground by $V_{CC}/2 = 15$ V. 0.5 Ω resistors are added in series with the composite emitters to ensure the composite transistors turn off on alternate half-cycles.

BIASING. Assuming all transistors are silicon and $\beta_0 = \beta_F = 50$ for each, the required dc voltage levels are shown in squares in Fig. 9-22. $R_2$ is adjusted for a dc idle current in the output collectors of 50 mA. $R_{B1}$ and $R_{B2}$ establishes $I_{BA}$, $I_{CA}$, and consequently the dc voltage levels. $I_{CA} = 2$ mA, and is selected to be large with respect to the 20-$\mu$A base currents of $Q_1$ and $Q_2$. Resistor $R_1$ is calculated to be

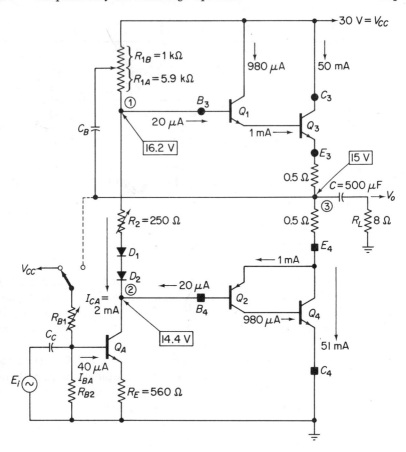

**Fig. 9-22.** A practical quasicomplementary amplifier.

$$R_1 = \frac{30 - 16.2}{(2 + 0.02)10^{-3}} = 6.9 \text{ k}\Omega$$

When $E_i$ goes positive in Fig. 9-22 the collector of $Q_A$ goes negative to develop a signal voltage $V_a$ across the ac input resistance $R_{i34}$ of the quasicomplementary stage. Under this condition the composite pnp $Q_2$-$Q_4$ turns on and the composite npn $Q_{13}$ turns off. The ac model in Fig. 9-23 shows how bootstrap capacitor $C_B$ mitigates against the shunting effect of $R_1$.

We can assume a typical voltage gain of 0.95 for the $Q_2$-$Q_4$ composite common collector in Fig. 9-23. The ac short circuit due to $C_B$ forces the wiper of $R_1$ to stay at $0.95E_i$ so that $0.05E'_i$ appears between point 2 and the wiper of $R_1$. If the wiper is moved up to point 1 then the shunt signal $I_s$ current will be

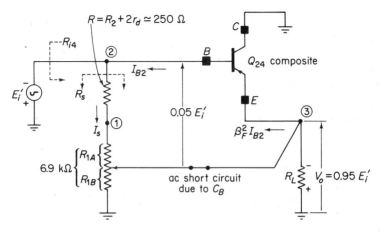

**Fig. 9-23.** Bootstrapping action of $C_B$ for positive going $E_i$.

$$I_s = \frac{0.05 E'_i}{250}, \quad \frac{E'_i}{I_s} = R_s = \frac{250}{0.05} = 5 \text{ k}\Omega$$

Let the wiper be moved down so that $R_{1A} = 1000 \ \Omega$, then

$$\frac{E'_i}{I_s} = R_s = \frac{1250}{0.05} = 25 \text{ k}\Omega$$

Thus the shunt resistance $R_s$ presented to $E'_i$ increases as the wiper is moved down $R_1$. Resistor $R_{1A}$ cannot be made too large or $R_{1B}$ will become comparable to $R_L$ and divert some of the load current. We therefore select $R_{1B}$ to be approximately 100 times $R_L$ or 1 k$\Omega$ and $R_{1A} = 5.9$ k$\Omega$.

The bootstrapping equations will not be developed in detail, but looking from the collector of $Q_A$ toward point 2 in Fig. 9-22 we would see an ac resistance during either half-cycle (assume all transistors have equal $\beta$) of

$$R_{i34} \cong \beta_F^2 R_L = (50)^2(8) = 20 \text{ k}\Omega \quad (9\text{-}30)$$

We can establish the gain of $Q_A$ to be approximately 34 as in the design example of Section 9-8. Evaluate $g_m = 2/25 = 0.08$ mho, $r_\pi = 50/0.08 = 625 \ \Omega$. The ac load is $R_L = R_{i34} = 20$ k$\Omega$ so that

$$\frac{V_o}{E_i} = \frac{\beta_0 R_L}{r_\pi + (\beta_0 + 1) R_E} = 34 = \frac{50(200{,}000)}{625 + 51 R_E}$$
$$R_E = 560 \ \Omega$$

$R_{B1}$ and $R_{B2}$ can be calculated to bias $Q_A$ at 2 mA, but the dashed line shown in Fig. 9-22 is a feedback biasing connection and we can gain operating-point stability through negative feedback by removing the connection from $R_{B1}$ to $V_{CC}$ and making the dashed line connection. Recalculate $R_{B1}$ and $R_{B2}$ to bias $Q_A$ at 2 mA but use the supply voltage at point 3 of 15 V. If the voltage at point 3 increases due to a supply voltage increase, bias and collector current of $Q_A$ increase, increasing the drop across $R_1 + R_2$. This lowers the voltage of point 1, and consequently point 3, stabilizing it at 15 V. Thus point 3 is monitored by $Q_A$ and an adjustment (negative feedback) is made by the circuit to oppose any slow change in voltage.

$C_B$ sees essentially $R_L$ in series with $R_{iB}$ and for a lower cutoff frequency of

$$C_B = \frac{1}{\omega_L(R_{1B} + R_L)} = \frac{1}{10(1008)} = 100 \ \mu F$$

## PROBLEMS

**9-1** In Example 9-1, what changes in the solution occur if $V_{BE2} = 0.6$ V and $\beta_{F1} = \beta_{F2} = 50$?

**9-2** Revise Fig. 9-4 for the case where both complementary output transistors are made of silicon.

**9-3** With $R_2$ retained at 50 in Example 9-2, recalculate $R_3$ and $R_1$ for $V_{BE1} = V_{BE2} = 0.6$ V.

**9-4** What changes occur in the power levels of Example 9-3 when $V_{CC}$ is reduced to 24 V?

**9-5** Given $I_{B3} = 1$ mA, $\beta_{F3} = 49$, and $\beta_{F4} = 99$ in Fig. 9-3. Show that $I_{C4} - I_{C3} = 0.5$ mA.

**9-6** Calculate bleeder current values in Fig. 9-5 for $R_2 = 25, 50$, and $100$ $\Omega$, with corresponding values of $R_1$ and $R_3$.

**9-7** $R_2$ is doubled to 100 $\Omega$ in Example 9-4. (a) Find the new values of peak output voltages. (b) Repeat for $R_2 = 25$ $\Omega$.

**9-8** On one graph plot (a) bleeder current versus $R_2$ and (b) peak output voltages versus $R_2$ from the data of Problems 9-6 and 9-7. (c) Add a plot of $R_{in3}$ and $R_{in4} + R_2 = R$ versus $R_2$. Based on the results make a choice for $R_2$.

**9-9** Retaining the same resistor values in Example 9-6, double $\beta_F$ to 120 for each transistor and find the new values of $R_{i3}$ and $R_{i4}$, with and without bootstrapping.

**9-10** Employing the results of Problem 9-9, how is $V_{o3p}$ and $V_{o4p}$ changed in Example 9-7 with $\beta_F = 120$ for each transistor?

Chap. 9　　　　　　　　　　　　　　　　　　　　　　　　Problems　297

*9–11*　In Fig. 9-12, what effect does doubling $\beta_{F2}$ to 200 have on $R_4$, and $R_{i2}$?

*9–12*　In order to match the ac and dc load lines in Fig. 9-14, assuming $\beta_{F3} = \beta_{F4} = 60$, what would be a good choice for $R_3$? Would $R_2$ have to be recalculated?

*9–13*　Revise the design example in Section 9-8 for a supply voltage of 24 V, retaining the other specifications.

*9–14*　If $Q_2$ were removed in Fig. 9-18, what would the single-stage gain of $Q_1$ be?

*9–15*　Eliminating $Q_2$ in Fig. 9-18 and connecting the collector of $Q_1$ to the base of $Q_3$ would not drastically change the dc bias conditions. What would the voltage gain of $Q_1$ be? Compare the results with Problem 9-14 and Design Example 9-8.

*9–16*　What change in output resistance results in Fig. 9-18 if $R_C$ is doubled to 14 kΩ? How does this change affect the required value of $C$ for $\omega_L = 30$ Hz?

*9–17*　Using Fig. 9-19 for guidance, revise Fig. 9-3 to eliminate output capacitor $C$.

*9–18*　Revise the dc voltage levels in Fig. 9-22 if $V_{CC}$ is changed to 24 V.

*9–19*　When $R_{1A}$ is set to 200 Ω in Fig. 9-23, approximately what resistance is attained by $R_S$?

*9–20*　Calculate values for $R_1$ and $R_2$ in Fig. 9-22 for a supply voltage of 15 V, assuming $R_B = 20R_E = 11.2$ kΩ.

# Chapter 10

| | | |
|---|---|---|
| **10-0** | INTRODUCTION | 299 |
| **10-1** | BASIC DIFFERENTIAL AMPLIFIER CIRUIT ANALYSIS | 300 |
| **10-2** | COMMON-MODE REJECTION RATIO | 304 |
| **10-3** | DIFFERENTIAL OUTPUT VOLTAGE | 308 |
| **10-4** | CONSTANT CURRENT BIASING | 309 |
| **10-5** | PHASE INVERTER | 310 |
| **10-6** | VOLTAGE-GAIN CONTROL | 311 |
| **10-7** | GAIN CONTROL BY OFFSET VOLTAGE | 313 |
| **10-8** | LINEAR MULTIPLIER | 314 |
| **10-9** | AMPLITUDE MODULATION | 316 |
| **10-10** | BALANCED AND SINGLE SIDE BAND MODULATORS | 320 |
| **10-11** | FREQUENCY SHIFTING | 321 |
| **10-12** | THE COMPARATOR | 321 |
| | PROBLEMS | 325 |

# Selected Communication and Control Circuits

## 10-0 Introduction

A balanced differential amplifier is the basic circuit configuration in most of the integrated circuits manufactured for use as audio, intermediate-frequency, video, tuned, and operational amplifiers. While these integrated circuits may differ in that certain characteristics are optimized, they are all related by the fact that their performance in a specific application is determined primarily by the method of connecting external circuit elements.

The basic differential amplifier in Fig. 10-1 has two input terminals and two output terminals. Input sources $E_{g1}$ and $E_{g2}$ must be able to pass the small dc base bias currents. If $E_{g1}$ and $E_{g2}$ cannot pass a dc current, each

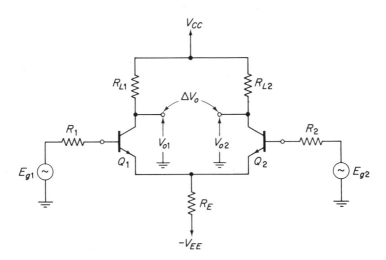

Fig. 10-1. Basic differential amplifier.

must be coupled through a capacitor to a base, and biasing resistors must be added between each base and ground to provide a return for the bias currents. Output voltage may be taken from either collector to ground $V_{o1}$ or $V_{o2}$, or else we may take their difference as the differential output voltage $\Delta V_o$. With respect to the inputs, there are two single-ended voltages $E_{g1}$ and $E_{g2}$ and we could also measure their difference as a differential input voltage $\Delta E_g$. It is possible to compare any one of the three output voltages to any one of the three input voltages, but initially we are interested primarily in $V_{o2}$ and how it depends on $E_{g1}$. Straight circuit analysis is therefore applied in Section 10-1 to study this relationship.

## 10-1 Basic Differential Amplifier Circuit Analysis

A dc model of Fig. 10-1 is given in Fig. 10-2. The circuit is assumed to

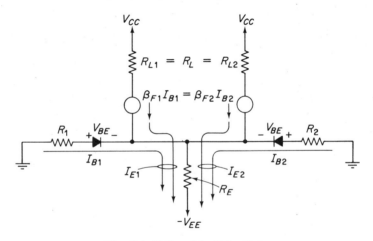

**Fig. 10-2.** DC model of Fig. 10-1.

be made symmetrical so that transistors are matched, $R_2 = R_1 = R$, and $R_{L1} = R_{L2} = R_L$. Thus base, emitter, and collector currents will be equal for each transistor and we can write an identical equation for either input loop in terms of $I_E$:

$$V_{EE} - V_{BE} = I_E\left[2R_E + \frac{R}{\beta_F + 1}\right] \tag{10-1}$$

Normally $R/(\beta_F + 1) \ll 2R_E$.

$$I_E \cong \frac{V_{EE} - V_{BE}}{2R_E} \cong I_C \tag{10-2}$$

Sec. 10-1      Basic Differential Amplifier Circuit Analysis      301

We can conclude that, with respect to ground, both bases are almost at 0 V and both emitters are negative by $V_{BE} \cong 0.6$ V. Most of $V_{EE}$ appears across $R_E$ and voltage drops across the equal load resistors are

$$V_{RL} = I_C R_L \cong \frac{R_L}{2R_E}(V_{EE} - V_{BE})$$

Letting $E_{g2} = 0$, expressions for voltage gain and input resistance are developed by noting that $Q_1$ operates as a common collector and $Q_2$ as a common base. Looking to the right from the emitter of $Q_1$, we see $R_{in2}$, which is $R_E$ in parallel with input resistance $R_{i2}$ of $Q_2$ as shown in Fig. 10-3(a). Then,

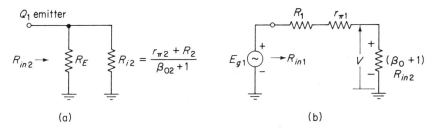

Fig. 10-3. Input resistance of $Q_1$ and $Q_2$ in Fig. 10-1 for $E_{g2} = 0$.

looking into the base of $Q_1$ we see $R_{in1}$ as shown in Fig. 10-3(b), where

$$R_{in1} = R_1 + r_{\pi 1} + (\beta_{01} + 1)R_{in2} \qquad (10\text{--}4a)$$

where

$$R_{in2} = R_E \left\| \frac{r_{\pi 2} + R_2}{\beta_{02} + 1} \right. = R_E \| R_{i2} \qquad (10\text{--}4b)$$

Assuming symmetry so that $r_{\pi 1} = r_{\pi 2} = r_\pi$, $R_1 = R_2 = R$, $\beta_{01} = \beta_{02} = \beta_0$ and assuming that $R_E \gg R_{i2}$ reduces Eq. (10-4) to

$$R_{in1} \cong 2(R + r_\pi) \qquad \text{for } R_E \gg R_{i2} \qquad (10\text{-}5a)$$
$$R_{in2} \cong R_{i2} \qquad \text{for } R_E \gg R_{i2} \qquad (10\text{-}5b)$$

Signal voltage $V$ in Fig. 10-3(b) is the input voltage to $Q_2$ and is expressed exactly by

$$V = E_{g1}\frac{(\beta_{01} + 1)R_{in2}}{R_{in1}} \qquad (10\text{--}6)$$

Substituting the approximations for Eq. (10-5) and employing symmetry yields

$$V = E_{g1} \frac{\frac{(\beta_0 + 1)(r_\pi + R)}{(\beta_0 + 1)}}{2(r_\pi + R)} = \frac{E_{g1}}{2} \tag{10-7}$$

We conclude from Eq. (10-7) that voltage gain of $Q_1$ will always be approximately 0.5 as long as symmetry is maintained and $R_E$ is large.

Voltage gain of $Q_2$ is found from Eq. (5-42) and the model of Fig. 10-4:

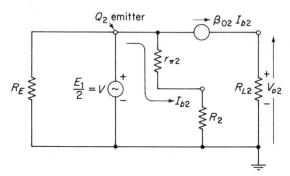

**Fig. 10-4.** Signal-voltage model for $Q_2$.

$$V_{o2} = \frac{V}{R_2 + r_{\pi 2}} R_{L_2} \beta_{0_2} \quad \text{and} \quad \frac{V_{o2}}{E_{g1}} = \beta_{02} \frac{R_{L2}}{2(R_2 + r_{\pi 2})} \tag{10-8}$$

If $R_1 = R_2 = 0$, Eq. (10-7) is unchanged and Eq. (10-8) simplifies to

$$\frac{V_{o2}}{E_{g1}} = \frac{g_{m2}}{2} R_{L2} \quad \text{for } R_1 = R_2 = 0 \tag{10-9}$$

**Example 10-1.** Given $\beta_{F1} = \beta_{F2} = \beta_{01} = \beta_{02} = \beta_F = 100$ in the circuit of Fig. 10-5. (a) Calculate $R_E$ to establish collector currents at 0.5 mA. (b) Find the voltages to ground at the collector, base, and emitter terminals of each transistor.

**Solution.** (a) $I_B = 0.5 \text{ mA}/100 = 5 \ \mu\text{A}$ and $I_E = 101(5) = 505 \ \mu\text{A}$. From Eq. (10-1),

$$R_E = \frac{V_{EE} - V_{BE}}{2I_E} - \frac{R}{2(\beta_F + 1)} = \frac{5.4}{1.01 \times 10^{-3}} - \frac{600}{202}$$
$$= 5340 - 3 = 5.34 \text{ k}\Omega$$

Note that the approximation from Eq. (10-2) is quite accurate.
(b) Base potentials are negative at $5 \ \mu\text{A} \times 600 = 3$ mV. Emitter potentials are negative at $-0.6$ V. The collector of $Q_1$ is at $+6$ V. Voltage drop across $R_{L2}$ is 0.5 mA $\times$ 6 k$\Omega$ = 3 V so that the collector of $Q_2$ is at $+3$ V.

Sec. 10-1      Basic Differential Amplifier Circuit Analysis   303

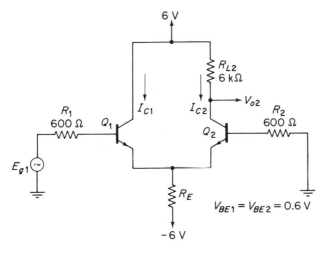

Fig. 10-5. Circuit for Examples 10-1 and 10-2.

*Example 10-2.* Employing the data in Example 10-1, calculate (a) $R_{in1}$, (b) $V/E_{g1}$, (c) $V_{o2}/E_{g1}$.

*Solution.* (a) Find $r_\pi$ for each transistor from $g_m = 0.5/25 = 0.02$ and $r_\pi = \beta_0/g_m = 5$ kΩ. From Eq. (10-4b),

$$R_{i2} = \frac{5000 + 600}{101} = 55.5 \text{ Ω} \qquad R_{in2} = 5340 \| 55.5 \cong 55 \text{ Ω} = R_{i2}$$

From Eq. (10-4a), $R_{in1} = 600 + 5000 \pm 101(55) = 11.1$ kΩ. Check against Eq. (10-5a): $R_{in1} = 11.2$ kΩ.
(b) From Eq. (10-7) or (10-6). $V = E_{g1}/2$.
(c) From Eq. (10-8),

$$\frac{V_{o2}}{E_{g1}} = 100\frac{6000}{2(5600)} = 53.5$$

From Eq. (10-9),

$$\frac{V_{o2}}{E_{g1}} \cong 60$$

It is evident from the models in Figs. 10-3 and 10-4 that there is no phase shift between $E_{g1}$, $V$, and $V_{o2}$. $R_{L1}$ was omitted so that *no* signal voltage can be developed at the collector of $Q_1$ for coupling back through $C_{\mu 1}$ to the base of $Q_1$. Also, $R_2$ can be shunted with a capacitor so that any signal voltage fed back from the collector of $Q_2$ through $C_{\mu 2}$ will be bypassed to ground. This configuration consequently exhibits outstanding performance at high frequencies. Also, when $R_{L2}$ is replaced by a tuned circuit, the need for neutralization is eliminated because of negligible internal feedback.

## 10-2 Common-Mode Rejection Ratio

It is useful to have a figure of merit for comparing one differential amplifier with another. One such figure of merit is called the *common-mode rejection ratio* CMRR and is a measure of how well $V_{o2}$ in Fig. 10-6 depends

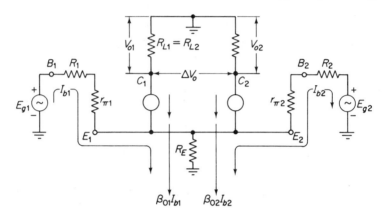

**Fig. 10-6.** Small-signal model of a differential amplifier.

on the *difference* between $E_{g1}$ and $E_{g2}$ and not on their individual magnitudes. To understand this concept we develop general expressions for $V_{o1}$ and $V_{o2}$ in terms of both $E_{g1}$ and $E_{g2}$ and then examine special cases for the relative sizes of $E_{g1}$ and $E_{g2}$. Begin by writing both input loop equations for Fig. 10-6:

$$E_{g1} = I_{b1}[R_1 + r_{\pi 1} + (\beta_{01} + 1)R_E] + I_{b2}(\beta_{02} + 1)R_E \quad (10\text{-}10a)$$
$$E_{g2} = I_{b1}(\beta_{01} + 1)R_E + I_{b2}[R_2 + r_{\pi 2} + (\beta_{02} + 1)R_E] \quad (10\text{-}10b)$$

Employ symmetry to simplify Eq. (10-10), where

$$R = (R_1 + r_{\pi 1}) = (R_2 + r_{\pi 2})$$

and

$$B = (\beta_{01} + 1)R_E = (\beta_{02} + 1)R_E$$
$$E_{g1} = I_{b1}(R + B) + I_{b2}B \quad (10\text{-}11a)$$
$$E_{g2} = I_{b1}B + I_{b2}(R + B) \quad (10\text{-}11b)$$

Solving Eq. (10-11) for $I_{b1}$ and $I_{b2}$,

$$I_{b1} = \frac{E_{g1}(R + B) - E_{g2}B}{R(R + 2B)} \qquad I_{b2} = \frac{E_{g2}(R + B) - E_{g1}B}{R(R + 2B)}$$

From the assumed current directions in Fig. 10-6, $V_{o1} = -\beta_{01}I_{b1}R_{L1}$ and

$V_{o2} = -\beta_{02}I_{b2}R_{L2}$. Substituting for $I_{b1}$ and $I_{b2}$ from above and letting $R_{L1} = R_{L2} = R_L$, $\beta_{01} = \beta_{02} = \beta_0$,

$$V_{o1} = -\beta_0 R_L \frac{E_{g1}(R+B) - E_{g2}B}{R(R+2B)} \tag{10-12a}$$

$$V_{o2} = -\beta_0 R_L \frac{E_{g2}(R+B) - E_{g1}B}{R(R+2B)} \tag{10-12b}$$

Equations (10-12a) and (10-12b) are perfectly general but we now examine special cases.

CASE 1. Let $E_{g1}$ be equal to and in phase with $E_{g2}$ so that a *common-mode* voltage $E_C$ is applied to both inputs, where

$$E_{g1} = E_{g2} = E_C \tag{10-13}$$

Substituting Eq. (10-13) into Eq. (10-12b) yields the common-mode gain $A_C$:

$$A_C = \frac{V_{o2}}{E_C} = -\frac{\beta_0 R_L}{R+2B} = \frac{V_{o1}}{E_C} \tag{10-14}$$

Common-mode voltage gain may be pictured as in Fig. 10-7, and input

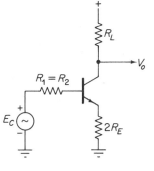

Fig. 10-7. Common-mode voltage gain. $E_C = E_{g1} = E_{g2}$ so that base currents are equal in Fig. 10-6.

resistance seen by $E_C$ is $R + 2B = R_1 + r_\pi + 2(\beta_0 + 1)R_E$.

CASE 2. Let $E_{g1} = E_d/2$ and $E_{g2} = -E_d/2$ so that the differential input voltage $E_d$ in Fig. 10-8 is

$$E_{g1} - E_{g2} = E_d \tag{10-15}$$

$E_{g1}$ will always be positive by exactly the same amount as $E_{g2}$ is negative so that the emitter terminals in Fig. 10-8 are always at signal ground potential and $R_E$ conducts no signal current. Substituting for $E_{g1}$ and $E_{g2}$ into Eq. (10-12b) gives the differential voltage gain $A_d$:

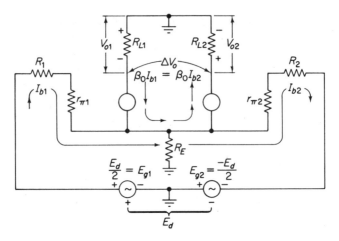

**Fig. 10-8.** Differential voltage gain.

$$A_d = \frac{V_{o2}}{E_d} = \frac{\beta_0 R_L}{2R} = \frac{V_{o1}}{E_d} \qquad (10\text{-}16)$$

In Eq. (10-16) the input resistance seen by $E_d$ is $2(R_1 + r_{\pi1})$ and the minus sign is dropped because of the reverse direction of $I_{b2}$ in Fig. 10-8.

CMRR is now defined specifically as the ratio of differential gain, Eq. (10-16), to common-mode gain, Eq. (10-14):

$$\text{CMRR} = \frac{A_d}{A_c} = \frac{R + 2B}{2R} \qquad (10\text{-}17)$$

where $R = R_1 + r_{\pi1}$ and $B = (\beta_0 + 1)R_E$.

Output voltage should depend only on the difference input signal $E_d$ and not on their *average level* $E_c$. From Fig. 10-9,

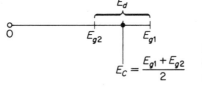

**Fig. 10-9.** Common-mode and differential input voltage in terms of the input signals.

$$E_{g1} = E_c + \frac{E_d}{2} \qquad E_{g2} = E_c - \frac{E_d}{2} \qquad (10\text{-}18)$$

Substituting Eq. (10-18) into Eq. (10-12a) or (10-12b) yields

$$V_{o2} = -\frac{\beta_0 R_L}{R + 2B}E_c + \frac{\beta_0 R_L}{2R}E_d \qquad (10\text{-}19)$$

Sec. 10-2                         Common-Mode Rejection Ratio     307

Comparing Eq. (10-19) with Eqs. (10-14) and (10-16) shows

$$V_{o2} = A_c E_c + A_d E_d \qquad (10\text{-}20)$$

From Eq. (10-20), $A_C$ should be zero to make $V_{o2}$ independent of the common-mode voltage. By reference to Eq. (10-14) we can decrease $A_c$ through increasing $B$ by increasing the size of $R_E$. The CMRR should be large and will also be increased by increasing $R_E$ as seen in Eq. (10-17).

*Example 10-3.* Assume symmetry in the circuit of Fig. 10-1, and $R_L = 6\,\text{k}\Omega$, $R_E = 5.34\,\text{k}\Omega$, $r_\pi = 5\,\text{k}\Omega$, $R_1 = R_2 = 600$, $\beta_0 = 100$. Find (a) $A_c$, (b) $A_d$, (c) CMRR.

*Solution.* (a) From Eq. (10-14),

$$A_c = -\frac{100(6000)}{5600 + 202(5340)} = -0.556$$

(b) From Eq. (10-16),

$$A_d = \frac{100(6000)}{2(5600)} = 53.5$$

(c) From Eq. (10-17),

$$\text{CMRR} = \frac{53.5}{0.556} = 96.3$$

*Example 10-4.* Employing the values and circuit of Example 10-3, evaluate $V_{o1}$ in millivolts for the following input voltages (in millivolts).
(a) $E_{g1} = 0.5 \qquad E_{g2} = -0.5$
(b) $E_{g1} = 1 \qquad E_{g2} = 0$
(c) $E_{g1} = 10.5 \qquad E_{g2} = 9.5$
(d) $E_{g1} = 96.8 \qquad E_{g2} = 95.8$

*Solution.* From Eq. (10-18), $E_c = (E_{g1} + E_{g2})/2$, $E_d = E_{g1} - E_{g2}$. Employ Eq. (10-20) and tabulate results, as shown in Table 10-1.

Table 10-1

|     | $E_c$ (mV) | $E_d$ (mV) | $A_C E_C$ (mV) | $A_d E_d$ (mV) | $V_{o2} = A_c E_c + A_d E_d$ (mV) |
|-----|------------|------------|----------------|----------------|-----------------------------------|
| (a) | 0          | 1          | — 0.00         | — 53.5         | — 53.5                            |
| (b) | 0.5        | 1          | — 0.28         | — 53.5         | — 53.8                            |
| (c) | 10         | 1          | — 5.5          | — 53.5         | — 59.0                            |
| (d) | 96.3       | 1          | — 53.5         | — 53.5         | — 107                             |

Example 10-4 clarifies the relationships between $E_{g1}$, $E_{g2}$, $E_c$, and $E_d$ and shows how $V_{o1}$ is not dependent solely on $E_d$. In (d) the common-mode voltage is greater than $E_d$ by the CMRR so that $E_c$ has the same affect on $V_{o1}$

as $E_d$. This point is illustrated by substituting Eq. (10-17) into (10-20) and tabulating results.

$$V_{o2} = A_d\left(E_d + \frac{E_c}{\text{CMRR}}\right) \qquad (10\text{-}21)$$

## 10-3 Differential Output Voltage

In a symmetrical differential amplifier the differential output voltage is independent of any common-mode voltage. An expression for differential output voltage $\Delta V_o$ in Figure 10-6 or 10-7 is found by subtracting Eq. (10-12b) from Eq. (10-12a) to show the independence as

$$\Delta V_o = \Delta E_g\left(\frac{\beta_0 R_L}{R}\right) \quad \text{where} \quad \Delta E_g = E_{g1} - E_{g2} \qquad (10\text{-}22)$$

We can exploit this unique advantage to extract a small transducer signal which would otherwise be lost in the presence of large spurious signals. In Fig. 10-10, $E_{g1}$ and $E_{g2}$ represent small signals from transducers such as solar cells in a location remote from the measuring equipment. $E_c$ represents a spurious voltage (up to several volts) representing the sum of electromagnetically induced, electrolysis, or ground-loop voltages. Resistors $R_B$ are adjusted

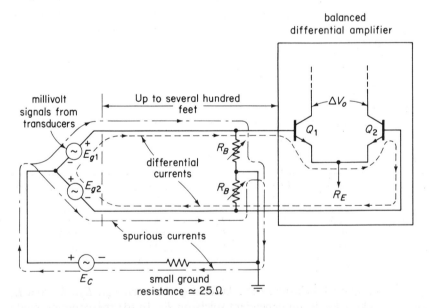

Fig. 10-10. Differential output voltage depends only on differential input voltage.

to equalize the spurious currents and the common-mode voltage at bases of $Q_1$ and $Q_2$. $\Delta V_o$ will respond to $E_{g1}$ and $E_{g2}$ only and will not be affected by unwanted spurious voltages. If $R_E$ can be made very large, so that none of the spurious currents can enter the differential amplifier, then $V_{o2}$ will depend only on $E_{g1}$ and $E_{g2}$.

## 10-4 Constant Current Biasing

Replacing $R_E$ by a current source transistor $Q_3$ in Fig. 10-11 improves CMRR. Source current $I_o$ divides equally between $Q_1$ and $Q_2$ so that their operating points depend primarily on $I_o$ and indirectly on biasing resistors $R_3$ and $R_4$. Any common mode voltages cannot force dc or incremental currents into the high impedance collector terminal of $Q_3$ and effectively we have created an infinite emitter resistance.

From Eq. (4-35a), constant bias current $I_o$ is expressed by

$$2I_{C1} = 2I_{C2} = I_o \cong I_{E3} = \frac{V_{EE} - V_{BE3}}{R_3 + \dfrac{R_4}{\beta_{F3} + 1}} \qquad (10\text{-}23)$$

Since the collector of $Q_3$ is clamped to approximately $-0.6$ V, $R_3$ must not be made too large or $Q_3$ will saturate. Also, when a negative dc common-mode

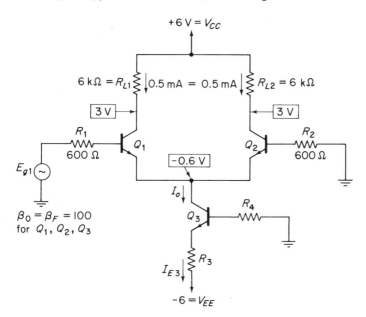

**Fig. 10-11.** Constant current biasing.

voltage of $-2$ V forces the collector of $Q_3$ to $-2.6$ V, we would still want $Q_3$ to be in its linear region. But a large value of $R_3$ gives good operating-point stability so we compromise and choose $R_E$ to drop $V_{EE}/2$ volts.

*Example 10-5.* Calculate values for $R_3$ and $R_4$ to bias $Q_1$ and $Q_2$ as shown in Fig. 10-11.

*Solution.* $I_o = 2(0.5) = 1$ mA $\cong I_{E3}$. Choose $R_3$ to drop $V_{EE}/2 = 3$ V or $R_3 = 3/0.001 = 3$ k$\Omega$. From Eq. (10-22),

$$R_4 = (\beta_{F3} + 1)\left[\frac{V_{EE} - V_{BE}}{I_{E3}} - R_3\right] = 101\left(\frac{5.4}{0.001} - 3000\right) = 240 \text{ k}\Omega$$

## 10-5 Phase Inverter

A phase inverter receives a signal at its input and delivers two output signals that are equal in amplitude and opposite in polarity. A small-signal model of Fig. 10-11 is drawn in Fig. 10-12 to show how the differential

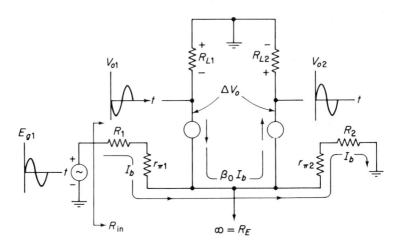

**Fig. 10-12.** Phase inverter.

amplifier performs phase inversion. Due to the infinite emitter resistance, input resistance $R_{in}$ is

$$R_{in} = 2R \quad \text{where} \quad R = R_1 + r_{\pi 1} = R_2 + r_{\pi 2} \quad (10\text{-}24)$$

and voltage gain is $A_d$ so that $V_{o2}$ is equal but opposite in phase with $V_{o1}$, or

$$V_{o2} = -V_{o1} = \frac{\beta_0 R_L}{2R}E_{g1} = \frac{1}{2}\Delta V_o \quad (10\text{-}25)$$

## 10-6 Voltage-Gain Control

Voltage gain of a differential amplifier is readily controllable by controlling the constant-current generator output current. Battery $V_G$ represents a variable-gain controlling voltage in Fig. 10-13. Assume $V_G$ is initially

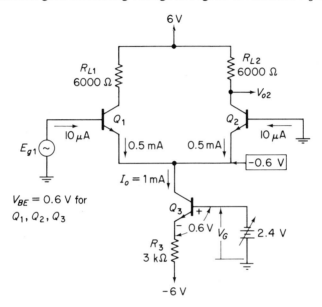

**Fig. 10-13.** Gain control by constant current supply.

set to $-2.4$ V, corresponding with $I_o = 1$ mA in Example 10-5. Resistances $R_1$ and $R_2$ must be negligible with respect to $r_\pi$ because gain will be controlled by varying collector current and $g_m$. From Eqs. (10-24) and (10-25),

$$V_{o2} = \frac{\beta_0 R_L}{2(R_1 + r_\pi)} = \frac{R_L}{2\left(\frac{R_1}{\beta_0} + \frac{1}{g_m}\right)} E_{g1} \tag{10-26a}$$

$\beta_0$ will remain fairly constant with collector current and a large $R_1$ will swamp out any variation in $g_m$. Assuming $R_1$ and $R_2$ are less than 10% of $r_{\pi 1}$ and $r_{\pi 2}$, we can simplify Eq. (10-26a) to

$$V_{o2} = \frac{g_m R_L}{2} E_{g1} \tag{10-26b}$$

Assuming symmetry, $I_{C1} = I_{C2} = I_C$ so that $g_{m1} = g_{m2} = g_m$, and since $I_o = 2I_C$

$$g_m = \frac{I_C}{\frac{kT}{q}} = \frac{I_o}{2(0.025)} = \frac{I_o}{0.05} \tag{10-27}$$

Substituting Eq. (10-27) into Eq. (10-26b) shows the dependence of voltage gain on operating-point current and temperature,

$$A_d = \frac{V_{o2}}{E_{g1}} = I_o \frac{R_L}{4\frac{kT}{q}} = \frac{I_o R_L}{0.1} \qquad (10\text{-}28)$$

Finally, $I_o$ can be expressed in terms of $V_G$ from Fig. 10-13 by

$$I_{E3} \cong I_o = \frac{V_{EE} - (V_G + V_{BE3})}{R_3} \qquad (10\text{-}29)$$

*Example 10-6.* What are the maximum and minimum allowable values for $V_G$ in Fig. 10-13?

*Solution.* The collector of $Q_3$ is clamped at $-0.6$ V by $V_{BE}$ of $Q_1$ or $Q_2$. Allowing a minimum of 1 V for $V_{CE3}$ to keep $Q_3$ out of saturation, the emitter of $Q_3$ must never be more positive than $-1.6$ V. Thus $V_G$ must never allow the base of $Q_3$ to go above $-1.6 + 0.6 = -1.0$ V. We check the situation of $Q_1$ and $Q_2$ by calculating $I_o$ from Eq. (10-29):

$$I_o = \frac{6 - (1 + 0.6)}{3000} = 1.46 \text{ mA}$$

Then $I_C = 1.46/2 = 0.73$ mA, $V_{RL} = 6000 \times 0.73 \times 10^{-3} = 4.4$ V.
Voltage at the collectors of $Q_1$ and $Q_2$ is $6 - 4.4 = 1.6$ V, and $V_{CE1} = V_{CE2} = 1.6 + 0.6 = 2.2$ V.
Any value of $V_G$ more negative than $-5.4$ V will not forward bias $Q_3$ and $Q_3$ will cut off.

*Example 10-7.* Plot voltage gain $V_{o2}/E_{g1}$ versus $V_G$ for values of $V_G$ between $-1$ and $-5$ V.

*Solution.* Substitute Eq. (10-29) into (10-28) and then solve for the variables indicated.

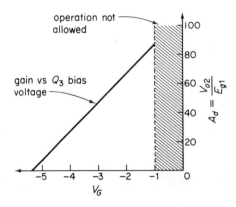

Fig. 10-14. Gain control by bias supply.

## Sec. 10-7      Gain Control by Offset Voltage

$$A_d = \frac{V_{EE} - V_{BE3} - V_G}{R_3}\left(\frac{R_L}{0.1}\right) = (5.4 - V_G)\frac{6000}{3000(0.1)}$$
$$= 108 - 20V_G.$$

When $V_G = 1$, $A_d = 88$; and when $V_G = -5$, $A_d = 8$. The curve is linear and is plotted in Fig. 10-14.

## 10-7 Gain Control by Offset Voltage

An additional control can be placed on collector current by applying a small offset voltage $V_F$ in Fig. 10-15(a). $I_o$ establishes equal collector cur-

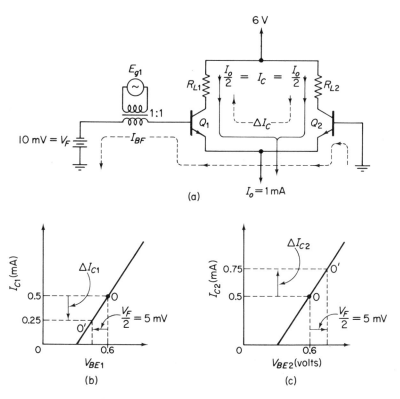

Fig. 10-15. Gain control by offset voltage.

rents of 0.5 mA when $V_F = 0$. Operating points for this condition are shown in Figs. 10-15(b) and (c). $V_F$ is a tiny voltage of 10 mV that divides equally between $Q_1$ and $Q_2$ to decrease the 0.6 V forward bias on $Q_1$ by 5 mV and increase the 0.6-V forward bias on $Q_2$ by 5 mV. From the $I_C - V_{BE}$ characteristics in Figs. 10-15(b) and (c), collector current of $Q_1$ is decreased by

0.25 mA and collector current of $Q_2$ is increased by 0.25 mA. Thus with a steady $V_F = 10$ mV, the new operating points are located at $0'$ and $g_m$ and $r_\pi$ will differ for each transistor. Assuming $\beta_0$ does not change significantly with collector current and since $R_1 = R_2 = 0$ we can rewrite Eq. (10-26a) and divide numerator and denominator by $\beta_0$ to get

$$\frac{V_{o2}}{E_{g1}} = \frac{\beta_0 R_L}{r_{\pi 1} + r_{\pi 2}} = \frac{R_L}{\frac{1}{g_{m1}} + \frac{1}{g_{m2}}} = R_L(g_{m1} \| g_{m2}) \qquad (10\text{-}30)$$

where

$$g_{m1} = \frac{I_{C1}}{C} \qquad g_{m2} = \frac{I_{C2}}{C} \qquad C = \frac{kT}{q}$$

$$I_{C1} = \frac{I_o}{2} - \Delta I_C \qquad I_{C2} = \frac{I_o}{2} + \Delta I_C$$

**Example 10-8.** Calculate voltage gain $V_{o2}/E_{g1}$ for Fig. 10-15(a) with $V_F = 0$.

**Solution.** Since $V_F = 0$, $I_{C1} = I_{C2}$, $g_{m1} = g_{m2} = g_m$ and $g_m = 0.5/25 = 0.02$ mho. Eq. (10-30) reduces to

$$V_{o2} = R_L(g_m \| g_m) = \frac{g_m R_L}{2} = \frac{0.02(6000)}{2} = 60$$

**Example 10-9.** Calculate voltage gain $V_{o2}/E_{g1}$ for Fig. 10-15(a) with $V_F = 10$ mV.

**Solution.** From Figs. 10-15(b) and (c), $I_{C1} = 0.25$ mA and $I_{C2} = 0.75$ mA, so

$$g_{m1} = \frac{0.25}{25} = 0.01 \qquad g_{m2} = \frac{0.75}{25} = 0.03 \qquad g_{m1} \| g_{m2} = 0.0075$$

From Eq. (10-30),

$$\frac{V_{o2}}{E_{g1}} = (6000)(0.0075) = 45$$

Examples 10-8 and 10-9 show that very small offset voltages give large changes in voltage gain. Automatic gain control can be established by rectifying a portion of $V_{o2}$ and feeding the resulting dc voltage back to control the offset voltage. As signal input is increased the rectified dc voltage increases to increase $V_F$, reducing gain and holding $V_{o2}$ relatively constant.

## 10-8 The Linear Multiplier

In Fig. 10-16, output $V_{o2}$ is proportional to the product of inputs $E_1$ and $E_3$. If $E_3 = 0$, then $V_{o2}$ is proportional to $E_1$ or $g_m$ in Eq. (10-26b) and $g_m$ is controlled by $I_o$ in Eq. (10-27). Both equations combine into Eq.

Sec. 10-8                                       The Linear Multiplier    315

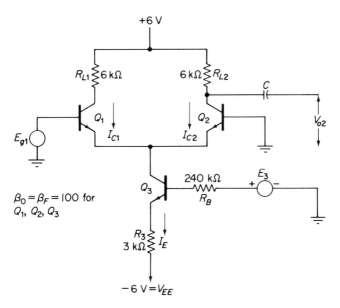

Fig. 10-16. Basic multiplier circuit.

(10-28). As shown by Eq. (10-29), $I_o$ is controllable by $E_3$ and consequently the gain of $V_{o2}/E_{g1}$. In Fig. 10-16, when $E_3 = 0$, $I_{E3}$, is found from

$$I_{E3} = \frac{V_{EE} - V_{BE}}{R_3 + \dfrac{R_B}{\beta_F + 1}} \tag{10-31}$$

But if $E_3$ is an incremental voltage it will cause an incremental component of emitter current $i_{e3}$ to be superimposed on the dc component $I_{E3}$, where

$$i_{e3} = \frac{E_3}{R_3 + \dfrac{r_{\pi 3} + R_B}{\beta_{03} + 1}} \tag{10-32}$$

The total instantaneous collector current of $Q_3$, $i_{C3}$ is now approximately equal to $i_E$, where

$$i_{C3} = I_{E3} + i_{e3} \tag{10-33}$$

But $i_{C3}$ divides equally between $i_{C1}$ and $i_{C2}$ so that both collector currents of $Q_1$ and $Q_2$ are always equal but now have an incremental component as well as a dc component. Substituting Eqs. (10-31) to (10-33) into Eq. (10-28) yields a product term to show multiplication

$$V_{o2} = E_{g1}(A_1 + A_2 E_3) \tag{10-34}$$

where

$$A_1 = \frac{(V_{EE} - V_{BE})R_L}{0.1\left(R_3 + \dfrac{R_B}{\beta_F + 1}\right)} \qquad A_2 = \frac{R_L}{0.1\left(R_3 + \dfrac{r_{\pi 3} + R_B}{\beta_{03} + 1}\right)}$$

**Example 10-10.** Calculate $V_{o2}/E_{g1}$ in Fig. 10-16 at (a) $E_3 = 0$, (b) a peak change in $E_3$ of 2.7 V, (c) a peak change in $E_3$ of $-2.7$ V.

**Solution.** For Eq. (10-34) Calculate $A_1$ and $A_2$. Use $r_{\pi 3} = 2500\ \Omega$ from $I_o \cong I_{E3}$ in Example 10-5.

$$A_1 = \frac{(6 - 0.6)6000}{0.1\left(3000 + \dfrac{240{,}000}{101}\right)} = 60$$

$$A_2 = \frac{6000}{0.1\left(3000 + \dfrac{242{,}500}{101}\right)} = 11$$

Eq. (10-34) can now be expressed by $V_{o2} = E_{g1}(60 + 11E_3)$ and voltage gains for $V_{o2}/E_{g1}$ are (a) 60, (b) 90 (c) 30. We see that $E_{g1}$ is multiplied by $\pm 30$ at the respective peak values of $E_3$.

## 10-9 Amplitude Modulation

The circuit of Fig. 10-16 can also function as an *amplitude modulator* since amplitude modulation is a multiplication process. To illustrate this application, let $E_{g1}$ be a sinusoidal signal of 10,000 Hz with a peak amplitude of 2 mV. $E_{g1}$ is shown for an interval of 2 msec in Fig. 10-17 with a total of 20 cycles. Let $E_3$ be zero during the first millisecond, and have a frequency of 1000 Hz with a peak amplitude of 2.7 V during the second millisecond. Only one cycle of $E_3$ will occur during this second millisecond.

The resulting output voltage in Fig. 10-17 can be explained quite simply by reference to the solution of Example 10-10. During the first millisecond, $E_3 = 0$ and the gain is 60 so that 10 cycles of $E_1$ are amplified to peak values of $2(60) = 120$ mV. During the first quarter of the second millisecond $E_3$ increases sinusoidally to 2.7 V and the amplifier's gain increases sinusoidally to 90 so that $V_{o1}$ increases sinusoidally to a peak of $2(90) = 180$ mV. Since $E_2$ decreases sinusoidally to $-2.7$ V during the second and third quarters of the second millisecond, voltage gain and $V_{o2}$ decrease sinusoidally to a peak of $2(30) = 60$ mV. During the last quarter cycle $V_{o2}$ increases sinusoidally back to a peak of 120 mV.

$V_{o2}$ in Fig. 10-16 is an amplitude-modulated wave represented by Eq. (10-34), repeated here for convenience.

$$V_{o2} = A_1 E_{g1} + A_2 E_{g1} E_3 \tag{10-34}$$

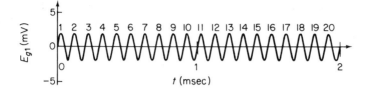

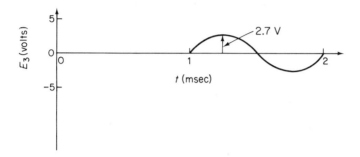

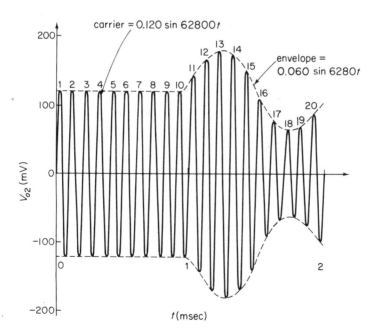

**Fig. 10-17.** Output voltage $V_{o2}$ for sinusoidal inputs $E_{g1}$ and $E_3$ in Fig. 10-16.

To introduce the concepts and terminology of amplitude modulation, stipulate that $E_{g1}$ is a *carrier* signal and $E_3$ is a *modulating* signal, both of which are sine waves

$$E_{g1} = E_{cp} \sin \omega_c t = \text{carrier signal} \tag{10-35}$$

$$E_3 = E_{mp} \sin \omega_m t = \text{modulating signal} \tag{10-36}$$

Substituting Eqs. (10-35) and (10-36) into Eq. (10-34) gives one method of representing the *modulated output voltage* $V_{o2}$ in Fig. 10-17.

$$\begin{aligned} V_{o2} &= A_1 E_{cp} \sin \omega_c t + A_2 (E_{cp} \sin \omega_c t)(E_{mp} \sin \omega_m t) \\ &= \text{carrier output} + \text{product term} \end{aligned} \tag{10-37}$$

The product term in Eq. (10-37) is the product of two sines and may be expanded by the trigonometric identity

$$(\sin a)(\sin b) = \tfrac{1}{2}[\cos(a-b) + \cos(a+b)]$$

so that Eq. (10-37) may be rewritten as

$$\begin{aligned} V_{o2} &= C_1 \sin \omega_c t - C_2 \cos(\omega_c + \omega_m)t + C_2 \cos(\omega_c - \omega_m)t \\ &= \text{carrier} + \text{sum frequency} + \text{difference frequency} \end{aligned} \tag{10-38}$$

where $C_1 = A_1 E_{cp}$ and $C_2 = A_2 E_{cp} E_{mp}/2$.

Equations (10-37) and (10-38) are different descriptions of the same modulated wave. However from Eq. (10-38) we can identify three distinct signal frequencies. The carrier frequency $\omega_c$ is usually 10 or more times greater than the modulating frequency $\omega_c$. One output voltage component is higher than the carrier by $\omega_m$ and is called the *upper side frequency* or sum frequency. The other voltage component is less than the carrier by $\omega_m$ and is called the *lower side frequency* or difference frequency. Any modulating frequencies applied to $Q_3$ will not appear directly in the output but will appear as sum and difference frequencies. If many different frequencies were applied to $Q_3$, the sum and difference of each would appear in the output to form a band of frequencies on either side of the carrier frequency. These groups of frequencies are appropriately identified as the upper and lower *side band*.

*Example 10-11.* Let $E_1 = 0.002 \sin 2\pi(10000t)$ and $E_3 = 2.7 \sin 2\pi(1000t)$ in the circuit of Fig. 10-16. Express the output voltage in two forms.

*Solution.* From the results of Example 10-10 substitute values into Eq. (10-34)

$$\begin{aligned} V_{o2} &= 0.002 \sin 62800t \, [60 + 11(2.7) \sin 6280t] \\ &= 0.120 \sin 62800t + 0.060(\sin 6280t)(\sin 62{,}800t) \end{aligned}$$

Observe from the product term that carrier signal ($\sin 62{,}800t$) has its amplitude varied to form the envelope in Fig. 10-17 of

$$\text{Envelope equation} = A_2 E_{cp} E_3 = 0.060 \sin 6280t$$

From Eq. (10-38). $C_1 = 60(0.002) = 0.120$ and $C_2 = 11(0.002)(2.7/2) \cong 0.030$. From Eq. (10-38) $V_{o2} = 0.120 (\sin 62{,}800t) - 0.030 (\cos 2\pi\, 11{,}000t) + 0.030 \cos(2\pi\, 9000t)$.

Results of Example 10-11 are plotted in Fig. 10-18(a) to illustrate relationships between carrier and side frequency components of $V_{o2}$. Since

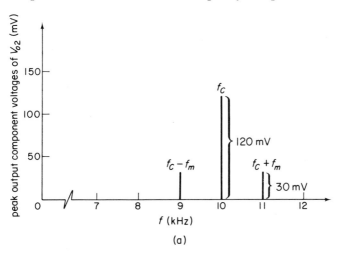

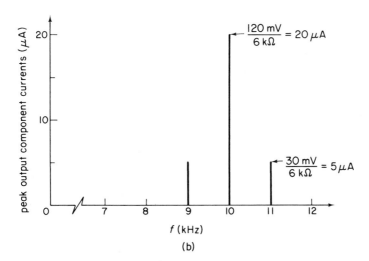

**Fig. 10-18.** Carrier and side frequencies for Example 10-11.

$V_{o2}$ is developed across $R_{L2} = 6$ kΩ we can also show the signal components of $Q_2$ collector current in Fig. 10-18(b) from $V_{o2}/R_{L2}$.

The modulation process thus impresses information contained in a modulating wave on a carrier wave whose envelope is varied according to the amplitude and frequency of the modulating wave. The carrier wave carries the information to a receiver where the evelope is recovered to duplicate the original information.

## 10-10 Balanced and Single Side Band Modulators

A tuned circuit can replace $R_{L2}$ in Fig. 10-16 as the load on a differential amplifier. The tuned circuit can be constructed of a parallel capacitor and inductor tuned to be resonant at a desired frequency. For example, the tuned circuit can be resonated to develop a very high impedance at $(\omega_c + \omega_m)$. At frequencies $\omega_c$ and $(\omega_c - \omega_m)$ the tuned circuit will have low impedance. The same currents of Fig. 10-18(b) will flow through the tuned circuit but only the sum frequency current will develop any appreciable output voltage. The other two current components will be unchanged in magnitude but will develop much smaller output voltage components across the low resistance of the tuned circuit. The tuned circuit may be designed to develop a large impedance over a bandwidth that corresponds with the upper side band. Only signal currents within this band would develop significant output voltages. This arrangement is called *single side band transmission*, because only the sum frequencies lying within the band are transmitted from the modulator.

It also may be required that the carrier frequency be eliminated to save power. Low-frequency modulating signals generate sum and difference signals that are very close to the carrier frequency and requirements for a tuning circuit to eliminate the carrier but not the side bands are stringent. A far better arrangement is to use the differential amplifier to balance out the carrier.

A *balanced modulator* (balance out the carrier) is made by connecting the carrier signal to the input of $Q_3$ and the modulating signal to the differential input. For example, a 1000-Hz signal into the differential input and a 10,000-Hz signal into $Q_3$ would generate output signals of 1000 Hz, 11,000 Hz, and 9000 Hz. The 10,000-Hz signal has been balanced out and it is relatively easy to tune out the 1000-Hz signal. A single side band, or two sidebands with the carrier balanced out, contain enough data to reconstruct the original modulation. At the receiver, the carrier is put back and the envelope detected.

## 10-11 Frequency Shifting

Frequency shifting is accomplished by the same modulation circuit of Fig. 10-16. Any signal applied to the differential input is shifted to a new frequency by a signal applied to current source transistor $Q_3$. The differential amplifier does the mixing and a tuned circuit selects the sum or difference frequency. The standard AM broadcasting system broadcasts carrier frequencies in a range 550–1600 kHz. The standard AM superheterodyne receiver employs frequency shifting to shift the radio carrier frequency to the commonly used *intermediate frequency* (IF) of 455 kHz. For example, a radio carrier frequency of 800 kHz with side bands of $\pm 5$ kHz is selected by a tuned RF amplifier and applied to the differential input. A local oscillator is tuned to 1255 kHz and the tuned load selects the difference frequencies of $455 \pm 5$ kHz.

The carrier frequency has been *shifted* from 800 kHz to the intermediate frequency of 455 kHz. Every carrier frequency is shifted to the IF frequency and this allows the design of the IF amplifier stages to be optimized. All the original modulation frequencies are still present in the modulated IF signal. The envelope shape at the output of the RF amplifier is identical to that at the output of the IF amplifier.

## 10-12 The Comparator

The differential amplifier functions as a *comparator* by comparing two input voltages. There will be no differential output voltage when the voltages at each of the differential inputs are identical. One of the inputs may be connected to a reference voltage while the other *measuring input* is connected to a point which may have to be held constant at the reference voltage. Any difference between these two voltages constitutes an error signal to be amplified so that the resulting output voltage actuates a control circuit to restore the measuring input to the reference voltage.

The comparator can be designed to watch for a specific voltage. A rising ramp voltage is applied to the measuring input, and when the ramp voltage reaches the reference voltage, the differential amplifier produces a signal. Alternatively, a precision ramp can be connected to the reference terminal and the measuring input to an unknown voltage. If a digital counter is set to count up, in step with the ramp, the count can be stopped when the two inputs are identical and the unknown voltage has been measured. This is one principle underlying the operation of a digital voltmeter.

Refer to Fig. 10-19 for a very basic comparator. The voltages shown are given for balanced conditions. At balance, the thermistor has a resistance

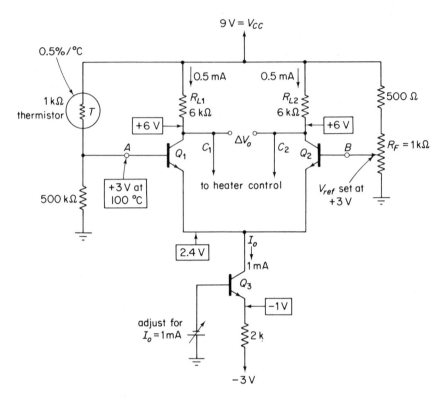

**Fig. 10-19.** Control applications for a comparator.

of 1 kΩ at 100 °C. Potentiometer $R_f$ is adjusted so that $V_o = 0$. Points $A$ and $B$ will be at $+3$ V and the collectors at $+6$ V with respect to ground. Assume that the thermistor is monitoring the temperature of an oven. When oven temperature drops, the resistance of the thermistor increases, causing point $A$ to go negative. A base-current increment is injected into the base of $Q_2$ and flows out of the base of $Q_1$ so that collector current of $Q_2$ increases and collector current of $Q_1$ decreases. The total voltage drop across $R_{L2}$ increases so that $C_2$ goes negative. $C_1$ goes positive since the drop across $R_{L1}$ decreases. Restated, the temperature drop caused a differential input voltage change which was amplified and applied to the heater control. The heater is activated, raising the temperature to 100 °C, restoring the voltage at $A$ to $+3$ V, whereupon $V_o = 0$ and the heater is turned off. A 1° temperature change will generate a differential input voltage of approximately 30 mV. The gain of the amplifier is 120 so that about 3.6 V are available to activate the heater turn-on control.

The comparator in Fig. 10-20 has ground as a reference voltage. When $E_1$ goes positive by a few millivolts, $Q_5$ should turn off to detect crossing of

## Sec. 10-12 The Comparator 323

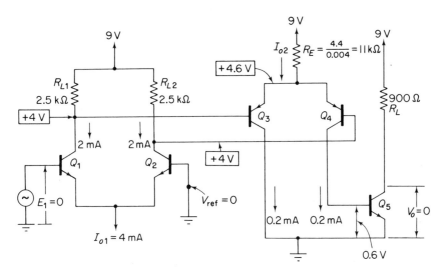

**Fig. 10-20.** Comparator application as a zero-crossing detector. $\beta_0 = \beta_F = 50$ for all transistors.

the zero reference in the positive direction. Conversely, when $Q_5$ turns on, the output goes to $V_{CE(\text{sat})}$ (of $Q_5$) to indicate the input has crossed zero in the negative direction. The known facts in this circuit are that $I_{o1} = 4$ mA and $I_{o2} = 0.4$ mA. At balance, the input voltage is zero, so that $I_{C1} = I_{C2} = 2$ mA, and $I_{C3} = I_{C4} = 0.2$ mA.

DC ANALYSIS. Base bias currents of $Q_3$ and $Q_4$ are 0.2 mA/50 = 4μA and do not affect the voltage drops across $R_{L1}$ and $R_{L2}$ of 2 mA × 2.5 kΩ = 5 V. Collectors of $Q_1$ and $Q_2$ are at $9 - 5 = 4$ V and emitters of $Q_3$ and $Q_4$ are higher by 0.6 V at 4.6 V. $R_E$ carries 0.4 mA and is set at 11 kΩ to drop 4.4 V. Collector current of $Q_4$ is the base current of $Q_5$, driving $Q_5$ into saturation since $I_{C5} = 0.2$ mA × 50 = 10 mA and $V_{RL} = 900 \times 10$ mA = 9 V. Thus $V_o = 0$ when $E_1 = 0$.

$Q_5$ TURNOFF. $Q_1 - Q_2$ must develop a minimum differential output voltage $V_{o1}$ to cause a base current increment $I_b$ in Fig. 10-21 that is sufficient to cause a change in collector currents of $Q_3 - Q_4$ of $\Delta I_c = 0.2$ mA. Collector current of $Q_4$ will go to zero, cutting off $Q_5$ and $V_o$ will go towards +9 V. If $\Delta V_{o1}$ is driven to the opposite polarity, $I_{C4}$ will increase to 0.4 mA and drive $Q_5$ deeper into saturation.

MINIMUM PLUS CROSSING VOLTAGE. We employ a small-signal model of $Q_1$ and $Q_2$ in Fig. 10-22 to find the required value of $\Delta V_{o1}$ to turn off $Q_5$. $I_b$ of $Q_3 - Q_4$ must be $I_c/\beta_0 = 0.2$ mA/50 = 4 μA. Calculate $r_{\pi 3}$ and $r_{\pi 4}$

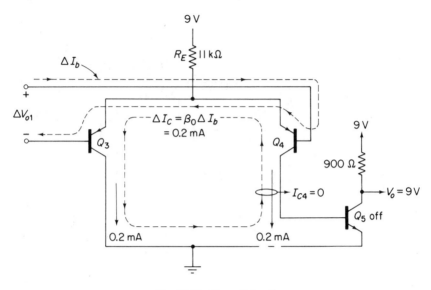

**Fig. 10-21.** Turnoff for $Q_5$.

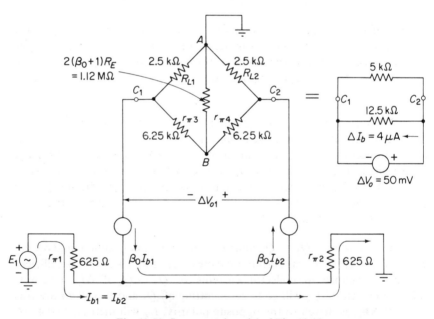

**Fig. 10-22.** Incremental model of Fig. 10-20.

from $g_{m3} = g_{m4} = 0.2/25 = 0.008$ and $r_{\pi 3} = r_{\pi 4} = 50/0.008 = 6250$. Calculate $r_{\pi 1} = r_{\pi 2} = (25)(50)/2 = 625 \, \Omega$ and $2(\beta_0 + 1)R_E = 102(11) \, \text{k}\Omega = 1.02 \, \text{M}\Omega$ to complete the model.

By drawing the load on $Q_1 - Q_2$ in the bridge configuration in Fig.

10-22 it is clear that, due to symmetry, points $A$ and $B$ are at the same potential so that the equivalent emitter resistance draws no current. The differential load presented to $\Delta V_{o1}$ is shown in the simplified model of Fig. 10-22 to be $R_L = 5 \| 12.5 = 3.57$ k$\Omega$. From Eq. (10-22),

$$\frac{V_{o1}}{E_1} = \frac{\beta_0 R_L}{2r_\pi} = \frac{50(3570)}{1250} = 142$$

The value of $\Delta V_{o1}$ required to cause a $\Delta I_b$ of 4 $\mu$A is shown to be 50 mV in Fig. 10-22 so that the minimum positive value of $E_1$ that will be detected by the cutoff of $Q_5$ will be

$$E_{1(\text{min})} = \frac{\Delta V_{o1}}{142} = \frac{0.050}{142} \cong 0.3 \text{ mV}$$

## PROBLEMS

*10-1* In Fig. 10-5, $R_E = 6$ k$\Omega$ and $\beta_0 = \beta_F = 100$ for both transistors. At what potential is the collector of $Q_2$ with respect to ground? Repeat the problem for $\beta_0 = \beta_F = 50$ to observe the small change in collector potential.

*10-2* Recalculate $R_E$ in Example 10-1 if transistors with $\beta_0 = \beta_F = 50$ are substituted.

*10-3* What changes occur in the results of Example 10-2 if transistors with $\beta_0 = \beta_F = 50$ are substituted?

*10-4* A differential amplifier with CMRR = 100 is driven by a differential input voltage of 2 mV. What common-mode voltage will cause an output voltage equal to that of the differential input voltage?

*10-5* Check values of $V_{o2}$ in Example 10-3 by employing Eq. (10-21).

*10-6* What are the differential output voltages in Example 10-4 for each combination of input voltages?

*10-7* What is the common-mode voltage when voltages to ground at each input of a differential amplifier are (a) 10 and 20 mV; (b) 5 and −5 mV?

*10-8* In Fig. 10-11, $R_3 = 2$ k$\Omega$ and $R_4 = 220$ k$\Omega$. Find the voltages to ground for the collectors of both transistors.

*10-9* If $R_{L1} = 7$ k$\Omega$ and $R_{L2} = 5$ k$\Omega$ in Example 10-5, what changes occur in $I_0$ and $V_{CE}$ of each transistor?

*10-10* Calculate $V_{o2}/E_{g1}$ in Fig. 10-13 when $V_G = -2.4$ V. Check your results with Fig. 10-14 and by Eq. (10-26b).

**10–11** $R_{L1} = R_{L2} = 3$ kΩ replaces the load resistors in Example 10-7. Plot $V_o/E_{g1}$ versus $V_G$ in Fig. 10-14.

**10–12** Plot $V_{o2}/E_{g1}$ versus $E_3$ for Example 10-8. On the same graph make the same plot for equal-load resistors of 3 kΩ (replacing the 6-kΩ resistors in Fig. 10-16).

**10–13** In Fig. 10-17, the peak value of $E_3$ in the second cycle is increased to 5.4 V. Sketch the resulting effect on $V_{o2}$.

**10–14** (a) Write the equation for $V_{o2}$ in Problem 10-13 both in the format of Eq. (10-34) and Eq. (10-38) with values. (b) Plot the output-voltage carrier side frequencies in Fig. 10-18(a), (with correct amplitude).

**10–15** Plot the output-voltage side frequencies in Fig. 10-18(a) for Example 10-11, assuming (a) $E_3 = 2.7 \sin 2\pi(2000t)$ and $E_1$ unchanged; (b) $E_1 = 0.002 \sin 2\pi(15000t)$ and $E_3$ unchanged; (c) $E_1 = 0.004 \sin 2\pi (10,000t)$ and $E_3$ unchanged.

**10–16** In the amplitude modulator circuit of Fig. 10-16, $E_{g1}$ is a carrier frequency of 1 MHz. Sine waves with frequencies of 1 and 5 kHz are *linearly* mixed and applied to $E_3$. What frequencies are present in the modulated output voltage $V_{o2}$?

**10–17** If the carrier signal is applied to $E_3$ and the sine waves to $E_{g1}$ in Problem 10-16, what frequencies are present in the output voltage?

**10–18** A broadcasting carrier signal of 550 kHz is applied to $E_{g1}$ in Fig. 10-16. What frequency or frequencies can be applied to $E_3$ so that $V_{o2}$ will contain a frequency of 455 kHz?

**10–19** Assuming all transistors in Figs. 10-20 to 10-22 have $\beta_0 = \beta_F = 100$, what are the new values of $V_o$ and minimum value of $E_i$ to switch $Q_5$?

# Chapter 11

**11-0** INTRODUCTION .............................329
**11-1** TUNED CIRCUIT REVIEW ......................329
**11-2** INDUCTANCE MEASUREMENT AND QUALITY FACTOR ..334
**11-3** TRANSFORMER REVIEW ........................336
**11-4** THE TAPPED INDUCTOR .......................338
**11-5** THE TUNED AMPLIFIER .......................342
**11-6** BANDWIDTH REDUCTION IN CASCADE .............348
**11-7** MULTIPLE-TUNED-CIRCUIT AMPLIFIER ...........350
      PROBLEMS ..................................356

# Tuned Circuits

## 11-0 Introduction

Each antenna continuously receives signals broadcast anywhere within a radius of about 100 mi. In addition, signals from other parts of the world may reflect from one or more ionized layers in the upper limits of our atmosphere to excite the antenna. Data signals from radar sets or satellites, time or location (LORAN) signals, television or radio are all present to some degree on every antenna. It is the job of the tuned circuit to create order out of this apparent chaos. The tuned circuit selects only the desired signal and rejects all others. Tuned circuits perform as frequency selectors in many applications that do not involve antennae. Oscillators are made to generate signals at a frequency determined by a tuned circuit. Many different frequencies can be sent simultaneously over a single communications link and tuned circuits sort them out. Since the performance of any tuned amplifier or oscillator depends on the tuned circuit elements, we begin with a review of tuned circuit behavior, measurement of inductance and the autotransformer.

## 11-1 Tuned Circuit Review

Our study of tuned circuits will be concerned with the tank circuit or parallel inductance, resistance, and capacitance. This $RLC$ combination resonates at a resonant frequency $\omega_R$ and selects a band of frequencies centered around the resonant frequency. Ideally the tuned circuit would filter out or pass a desired band of frequencies with no change in their phase angles and reject all others completely. The $RLC$ tuned circuit approaches the ideal if the band of frequencies to be passed is small with respect to the resonant frequency. Our study is therefore restricted to *narrow-band* tuned circuits where the bandwidth of frequencies selected is $\frac{1}{10}$ or less than the resonant frequency.

For most practical applications the behavior of an *RLC* tuned circuit can be described completely and simply by two equations and three statements.

$$\omega_R = \frac{1}{\sqrt{LC}} \tag{11-1}$$

$$B = \frac{1}{RC} \tag{11-2}$$

1. An *RLC* network is tuned to a resonant frequency $\omega_R$, where its impedance $Z$ is equal to $R$ and is maximum.
2. Bandwidth $B$ lies between an upper frequency limit $\omega_H$ and a lower frequency limit $\omega_L$, where $\omega_H$ and $\omega_L$ are equidistant from $\omega_R$.
3. At $\omega_L$ and $\omega_H$ the magnitude of $Z$ is 0.707 times its maximum value or $0.707R$, and has phase angles of 45 and $-45°$, respectively.

The remainder of this section is concerned with proving Eqs. (11-1) and (11-2). In Fig. 11-1, $E$ is a constant-voltage generator whose frequency

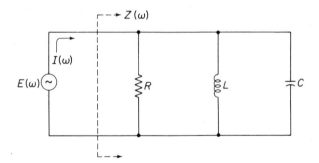

**Fig. 11-1.** *RLC* tuned circuit.

can be varied. The ratio of $E$ to the resulting current $I$ allows a study of how impedance $Z$ varies with frequency. Since $E$ is constant the product $IZ$ must be constant and may be expressed in several ways.

$$E = IZ = I(R \| X_L \| X_C) = \frac{I}{\dfrac{1}{R} + \dfrac{1}{jX_L} + \dfrac{j}{X_c}} \tag{11-3}$$

It is convenient to substitute conductance $G$ for $1/R$, $\omega L$ for $X_L$, and $1/\omega C$ for $X_C$ and present $Z$ as

Sec. 11-1                           Tuned Circuit Review    331

$$Z = \frac{E}{I} = \frac{1}{G + \frac{1}{j\omega L} + j\omega C} \quad (11\text{-}4)$$

it is clear from Eq. (11-4) that $Z$ varies with $\omega$ as a complex variable with a magnitude and phase angle depending on $\omega$. There are three frequencies where we are interested in evaluating $Z$. First is the resonant frequency $\omega_R$, defined when the magnitude of the $j$ term equals zero in Eq. (11-4).

$$\text{At } \omega = \omega_R, \quad \frac{1}{j\omega_R L} + j\omega_R C = 0$$

Solving for $\omega_R$,

$$\omega_R = \frac{1}{\sqrt{LC}} \quad (11\text{-}5)$$

Evaluate $Z$ at $\omega_R$ by substituting Eq. (11-5) into Eq. (11-4).

$$\text{At } \omega = \omega_R, \quad Z = \frac{1}{G} = R\,\underline{/0°} \quad (11\text{-}6)$$

The second frequency of interest is $\omega_H$, where $j\omega_H C$ is larger than $1/j\omega_H L$ and the two add to equal $jG$ in Eq. (11-4), or,

$$\text{At } \omega = \omega_H, \quad Z = \frac{1}{G + jG} = 0.707R\,\underline{/-45°} \quad (11\text{-}7)$$

The last frequency is $\omega_L$, where $1/j\omega_L L$ is larger than $j\omega_L C$ and the two add to equal $-jG$ in Eq. (11-4), or

$$\text{At } \omega = \omega_L, \quad Z = \frac{1}{G - jG} = 0.707R\,\underline{/45°} \quad (11\text{-}8)$$

From Eqs. (11-6) to (11-8) it is evident that the phase angle of $Z$ varies from inductive through zero to capacitive as $\omega$ is varied from $\omega_L$ to $\omega_H$. The variation in magnitude of $Z$ with $\omega$ is shown in Fig. 11-2 for tuned circuits with identical $L$ and $C$ but with different values of $R$. $R_1$ is greater than $R_2$, making the magnitude of $Z_1$ greater than $Z_2$ at resonance. The higher impedance curve, however, has a smaller bandwidth.

Finding the bandwidth is simplified if we work in terms of deviation from resonance $\Delta\omega$, where

$$\Delta\omega = \omega_H - \omega_R = \omega_R - \omega_L \quad (11\text{-}9)$$

Bandwidth $B$ equals twice the frequency deviation:

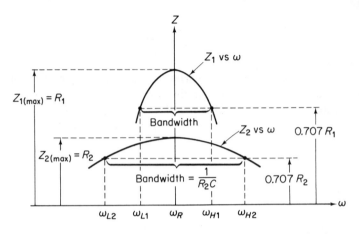

**Fig. 11-2.** Impedance variation with frequency of an *RLC* circuit. $L$ and $C$ are identical for each curve but $R_1 = 2R_2$.

$$B = 2\Delta\omega \tag{11-10}$$

Substituting for $\omega = \omega_H = \Delta\omega + \omega_R$ in Eq. (11-4),

$$\text{At } \omega_H \quad Z = \frac{1}{G + \dfrac{1}{j(\omega_R + \Delta\omega)L} + j(\omega_R + \Delta\omega)C} \tag{11-11}$$

It can be shown that Eq. (11-11) is approximated by

$$\text{At } \omega_H \quad Z = \frac{1}{G + \dfrac{1}{j\omega_R L} + j\omega_R C + j2\Delta\omega C} \tag{11-12}$$

The resonant terms add to zero, substituting for $B$ and comparing with Eq. (11-7)

$$\text{At } \omega = \omega_H \quad Z = \frac{1}{G + j\omega CB} = \frac{1}{G + jG} \tag{11-13}$$

And by inspection of Eq. (11-13),

$$G = CB \quad \text{or} \quad B = \frac{1}{RC} \tag{11-14}$$

The denominator term of Eq. (11-11) may be simplified by manipulating the $L$ term as follows:

Sec. 11-1    Tuned Circuit Review    333

$$\frac{1}{\omega_R L + L\Delta\omega} = \omega_R L + L\Delta\omega)\overline{\begin{array}{l}(\omega_R L)^{-1} - (\omega_R L)^{-2} L\Delta\omega + \cdots \\ 1 \\ \underline{1 + (\omega_R L)^{-1} L\Delta\omega} \\ -(\omega_R L)^{-1} L\Delta\omega \\ \underline{-(\omega_R L)^{-1} L\Delta\omega - (\omega_R L)^{-2}(L\Delta\omega)^2} \\ (\omega_R L)^{-2}(L\Delta\omega)^2\end{array}}$$

The result of the long division is approximated by the first two terms, since each succeeding term is $\omega/\omega_R$ times the preceeding term:

$$\frac{1}{\omega_R L + L\Delta\omega} \simeq \frac{1}{\omega_R L} - \frac{L\Delta\omega}{(\omega_R)^2 L^2}$$

Substituting for $(\omega_R)^2$,

$$\frac{1}{\omega_R L} - \frac{L\Delta\omega}{\left(\frac{1}{LC}\right)L^2} = \frac{1}{\omega_R L} - C\Delta\omega \qquad (11\text{-}15)$$

Substituting Eq. (11-15) into the denominator of Eq. (11-11) yields Eq. (11-12).

The tuned circuit is fully described for our purposes by Eqs. (11-1) and (11-2). Assuming $C$ is fixed, $L$ determines the resonant frequency and $R$ determines the sharpness of its impedance variation with frequency or selectivity.

*Example 11-1.* Given $C = 136$ pF in Fig. 11-1. (a) Calculate $L$ to resonate at 455 kHz and (b) $R$ for a bandwidth of 10 kHz.

*Solution.* (a) Convert frequency from hertz to radians by multiplying by $2\pi$ so that $\omega_R = 2.86 \times 10^6$ rad/sec and $B = 2\pi \times 10^4 = 6.28 \times 10^4$ rad/sec. From Eq. (11-1),

$$L = \frac{1}{(2.86 \times 10^6)^2(136 \times 10^{-12})} = 900 \ \mu\text{H}$$

From Eq. (11-2),

$$R = \frac{1}{6.28 \times 10^4 \times 136 \times 10^{-12}} = 117 \ \text{k}\Omega$$

*Example 11-2.* If $R$ is decreased by 50% to 58.5 k$\Omega$ in Example 10-1, what is the effect on (a) resonant frequency, (b) bandwidth?

*Solution.* (a) No effect. (b) From Eq. (11-2), $B$ doubles. Fig. 11-2 is a graph to compare Examples 10-1 and 10-2.

334   Tuned Circuits                                              Chap. 11

**Example 11-3.** If $R$ is a load resistor of 1000 $\Omega$, what values of $C$ and $R$ are required to meet the tuning requirements of Example 10-1?

**Solution.** From Eq. (11-2),

$$C = \frac{1}{10^3(6.28 \times 10^4)} = 15900 \text{ pF}$$

From Eq. (11-1),

$$L = \frac{1}{\omega_R^2 C} = \frac{1}{(8.18 \times 10^{12})(15900 \times 10^{-12})} = 7.7 \text{ }\mu\text{H}$$

## 11-2 Inductance Measurement and Quality Factor

In real inductors there are losses that may be represented by a resistance $R_p$ connected across the inductance. If a load resistor $R_L$ parallels $R_p$, the resultant $R = R_L \| R_p$ may give a bandwidth that is too broad or too narrow. We can add another resistor across the tuned circuit to lower the net $R$ and increase bandwidth. It is imperative therefore that we be able to measure $R$ and control it, in order to control the bandwidth of a tuned circuit.

Inductance is measured with a bridge or $Q$ meter and indicates the measurements in terms of $Q$, $\omega$, and $L$. Depending on the type of bridge these readings can model the inductor with either a series resistance $R_s$ and inductance $L_s$, or with a parallel resistance $R_p$ and inductance $L_p$. For a series model $R_s$ is calculated from the definition of *quality factor* $Q_s$.

$$Q_s = \frac{\text{Reactive power}}{\text{Real power}} = \frac{I^2 X_{LS}}{I^2 R_s} = \frac{\omega L_s}{R_s} \quad (11\text{-}16)$$

For a parallel model of the inductor,

$$Q_p = \frac{\frac{E^2}{X_{Lp}}}{\frac{E^2}{R_p}} = \frac{R_p}{\omega L_p} \quad (11\text{-}17)$$

These models are shown in Fig. 11-3. Connecting a capacitor $C$ across the real inductor gives a minimum possible bandwidth of

$$B = \frac{1}{R_p C} \quad (11\text{-}18)$$

and a resonant frequency of $\omega_R = (L_p C)^{-1/2}$. Substituting for $\omega_R$ in Eq. (11-17) and the resulting expression for $R_p$ into Eq. (11-18) yields

Sec. 11-2    Inductance Measurement and Quality Factor    335

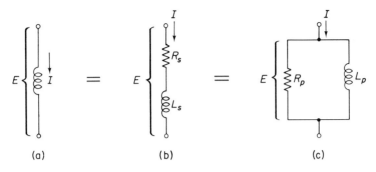

(a)   (b)   (c)

**Fig. 11-3.** The real inductor in (a) is represented by either a series model in (b) or a parallel model in (c).

$$B = \frac{\omega_R}{Q_p} \qquad (11\text{-}19)$$

Often the inductor is measured in terms of $Q_s$, $L_s$, and $\omega$ to yield $R_s$ from Eq. (11-16). $R_s$ must be converted to an equivalent $R_p$, *at the same frequency*, by equating the admittances in Figs. 11-3(b) and (c).

$$\frac{1}{R_s + j\omega L_s} = \frac{1}{R_p} + \frac{1}{j\omega L_p}$$

Conjugating the series term and equating real terms and $j$ terms, respectively, yields

$$R_p = R_s(1 + Q_s)^2 \qquad (11\text{-}20)$$

$$L_p = L_s\left(1 + \frac{1}{Q_s^2}\right) \qquad (11\text{-}21)$$

$Q_p$ equals $Q_s$ for values of $Q$ greater than 10 so that conversion equations are simplified to

$$R_p = R_s Q^2 \qquad (11\text{-}22a)$$

$$L_p = L_s \qquad (11\text{-}22b)$$

$$B = \frac{\omega_R}{Q} \qquad (11\text{-}22c)$$

$$Q = \frac{\omega L_s}{R_s} = \frac{R_p}{\omega L_p} \qquad (11\text{-}22d)$$

When a resistor $R$ is connected across a real inductor the resulting quality factor will be designated $Q_L$ to signify loading and will be referred to as *loaded Q*. $Q_L$ will be lower than the natural $Q$ of the inductor.

*Example 11-4.* The primary of an IF transformer is measured by a $Q$ meter to be $C = 136.5$ pF and $Q = 100$ at 455 kHz. Assuming the $Q$ meter indicates a parallel model of the primary, find $R_p$. (Note $C$ is a calibrated capacitor in the $Q$ meter that resonates with the unknown inductor at the measuring frequency).

*Solution.* From Eq. (11-1),

$$L_p = \frac{1}{(\omega_R)^2 C} = \frac{1}{(6.28 \times 455 \times 10^3)^2 (136 \times 10^{-12})} = 900 \text{ }\mu\text{H}$$

From Eq. (11-22d),

$$R_p = L_p \omega Q = 100(6.28 \times 455 \times 10^3) \times 900 \times 10^{-6} = 257 \text{ k}\Omega.$$

*Example 11-5.* Assume that the $Q$ meter readings in Example 11-4 represent a series model and calculate $R_p$.

*Solution.* From Eq. (11-22d),

$$R_s = \frac{\omega L_s}{Q_s} = \frac{6.28 \times 455 \times 10^3 \times 900 \times 10^{-6}}{100} = 25.7 \text{ }\Omega$$

From Eq. (11-22a),

$$R_p = R_s Q^2 = 25.7 \times 10^4 = 257 \text{ k}\Omega.$$

*Example 11-6.* If the inductor in Example 11-4 were connected in parallel with 136.5 pF, what would be the bandwidth of the resulting tuned circuit?

*Solution.* From Eq. (11-22c).

$$B = \frac{455 \times 10^3 \times 2\pi}{100} = 28.6 \text{ krad}$$

From Eq. (11-2),

$$B = \frac{1}{R_p C} = \frac{1}{257 \times 10^3 \times 136 \times 10^{-12}} = 28.6 \text{ krad} = 4.55 \text{ kHz}.$$

## 11-3 Transformer Review

Inductance is a measure of the flux linkages produced by a current and may be calculated for a particular inductor from

$$L = c\mu N^2 \tag{11-23}$$

where $c$ = a constant depending on inductor geometry,
$\mu$ = permeability of the magnetic circuit, and
$N$ = number of turns of wire in the inductor.

Sec. 11-3                                    Transformer Review   337

In an *ideal transformer*, $c$ and $\mu$ are identical for both primary winding $L_1$ and secondary winding $L_2$, so that

$$\frac{L_1 = c\mu N_1^2}{L_2 = c\mu N_2^2} \quad \text{or} \quad \frac{L_1}{L_2} = \frac{N_1^2}{N_2^2} \tag{11-24}$$

In some transformers a *tuning core* made of compressed powered iron and binder allows adjustment of permeability and consequently inductance. These tuning cores or *slugs* are threaded for adjustment with an alignment tool, and inserted into the inductor. Lowering the slug increases inductance. Brass or aluminum cores are nonmagnetic and lowering these cores decreases inductance. IF transformers and other types of transformers with metallic cores approximate the ideal transformer and almost all flux generated by primary and secondary currents link the primary and secondary windings. For an ideal transformer the volts per turn of the primary equals the volts per turn of the secondary since both depend on the same flux, and from Fig. 11-4(a),

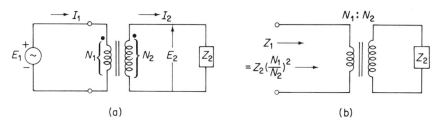

**Fig. 11-4.** Transformer current-voltage relationships in (a) form the impedance relationships in (b).

$$\frac{E_1}{N_1} = \frac{E_2}{N_2} \tag{11-25}$$

Ampere turns of both primary and secondary depend on the same flux, so

$$I_1 N_1 = I_2 N_2 \tag{11-26}$$

Dividing Eq. (11-25) by Eq. (11-26) yields the impedance-transforming relationship shown in Fig. 11-4(b):

$$\frac{E_1}{I_1 N_1^2} = \frac{E_2}{I_2 N_2^2} \quad \text{or} \quad \frac{Z_1}{N_1^2} = \frac{Z_2}{N_2^2} \tag{11-27}$$

*Example 11-7.* A primary winding $L_1$ is wound with 144 turns of #28 enameled wire on a nylon core. A secondary winding $L_2$ is wound with 72 turns of the same wire on top of $L_2$. The ends are burned, scraped, and brought out. The core is enclosed by a holder made of powdered iron and epoxy to contain the flux. $L_1$ is measured to be 88 mH and $L_2 = 21.9$ mH at 1000 Hz. (a) Check the validity of Eq. (11-24). (b) What is the turns ratio $N_1/N_2$? (c) If $Z_2 = 100 \, \Omega$ in Fig. 11-4, what is $Z_1$?

**Solution.** (a)

$$\frac{L_1}{L_2} = 4.01, \qquad \frac{N_1^2}{N_2^2} = 4.0$$

(b) 2/1.
(c) From Eq. (11-27),

$$Z_1 = 100(4) = 400 \, \Omega.$$

## 11-4 The Tapped Inductor

In Fig. 11-5, 11 turns are wound around a metallic core so that coil losses are negligible. To see how $R_s$ is transformed to appear between the input terminals, define $R_p$ as the resistance presented to $E$. Voltage per turn of input and output are equal, or

$$\frac{E}{N} = \frac{V_s}{N_s} \qquad (11\text{-}28)$$

Power supplied to the input of $E^2/R_p$ equals power supplied to the output (assuming no losses), or

$$\frac{E^2}{R_p} = \frac{V_s^2}{R_s} \qquad (11\text{-}29)$$

Solving Eq. (11-28) for $V_s$ and substituting into Eq. (11-29),

$$R_p = R_s \left(\frac{N}{N_s}\right)^2 = R_s n^2 \qquad (11\text{-}30)$$

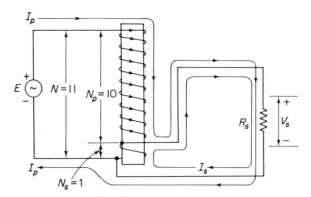

**Fig. 11-5.** Tapped inductor.

Where $n$ is the *turns ratio* of

$$n = \frac{N}{N_s} = \frac{N_p + N_s}{N_s} = \frac{N_p}{N_s} + 1 = n' + 1 \qquad (11\text{-}31)$$

and $n'$ is the ratio of $N_p$ to $N_s$ in Fig. 11-5.

*Example 11-8.* In Fig. 11-5, $R_s = 10/11\ \Omega$ and $E = 110$ V. Find (a) resistance presented to $E$, (b) voltage across load resistor $R_s$, (c) current through the load resistor.

*Solution.* (a) From Eq. (11-30),

$$R_p = \frac{10}{11}\left[\frac{11(11)}{1}\right] = 110\ \Omega$$

(b) From Eq. (11-28),

$$V_s = \frac{1}{11}(110) = 10\text{ V}$$

(c) Current through $R_s$ is $I = V_s/R_s = 10 \div 10/11 = 11$ A. Check: $I_p = E/R_p = 1$ A, from $I_s N_s = I_p N_p$, $I_s = 1(10) = 10$ A. $I = I_p + I_s = 11$ A.

*Example 11-9.* A capacitor of 121 pF replaces $R_s$ in Fig. 11-5. How does this capacitor appear to $E$?

*Solution.* Employing impedances in Eq. (11-30),

$$Z_p = \frac{1}{\omega C_p} = n^2 Z_s = n^2 \frac{1}{\omega C_s}$$

or

$$C_p = \frac{C_s}{n^2} = \frac{121}{121} = 1\text{ pF}$$

One useful property of the tapped inductor is demonstrated by varying the frequency of current source $I$ in Fig. 11-6(a) until resonance occurs at the natural frequency determined by $L$ and $C$. When the current source is moved to excite the tank between two other points in Fig. 11-6(b), the resonant frequency will not change. But from Eq. (11-30), the current source in Fig. 11-6(b) will see a capacitance as shown in Fig. 11-6(c) of

$$\frac{1}{C_p(N_1)^2} = \frac{Z_p}{N_1^2} = \frac{Z_s}{N^2} = \frac{1}{C_s N^2}$$

or

$$C_p = C_s \frac{N^2}{N_1^2} \qquad (11\text{-}32)$$

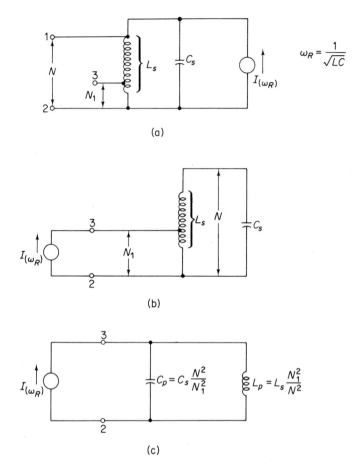

**Fig. 11-6.** The tuned circuit resonates at the same frequency in (a) or (b). Both $L_s$ and $C_s$ are transformed in (b) to the values shown in (c).

But $C_p$ must tune with some other *transformed* inductance $L_p$ since the tuned circuit frequency cannot change, or

$$\omega_R^2 = \frac{1}{L_s C_s} = \frac{1}{C_p L_p} \tag{11-33}$$

Substituting for $C_p$ from Eq. (11-32),

$$L_p = \left(\frac{N_1^2}{N^2}\right) L_s \tag{11-34}$$

***Example 11-10.*** A tuned circuit is required with 5 $\mu$H tuned with a capacitance to 455 kHz. However, only 500-$\mu$H inductors are available. Where should the inductor be tapped and what capacitor should be connected across it?

Sec. 11-4                                          The Tapped Inductor    341

*Solution.* From Eq. (11-34), $L_p = 5 \ \mu H$ and $L_s = 500 \ \mu H$:

$$5 = 500 \frac{N_1^2}{N^2} \quad \text{and} \quad N = 10 N_1$$

Capacitor $C_s$ must resonate with $L_s$ and, from Eq. (11-1),

$$C_s = \frac{1}{L_s \omega_R^2} = \frac{1}{(500 \times 10^{-6})(2\pi \times 455 \times 10^3)^2} = 244 \text{ pF}$$

The actual and equivalent tuned circuits are shown in Figs. 11-7(a) and (b) respectively.

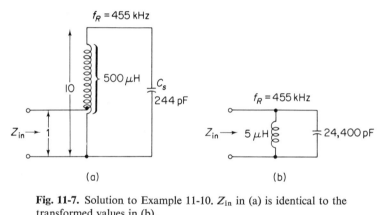

**Fig. 11-7.** Solution to Example 11-10. $Z_{in}$ in (a) is identical to the transformed values in (b).

*Example 11-11.* In the circuit of Fig. 11-8(a), find the effective capacitance and resistance appearing between terminals $AA'$. The transformer has a turns ratio of 10:1 with primary tap at 10:2.

*Solution.* (a) Transform $C_i$ to appear across $AA'$ as $C_i'$ from Eq. (11-27):

$$\frac{1}{C_i}\left(\frac{1}{1^2}\right) = \frac{1}{C_i'}\left(\frac{1}{N^2}\right) \quad \text{or} \quad C_i' = \frac{500}{100} = 5 \text{ pF}$$

(b) Transform $C_o$ as $C_o'$ by Eq. (11-32):

$$\frac{1}{C_o}\left(\frac{1}{A^2}\right) = \frac{1}{C_o'}\left(\frac{1}{N^2}\right) \quad \text{or} \quad C_o' = 100\left(\frac{4}{100}\right) = 4 \text{ pF}$$

(c) Effective capacitance across terminals $AA'$ is known to be $C = C_T + C_o' + C_i' = 109 \text{ pF}$ in Fig. 11-8(b).
(d) $R_L$ is transformed as $R_L'$ from Eq. (11-30):

$$\frac{R_L'}{N^2} = \frac{R_L}{1} \quad \text{or} \quad R_L' = 100(1) \text{ k}\Omega = 100 \text{ k}\Omega$$

The resonant frequency of Fig. 11-8(a) is found from $\omega_R = 1/\sqrt{L_1 C}$ and bandwidth from $1/(R_L' C)$.

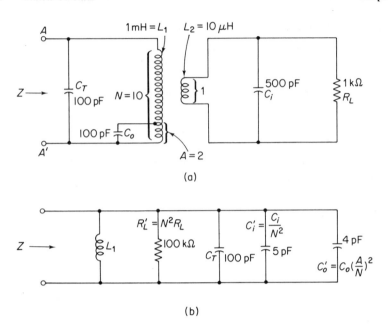

**Fig. 11-8.** Solution to Example 11-11. Impedance $Z$ is identical in (a) and (b).

## 11-5 The Tuned Amplifier

When a tuned circuit is the load on an amplifier the resulting *tuned amplifier* will provide gain for those frequencies selected by the tuned circuit and, ideally, will reject all others. Our studies will employ the hybrid-pi model and will be valid for narrow-band tuned amplifiers in the frequency range between dc and $\omega_\beta$.

A basic low-frequency tuned amplifier is introduced in Fig. 11-9(a) and modeled in Fig. 11-9(b). Voltage gain will be maximum at resonant frequency $\omega_R$ and is

$$A_v = \frac{V_o}{E_i} = -\frac{\beta_0 R}{r_x + r_\pi} \quad \text{at } \omega_R \quad (11\text{-}35)$$

where $R = R_L \| R_p$, and $R_p$ is found from Eq. (11-17). Bandwidth is given by Eq. (11-2). Voltage gain is frequency dependent on $Z_L = R \| X_L \| X_c$ and not on $g_m V$. Current $g_m V$ will remain constant as long as $\omega_R$ is below that frequency range where we must account for $C_\pi$ and $C_\mu$. Voltage gain may be stabilized against transistor-parameter variation and operating-point stability enhanced by adding an unbypassed emitter resistor. Voltage gain at resonance will then be approximately $R/R_E$.

Sec. 11-5                                   The Tuned Amplifier   343

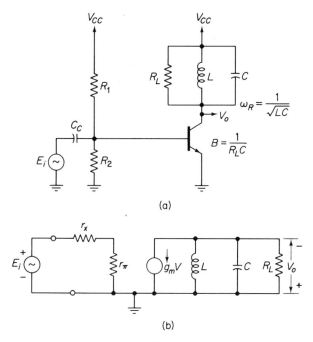

Fig. 11-9. The tuned amplifier in (a) is modeled in (b).

Load $Z_L$ will affect the input circuit by feedback through $C_\mu$ as shown in Fig. 11-10. $Z_L$ will be a maximum of $R_L$ at $\omega_R$ and be negligible at frequencies

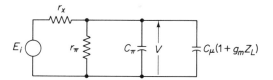

Fig. 11-10. Equivalent input model of Fig. 11-9(a) at high frequencies.

more than 10% of $\omega_R$ on either side of resonance. Thus as long as $\omega_R$ is less than approximately 90% $\omega_H$, the tuned circuit and not the transistor will control variation of voltage gain with frequency. Since any signal generator resistance will lower $\omega_H$ it is good practice to drive a tuned circuit from a low-impedance source.

In Fig. 11-9(a) inductor $L$ not only tunes with $C$ but also with the output capacitance of the transistor as in Fig. 11-11(a). Since $r_x$ is small with respect to $r_\pi$, voltage $V$ approximately equals $I_\mu r_x$ and, from the outside loop equation,

$$I = \frac{E_T}{r_x + \dfrac{1}{j\omega C}} \qquad (11\text{-}36)$$

**344** Tuned Circuits   Chap. 11

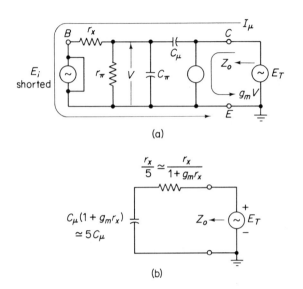

**Fig. 11-11.** The test circuit in (a) yields the approximate output capacitance, in (b), presented to the tuned circuit in Fig. 11-9.

But $Z_o$ equals test voltage $E_T$ divided by $I + g_m V$ and substituting Eq. (11-36),

$$Z_o = \frac{E_T}{I + g_m V} = \frac{E_T}{I(1 + g_m r_x)} = \frac{1}{(1 + g_m r_x)}\left(r_x + \frac{1}{j\omega C}\right) \quad (11\text{-}37)$$

Equation (11-37) is modeled in Fig. 11-11(b). Typical values are $g_m = 0.04$, $C_\mu = 2$ pF, and $r_x = 100\,\Omega$ so that the transistor presents a shunt impedance to the tuned circuit of 20 $\Omega$ in series with 10 pF. As long as $C$ is 10 times (or more) greater than 10 pF we may safely ignore the affects of $C_\mu$ on $\omega_R$. This approximation holds for low values of source resistance and at frequencies below $\omega_\beta$.

A gain-bandwidth product can be derived by rewriting Eq. (11-35) as $g_m R_L$ and multiplying by Eq. (11-2) to obtain

$$A_v B = \frac{g_m}{C} \quad (11\text{-}38)$$

Equation (11-38) is useful in predicting whether the operating point will give an adequate $g_m$.

*Example 11-12.* Calculate (a) $\omega_R$, (b) bandwidth, and (c) voltage gain at $\omega_R$ for the circuit of Fig. 11-9(a) with values given as follows: $L_1 = 0.5$ mH, $Q = 100$, $r_x = 10\% \, r_\pi$, $I_C = 2$ mA, $C = 0.2\,\mu$F, $\beta_0 = 100$, and $R_L = 1$ k$\Omega$.

Sec. 11-5                              The Tuned Amplifier    345

*Solution.* (a) From Eq. (11-1), $\omega_R = (0.5 \times 10^{-3} \times 0.2 \times 10^{-6})^{-1/2} = 100$ krad/sec.
(b) From Eq. (11-17), $R_p = \omega L_p Q = (10^5)(0.5 \times 10^{-3})(100) = 5$ k$\Omega$. The actual resistance across $L$ is $R = R_p \| R_L = 833\ \Omega$ and, from Eq. (11-2),

$$B = \frac{1}{RC} = \frac{1}{833 \times 0.2 \times 10^{-6}} = 6000 \text{ rad/sec}$$

(c) $g_m = 2/25 = 0.08$, $r_\pi = 100/0.08 = 1250\ \Omega$, $r_x = 0.1 r_\pi = 125\ \Omega$ and from Eq. (11-35),

$$\frac{V_o}{E_i} = -\frac{100(833)}{125 + 1250} = 60$$

*Example 11-13.* (a) Find $\omega_H$ for Example 11-12 assuming $C$ and $L$ are not present and $C_\pi = 100$ pF, $C_\mu = 6$ pF. (b) How does $C_\mu$ affect the effective capacitance of the tuned circuit?

*Solution.* (a) $r_x \| r_\pi = 0.1 r_\pi / 1.1 = 113\ \Omega$,

$$C_\pi + C_\mu(1 + g_m R_L) = 100 + 6(1 + 80) = 586 \text{ pF}$$

$$\omega_H = \frac{1}{(113)(586 \times 10^{-12})} = 15.1 \text{ Mrad/sec}$$

(b) From Fig. 11-11(b), $C_\mu$ is transformed across $C$ as $C'_\mu$, where

$$C'_\mu = C_\mu(1 + g_m r_x) = 6[1 + 0.08(125)] = 66 \text{ pF}$$

Clearly $C'_\mu$ is negligible with respect to $C$ and since $\omega_H \gg \omega_R$ the tuned circuit alone determines $\omega_R$. Observe that had we neglected $R_p$ in Example 11-12, $B$ would seem lower at 5000 rad/sec and gain higher at almost 73.

*Design Example 11-14.* The circuit of Fig. 11-12(a) is to be designed for a pass band of 10 kHz = 62.8 krad/sec at a resonant frequency of $\omega_R = 455$ kHz = 2.86 Mrad/sec with a voltage gain of 30 or 24 dB. Neglect coil losses and assume a low-resistance source.

*Solution.* From Eq. (11-2),

$$C = \frac{1}{62.8 \times 10^3 \times 10^3} = 15{,}900 \text{ pF}$$

$L$ is found from Eq. (11-1):

$$L = \frac{1}{(2.86 \times 10^6)^2 \times 15900 \times 10^{-12}} = 7.7\ \mu\text{H}$$

$L$ is too small for stock inductors and $C$ is too large for low-loss mica or

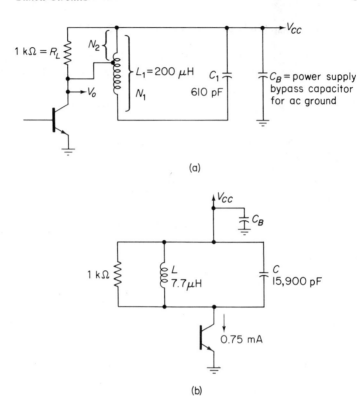

**Fig. 11-12.** The practical tuned circuit in (a) is viewed by the transistor as the model in (b).

ceramic capacitors. Select a convenient inductor of $L_1 = 200 \ \mu H$ and tap it so that $R_L$ sees 7.7 $\mu H$ through the turns ratio of

$$\left(\frac{N_1}{N_2}\right)^2 = \frac{L_1}{L} = \frac{200}{7.7} = 26.0 \qquad \frac{N_1}{N_2} = 5.1$$

$R_L$ must also see 15,900 pF through the turns ratio of 5.1, so we connect a capacitor $C$ of

$$C_1 = C\left(\frac{N_2}{N_1}\right)^2 = \frac{15,900}{26} = 610 \ \text{pF}$$

To establish the voltage gain find $g_m$ from Eq. (11-38).

$$g_m = A_v BC = 30 \times 62.8 \times 10^3 \times 15,900 \times 10^{-12} = 0.03$$

Collector bias current must be set at $I_C = 25g_m = 0.75$ mA. The equivalent tuned circuit is modeled in Fig. 11-12(b). In practice $L_1$ would have an adjustable core or a small adjustable capacitor would be connected across $C_1$ for purposes of fine tuning.

It is convenient to think of $L_1$ and $C_1$ being transformed to appear as loads to the transistor between collector and ac ground. In Design Example 11-15 we remove resistor $R_L$ but apply a real load on the tuned circuit as the input of another transistor.

**Design Example 11-15.** It is required to deliver a voltage gain of 30 at $\omega_R = $ 455 kHz with a pass band of 10 kHz to the input of a second stage with an equivalent input resistance of $R_i = 400\ \Omega$ and $C_i = 100$ pF. We retain the tapped inductor of Example 11-14 as the primary of an auto transformer to allow a comparison, but in this example we will account for inductor losses by assuming $Q = 100$ for $L_1$.

**Solution.** In Fig. 11-13(a), $R_i$ and $C_i$ are the equivalent inputs of the next stage. They will be transformed across $L_1$ by $N_1$ and $N_3$ as $R_i'$ and $C_i'$, respectively, in Fig. 11-13(b). Losses of $L_1$ are found from Eq. (11-17) as $R_p = \omega_R L_1 Q = 57.2$ k$\Omega$. But the parallel combination of $R_p$ and $R_i'$ must be transformed by $N_1$ and $N_2$ to yield $R_L = 1$ k$\Omega$. Since $N_1$ and $N_2$ are fixed by the tap we can solve for the required value of

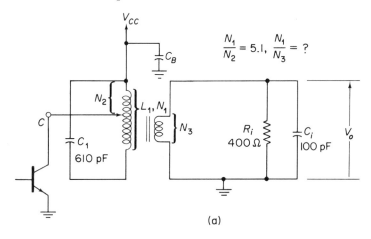

(a)

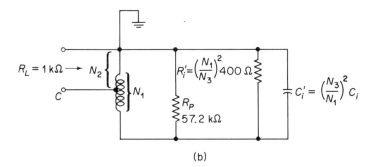

(b)

**Fig. 11-13.** Secondary load $R_i$ and $C_i$ on the IF transformer in (a) is transformed across the primary in (b).

$$R_p \| R_i' = R_L \left(\frac{N_1}{N_2}\right)^2 = 1000(26) = 26 \text{ k}\Omega$$

Now $R_i'$ can be found from $57.2R_i'/(57.2 + R_i') = 26$, and $R_i' = 48$ k$\Omega$. The required turns ratio of $N_1/N_3$ must transform $R_i'$ from 400 $\Omega$ to 48 k$\Omega$, or

$$\frac{R_i'}{N_1^2} = \frac{R_i}{N_3^2} \quad \text{and} \quad \frac{N_1}{N_3} = \left(\frac{48000}{400}\right)^{1/2} \cong 11$$

From Fig. 11-13(b), the capacitance of $C_i'$ is negligible at $100/(11 \times 11) \cong 1$ pF and the value of $C_1 = 610$ pF is retained from Example 11-14. If $C_i'$ were not negligible we would merely choose a smaller $C_1$ so that $C_1 + C_i' = 610$ pF.

Output voltage is developed across $R_i$ in Fig 11-13 and not the tuned circuit. Output voltage and voltage gain at the collector will be stepped up through $N_2$ to $N_1$ by a factor of $N_1/N_2 = 5.1$, then down again by $N_1$ and $N_3$ by a factor of $N_3/N_1 = 1/11$. Thus voltage gain from transistor output to $R_i$ will be $5.1/11 = 0.47$ so the transistor's gain must be raised to $30/0.47 = 64$ by increasing $I_C$ to $0.75/0.47 = 1.6$ mA.

The remarkable impedance-transforming properties of the tapped inductor and transformer were demonstrated in Example 11-14 and 11-15. Further examples will be studied after we consider the effects of more than one tuned circuit.

## 11-6 Bandwidth Reduction in Cascade

When a band of frequencies is passed through more than one tuned circuit, there is an uneven attenuation of the pass band that tends to shrink the net bandwidth. For example, suppose two identical tuned amplifiers could each pass a bandwidth of 10 kHz, with a gain of 1 at a resonant frequency of 455 kHz. Individual characteristics of each amplifier are shown in Fig. 11-14

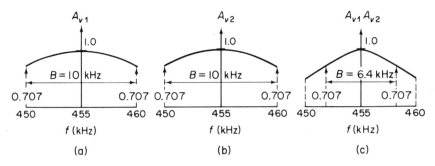

Fig. 11-14. Signals passing through identical tuned circuits in (a) and (b) are unevenly attenuated to reduce net bandwith in (c).

Sec. 11-6    Bandwidth Reduction in Cascade    349

(a) and (b). A 10-V signal at 455 kHz would pass through amplifiers 1 and 2 unattenuated to develop an output voltage of 10 V. Next, a signal of 10 V at 450 kHz passes through amplifier 1 and exits at 7.07 V, then enters amplifier 2 and exits at $7.07(0.707) = 5$ V. Since 5 V is less than $0.707V_o$ at $\omega_H$, clearly 450 kHz is not within the net pass band and bandwidth has been reduced by cascading.

As was shown in Chapter 6, voltage gain at high frequencies, $A_H$, can be expressed in terms of the upper cutoff frequency $\omega_H$ and midfrequency gain $A_{mid}$ by

$$A_H = \frac{A_{mid}}{1 + j\frac{\omega}{\omega_H}} \tag{11-39}$$

The magnitude of $A_H$ is

$$|A_H| = \frac{A_{mid}}{\sqrt{1 + \left(\frac{\omega}{\omega_H}\right)^2}}$$

If $n$ identical amplifiers are connected in cascade, net voltage gain will equal the product of each identical stage gain, or

$$A_{cascade} = (A_H)^n = \left[\frac{A_{mid}}{\sqrt{1 + \left(\frac{\omega}{\omega_H}\right)^2}}\right]^n \tag{11-40}$$

At the cascade cutoff frequency $\omega_{HC}$, the value of $A_{cascade}$ will be reduced by 0.707 or, divided by $\sqrt{2}$,

At $\omega_{HC}$

$$A_{cascade} = \frac{(A_{mid})^n}{\sqrt{2}} \tag{11-41}$$

Substituting Eq. (11-41) into Eq. (11-40) and solving for $\omega_{HC}$ gives

$$\omega_{HC} = \omega_H(2^{1/n} - 1)^{1/2} \tag{11-42}$$

In a similar fashion it can be shown that the single-stage gain with a lower cutoff frequency of $\omega_L$ expressed by

$$A_L = A_{mid}\left(1 + j\frac{\omega_L}{\omega}\right)$$

will have a cascaded lower cutoff frequency $\omega_{LC}$ of

$$\omega_{LC} = \omega_L(2^{1/n} - 1)^{1/2} \tag{11-43}$$

Bandwidth $B_C$ of the cascade is found in terms of the individual stage bandwidth $B$ by subtracting Eq. (11-43) from Eq. (11-42) to obtain

$$B_C = \omega_{HC} - \omega_{LC} = B(2^{1/n} - 1)^{1/2} \qquad (11\text{-}44)$$

where $B = \omega_H - \omega_L$ of the individual, identical stages.

If one tuned circuit with a bandwidth of 10 kHz formed the load on an amplifier and a second identical tuned circuit were present at the amplifier's input, the resulting bandwidth (see Fig. 11-14) is reduced, because $n = 2$, in Eq. (11-44) to

$$B_C = 10[2^{1/2} - 1]^{1/2} = 6.4 \text{ kHz}$$

Interaction can occur between these tuned circuits because of internal feedback through $C_\mu$ of the transistor and produce oscillation through a negative input resistance. This point is investigated in Chapter 13. It is possible to cancel the negative input resistance by introducing feedback with an external neutralizing capacitor from output to input with a technique called *neutralization*. A phase reversal is necessary at the output by a transformer so that external neturalizing current can cancel internal feedback current. However, neutralization is effective over a rather small frequency range and may well be unreliable at all volume levels, particularly where automatic volume control is featured.

A better engineering solution to the problems of internal feedback is the integrated circuit common-collector to common-base or common-emitter to common-base configuration. These low cost, reliable, and space-saving packages practically eliminate internal feedback and do not require neutralization. A significant advantage of the configurations is that one tuned circuit can be adjusted without affecting another.

## 11-7 Multiple-Tuned-Circuit Amplifier

A single-tuned common-collector to common-base amplifier is given in Fig. 11-15(a) to show how internal feedback is minimized. In the model of Fig. 11-15(b), feedback currents through $C_{\mu 2}$ are bypassed through $r_x$ to ground and $C_{\mu 2}$ will have negligible affect on $C$ of the tuned circuit. The collector of $C_1$ is at ac ground because of $C_B$ or the power supply so that no voltage is fed back through $C_{\mu 1}$. Thus output and input are isolated.

To learn what input capacitance is seen by $E_i$, refer to the model in Fig. 11-16. Neglecting $r_x$ and assuming that $Q_3$ offers infinite resistance to ground, $E_i$ sees $C_{\mu 1}$ in parallel with the series combination of $C_{\pi 1}$ and $C_{\pi 2}$.

Sec. 11-7        Multiple-Tuned Circuit Amplifier    351

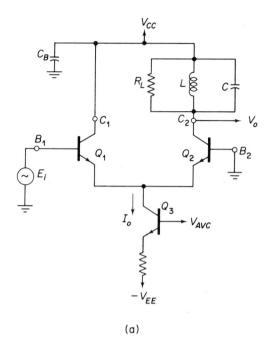

(a)

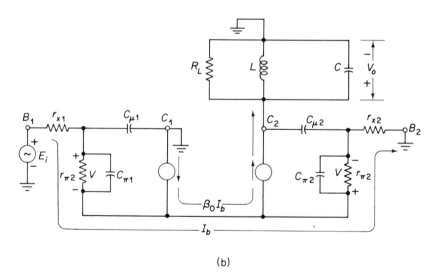

(b)

**Fig. 11-15.** The tuned common-collector to common-base amplifier in (a) is modeled in (b).

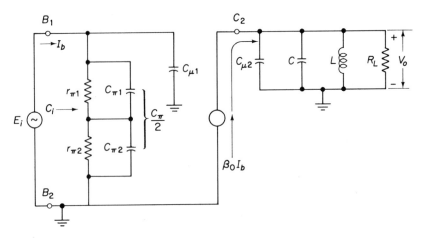

**Fig. 11-16.** Simplified model of Fig. 11-15(b) to show input capacitance $C_i = C_\mu + C_\pi/2$.

For identical $Q_1$ and $Q_2$, $C_{\pi 1} = C_{\pi 2} = C_\pi$ so that the series combination equals $C_\pi/2$. Thus input capacitance of the differential pair is lower than that of a single transistor and input capacitance $C_i$ is

$$C_i = C_\mu + \frac{C_\pi}{2} \tag{11-45}$$

**Example 11-16.** The amplifier of Fig. 11-17 is to develop a voltage gain of $V_o/E_g$

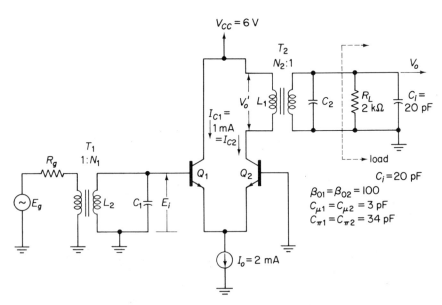

**Fig. 11-17.** Differential amplifier with two tuned circuits.

Sec. 11-7  Multiple-Tuned-Circuit Amplifier  353

$= 40$ at $\omega_R = 45$ MHz $= 283$ Mrad/sec with a net bandwidth of 4.5 MHz $= 28.3$ Mrad/sec. Transformers $T_1$ and $T_2$ are in tuned circuits that must be broad banded because they are in cascade. Since $n = 2$, employ Eq. (11-14) to find that the bandwidth of each tuned circuit must be $28.3/0.64 = 44.2$ Mrad/sec. Design the input tuned circuit only by (a) calculating values for $C_1$, $L_2$, and $N_1$ and (b) finding gain $E_i/E_g$. Disregard coil losses.

*Solution.* (a) Calculate $g_m = I_C/25 = 0.040$ and $r_{\pi 1} = r_{\pi 2} = \beta_0/g_m = 2500 \, \Omega$. Input capacitance is found from Eq. (11-45):

$$C_i = \frac{C_\pi}{2} + C_\mu = 20 \text{ pF}$$

A Norton equivalent of $E_g - R_g$ is formed in Fig. 11-18(a), where $R_i = 2r_\pi = 5$ k$\Omega$ to complete the known elements connected to $T_1$. Choose $C_1$ to be 30 pF (variable) so that the total tuning capacitance is $C = 50$ pF. From Eq. (11-2), find the net resistance required across $L_2$.

$$R_p = \frac{1}{BC} = \frac{1}{44.2 \times 10^6 \times 50 \times 10^{-12}} = 452 \, \Omega$$

$R_g$ will transform across $L_2$ as $R_g N_1^2$ in Fig. 11-18(b) and the parallel combination of $R_i$ and $R_g N_1^2$, must equal $R_p$. Solving for $N_1$,

$$452 = 5000 \, || \, (50 N_1)^2$$
$$464 = 50 N_1^2 \quad \text{and} \quad N_1 = 3.06$$

Evaluate $L_2$ from Eq. (11-1):

$$L_2 = \frac{1}{(283 \times 10^6)^2 \times 50 \times 10^{-12}} = 0.25 \, \mu\text{H}$$

(b) As shown in Fig. 11-18(c), $R_i$ is transformed into the primary of $T_1$ as $5000/N_1^2 = 540 \, \Omega$, forming a voltage divider so that $0.915 E_g$ is applied to the primary of $T_1$. Its secondary signal voltage $E_i$ is higher by the turns ratio and, from Fig. 11-18(d), is

$$E_i = N_1(0.915 E_g) = 2.8 E_g$$

*Example 11-17.* In the circuit of Fig. 11-17 we want $V_o/E_g = 40$ and have found $E_i/E_g = 2.8$ in Example 11-16. Voltage gain $V_o/E_i$ will depend on $g_m$, and the transformed resistance of $R_L$ into the primary as well as $T_2$. We arbitrarily pick a convenient value of $N_2 = 2$ to fix one of the variables so that $V_o/V_o' = \frac{1}{2}$ and solve for $V_o'/E_i$ from

$$\frac{V_o}{E_g} = 40 = \frac{E_i}{E_g}\left(\frac{V_o'}{E_i}\right)\left(\frac{V_o}{V_o'}\right) = 2.8 \left(\frac{V_o'}{E_i}\right)\left(\frac{1}{2}\right)$$

$$\frac{V_o'}{E_i} = 28.6$$

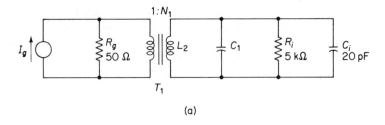

(a)

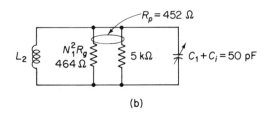

(b)

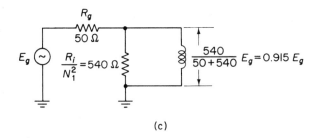

(c)

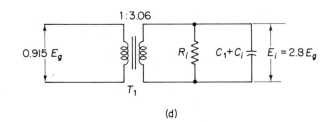

(d)

**Fig. 11-18.** The input of Fig. 11-17 is modeled in (a). $R_g$ is transformed into the secondary in (b) to find $N_1$. In (c) the actual primary signal voltage is less than $E_g$ but is transformed to the secondary by $N_1$ in (d).

(a) Calculate the net effective resistance $R_p$ required across $L_1$ to develop $V'_o/E_i = 28.6$ (b) What shunt resistance must be connected across the primary to realize $R_p$? (c) What effective capacity must exist across the primary to give a bandwidth of 44.2 Mrad? (d) Calculate $L_2$ for $\omega_R = 283$ Mrad/sec. (e) What value of capacitor $C_2$ should be added to fulfill the requirements of (c)?

Sec. 11-7  Multiple-Tuned-Circuit Amplifier  355

Solution. (a) $\dfrac{V'_o}{E_i} = 28.6 = \dfrac{\beta_0 R_p}{2r_\pi} = \dfrac{100 R_p}{5000}$, $R_p = 1430$

(b) $R_L$ will be transformed across the primary as $N_2^2 R_L = 4(2000) = 8000$. An external shunt resistor $R_A$ must be added across $L_1$ so that

$$1430 = R_p = R_A \| 8000 \quad \text{or} \quad R_A = 1740 \ \Omega$$

(c) Effective capacitance $C$ required across $L_1$ is

$$C = \dfrac{1}{BR_p} = \dfrac{1}{44.2 \times 10^6 \times 1430} = 15.8 \ \text{pF}$$

(d) $L_1 = \dfrac{1}{\omega_R^2 C} = \dfrac{1}{(283 \times 10^6)^2 \times 15.8 \times 10^{-12}} = 0.79 \ \mu\text{H}$

(e) $C$ transforms into the secondary of $T_2$ as $N_2^2 C = 4(15.8) = 63 \ \text{pF}$. Since 20 pF is already present in the form of $C_i$, add $63 - 20 = 43 \ \text{pF}$ as $C_2$. The resulting connections on $T_2$ are shown in Fig. 11-19.

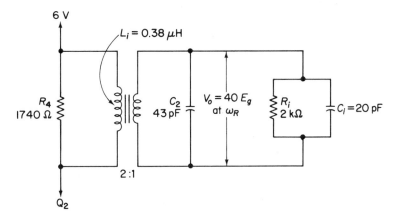

Fig. 11-19. Solution to Example 11-17.

Examples 11-16 and 11-17 illustrate the general technique in designing or analyzing tuned amplifiers. Capacitance and resistance is transformed to one side of the transformer for calculating effective values to meet resonant frequency and bandwidth stipulations. Effective values may be retransformed to the other side of the transformer if convenient. Existing capacitance or resistance is combined with added capacitance or resistance to make up the effective values. Any turns ratio may be selected as in Example 11-17 and a design carried through. Experience gained from the design will allow a better selection of turns ratio to minimize added components.

## PROBLEMS

**11-1** In the circuit of Fig. 11-1, $C = 150$ pF. Calculate $L$ and $R$ for a resonant frequency of 600 kHz with a bandwidth of 6000 Hz.

**11-2** Given a high-$Q$, 1-mH inductor. (a) Design a parallel tuned circuit to resonate at 455 kHz with a bandwidth of 10 kHz. (b) What is the tuned circuit's impedance at the resonant frequency?

**11-3** A $Q$-meter measurement of an inductor gives $Q = 80$, $C = 150$ pF at 600 kHz. (a) Draw a parallel and series model of the inductor based on these measurements. (b) If we connect $C = 150$ pF across the inductor, what is the bandwidth of the resulting tuned circuit?

**11-4** A resistor of 257 k$\Omega$ is connected across the tuned circuit of Example 11-6. What is the resulting bandwidth? Repeat for a resistor of 25.7 k$\Omega$.

**11-5** In Fig. 11-4(a), the ratio of $N_1$ to $N_2$ is 5; $Z_2$ is a resistor equal to 10 $\Omega$; inductance of $N_2$ is measured (with $Z_2$ and $E_1$ removed) as 20 mH. (a) What is the self-inductance of $N_1$, assuming an ideal transformer? (b) What value of $E_1$ is required to yield $E_2 = 1$ V, and what will be the values of $I_1$ and $I_2$? (c) What is the value of $Z_1$?

**11-6** In Fig. 11-5, $R_s = 1$ k$\Omega$ and $N_p/N_s = 9$. (a) What resistance is presented to $E$? (b) If $E = 10$ V, find $V_s$, $I_p$, and $I_s$.

**11-7** Connect $E$ in place of $R_s$ and $R_s$ in place of $E$ in Fig. 11-5. With $R_s = 110$ $\Omega$ and $E = 10$ V, find (a) resistance presented to $E$, and (b) voltage developed across $R_s$. Compare your results with Example 11-8.

**11-8** A capacitor of 10 pF replaces $E$ in Fig. 11-5. What capacitance is presented by the secondary terminals?

**11-9** Only 250-$\mu$H inductors are available for the problem in Example 11-10. Where should the tap be placed and what value of $C_s$ is required? Sketch the resulting circuit as in Fig. 11-7.

**11-10** Find the resonant frequency and bandwidth of the circuit in Fig. 11-8.

**11-11** $R_L$ is changed to 500 $\Omega$, $C_i$ to 2500 pF, and $C_o$ to 500 pF in Fig. 11-8(a). Find the effective capacitance and resistance appearing across terminals $AA'$.

**11-12** If $Q$ of the inductor in Example 11-12 equals 50, what changes would result in bandwidth and voltage gain?

**11-13** Revise Design Example 11-14 for a pass band of 20 kHz at the same resonant frequency and voltage gain.

*11-14* What changes are required to meet the specifications of Design Example 11-15 if $C_i = 200$ pF and $R_i = 800\,\Omega$ in Fig. 11-13(a)?

*11-15* Three identical tuned amplifiers, each with a bandwidth of 10 kHz and voltage gain of 10 are cascaded. What is the overall (a) bandwidth, (b) voltage gain at $\omega_R$, (c) voltage gain at the lower and upper cutoff frequencies?

*11-16* What changes occur in Example 11-16 if $\omega_R = 455$ kHz and the bandwidth $= 10$ kHz?

*11-17* What changes occur in Example 11-17 if $\omega_R = 455$ kHz and the bandwidth $= 10$ kHz?

# Chapter 12

- **12-0** INTRODUCTION ...............................359
- **12-1** NEGATIVE FEEDBACK AND VOLTAGE GAIN ........360
- **12-2** FUNDAMENTAL CONCEPTS OF NEGATIVE FEEDBACK..363
- **12-3** SINGLE-STAGE AMPLIFIERS WITH NEGATIVE FEEDBACK ...................................365
- **12-4** REDUCTION OF DISTORTION BY NEGATIVE FEEDBACK...368
- **12-5** MULTISTAGE PARALLEL-OUTPUT TO SERIES-INPUT NEGATIVE FEEDBACK ..........................371
- **12-6** MULTISTAGE SERIES-OUTPUT TO PARALLEL-INPUT NEGATIVE FEEDBACK ..........................375
- **12-7** MULTISTAGE PARALLEL-OUTPUT TO PARALLEL-INPUT NEGATIVE FEEDBACK ............381
- **12-8** MULTISTAGE SERIES-OUTPUT TO SERIES-INPUT NEGATIVE FEEDBACK ..........................385
- **12-9** EFFECT OF NEGATIVE FEEDBACK ON RESISTANCE LEVELS ............................389
- **12-10** NEGATIVE FEEDBACK AND FREQUENCY RESPONSE ..390
- **12-11** DESIGN AND ANALYSIS EXAMPLES ................391
- **12-12** OSCILLATION IN AMPLIFIERS ....................396
- **12-13** OSCILLATORS ..............................398
- **12-14** PHASE-SHIFT OSCILLATORS .....................399
- **12-15** TUNED-CIRCUIT OSCILLATORS ..................403
- PROBLEMS ...................................409

# Negative and Positive Feedback

## 12-0 Introduction

Amplifier characteristics, such as voltage gain, depend on transistor parameters and supply voltage. Variations in supply voltage and wide variations in transistor parameters make for unpredictable performance unless constant manual control is excercised from outside the amplifier or automatic control is built into the amplifier.

Feedback is a fundamental method of introducing stability or predictability to an amplifier by feeding a fraction of the outback back to the input so that the amplifier's input is driven by a combination of signals from output and signal source. If signals from the output oppose or subtract from the source signals, the amplifier experiences *negative feedback*. On the other hand, signals from the output can be connected to aid, or even to replace the source signals and cause *positive feedback*, a condition normally associated with oscillation.

We shall be concerned primarily with feedback amplifiers employing a *significant* amount of feedback or the amount required for amplifier performance to depend on precision feedback elements rather than transistor characteristics. Since the feedback elements are usually linear passive resistors, gain expressions of amplifier performance will be both precise and simple.

When considering gain we can select for output variables either output voltage or output current and can select for input variables either input voltage or input current. Since gain is an output-to-input ratio we can choose four possible combinations of gain. Usually we work with voltage gain or current gain but could just as well work with the ratio of output voltage to input current (transresistance) or output current to input voltage (transconductance). All of the improvements obtained from the addition of negative feedback are paid for by loss of gain, so our introduction begins with a study of the effect of negative feedback on the more familiar voltage gain.

## 12-1 Negative Feedback and Voltage Gain

Gain between source and load in Fig. 12-1(a) depends on amplifier gain $A$, and $A$ may be both unpredictable and variable. It will be shown from Fig. 12-1($b$) that these deficiencies in $A$ do not matter, so long as $A$ is large enough, and that gain $A_f$ between source and load depends almost entirely on the feedback network. Initially we will stipulate that (1) no significant amount of $E_i$ is transmitted directly from left to right through $f$ to affect the output, and (2) the feedback network's loading on the output is negligible.

In Fig. 12-1(b), $f$ consists of a precision resistor network. Output voltage is sampled by the parallel connection of load, amplifier output, and feedback network. A precise fraction of $V_o$, or $fV_o$ is fed back and connected in series with source and amplifier input. Source voltage is opposed by $fV_o$, so the feedback is negative and the amplifier is driven by their difference:

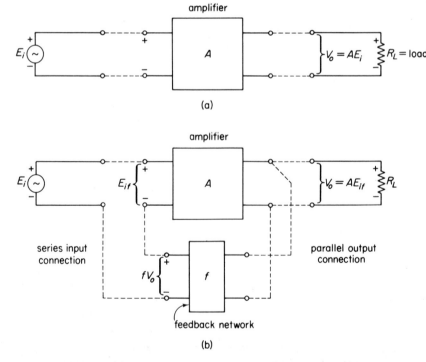

Fig. 12-1. Voltage gain $V_o/E_i$ in (a) is modified by the feedback network in (b).

Sec. 12-1   Negative Feedback and Voltage Gain   361

$$E_{if} = E_i - fV_o \qquad (12\text{-}1)$$

But $V_o = AE_{if}$ and substituting for $E_{if}$ from Eq. (12-1),

$$V_o = AE_{if} = A(E_i - fV_o) \qquad (12\text{-}2)$$

Solving for source-to-load gain with feedback $A_f$,

$$A_f = \frac{V_o}{E_i} = \frac{A}{1+fA} \qquad (12\text{-}3)$$

$E_{if}$ is multiplied by $A$ and again by $f$ in transmission around the feedback loop of Fig. 12-1(b). *Loop gain* or *return ratio* are names for the product $fA$, and its limits are of interest in Eq. (12-3). If $fA$ is small with respect to 1, $A$ approaches $A_f$ and $E_i$ approaches $E_{if}$ to signify negligible feedback. But if $fA \gg 1$, then Eq. (12-3) approaches

$$A_f = \frac{V_o}{E_i} \cong \frac{1}{f} \qquad fA \gg 1 \qquad (12\text{-}4)$$

or

$$fV_o \cong E_i \qquad fA \gg 1 \qquad (12\text{-}5)$$

A most important and fundamental conclusion is drawn from Eq. (12-5). If $fA \gg 1$, not only is the system output related to system input by precision feedback fraction $f$, but more important, *system input is approximately equal to the fed-back fraction of the output*. Normally we want $fA$ to be large and practically we make $fA$ large by making $A$ large. Then if we can determine $f$ easily we know $V_o/E_i$. This conclusion applies in principle to all of the four possible gain relations. We shown how the validity of Eqs. (12-4) and (12-5) depend on $fA$ and $A$ in example 12-1.

*Example 12-1.* In the circuit of Fig. 12-1(b), $E_i$ is held constant at 1 mV and $f$ is adjusted to 0.01. Amplifiers with gains ranging from $A = 1$ to $A = 10,000$ are inserted. (a) Plot $A_f$ versus $A$. (b) Calculate $fV_o$ to see how it approaches $E_i$ as $A$ increases. (c) Calculate $V_o$ to see how $V_o = AE_i$ at small values of $A$ and approaches $E_i/f$ at large values of $A$.

*Solution.* Calculations are arranged in Table 12-1. All voltages are in millivolts. Gain and $f$ are without units. Note that since $E_i = 1$ mV, values of $V_o$, in millivolts, and $A_f$ are equal in the fifth column.

These calculations should be studied with care. Observe that as $A$ increases, $E_{if}$ decreases just enough for $V_o$ to equal $AE_{if}$ as well as $V_o = E_i A_f$. When $fA = 10$, $fV_o$ is equal to 90% $E_i$ and when $fA = 100$, $fV_o = 99\% E_i$. The solution to (a) is plotted as a solid line in Fig. 12-2. For pur-

Table 12-1

| $E_i$ | $A$ | $fA$ | $1+fA$ | $A_f = \dfrac{A}{1+fA}$ or $V_o = A_f E_i$ | $fV_o$ | $E_{if} = E_i - fV_o$ |
|---|---|---|---|---|---|---|
| 1 | 1 | 0.01 | 1.01 | 0.99 | 0.0099 | 0.99 |
| 1 | 10 | 0.1 | 1.1 | 9.1 | 0.09 | 0.91 |
| 1 | $10^2$ | 1.0 | 2 | 50 | 0.5 | 0.5 |
| 1 | $10^3$ | 10 | 11 | 90.9 | 0.909 | 0.0909 |
| 1 | $10^4$ | 100 | 101 | 99 | 0.990 | 0.0099 |

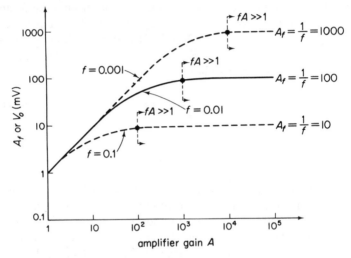

**Fig. 12-2.** Solution to Example 12-1. $A_f$ approaches $1/f$ as $fA$ gets large with respect to 1.

poses of comparison the same gain variation in $A$ is plotted as dashed lines for $f = 0.001$ and $f = 0.1$. In all three curves, a heavy dot locates $fA = 10$ to show where $A_f$ approaches $1/f$.

FEEDBACK IN DECIBELS. Degree of feedback may be expressed in decibels (dB) and is defined as the ratio of gain with feedback, $A_f$, to gain without feedback, $A$, or

$$\text{dB} = 20 \log \frac{A_f}{A} = 20 \log \frac{1}{1+fA} \qquad (12\text{-}6)$$

In Example 12-1, when $A = 10^4$ and $A_f = 99$, decibels of feedback is $20 \log (1/101) = -20(2.0) = -40$ dB.

GAIN STABILIZATION BY FEEDBACK. While Example 12-1 illustrates that gain $V_o/E_i$ is stabilized against variations in $A$ by feedback, it is informative to differentiate Eq. (12-3) to obtain

$$dA_f = \frac{dA}{(1+fA)^2} \quad (12\text{-}7)$$

Dividing Eq. (12-7) by Eq. (12-3) yields the fractional change in gain $A_f$ caused by a change in $A$:

$$\frac{dA_f}{A_f} = \frac{1}{(1+fA)}\frac{dA}{A} \quad (12\text{-}8)$$

For example, in Example 12-1, let $A$ increase from 1000 to 3000 for a fractional change of $2000/3000 = 0.67 = dA/A$. The fractional change in $A_f$ (for $f = 0.01$) will be

$$\frac{dA_f}{A_f} = \frac{1}{1+(0.01)3000}0.67 = 0.021$$

## 12-2 Fundamental Concepts of Negative Feedback

The feedback concept is superior to basic circuit analysis only if there is a simple method of obtaining the expression for $f$. Generally $f$ is a precise sample of the output variable (either output current or voltage) so that the fedback variable opposes the input variable (either source voltage or current). Gain will be the ratio of output variable to input variable and will equal $1/f$ *if the fed-back variable almost equals the input variable*. These statements are expressed generally by

$$f \times \text{sampled output} = \text{fed-back variable} \cong \text{input variable} \quad (12\text{-}9a)$$

$$A_f = \frac{1}{f} = \frac{\text{output variable}}{\text{input variable}} \quad (12\text{-}9b)$$

How do we know if output voltage or output current is the sampled output variable? Is the fed-back variable or input variable the source voltage or source current? The answers to these questions are found by focusing attention on the method by which the feedback network is connected to the output circuit and to the input circuit. If the feedback network is connected in *parallel* with load and amplifier output, *output voltage is sampled*. $V_o$ is the output variable that will be stabilized, and $fV_o$ is the fed-back variable.

If the feedback network is connected *in series* with load and amplifier, *output current is sampled*. $I_o$ is the output variable that will be stabilized, and $fI_o$ is the fed-back variable.

With respect to the input, the feedback network can be connected in series with signal source and amplifier input so that the *fed-back variable is a voltage* $E_f$ that opposes the signal source. The input source voltage $E_i$ is then the input variable. If the feedback network is connected in parallel with signal source and amplifier input, then the fed-back variable is a current $I_f$ to oppose the signal source current. Input source current $I_i$ is the input variable.

What is stabilized by the feedback connections? Again the answer is found from the method of connecting the feedback network. Parallel-output to series-input (PO-SI) feedback samples output voltage and feeds back a voltage to stabilize voltage gain $V_o/E_i = A_{fv}$.

$$fV_o = E_f \cong E_i \quad \text{and} \quad A_{fv} = \frac{1}{f} = \frac{V_o}{E_i} \quad \text{PO-SI} \quad (12\text{-}10)$$

Parallel-output to parallel-input (PO-PI) samples output voltage and feeds back a current to stabilize transresistance $V_o/I_i = A_{fr}$ so that

$$fV_o = I_f \cong I_i \quad \text{and} \quad A_{fr} = \frac{V_o}{I_i} = \frac{1}{f} \quad \text{PO-PI} \quad (12\text{-}11)$$

Series-output to parallel-input (SO-PI) samples output current and feeds back a current to stabilize current gain $A_{fi}$ so that

$$fI_o = I_f \cong I_i \quad \text{and} \quad A_{fi} = \frac{I_o}{I_i} = \frac{1}{f} \quad \text{SO-PI} \quad (12\text{-}12)$$

Finally, series-output to series-input (SO-SI) samples output current and feeds back a voltage to stabilize transconductance $A_{fg}$, or

$$fI_o = E_f \cong E_i \quad \text{and} \quad A_{fg} = \frac{I_o}{E_i} = \frac{1}{f} \quad \text{SO-SI} \quad (12\text{-}13)$$

How is $f$ obtained? (1) Make a test circuit by applying $V_o$ or $I_o$ to the feedback network, depending on whether $V_o$ or $I_o$ is sampled. (2) With respect to the input end of the feedback network, (a) short circuit a paralleled input to measure $I_f$ and find $f = V_o/I_f$ or $I_o/I_f$; (b) open circuit a series input to measure $E_f$ and find $f = V_o/E_f$ or $I_o/E_f$. The test circuit of (a) and (b) stops feedback at the input so that we can measure the fed-back variable. These fundamental concepts will be applied in the next section, and are summarized in Table 12-2.

Table 12-2

| Feedback Network Connections | | Output Variable Sampled | Feedback and Input Variable | Feedback Stabilized Relationship |
|---|---|---|---|---|
| Output | Input | | | |
| parallel | parallel | $V_o$ | $I_f, I_i$ | $A_{fr} = V_o/I_i$ |
| parallel | series | $V_o$ | $E_f, E_i$ | $A_{fv} = V_o/E_i$ |
| series | parallel | $I_o$ | $I_f, I_i$ | $A_{fi} = I_o/I_i$ |
| series | series | $I_o$ | $E_f, E_i$ | $A_{fg} = I_o/E_i$ |

## 12-3 Single-Stage Amplifiers with Negative Feedback

PO-SI. The familiar emitter follower in Fig. 12-3(a) is modeled to emphasize the nature of the feedback connections in Fig. 12-3(b). $V_o$ is sampled and fed back, unattenuated by the feedback network, as $fV_o = E_f = V_o$ to oppose and almost equal $E_i$. Therefore, $V_o/E_i$ is stabilized. In Fig.

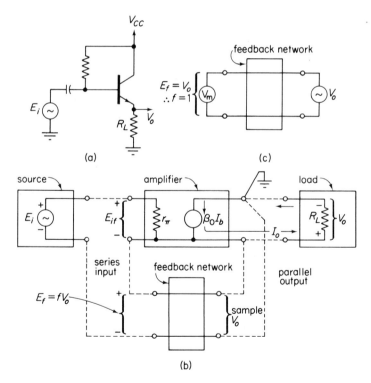

**Fig. 12-3.** The emitter follower in (a) is modeled in (b) to show PO-SI feedback connections. In (c) the test circuit shows $f = 1$.

12-3(c) test voltage $V_o$ is applied to the feedback network; the input loop is opened and since $fV_o = E_f = V_o, f = 1$. From Eq. (12-10),

$$A_{fv} = \frac{V_o}{E_i} = \frac{1}{1} = 1 \qquad (12\text{-}14)$$

**SO-SI.** A common-emitter amplifier in Fig. 12-4(a) is modeled in Fig. 12-4(b) to illuminate the feedback connections. Output current is sam-

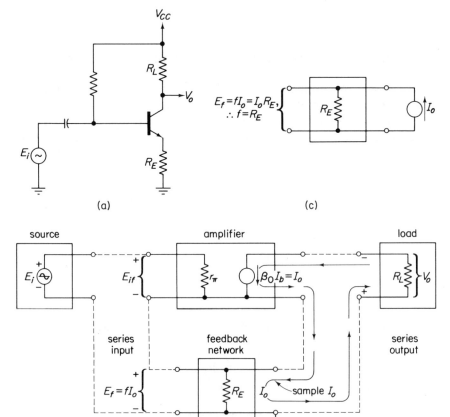

**Fig. 12-4.** The common emitter in (a) is modeled in (b) to show SO-SI feedback connections. In (c) the test circuit shows $f = R_E$.

pled and fed back as a voltage to oppose and almost equal $E_i$. Since the sampled variable is current, $I_o$ is applied to the feedback network in Fig. 12-4(c). Feedback is stopped by opening the input series connection. The fed-back variable $fI_o$ is measured to be $I_o R_E$ so that $f = R_E$ and, from Eq. (12-13),

Sec. 12-3   Single-Stage Amplifiers with Negative Feedback   367

$$A_{fg} = \frac{I_o}{E_i} = \frac{1}{R_E} \quad (12\text{-}15)$$

Conversion of this unfamiliar transconductance gain to the familiar voltage gain is accomplished by noting that $I_o = V_o/R_L$ in Fig. 12-4. Substituting for $I_o$ in Eq. (12-15),

$$\frac{I_o}{E_i} = \frac{V_o}{R_L}\left(\frac{1}{E_i}\right) = \frac{1}{R_E} \quad \text{and} \quad \frac{V_o}{E_i} = \frac{R_L}{R_E} = A_{fv} \quad (12\text{-}16)$$

PO-PI. The collector-base resistor $R_F$ in Fig. 12-5(a) samples output voltage and feeds back a current because of parallel connections at the input nodes. Source resistance must always be included in any analysis involving parallel connected inputs. Otherwise, as is evident in the model of Fig. 12-5(b), all of $I_f$ would be shunted through $E_g$ (if $R_g = 0$) and no feedback action

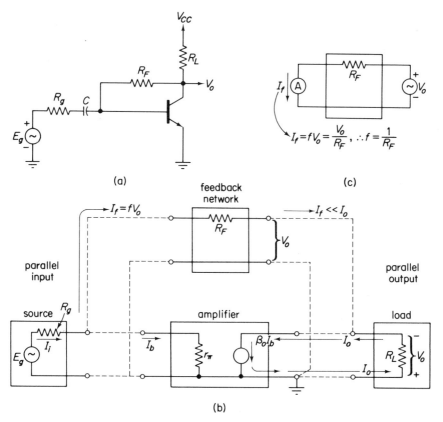

Fig. 12-5. Feedback in (a) is shown to be PO-PI in the model of (b). In (c) the test circuit shows $f = 1/R_F$.

would occur. $V_o$ is sampled and fed back as a current $I_f = fV_o$ that almost equals source current $I_i$, to stabilize $V_o/I_i = A_{fr}$. In Fig. 12-5(c), test voltage $V_o$ is applied to the feedback network; the input nodes are short circuited through an ammeter to stop feedback and measure $I_f = V_o/R_F$. But since $I_f$ also equals $fV_o$, $f = 1/R_F$. Accordingly, from Eq. (12-11)

$$A_{fr} = \frac{V_o}{I_i} = R_F \qquad (12\text{-}17)$$

Equation (12-17) is an unfamiliar transresistance gain but can be converted to the familiar voltage-gain form by redrawing the essential portions of Fig. 12-5(b) in Fig. 12-6(a). Fed-back variable $I_f$ must approximately

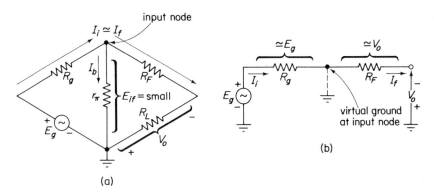

Fig. 12-6. In (a) the simplified model of Fig. 12-5(b) shows that $E_{if}$ is small so that the input node is virtually at ground potential in (b).

equal input variable $I_i$ and therefore $I_b$ is small. The product of a small $I_b$ and $r_\pi$ makes for a small voltage $E_{if}$ so that the input node is almost at ground potential, as shown in Fig. 12-6(b). $E_g$ must drop across $R_g$ and $V_o$ across $R_F$, or $I_i R_g = E_g$ and $I_f R_F = V_o$. Substituting for $I_i = E_g/R_g$ in Eq. 12-17 yields

$$\frac{V_o}{I_i} = V_o\left(\frac{R_g}{E_g}\right) = R_F \quad \text{and} \quad \frac{V_o}{E_g} = \frac{R_f}{R_g} \qquad (12\text{-}18)$$

SO-PI feedback analysis will be covered in Section 12.6 on multistage feedback.

## 12-4 Reduction of Distortion by Negative Feedback

If distortion is introduced in the final stages of an amplifier and the preceeding stages have both negligible distortion and high gain, then distortion may be reduced by negative feedback. In practice a preamplifier and

## Sec. 12-4    Reduction of Distortion by Negative Feedback

voltage amplifier introduce negligible distortion and a complementary output stage furnishes large signal current and voltage swings extending into nonlinear regions of the power transistor's characteristic. For our purposes, the difference between what the undistorted output signal should be and what it actually is, will be studied by thinking of the difference as a distortion voltage $D$ introduced as some percentage of the output voltage.

In Fig. 12-7(a), distortion voltage $D$ is inserted into the output stage and in the absence of feedback, output voltage *including distortion* is $V'_o$, where

$$V'_o = V_o + DA_{v2} \qquad (12\text{-}19)$$

But $V_o = A_{v1}A_{v2}E_i$ and, substituting into Eq. (12-19),

$$V'_o = A_v E_i + A_{v2} D \qquad (12\text{-}20)$$

where $A_v = A_{v1}A_{v2}$. PO-SI negative feedback is introduced in Fig. 12-7(b). Looking into terminals $yy$ we see $V_o + D$. To simplify circuit analysis we can transform $D$ back to the amplifier's input as $D/A_{v1}$ in Fig. 12-7(c) and still see $V_o + D$ between terminals $yy$.

$V'_{of}$ in Figs. 12-7(b) and (c) represents the sum of (a) undistorted output voltage $V_{of}$ plus (b) distortion component $D/A_{v1}$ interacting with a portion of the distortion present in the output. $E_{if}$ equals source input $E_i$ minus (1) distortion component $D/A_{v1}$ and (2) fraction $f$ of $V'_{of}$ or from Fig. 12-7(c):

$$E_{if} = E_i + \frac{D}{A_{v1}} - fV'_{of} \qquad (12\text{-}21)$$

Substituting for $E_{if}$ from $V'_{of} = A_{v1}A_{v2}E_{if}$ and solving for $V'_{of}$

$$V'_{of} = \frac{A_v E_i}{1 + fA_v} + \frac{DA_{v2}}{1 + fA_v} \qquad (12\text{-}22)$$

where $A_v = A_{v1}A_{v2}$. Compare Eqs. (12-20) and (12-22) to see that not only is voltage gain reduced by 1 plus loop gain $fA_v$ but distortion $DA_{v2}$ has also been reduced by the same factor. The expression 1 *plus loop gain* $(1 + fA)$ will be called the *improvement factor* because it shows clearly the extent of improvement by negative feedback.

*Example 12-2.* Assume no distortion is present in both amplifiers of Fig. 12-7, $A_{v1} = 100$, $A_{v2} = 1$, and $f = 0.2$. (a) What value of $E_i$ is required for $V'_o$ to equal 2.5 V without feedback? (b) What value of $E_i$ is required to give $V'_{of} = 2.5$ V with feedback?

*Solution.* (a) Without feedback, $E_i = V'_o/A_v = 2.5/100 = 25$ mV. (b) With feedback, $E_{if}$ must equal the same value of 25 mV and, from Eq. (12-1),

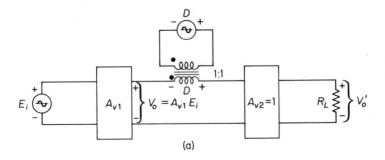

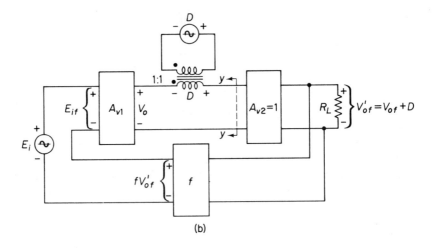

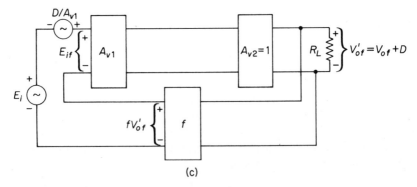

**Fig. 12-7.** The same distortion voltage is introduced into the nonfeedback and feedback circuits of (a) and (b), respectively. $D$ is transformed to the input loop in (c) for simplication.

$$E_i = E_{if} + fV_{of} = 0.025 + (0.2)(2.5) = 0.525 \text{ V}$$

Substituting for $V_{of} = A_v/E_{if}$, $E_i$ must be increased by the improvement factor, $(1 + fA_v)$ to make up for gain lost by feedback, or

$$E_i = E_{if}(1 + fA_v) = 0.025(1 + 20) = 0.525 \text{ V}$$

*Example 12-3.* A distortion voltage of $10\% \, V_o$ or 0.25 V is introduced into both circuits of Example 12-2. What distortion-voltage component is present in the output of each?

*Solution* Without feedback, from Eq. (12-20),

$$V'_o = 100(0.025) + 1(0.25) = 2.5 \text{ V} + 0.25 \text{ V}$$

With feedback, from Eq. (12-22), distortion is reduced by the improvement factor

$$V'_{of} = \frac{100(0.525)}{1 + 20} + \frac{(0.25)(1)}{1 + 20} = 2.5 + 0.0119 \text{ V}$$

*Example 12-4.* Evaluate the total distortion component present at the amplifier's input with the feedback circuit of Examples 12-2 and 12-3.

*Solution.* Substitute into Eq. (12-21) and segregate the distortion terms:

$$\begin{aligned} E_{if} &= 0.525 + \frac{0.25}{100} - 0.2(2.5 + 0.0119) \\ &= (0.525 - 0.50) + (0.0025 - 0.00238) \\ &= 0.025 \text{ V} + 0.00012 \text{ V} \quad \text{Distortion component} = 0.00012 \text{ V} \end{aligned}$$

Check that dividing $V'_{of}$ by $A_v$ gives the same components.

## 12-5 Multistage Parallel-Output to Series-Input Negative Feedback

The multistage amplifier in Fig. 12-8(a) is modeled in Fig. 12-8(b) to show PO-SI negative feedback. Output voltage $V_o$ is sampled and fed back as a voltage $fV_o$ that opposes and almost equals $E_i$. To find $f$ we open the input loop, apply $V_o$ to the feedback network, and measure

$$E_i \cong fV_o = \frac{R_E}{R_E + R_F}V_o \qquad f = \frac{R_E}{R_E + R_F} \qquad (12\text{-}23a)$$

The stabilized output-input variables are

**372** Negative and Positive Feedback  Chap. 12

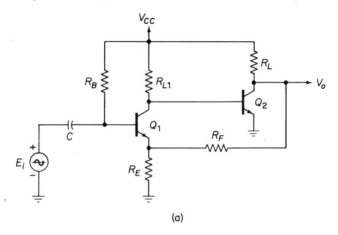

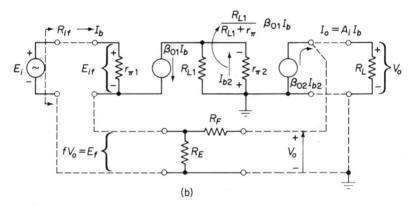

**Fig. 12-8.** Multistage amplifier in (a) is modeled in (b) to show PO-SI negative feedback.

$$\frac{V_o}{E_i} = A_{fv} = \frac{1}{f} = \frac{R_E + R_F}{R_E} \quad (12\text{-}23b)$$

In the event $f$ does not entirely control the output-input relationship we derive the general feedback output-input equation and force it into the format of Eq. (12-3). From this result we shall learn to form the $A$ terms from a simpler amplifier comprised of *the original amplifier without feedback but with loading on input and output by the feedback elements*. The derivation of $V_o/E_i$ is simplified by writing the input loop equation of Fig. 12-8(b) from the equivalent input model of Fig. 12-9.

$$E_i = I_b R_i + \frac{R_E}{R_E + R_F} V_o = I_b R_i + f V_o \quad (12\text{-}24a)$$

where

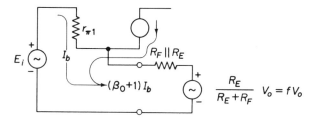

**Fig. 12-9.** Thévenin equivalent of the feedback circuit presented to $E_i$ in Fig. 12-8.

$$R_i = r_{\pi 1} + (\beta_{01} + 1)(R_E \| R_F) \tag{12-24b}$$

In Fig. 12-8(b), $E_i$ has no significant direct effect through the feedback network on $V_o$. That is, if we removed $r_{\pi 2}$ to disable the second stage, an output voltage of approximately $R_L E_i/(R_F + R_E)$ would be developed across $R_L$. Usually $R_F \gg R_L$ and this *direct-feed forward* component is negligible. We can therefore reason that $V_o$ is developed by $I_o$ flowing through the parallel combination of $R_L$ and $R_F + R_E$, or

$$V_o = A_i I_b R'_L \tag{12-25a}$$

where

$$A_i = \beta_{01}\beta_{02}\frac{R_{L1}}{R_{L1} + r_{\pi 2}} \tag{12-25b}$$

and

$$R'_L = R_L \|(R_F + R_E) \tag{12-25c}$$

Solve Eq. (12-24a) for $I_b$ and substitute into Eq. (12-25a) to yield

$$A_{vf} = \frac{V_o}{E_i} = \frac{A_i \dfrac{R'_L}{R_i}}{1 + \left(\dfrac{R_E}{R_E + R_F}\right) A_i \dfrac{R'_L}{R_i}} = \frac{A_v}{1 + fA_v} \tag{12-26a}$$

where

$$A_v = A_i \frac{R'_L}{R_i} = \frac{V_{oo}}{E_i} = \text{open-loop gain} \tag{12-26b}$$

and

$$f = \frac{R_E}{R_E + R_F} \tag{12-26c}$$

Equation (12-26a) is in the feedback format of an open-loop (no feedback) gain term, over 1 plus $f$ times the open-loop gain. Closer examination of Eq. (12-26b) shows that this equation describes the original amplifier $A_i$ with an open-loop output voltage $V_{oo}$ developed across $R_L$ and a parallel loading by the feedback network of $R_F + R_E$. Input voltage $E_i$ sees $R_i$ or the input of the amplifier, loaded by the feedback network but *without* a feedback voltage.

We can write two simple *rules* to identify the open-loop gain circuit. (1) To find the input loading, stop the feedback by (a) shorting the amplifier's output for PO feedback. (b) Remove connections to the amplifier's output for SO feedback. (2) To find output loading, stop feedback by (a) shorting the amplifier's input for PI feedback or (b) removing connections to the amplifier's input for SI feedback. For example, in Fig. 12-8(b), short circuit $R_L$ and the emitter of $Q_1$ sees $R_E \| R_F$ to ground, as input loading. For output loading, break the connections to $r_{\pi 1}$ and $V_{oo}$ sees $R_L \| (R_F + R_E) = R_L'$. The open-loop gain circuit is given in Fig. 12-10(a), and $f$ is found from Fig. 12-10(b).

SUMMARY. Beginning with the feedback circuit of Fig. 12-8 we draw the open-loop circuit of Fig. 12-10, which represents amplifier gain without

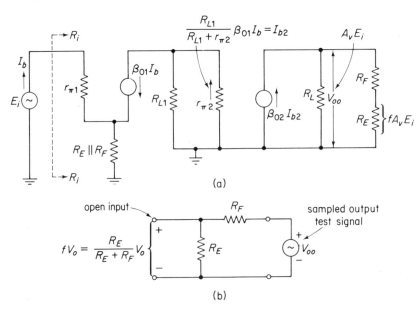

**Fig. 12-10.** Open-loop gain for Fig. 12-8 is developed in (a) and $f$ is found from the test circuit in (b).

feedback, but with loading by the feedback network and write the open-loop gain $A_v$ by inspection. Note: The gain can also be $A_r$, $A_g$, or $A_i$ depending on how the feedback connections are made. Next, obtain the expression for $f$ as instructed in Section 12-2: (1) Apply a signal to the feedback network identical with the sampled output ($V_o$ for PO and $I_o$ for SO). (2) Stop feedback at the input to measure the fed-back variable by shorting a PI and opening an SI. Finally, substitute for $A$ and $f$ in Eq. (12-3) to construct the feedback equation for $A_f$.

INPUT RESISTANCE. Input resistance with feedback is $R_{if} = E_i/I_b$ in Fig. 12-8. Substitute into Eq. (12-24a) for $V_o$ from Eq. (12-25a):

$$E_i = I_b(R_i + fA_iR'_L) = I_bR_i\left(1 + f\frac{A_iR'_L}{R_i}\right)$$

Now substituting for $A_i R'_L/R_i$ from Eq. (12-26b),

$$R_{if} = \frac{E_i}{I_b} = R_i(1 + fA_v) \qquad (12\text{-}27)$$

The series-input feedback connection feeds back a voltage to oppose $E_i$ and *increase* input resistance by the improvement factor. Distinguish clearly that $R_i$ is the input resistance, without a feedback voltage, but *with* input *feedback circuit* loading and including a local feedback component of $(\beta_0 + 1)(R_E \| R_F)$. We can generalize and say that any series-connected negative feedback will increase (loaded) input resistance without feedback, by the improvement factor. An example of this type of feedback will be studied as a design example in Section 12.11.

## 12-6 Multistage Series-Output to Parallel-Input Negative Feedback

Output current is sampled by the series-output connection in Fig. 12-11 and fed back as a current by the parallel-input connection. Assuming $fI_o$ almost equals input current $I_g$, current gain $A_{fi} = I_o/I_g$ is stabilized. A Norton equivalent signal source is used in Fig. 12-11(b) to show how generator resistance is included. Feedback factor $f$ is found by (1) applying current $I_o$ to the feedback network, (2) short circuiting the input node to ground and measuring $fI_o$. Since $I_o$ will divide between $R_E$ and $R_F$, $f$ is found from the divider.

$$fI_o = \frac{R_E}{R_E + R_F}I_o, \quad f = \frac{R_E}{R_E + R_F} \qquad (12\text{-}28)$$

The output-input stabilized relationship is then

$$A_{fi} = \frac{I_o}{I_g} \simeq \frac{I_o}{fI_o} = \frac{1}{f} = \frac{R_E + R_F}{R_E} \qquad (12\text{-}29)$$

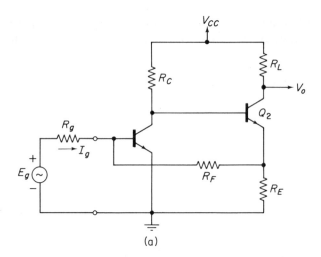

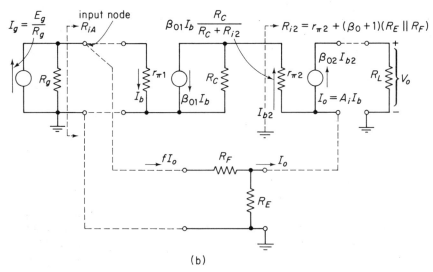

**Fig. 12-11.** The multistage amplifier in (a) is modeled in (b) to show SO-PI negative feedback.

To express voltage gain, $I_o$ flows through $R_L$ to develop $V_o$ and $I_o = V_o/R_L$. Substituting for $I_o$ in Eq. (12-29) and for $I_g = V_g/R_g$ from the Norton transformation in Fig. 12-11(b),

$$\frac{I_o}{I_g} = \frac{\frac{V_o}{R_L}}{\frac{E_g}{R_g}} = \frac{R_E + R_F}{R_E} \quad \text{and} \quad \frac{V_o}{E_g} = \left(\frac{R_E + R_F}{R_E}\right)\frac{R_L}{R_g} \qquad (12\text{-}30)$$

The exact output-input relationship will be derived to determine open-loop gain and input resistance, and to prove Eq. (12-29). To find $I_b$ in terms of $I_g$ and $fI_o$, first take a Norton equivalent of the feedback network as seen by the input nodes in Fig. 12-12. Both $I_g$ and $fI_o$ divide between $R_g \parallel (R_F + R_E)$ and $r_{\pi 1}$ according to current divider $D_i$, where

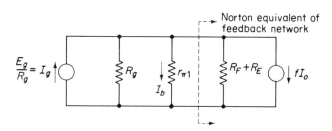

**Fig. 12-12.** Simplified model of the input circuit in Fig. 12-11(b).

$$D_i = \frac{R_g \parallel (R_F + R_E)}{r_{\pi 1} + R_g \parallel (R_F + R_E)} \tag{12-31a}$$

Use superposition to subtract component $D_i fI_o$ from $D_i I_g$ and express $I_b$ as

$$I_b = D_i(I_g - fI_o) \tag{12-31b}$$

Find $I_o$ in terms of $I_b$ from Fig. 12-11. $R_{i2}$ is found by neglecting any direct-feed forward voltage from $I_g$ so that the emitter of $Q_2$ sees essentially $R_E \parallel R_F$. $I_o$ is then expressed by

$$I_o = A_i I_b \tag{12-32a}$$

where

$$A_i = \beta_{01}\beta_{02}\frac{R_C}{R_C + R_{i2}} \tag{12-32b}$$

and

$$R_{i2} = r_{\pi 2} + (\beta_{02} + 1)(R_E \parallel R_F) \tag{12-32c}$$

Substitute for $I_b = A_i I_o$ in Eq. (12-31c) to obtain current gain with feedback,

$$A_{if} = \frac{I_o}{I_g} = \frac{D_i A_i}{1 + fD_i A_i} = \frac{A}{1 + fA} \tag{12-33}$$

where

$$A = D_i A_i = \frac{I'_o}{I_g}$$

$D_i A_i$ represents the open-loop gain $A$, without feedback *but with loading* on input and output by the feedback network. We could have ob-

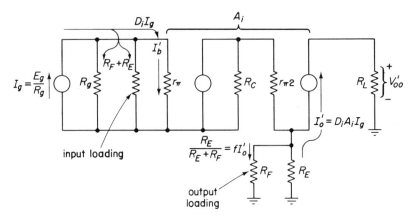

Fig. 12-13. Open-loop gain circuit for Fig. 12-11.

tained the expression for $D_i A_i$ quickly and simply by employing the two general rules of Section 12-5. This procedure is illustrated in Fig. 12-13, where we (1) remove the series connection to the amplifier's output to find input loading and (2) short the paralleled amplifier's input to obtain output loading. $D_i$ and $A_i$ can be written directly from Fig. 12-13, as well as $f$, to obtain Eq. (12-33).

INPUT RESISTANCE. The relative magnitudes of $R_g$ and $r_{\pi 1}$ determine both loop gain and input resistance so that $R_g$ must appear in both expressions. We modify Fig. 12-12 by replacing $I_g$ with test voltage $E$ in Fig. 12-14(a)

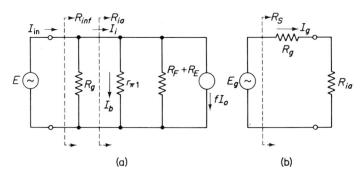

Fig. 12-14. $R_{\text{in } f}$ and $R_{ia}$ are derived from the test circuit in (a) to show $R_s$ in (b).

to measure $R_{\text{in } f}$ and input resistance of the amplifier $R_{ia}$. Neither of these resistances are identical with the actual resistance, $R_s$ presented to $E_g$, but are necessary steps in the development of $R_s$ as will be illustrated in Example 12-5.

## Sec. 12-6      Multistage SO-PI Negative Feedback

**Example 12-5.** In Fig. 12-14(a), removing $fI_o$ would result in an input resistance that would account for feedback network loading but without feedback action. This resistance would be

$$R_{in} = R_g \| r_{\pi 1} \| (R_F + R_E)$$

Let $fI_o$ be activated and derive an expression for (a) $R_{in\,f}$ in terms of $R_{in}$, (b) resistance of the amplifier input $R_{ia}$, and (c) resistance presented to $E_g$ in Fig. 12-11.

**Solution.** (a) By inspection of Fig. 12-14(a),

(1) $\quad\quad R_{in\,f} = \dfrac{I_{in}}{E} \quad$ and $\quad$ (2) $\quad I_{in} = \dfrac{R_{in}}{E} + fI_o$

$I_b$ is expressed in terms of $I_{in}$ from Eq. (12-31a):

(3) $\quad\quad\quad\quad I_b = D_i(I_{in} - fI_o)$

Substituting for $I_b$ from Eq. (12-32a),

(4) $\quad\quad\quad I_o = \dfrac{AI_{in}}{1+fA} \quad$ where $\quad A = D_i A_i$

Substituting for $I_o$ into (2) yields

(5) $\quad\quad\quad\quad \dfrac{E}{I_{in}} = R_{in\,f} = \dfrac{R_{in}}{1+fA}$

(b) Input resistance of the amplifier, $R_{ia}$, parallels $R_g$ in Fig. 12-14(a) so that

(6) $\quad\quad\quad\quad R_{ia} = \dfrac{R_g - R_{in\,f}}{R_g R_{in\,f}}$

(c) Equation (6) expresses $R_{ia}$ as shown in Fig. 12-14(a) and also in Fig. 12-11. Convert the Norton circuit of $I_g - R_g$ back to $E_g - R_g$ as shown in Fig. 12-14(b) to portray $R_s$ as

(7) $\quad\quad\quad\quad R_s = \dfrac{E_g}{I_g} = R_g + R_{ia}$

The significant equations of Example 12-5 are repeated and renumbered here for convenience and analysis:

$$R_{in} = R_g \| r_{\pi 1} \| (R_F + R_E) \quad\quad (12\text{-}34)$$

$$R_{in\,f} = \dfrac{R_{in}}{1+fA} \quad\quad (12\text{-}35)$$

$$R_{ia} = \frac{R_g R_{\text{in } f}}{R_g - R_{\text{in } f}} \quad (12\text{-}36)$$

$$R_s = R_g + R_{ia} \quad (12\text{-}37)$$

Two important conculsions are drawn from these equations. First, $R_{\text{in}}$ represents an input resistance that includes loading by the feedback network ($R_F + R_E$) and source ($R_g$). $R_{\text{in}}$ can be expressed directly from the circuit model of Fig. 12-11. Second, input resistance is reduced by the improvement factor of 1 plus the open-loop gain. *Reduced input resistance* is charcteristic of all *parallel-input negative feedback* connections. This is because the feedback network draws current from the input node that is much greater than would be drawn if feedback were not present.

*Example 12-6.* Disregard the bias resistor in Fig. 12-15 to calculate current gain

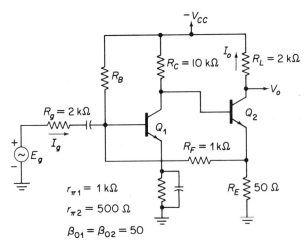

**Fig. 12-15.** Circuit for Examples 12-5 and 12-6.

$I_o/I_g$ from (a) the approximate relationship $1/f$, and (b) the exact current gain relationship. (c) Find voltage gain $V_o/E_g$.

**Solution.** (a) From Eq. (12-29),

$$A_{if} = \frac{1050}{50} = 21$$

(b) From Fig. 12-15 and Eq. (12-31a),

$$D_i = \frac{2000 \| 1050}{1000 + 2000 \| 1050} = 0.408$$

From Eq. (12-32),

$$A_i = \frac{50(50 \times 10^4)}{10^4 + 500 + (51)(50 \| 1000)} = 1940$$

From Eq. (12-29), $f = 50/1050 = 0.0476$. Calculate $A$ from $A = D_i A_i = 793$ and improvement factor from

$$1 + fA = 1 + (0.0476)(793) = 38.8$$

From Eq. (12-23),

$$A_{if} = \frac{793}{38.8} = 20.4$$

Results of (a) and (b) agree closely because $fA = 37.8 \gg 1$.
(c) From Eq. (12-30),

$$\frac{V_o}{E_g} = 21 \left(\frac{2000}{2000}\right) = 21$$

**Example 12-7.** Again disregarding $R_B$ and employing the results of Example 12-6, evaluate (a) $R_{\text{in}}$, $R_{\text{in}f}$, and resistance presented to $E_g$.
**Solution.** (a) From Eq. (12-34),

$$R_{\text{in}} = 2000 \| 1000 \| 1050 = 408 \ \Omega$$

(b) From Eq. (12-35),

$$R_{\text{in}f} = \frac{408}{38.8} = 10.5 \ \Omega$$

(c) From Eq. (12-37),

$$R_s = 2000 + 10.5 \cong 2 \ \text{k}\Omega$$

## 12-7 Multistage Parallel-Output to Parallel-Input Negative Feedback

The parallel-output connection in Fig. 12-16(a) samples $V_o$ and current $I_f = fV_o$ is fed back to the parallel input connection to almost equal input current $I_g$ and stabilize $V_o/I_g$:

$$fV_o = I_f \cong I_g \quad \text{and} \quad \frac{V_o}{I_g} = A_{fr} = \frac{1}{f} \quad (12\text{-}38a)$$

By applying $V_o$ to the feedback network and short circuiting the amplifier's input we measure $I_f = V_o/R_F = fV_o$ to conclude

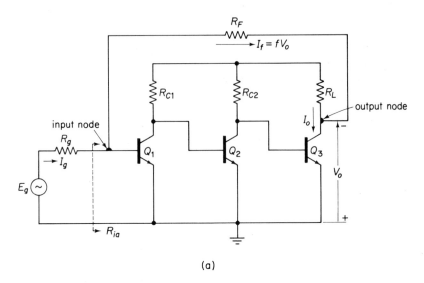

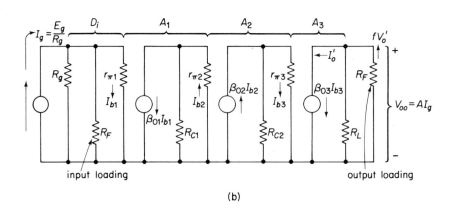

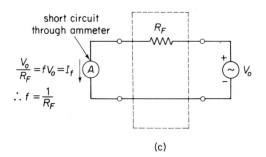

**Fig. 12-16.** The PO-PI feedback circuit in (a) has the open-loop gain model in (b). The test circuit in (c) shows $f = 1/R_F$.

Sec. 12-7                  Multistage PO-PI Negative Feedback     383

$$fV_o = I_f = \frac{V_o}{R_F} \quad \text{and} \quad f = \frac{1}{R_F} \quad (12\text{-}38\text{b})$$

Combining Eqs. (12-38a) and (12-38b),

$$A_{fr} = \frac{V_o}{I_g} = R_F \quad (12\text{-}38\text{c})$$

Equation (12-38c) can be written in terms of current gain because almost all of $I_o$ flows through $R_L$ in Fig. 12-16(a). Substituting for $V_o = I_o R_L$ in Eq. (12-38c),

$$\frac{I_o R_L}{I_g} = R_F \quad \text{and} \quad A_{fi} = \frac{I_o}{I_g} = \frac{R_F}{R_L} \quad (12\text{-}39)$$

As in the previous section, $R_{ia}$ will be small because of the parallel input connection and almost all of $E_g$ will drop across $R_g$. Substituting for $I_g = E_g/R_g$ in Eq. (12-38c) yields voltage gain:

$$\frac{V_o}{E_g} R_g = R_F \quad \text{and} \quad A_{fv} = \frac{V_o}{E_g} = \frac{R_F}{R_g} \quad (12\text{-}40)$$

Let's sketch the open-loop gain model directly from the circuit in order to write the complicated general feedback relationship of $V_o/I_g$ without deriving it. The precedure is as follows. (1) Make a Norton transformation of $E_g$ and $R_g$ to simplify the input. (2) To find input loading by the feedback network. Short the parallel-connected amplifier output terminals and the input node sees $R_F$ to ground. (3) For output loading, short the parallel-connected amplifier input terminals and the output node sees $R_F$ to ground. (4) Apply the rules from Section 12-5 to measure $f$. The resulting open-loop circuit and "$f$" circuit are given in Figs. 12-16(b) and (c).
The expression for current is long but straightforward and, by inspection of Fig. 12-16(b) is

$$A_i = \frac{I'_o}{I_g} = D_i A_1 A_2 A_3 \quad (12\text{-}41)$$

where the input current divider is

$$D_i = \frac{R_g \| R_F}{r_{\pi 1} + R_g \| R_F} = \frac{I_{b1}}{I_g}$$

Gain of $Q_1$    $A_1 = \dfrac{I_{b2}}{I_{b1}} = \beta_{01} \dfrac{R_{C1}}{r_{\pi 2} + R_{C1}}$

Gain of $Q_2$    $A_2 = \dfrac{I_{b3}}{I_{b2}} = \beta_{02} \dfrac{R_{C2}}{r_{\pi 2} + R_{C2}}$

Gain of $Q_3$    $A_3 = \dfrac{I'_o}{I_{b2}} = \beta_{03}$

384  Negative and Positive Feedback  Chap. 12

Output voltage is $V_{oo} = I'_o(R_L \| R_F)$ and substituting for $I'_o$ from Eq. (12-41),

$$\frac{V'_{oo}}{I_g} = (R_L \| R_F)A_i = A \tag{12-42}$$

The general expression is constructed from the feedback format of Eq. (12-3),

$$A_{fr} = \frac{V_o}{I_g} = \frac{A}{1 + fA} \tag{12-43}$$

Where expressions for $A$ and $f$ are given in Eqs. (12-42) and (12-38b), respectively. Voltage gain is found by substituting for $I_g = E_g/R_g$:

$$A_{fv} = \frac{V_o}{E_g} = \frac{A_{fr}}{R_g} \simeq \frac{R_F}{R_g} \tag{12-44}$$

Expressions for input resistance are identical with Eqs. (12-34) to (12-37) if we substitute $R_F$ for $R_F + R_E$ and employ values for $f$ and $A$ of this section.

**Example 12-8.** Values are given for the circuit of Fig. 12-16(a) as follows. $R_g = R_L = r_{\pi 3} = 500 \,\Omega$, $R_{C2} = r_{\pi 2} = 1 \,\text{k}\Omega$, $R_{C1} = r_{\pi 1} = 2 \,\text{k}\Omega$, $\beta_0$ of each transistor is 50 and $R_F = 20 \,\text{k}\Omega$. (a) Find $V_o/E_g$ by the approximate methods. (b) Evaluate $V_o/I_g$ from the feedback format.

**Solution.** (a) From Eq. (12-40), $A_{fv} = 20{,}000/500 = 40$.
(b) From Eq. (14-41),

$$D_i = \frac{500 \| 20{,}000}{2000 + 500 \| 20{,}000} = 0.196 \qquad A_3 = 50$$

$$A_1 = 50 \frac{2000}{1000 + 2000} = 33.3 \qquad A_2 = 50 \frac{1000}{500 + 1000} = 33.3$$

$$A_i = (0.196)(33.3)(33.3)(50) = 10{,}900$$

From Eq. (12-42), $A = (500 \| 20{,}000)(10900) = 5.32 \times 10^6$ and $fA = (1/20{,}000)(5.32 \times 10^6) = 266$. From Eq. (12-43),

$$A_{fr} = \frac{V_o}{I_g} = \frac{5.32 \times 10^6}{1 + 266} = 19{,}950$$

**Example 12-9.** Employing the results of Example 12-8, (a) find the resistance presented to $E_g$. (b) Find $A_{fv}$.

**Solution.** (a) From Eqs. (12-34), (12-35), and (12-36),

$$R_{in} = R_g \| r_{\pi 1} \| R_F = 500 \| 2000 \| 20{,}000 = 392 \,\Omega$$

$$R_{inf} = \frac{392}{1 + 266} = 1.46 \,\Omega$$

$$R_{ia} = \frac{500(1.46)}{500 - 1.46} \cong 1.46 \, \Omega$$

Resistance presented to $E_g$ is $R_s = 500 + 1.46 = 501 \, \Omega$.
(b) From (a) $E_g \cong I_g R_s = 501 I_g$. Substitute for $I_g$ into $A_{fr}$ obtained in (b) of Example 12-8:

$$A_{fr} = \frac{V_o(501)}{E_g} \qquad \frac{V_o}{E_g} = \frac{A_{fr}}{501} = \frac{19950}{501} = 39.8 = A_{fv}$$

## 12-8 Multistage Series-Output to Series-Input Negative Feedback

A three-stage amplifier in Fig. 12-17 illustrates negative feedback with output current sampled by the output-series connection. Feedback variable $fI_o$ is fed back as a voltage $E_f$ because of the series-input connection to almost-equal input voltage $E_i$. Thus the output-current to input-voltage relationship $A_{fg}$ is stabilized, or

$$fI_o = E_f \cong E_i \quad \text{and} \quad A_{fg} = \frac{I_o}{E_i} = \frac{1}{f} \tag{12-45}$$

To evaluate $f$ we assume collector and emitter current of $Q_3$ approximately equal $I_o$. Remove connections to the amplifier's input (base and emitter of $Q_1$): apply $I_o$ to the feedback network consisting of $R$ in parallel with $R_E + R_F$ and find the voltage $fI_o$ that is developed across $R_3$ as $E_f$. Thus $I_o$ develops a voltage across $R \| (R_E + R_F)$ that then divides between $R_E$ and $R_F$

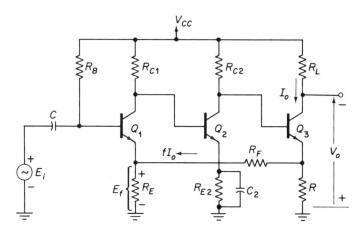

Fig. 12-17. Three-stage amplifier with negative feedback.

$$E_f = fI_o = I_o[R \| (R_E + R_F)]\frac{R_E}{(R_F + R_E)} = I_o\frac{RR_E}{(R + R_E + R_F)}$$

Solving for $f$,

$$f = \frac{RR_E}{R + R_E + R_F} \tag{12-46}$$

We can develope the open-loop circuit by opening the output loop in Fig. 12-17 to see that input loading by the feedback network places $R_F + R$ in parallel with $R_E$. Likewise, opening the input loop places $R_E + R_F$ in parallel with $R$ as output loading. Open-loop gain $I'_o/E_i = A$ is obtained from the resulting model in Fig. 12-18 by finding $E_i$ from the input loop and current gain $A_i$:

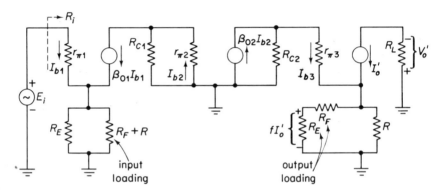

Fig. 12-18. Open-loop model of Fig. 12-17.

$$I_b = \frac{E_i}{R_i} \quad \text{where} \quad R_i = r_{\pi 1} + (\beta_{01} + 1)[R_E \| (R_F + R)] \tag{12-47}$$

$$A_i = \frac{I'_o}{I_b} = \left(\frac{R_{C1}}{R_{C1} + r_{\pi 2}}\right)\left(\frac{R_{C2}}{R_{C2} + r_{\pi 3}}\right)\beta_{01}\beta_{02}\beta_{03} \tag{12-48}$$

Substituting for $I_b$ from Eq. (12-47) into Eq. (12-48),

$$A = \frac{I'_o}{E_i} = \frac{A_i}{R_i} \tag{12-49}$$

Construct the general feedback equation

$$A_{fg} = \frac{A}{1 + fA} = \frac{I_o}{E_i} \tag{12-50}$$

*Example 12-10.* Employing the SO-SI feedback model in Fig. 12-19(a), derive (a) the general feedback equation for $I_o/E_i$ as given in Eq. (12-50), (b) the

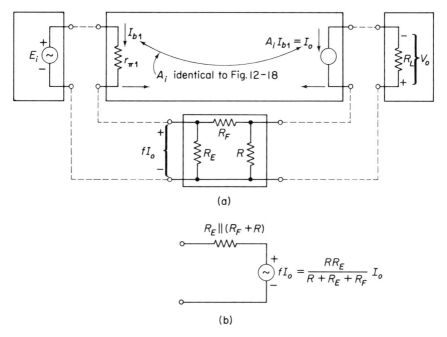

**Fig. 12-19.** The feedback model of Fig. 12-17 is shown in (a) and may be simplified by the Thévenin equivalent of the feedback network in (b).

expression for input resistance $R_{\text{in }f} = E_i/I_b$, (c) the expression for voltage gain $V_o/E_i$.

*Solution.* (a) Write an input-loop equation using the Thévenin input circuit simplification in Fig. 12-19(a):

(1) $$E_i = I_b R_i + fI_o$$

where $f$ and $R_i$ are given in Eqs. (12-46) and (12-47). Current gain $A_i = I_o/I_b$ is the same as in Eq. (12-48), so substitute for $I_b$ in Eq. (1):

(2) $$E_i = I_o \frac{R_i}{A_i} + fI_o$$

Substitute for $A_i/R_i = A$ from Eq. (12-49) and solve for $I_o/E_i$:

(3) $$\frac{I_o}{E_i} = A_{fg} = \frac{A}{1 + fA}$$

and is identical to Eq. (12-59).
(b) Substitute for $I_o = A_i I_b$ into Eq. (1):

(4) $$E_i = I_b(R_i + fA_i)$$

Now substitute for $A_i = R_i A$ from Eq. (12-49) and solve for $R_{\text{in } f}$

(5) $$\frac{E_i}{I_b} = R_{\text{in } f} = R_i(1 + fA)$$

(c) In Fig. 12-19, $V_o = I_o R_L$. Substituting for $I_o$ in Eq. (12-45),

(6) $$A_{fg} = \frac{V_o}{R_L}\left(\frac{1}{E_i}\right) = \frac{1}{f} \quad \text{and} \quad \frac{V_o}{E_i} = A_{fv} = R_L A_{fg}$$

*Example 12-11.* In the circuit of Fig. 12-17, $R_{C1} = R_{C2} = 5 \text{ k}\Omega$, $r_{\pi 1} = r_{\pi 2} = R_L = 1 \text{ k}\Omega$, $r_{\pi 3} = R_E = 500 \text{ }\Omega$, $R = 200 \text{ }\Omega$, $R_F = 10 \text{ k}\Omega$, $\beta_{01} = \beta_{02} = \beta_{03} = 50$. (a) Find $A_{fg}$ from both Eqs. (12-50) and (12-45). (b) Evaluate $A_{fv} = V_o/E_i$ and (c) $R_{\text{in } f}$.

*Solution.* (a) From Eq. (12-48),

$$A_i = \left(\frac{5000}{6000}\right)\left(\frac{5000}{5500}\right)(50)^3 = 94.6 \times 10^3$$

From Eq. (12-47),

$$R_i = 1000 + 51(500 \| 10,200) = 24.3 \text{ k}\Omega$$

From Eq. (12-49),

$$A = \frac{94.6 \times 10^3}{24.3 \times 10^3} = 3.9$$

From Eq. (12-46),

$$f = \frac{200(500)}{10,700} = 9.35$$

From Eq. (12-50),

$$A_{fg} = \frac{3.9}{1 + 9.35(3.9)} = 0.104 \text{ mho}$$

Compare with Eq. (12-45):

$$A_{fg} = \frac{1}{9.35} = 0.107 \text{ mho}$$

(b) From Eq. (6) of Example 12-9,

$$A_{fv} = R_L A_{fg} = 1000(0.104) = 104$$

(c) From Eq. (5) of Example 12-9.

$$R_{\text{in } f} = 24.3(1 + 9.35 \times 3.9) = 910 \text{ k}\Omega$$

## 12-9 Effect of Negative Feedback on Resistance Levels

It was shown in previous sections that when feedback was connected in series with the input voltage and amplifier, input resistance increased by the improvement factor. Parallel connections at the input circuit reduced input resistance by the improvement factor. These statements are true regardless of whether the output terminals are connected in series or in parallel. The same statements pertain to output resistance. Parallel connections reduce and series connections increase output resistance by the improvement factor. regardless of how feedback as applied to the input terminals.

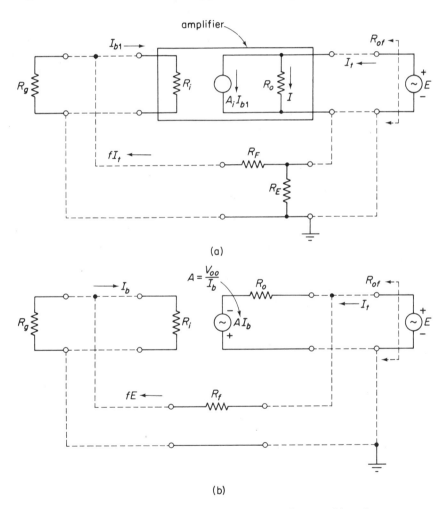

Fig. 12-20. Test circuits to measure output resistance with series-output feedback in (a) and parallel output feedback in (b).

390   Negative and Positive Feedback                               Chap. 12

Effect of feedback on output resistance is developed generally from Fig. 12-20. We apply test voltage $E$ in place of $R_L$ in Fig. 12-20 and measure $I_t$. $R_o$ represents output resistance of $Q_3$ and the signal generator is replaced by its internal resistance. Without feedback $E$ would see output resistance $R_o$, where $R_o \cong E/I$. With feedback, a fraction of $I_t$ is fed back as $R_E/(R_E + R_F)I_t = fI_t$ to excite the amplifier and current drawn from $E$ will be

$$I_t = I + A_i I_{b1} \tag{12-51}$$

Assuming $R_o \gg R_E$, then $I = E/R_o$. Assuming also that $fI_t \cong I_{b1}$, substitute into Eq. (12-51) and solve for $E/I_t$

$$\frac{E}{I_t} = R_{of} = R_o(1 + fA_i) \tag{12-52}$$

To find output resistance of the *PO* connection in Fig. 12-16(a), we employ a Thévenin model of the output in Fig. 12-20(b). $A$ is given by Eq. (12-42) and $R_o$ is the output resistance of $Q_3$. Current drawn by the feedback network is negligible, so $E$ supplies

$$I_t = \frac{E + AI_b}{R_o} \tag{12-53}$$

Assuming $I_b = fE$ substitute into Eq. (12-53) to show

$$\frac{E}{I_t} = R_{of} = \frac{R_o}{1 + fA}$$

## 12-10 Negative Feedback and Frequency Response

Without feedback an amplifier has an upper cutoff frequency $\omega_H$ and its gain $A$ varies with frequency as

$$A = \frac{A_o}{1 + j\dfrac{\omega}{\omega_H}} \tag{12-54a}$$

where $A_o$ = midfrequency gain. When feedback is added the new gain is expressed by

$$A_f = \frac{A}{1 + fA} = \frac{1 + \dfrac{A_o}{j\dfrac{\omega}{\omega_H}}}{1 + f\dfrac{A_o}{1 + j\dfrac{\omega}{\omega_H}}} = \frac{A_o}{1 + j\dfrac{\omega}{\omega_H} + fA_o} \tag{12-54b}$$

The new upper cutoff frequency $\omega_{fH}$ occurs when real and imaginary denominator terms are equal at $\omega = \omega_{fH}$, or

$$\omega_{fH} = \omega_H(1 + fA_o) \tag{12-54c}$$

Thus gain has been sacrificed to raise the upper break frequency by the improvement factor. Usually $\omega_H$ is approximately equal to the bandwidth $B$ and $\omega_H \cong B$ in Eq. (12-54).

Lower break frequency $\omega_L$ is given in Eq. (6-8)

$$A = \frac{V_o}{E_i} = \frac{A_o}{1 + \frac{\omega_L}{j\omega}} \tag{6-8}$$

With feedback the new gain is

$$A_f = \frac{A}{1 + fA} = \frac{A_o}{1 + fA_o + j\frac{\omega_L}{\omega}}$$

Denominator real and $j$ terms will be equal at $\omega_{fL} = \omega$,

$$1 + fA_o = \frac{\omega_L}{\omega_{fL}} \quad \text{and} \quad \omega_{fL} = \frac{\omega_L}{1 + fA_o} \tag{12-55}$$

We conclude that the lower break frequency is reduced by the improvement factor to improve low-frequency response.

*Example 12-12.* Voltage gain of an amplifier is $A_o = 100$ with $f_H = 50$ kHz and $f_L = 20$ Hz. Feedback of $f = 0.1$ is introduced. What is the resulting (a) improvement factor, (b) voltage gain, (c) $\omega_{fH}$, (d) $\omega_{fL}$?

*Solution.* (a) $\qquad (1 + fA_o) = 11.$

(b) $\qquad A_{fv} = \dfrac{A_o}{1 + fA_o} = 9.09$

(c) From Eq. (12-54),

$$\omega_{fH} = 50(11) = 550 \text{ kHz}.$$

(d) From Eq. (12-55),

$$\omega_{fL} = \frac{20}{11} = 1.8 \text{ Hz}.$$

## 12-11 Design and Analysis Examples

In Fig. 12-21 biasing resistors are shown by their parallel equivalent resistance to focus attention on small-signal behavior. Design and analysis of the low cutoff frequency is examined in Example 12-13.

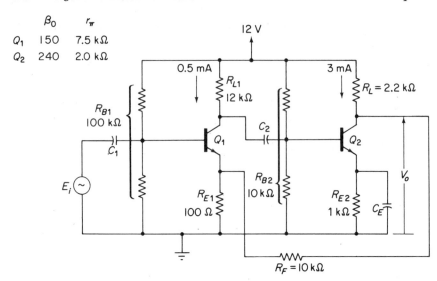

**Fig. 12-21.** PO-SI negative feedback circuit.

**Example 12-13.** (a) Draw the open-loop circuit of Fig. 12-21 and calculate its gain and input resistance. (b) Find $f$, the improvement factor, input resistance, and gain with feedback. (c) Draw the short-circuit time-constant circuit models seen by $C_E$, $C_2$, and $C_1$. (d) Calculate values for $C_E$, $C_2$, and $C_1$ to cause low break frequencies at 200, 20, and 20 rad/sec respectively. (e) What will be the value of $\omega_{fL}$?

**Solution.** (a) The open-loop gain circuit is shown in Fig. 12-22(a), where input resistance is $R_i = r_{\pi 1} + (\beta_{01} + 1)(R_E \| R_F) = 22.6 \text{ k}\Omega$. Voltage gain $V_{oo}/E_i$ is

$$A_{v1} = \frac{150(5.45 \| 2) \times 10^3}{22.6 \times 10^3} = 9.7 \qquad A_{v2} = \frac{240(2.2 \| 10.1) \times 10^3}{2 \times 10^3} = 217$$

$$A_v = 9.7(217) = 2100$$

(b) $f = R_{E1}/(R_{E1} + R_F) = 0.0099, (1 + fA_v) = 1 + (0.0099)(2100) = 21.8$.

$$A_{fv} = \frac{2100}{21.8} = 96$$

Without $R_{B1}$, $R_{if} = R_i(1 + fA_v) = 490 \text{ k}\Omega$. With $R_{B1}$, $R_{inf} = R_{if} \| R_{B1} = 83 \text{ k}\Omega$.

(c) Short-circuit time-constant circuits are shown in Figs. 12-22(b), (c), and (d).

(d) From Fig. 12-22(b),

$$C_E = \frac{1}{\omega_L R} = \frac{1}{(200)(30)} = 166 \text{ }\mu\text{F}$$

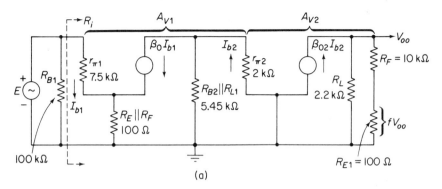

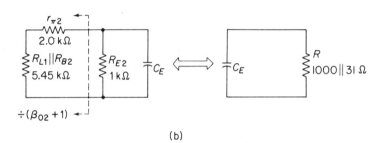

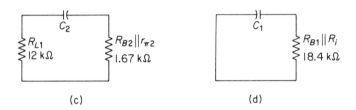

**Fig. 12-22.** Solutions to Example 12-12. Open-loop circuit in (a); time constants in (b), (c), and (d) for $C_E$, $C_2$, and $C_1$, respectively.

From Fig. 12-22(c),

$$C_2 = \frac{1}{(20)(13.6 \times 10^3)} = 3.7 \ \mu\text{F}$$

From Fig. 12-22(d),

$$C_1 = \frac{1}{20(18.4 \times 10^3)} = 2.7 \ \mu\text{F}$$

(e) From Eq. (12-55),

$$\omega_{fL} = \frac{200}{(1 + 20.8)} \cong 10 \ \text{Hz}$$

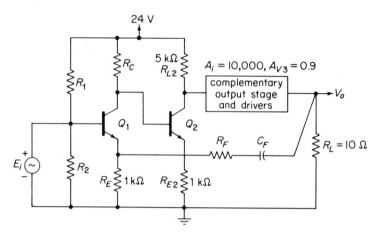

**Fig. 12-23.** Circuit for Example 12-13 ($\beta_{F2} = 65$, $\beta_{F1} = 65$).

**Example 12-14.** Complete the design of the amplifier in Fig. 12-23 to give $A_{vf} = V_o/E_i = 10$. Select an open-loop gain of 150.

**Solution.** (a) $R_F$ selection. Find $f$, then $R_F$ from

$$A_{fv} = 10 = \frac{A_v}{1 + fA_v} = \frac{150}{1 + f(150)} \qquad f = \frac{14}{150}$$

$$f = \frac{14}{150} = \frac{R_E}{R_E + R_F} = \frac{1000}{1000 + R_F} \qquad R_F = 9.7 \text{ k}\Omega$$

Improvement factor is $(1 + fA_v) = 15$.

(b) *Biasing* $Q_2$. The collector of $Q_2$ should be at $V_{CC}/2 = 12$ V, making $I_{C2} = 12/5000 = 2.4$ mA, $I_{B2} = 2.4/65 = 37$ μA. DC voltage across $R_E$ is $I_{C2}R_{E2} = 2.4$ V and voltage to ground at the base of $Q_2$ is $2.4 + 0.6 = 3.0$ V.

(c) *Voltage Gain of* $Q_1$. To obtain $A_v = 150$, $Q_1$ and $Q_2$ must develop a gain of $150/0.9 = 167$. $A_{v2}$ of $Q_2$ is $R_{L2}/R_{E2} = 5$ and $Q_1$ must furnish $A_{v1} = 167/5 = 33.4$. The ac load on $Q_1$ is $R_{L1} = R_C \| R_{i2}$. From $A_{v1} = 33.4 = R_{L1}/(R_E \| R_F) = R_{L1}/900$, $R_{L1} = 30.0$ kΩ. $R_{i2} = r_{\pi2} + (\beta_{02} + 1)R_{E2} = 67$ kΩ. Hence from $R_C \| 67$ kΩ $= 30.0$ kΩ, $R_c = 54.4$ kΩ.

(d) *Biasing* $Q_1$. In order to put the base of $Q_2$ at 3.0 V, the voltage drop across $R_C$ must be $24 - 3.0 = 21.0$ V. DC collector current of $Q_1$ is to be (neglecting $I_{B2}$)

$$I_{C1} = \frac{V_{RC}}{R_C} - I_{B2} = \frac{21}{54.4 \text{ k}} - 0.037 = 0.35 \text{ mA}, \quad I_{B1} = \frac{I_{C1}}{\beta_{F1}} \cong 5 \text{ μA}$$

The dc drop across $R_E$ is 0.38 V and $V_{CE1}$ is $V_{CC} - V_{RC} - V_{RE} = 2.6$ V $= V_{CE1}$. The base of $Q_1$ will be at a potential of $0.38 + 0.6 \cong 1$ V. A choice of $R_{B1} \cong 10R_E$ is suggested and calculations of $R_1$ and $R_2$ are left for the student.

Sec. 12-11                        Design and Analysis Examples    395

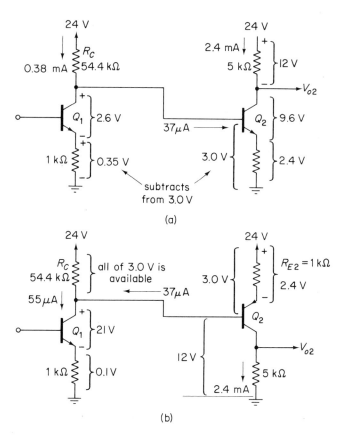

**Fig. 12-24.** When peak output voltage is limited by $Q_1$ in (a) $Q_2$ can be changed to a *pnp* in (b).

The dc voltage levels for $Q_1$ in Example 12-14 are shown in Fig. 12-24(a). Maximum output-voltage swing is limited by the low value of $V_{CE1}$. Ideally $Q_1$ could be driven for a maximum symetrical swing between $V_{CE1} = 0$ and $V_{CE1} = 5.2$ V, for a peak increment of 2.6 V. But this increment must divide between $R_E \| R_F$ and $R_{L1}$ so that peak output voltage of $Q_1$ is

$$V_{op1} = 2.6 \frac{R_{L1}}{R_{L1} + R_E \| R_F} = 2.5 \text{ V}$$

Introduce complementary operation by substituting a *pnp* transistor for $Q_2$ in Fig. 12-24(b). Biasing currents of $Q_2$ are unchanged. However, $I_{C1}$ must be reduced to $(55 + 37) \mu\text{A} = 92 \mu\text{A}$ in order to reduce the drop across $R_C$ to 3.0 V. The maximum peak output voltage is now limited by $V_{RC} = 3.0$ V and represents an increase of 20% at no additional cost. Choosing values from Fig. 12-24(b), assume a peak signal of $E_{ip} = 0.5$ V is applied to the input of the amplifier in Fig. 12-23. $V_{op}$ would be 5 V and $fV_o = 0.466$ V is fed back to oppose $E_i$ so that $Q_1$ is driven by the difference of 0.034 V. The gain of $Q_1$ develops $V_{op1} = 33.4(0.034) = 1.1$ V across the input to $Q_2$. $Q_2$ delivers 1.1 $A_{v2} = 5.5$ V to the output stage which reduces to 5.5 (0.9) $= 5.0$ V $= V_{op}$.

Perhaps the best way to review feedback action is to assume that the basic amplifier of Fig. 12-23, without feedback, has a distortion of 5% at 1000 Hz. With feedback, distortion would be reduced by the improvement factor of 15, in Example 12-13(a) to 0.3%. Suppose this amplifier also had $\omega_L = 150$ rad/sec, reduced to $\omega_{fL} = 10$ rad/sec by feedback. If we measure distortion below 150 rad/sec we would notice a sudden dramatic increase in distortion as frequency is decreased. This occurs because the gain of the basic amplifier decreases, reducing $fV_o$ until the difference $E_i - fV_o$ becomes large enough to overdrive one of the stages. Feedback is striving to maintain a constant output voltage and is working properly. One answer to this problem is to employ complementary transistors to increase the internal swing capacity of the basic amplifier.

## 12-12 Oscillation in Amplifiers

Focus attention on the phase relationships in Fig. 12-25(a). $A_v$ could consist of four common-emitter stages with 180° phase shift for each totaling 720° at 1 kHz. Total phase shift between $E_{if}$ and $V_o$ is a multiple of 360° and is effectively 0° as shown. There is a 180° phase shift in feedback network $f$ because $fV_o$ opposes $E_i$ from the input-loop equation $E_i = E_{if} + fV_o$. If we suddenly short circuit $E_i$ so that $E_i = 0$, then the input-loop equation is

$$E_{if} = -fV_o \qquad E_i = 0 \qquad (12\text{-}56)$$

Equation (12-56) signifies a net phase shift of 180° around the loop from $E_{if}$ to $V_o$ to $fV_o$. Thus when $E_i$ suddenly goes to zero, $fV_o$ will reverse the polarity of $E_{if}$ in Fig. 12-25(a), driving $V_o$ to a stable condition at zero.

Assume the phase shifts increases by 45° for each stage, at 100 kHz, for a total amplifier phase shift of 4(225°) $= 900°$ or an effective shift of 900° $- 4(180°) = 180°$. The feedback network is made of resistors and $f$ is not frequency dependent, retaining its shift of 180°. Phase shift around the loop from $E_{if}$ to $V_o$ is 180° plus 180° from $V_o$ to $fV_o$ for a total loop shift of 360° or 0°. The input-loop equation may now be written to show how $fV_o$ aids $E_i$ to yield positive feedback:

$$E_i = E_{if} - fV_o \qquad \text{or} \qquad E_{if} = fV_o \quad \text{when } E_i = 0 \qquad (12\text{-}57)$$

Equation (12-57) applies only at a frequency where phase shift in the loop

Sec. 12-12   Oscillation in Amplifiers   397

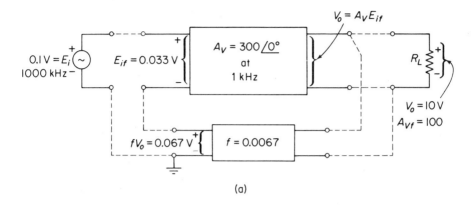

(a)

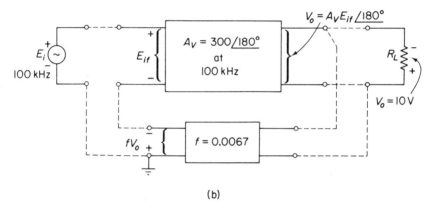

(b)

**Fig. 12-25.** The amplifier in (a) is stable at 1000 Hz but may oscillate as shown in (b) when $fV_o > E_{if}$.

$E_{if}$ to $V_o$ to $fV_o$ is zero and specifies at *what* frequency the amplifier will oscillate if it is going to oscillate. If we short circuit $E_i$, the amplifier will oscillate *if* $fV_o$ is equal to or greater than $E_{if}$, or

$$E_{if} = fV_o = fA_v E_{if} \quad \text{and} \quad fA_v = 1 \quad (12\text{-}58)$$

Equation (12-58) shows that the amplifier will oscillate at 100 kHz if $fA_v$ is equal to or greater than 1, at 100 kHz.

*Example 12-15.* Given $A_v = 300$ and $A_{vf} = 100$ for the amplifier in Fig. 12-25. This amplifier is built and oscillates at 100 kHz. (a) At what frequency does the loop phase shift reach zero? (b) by how much must $A_v$ be reduced to stop oscillation?

*Solution.* (a) 100 kHz. (b) From Eq. (12-3), $f = 0.0067$. From Eq. (12-58), $0.0067 A_v = 1$, or $A_v = 150$ to stop oscillation.

Reduction of amplifier gain at or before its natural frequency of oscillation may be accomplished experimentally by deliberate reducing the high-frequency response. Connect small capacitors across points of high gain (base-to-collector transistor terminals).

## 12-13 Oscillators

In the previous section our objective was to prevent undesired oscillations in an amplifier. Now our principle objective is to cause stable oscillations at a predictable frequency. When the feedback signal $fV_o$ exactly equals $E_{if}$, oscillations are self-sustaining and sinusoidal because only a sine (or exponential) wave can pass through an $RL$ or $RC$ network without changing shape. To analyze qualitatively the conditions for oscillation in Fig. 12-26,

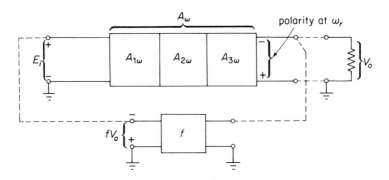

**Fig. 12-26.** Feedback oscillator.

assume each stage has an identical gain-frequency variation as expressed by Eq. (12-54a), so that amplifier gain without feedback is

$$A_\omega = \frac{A_o^3}{\left(1 + j\dfrac{\omega}{\omega_H}\right)^3} \tag{12-59a}$$

With feedback the gain is expressed from Eq. (12-54c)

$$\frac{V_o}{E_i} = A_{f\omega} = \frac{A_\omega}{1 + fA_\omega} \tag{12-59b}$$

The term $fA_\omega$ is complex with both magnitude and phase angle. The *Barkhausen criterion* states that oscillation will occur with *unity loop gain* when $-fA_\omega = 1$. In Eq. (12-59b) $A_{f\omega}$ approaches infinity under this criterion and implies that the phase of $-fA_\omega$ is zero with a magnitude of 1. Assuming the midfrequency phase shift of each stage in Fig. 12-26 is 180°, each stage must shift only 60° to bring the net amplifier phase shift to zero. A phase shift of 45° exists at $\omega_H$ of each stage so that oscillation will occur at some frequency above $\omega_H$. To find the frequency of oscillation, $\omega_r$, expand Eq. (12-59a).

$$-fA_\omega = f \frac{A_o^3}{1 - 3\left(\frac{\omega}{\omega_H}\right)^2 + j\frac{\omega}{\omega_H}(3) - \left(\frac{\omega}{\omega_H}\right)^3} \quad (12\text{-}60\text{a})$$

The $j$ terms of Eq. (12-60) will add to zero when $\omega = \omega_r$ to make the phase angle zero and, solving for $\omega_r$,

$$\omega_r = (3)^{1/2}\omega_H \quad (12\text{-}60\text{b})$$

The denominator's real terms are evaluated at $\omega = \omega_r$

$$1 - 3\left(\frac{\omega_r}{\omega_H}\right)^2 = 1 - 3\left[\frac{(3)^{1/2}\omega_H}{\omega_H}\right]^2 = -8$$

so that

$$fA_\omega = \frac{f(A_o)^3}{8} \quad \text{at } \omega = \omega_r \quad (12\text{-}60\text{c})$$

*Example 12-16.* Assuming each stage of the amplifier in Fig. 12-26 has a mid-frequency gain of 10, (a) what is the required value of $\omega_H$ for each stage and (b) value of $f$ required for oscillation at 100 kHz?

*Solution.* (a) From Eq. (12-60b),

$$f_H = 10^5/1.73 = 57.7 \text{ kHz.}$$

(b) From Eq. (12-60c),

$$fA\omega = 1 = \frac{f(10)^3}{8} \quad f = 0.008$$

In practice, loop gain of an oscillator is made larger than 1. Assume that $E_i$ is initially 1 mV in Fig. 12-26 and $fA_\omega = 1.1$, then $E_i$ will go to 1.1 mV after a trip around the loop, over an internal period, equal to $T$, or the oscillating frequency. The oscillator's amplitude will continue to increase until stabilized by some unpredictable nonlinearity in the circuit. For example $\beta_F$ decreases with increasing collector current, to limit $A$.

## 12-14 Phase-Shift Oscillators

Low-frequency oscillators are readily constructed from circuits similar to Fig. 12-27. $Q_1$ provides a gain $V_o/E_i$ sufficient to offset loss in the $RC$ feedback network. $Q_1$ has a phase shift of 180° from base to collector. The

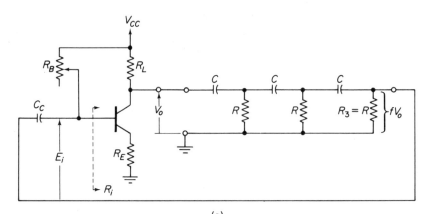

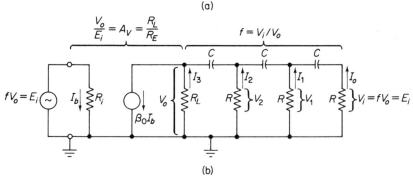

**Fig. 12-27.** The phase-shift oscillator in (a) is modeled in (b), where $E_i = V_i = fV_o$.

feedback network can be designed for an additional phase shift of 180° at a specific frequency $\omega_r$. Oscillator frequency and required amplifier gain are determined from an analysis of Fig. 12-27 in Example 12-17.

**Example 12-17.** In Fig. 12-27(b), $R_L$ is assumed small so that gain $V_o/E_i \cong R_L/R_E$ and is not reduced significantly by loading from the feedback network. $R_B$ and $R_i$ are large and do not affect the value of $R_3 = R$. $C_C$ is large enough to introduce negligible phase shift at $\omega_r$. (a) Find frequency $\omega_r$, where the feedback network's phase shift equals 180°. (b) What is the attenuation $f$ at $\omega_r$? (c) What value of $A_v = V_o/E_i$ is required so that $fA_v = 1$?

**Solution.** (a) Develop the general expression for $V_i/V_o = f$ in Fig. 12-27(b):

(1) $I_o = \dfrac{V_i}{R}$

(2) $V_1 = \dfrac{I_o}{j\omega C} + V_i = V_i\left(1 + \dfrac{1}{j\omega RC}\right)$

## Sec. 12-14    Phase-Shift Oscillators    401

(3) $I_1 = \dfrac{V_1}{R}$

(4) $V_2 = \dfrac{I_1 + I_o}{j\omega C} + V_1 = V_i\left(\dfrac{1}{j\omega RC}\right)^2 + \dfrac{3}{j\omega RC} + 1$

(5) $I_2 = \dfrac{V_2}{R}$

(6) $V_o = \dfrac{I_o + I_1 + I_2}{j\omega C} + V_2 = V_i\left(\dfrac{1}{j\omega RC}\right)^3 + 5\left(\dfrac{1}{j\omega RC}\right)^2 + \dfrac{6}{j\omega RC} + 1$

At $\omega_r$, the $j$ terms in Eq. (6) add to 0 for a 180° phase shift to solve for $\omega_r$:

(7) $\left(\dfrac{1}{j\omega_r RC}\right)^3 + \dfrac{6}{j\omega_r RC} = 0$   and   $\omega_r = \dfrac{1}{\sqrt{6}\,RC}$

(b) Attentation of the feedback network is found from the real terms of Eq. (6) when $\omega = \omega_r$

(8) $V_o = V_i\left[1 - 5\left(\dfrac{1}{\omega_r RC}\right)^2\right]$

Substituting for $\omega_r$ from Eq. (7),

(9) $\dfrac{V_i}{V_o} = f = \dfrac{1}{29}$

(c) Gain of the amplifier $A_v$ must be set at 29 or slightly greater to make $fA_v \cong 1$.

Phase shift of $Q_1$ and $Q_2$ are each 180° in Fig. 12-28. Phase shift around the loop will be effectively 360 or 0° when the phase shift of feedback network $R_1 R_2 C_1$ and $C_2$ is zero. Assuming that input resistance of $Q_1$ and $R_{B1}$ are large with respect to $R_2$, $V_o$ divides across the feedback network to give $fV_o = E_i$ as

$$\dfrac{1}{f} = \dfrac{V_o}{E_i} = 1 + \dfrac{R_1}{R_2} + \dfrac{C_2}{C_1} + j\omega R_1 C_2 + \dfrac{1}{j\omega C_1 R_2} \qquad (12\text{-}61\text{a})$$

The $j$ terms add to zero at $\omega = \omega_r$. Assuming $R_1 = R_2 = R$ and $C_1 = C_2 = C$,

$$\omega_r^2 = \dfrac{1}{R_1 C_1 R_2 C_2} \quad\text{and}\quad \omega_r = \dfrac{1}{RC} \qquad (12\text{-}61\text{b})$$

Substituting for $\omega_r$ into Eq. (12-61a) gives $f$ at $\omega_r$

$$\dfrac{V_o}{E_i} = 1 + \dfrac{R_1}{R_2} + \dfrac{C_2}{C_1} = \dfrac{1}{f} \qquad (12\text{-}61\text{c})$$

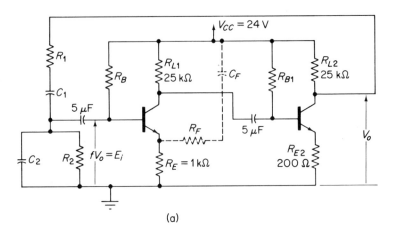

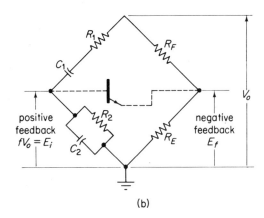

**Fig. 12-28.** Wien bridge phase-shift oscillator in (a) is modeled in (b) to show bridge arrangement.

When $R_1 = R_2$ and $C_1 = C_2$, Eq. (12-61c) simplifies to

$$\frac{V_o}{E_i} = 3 \quad \text{and} \quad f = \frac{1}{3} \tag{12-61d}$$

Gain of the amplifier must be set at 3 to ensure oscillation. Adding negative feedback by the $C_c - R_f$ network yields the *Wien bridge oscillator*, whose bridge configuration is emphasized in the model of Fig. 12-28(b). Amplifier gain is set by choosing $R_F$ from $A_v = 3 = (R_F + R_E)/R_E$. $C_c$ is a large capacitor acting as a short circuit at $\omega_r$.

An excellent, stable oscillator can be constucted quickly from an integrated circuit operational or differential amplifier in Fig. 12-29. *RC* networks determine $\omega_r$ and provide positive feedback. Negative feedback is furnished by $R_F$ and $R_i$, yielding a gain of $R_F/R_i$. The diodes ensure oscillation begins by disconnecting $R_F$ when $V_o$ is less than their forward bias voltage. Thus the amplifier applies its high open-loop gain to the differential input terminals 8 and 9 when $V_o$ is small.

Nonlinearities are also introduced by the diodes to limit amplitude of oscillation. Assume $V_o$ is going positive and dividing between $R_F$ and $R_i$ to make point 8 slightly more positive than point 9 in Fig. 12-29. Negative

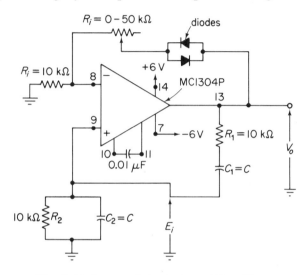

Fig. 12-29. Integrated circuit phase-shift oscillator.

feedback current through the top diode will increase diode voltage drop as feedback current increases until the positive potential at point 8 no longer exceeds that at point 9. Charges on $C_2$ and $C_1$ begin to leak off (because $V_o$ is no longer increasing) and begin a decrease of $V_o$ and the difference voltage between 8 and 9. Thus the natural time constants of $C_1$ and $C_2$ control frequency of oscillation and the diode characteristic, together with $R_F$ and $R_i$, controls amplitude. As in Eq. (12-61d), $R_F/R_i$ should be somewhat greater than 3.

## 12-15 Tuned-Circuit Oscillators

COLPITTS OSCILLATOR. The Colpitts oscillator in Fig. 12-30 must have a loop gain of unity and will have zero degree phase shift at the resonant frequency $\omega_r$ of the tuned circuit. The choke prevents $V_{cc}$ from short circuiting $C_1$ and $C$ is a short circuit at oscillating frequency $\omega_r$. $C_2$ may be modified to include effects of $C_\pi$ or $C_\mu$.

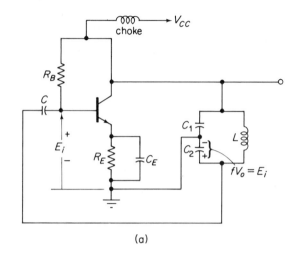

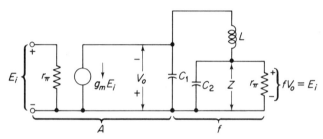

**Fig. 12-30.** Colpitts oscillator in (a) and circuit model in (b).

*Example 12-18.* (a) Derive the loop gain equation from the model in Fig. 12-30(b). (b) Find $\omega_r$ and (c) minimum gain required for oscillation.

*Solution.* (a) From Fig. 12-30(b), $V_o = g_m E_i [X_{C1} \| (X_L + Z)]$ and $fV_o$ is found from the division of $V_o$ between $Z$ and $X_L$, or

(1) $\quad fV_o = g_m E_i \left( \dfrac{Z X_{C1}}{X_{C1} + X_L + Z} \right) \quad$ where $Z = \dfrac{r_\pi}{1 + j\omega C_2 r_\pi}$

For oscillation $fV_o$ must equal $E_i$. Substituing for $fV_o = E_i$, $Z$, $X_L = j\omega L$, and $X_{C1} = 1/j\omega C_2$ yields

(2) $\quad 1 = \dfrac{g_m r_\pi}{1 + j\omega(C_1 r_\pi + C_2 r_\pi - \omega^2 L C_1 C_2 r_\pi) - \omega^2 L C_1}$

(b) Setting the denominator $j$ terms equal to 0 gives $\omega_r$:

(3) $\quad \omega_r^2 = \dfrac{1}{L\left( \dfrac{C_1 C_2}{C_1 + C_2} \right)}$

Sec. 12-15                              Tuned-Circuit Oscillators    405

The expression $C_1C_2/(C_1 + C_2)$ in Eq. (3) represents a capacitor $C$ that is the series equivalent of $C_1$ and $C_2$. Hence $\omega_r$ is determined by inspection of the resonant frequency of the tank circuit $LC$.
(c) At $\omega = \omega_r$, the real terms in Eq. (2) yield

(4) $\quad -\beta_0 = \dfrac{C_1}{C_2} \quad$ or $\quad -fA = -\beta_0 \dfrac{C_2}{C_1}$

In Eq. (4), $f = C_2/C_1$ and $A = \beta_0$. $\beta_0$ must be greater than $C_2/C_1$ for oscillation to occur.

CRYSTAL CONTROL OF OSCILLATORS. Some types of natural crystals exhibit a piezoelectric effect. If a voltage is applied to the crystal in Fig. 12-31

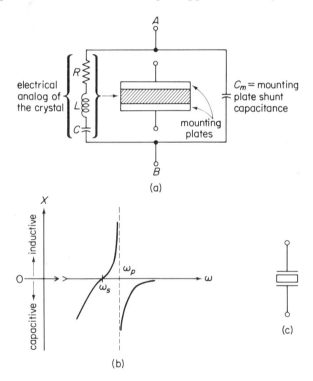

Fig. 12-31. Reactance of the crystal's electrical analog in (a) is shown in (b) and symbol in (c).

it will either compress or expand, depending on polarity. When voltage is removed the crystal will return to its original shape in a reaction to generate a voltage opposing the original voltage action. In fact the crystal overreacts to exhibit a damped oscillation similar to that of a compressed spring. An electrical analog of the crystal is represented by an $RLC$ circuit, where $R$

models losses, $C$ models restoring force, and $L$ models inertia. The crystal's natural mechanical frequency depends on its geometry and is very stable, with small losses and high values of $Q$.

From the electrical model of Fig. 12-31(a) we see that series resonance occurs at $\omega_s$ when series arm reactances are equal. Above $\omega_s$, the net inductance of the series arm enters into parallel resonance with $C_m$ and $\omega_p$. The resonant frequencies are drawn in Fig. 12-31(b) and approximated by

$$\omega_s = \frac{1}{\sqrt{LC}} \tag{12-62a}$$

$$\omega_p = \sqrt{\frac{1}{L}\left(\frac{1}{C_m} + \frac{1}{C}\right)} \tag{12-62b}$$

Typical values of $C$ are 0.05 pF and 2pF for $C_m$, so that $C_m \gg C$. Thus $\omega_p$ and $\omega_s$ are almost equal in Eq. (12-62). Furthermore any capacitance shunting $C_m$ makes $\omega_p$ approach $\omega_s$ even closer and does not affect the crystal's natural frequency. We can replace inductor $L$ in Fig. 12-30 to make a crystal-controlled Colpitts oscillator. $C_1$ and $C_2$ will not affect $\omega_r$ but will control the feedback factor $f$. $C_1$ and $C_2$ can only affect oscillator frequency when their series-equivalent capacitance approaches that of the crystal $C$.

HARTLEY OSCILLATOR. When the roles of $L$ and $C$ are interchanged in the Colpitts oscillator we have the Hartley oscillator modeled in Fig. 12-32 (a).

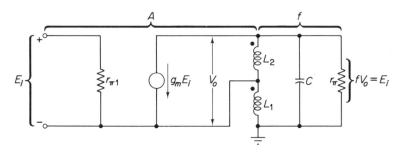

Fig. 12-32. Open-loop model of a Hartley oscillator.

Assuming no coupling between $L_1$ and $L_2$, resonance and oscillation will occur at a frequency of

$$\omega_r = \frac{1}{\sqrt{C(L_1 + L_2)}}, \quad \text{no coupling} \tag{12-63a}$$

It can be shown that $f = L_2/L_1$ and $A = \beta_0$ so that $\beta_0$ must exceed $L_1/L_2$ for oscillation. If close mutual coupling, $M$, does exist then

$$\omega_r = \frac{1}{\sqrt{LC}}, \quad \text{close coupling} \qquad (12\text{-}63b)$$

where $L = L_1 + L_2 + 2M$

Our final example of a tuned-circuit oscillator is given in Fig. 12-33(a).

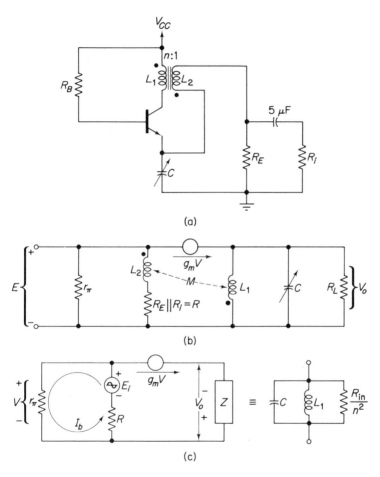

**Fig. 12-33.** The tuned-circuit oscillator in (a) is modeled in (b) and simplified in (c).

For purposes of analysis we open the feedback loop through $M$ in Fig. 12-33 (b) and assume voltage $E_i$ replaces $L_2$. $E_i$ causes $V_o$ in Fig. 12-33(c), which is dependent on open-loop gain. If we step $V_o$ down by a turns ratio between $L_1$ and $L_2$ to a value exactly equal to $E_i$, the circuit will oscillate. Load impedance $Z$ contains resistance $R_{in}/n^2$ reflected by $L_2$ across $L_1$.

*Example 12-19.* (a) Derive an expression for loop gain $E_i/V_o$ in Fig. 12-33(c). (b) Assume a turns ratio $n$, where $V_o = nE_i$ to find $\omega_r$ and (c) the minimum value of $\beta_0$.

*Solution.* (a) Writing the input loop to find $R_{in} = E_i/I_b$,

(1) $$E_i = I_b(r_\pi + R) - g_m V r_\pi$$

Substituting for $V = E_i - I_b r_\pi$ yields

(2) $$\frac{E_i}{I_b} = R_{in} = \frac{r_\pi}{\beta_0 + 1} + R$$

Output voltage $V_o$ equals $g_m V Z$. Substituting for $V$ and $I_b$ from (1) and (2) and for $Z$ from Fig. 12-33(c) yields

(3) $$\frac{E_i}{V_o} = \frac{r_\pi + (\beta_0 + 1)R}{\beta_0 Z} = \frac{r_\pi + (\beta_0 + 1)R}{\beta_0}\left(\frac{n^2}{R_{in}}\right) + j\omega C + \frac{1}{j\omega L_1}$$

(b) Set the $j$ terms of Eq. (3) equal to zero to find $\omega_r$:

(4) $$\omega_r = \frac{1}{\sqrt{L_1 C}}$$

(c) The fed-back fraction of $V_o$ is $fV_o$ and $fV_o$ must equal $E_i$ or $n = f$. Invert the real terms of Eq. (3) and substitute for $R_{in}$ to obtain

(5) $$A = \frac{V_o}{E_i} = \frac{\beta_0}{(\beta_0 + 1)n^2} = \frac{\alpha_0}{n^2} \cong \frac{1}{n^2}$$

Equation (5) represents transistor gain at $\omega_r$. Multiply $A$ by $f$ and the right-hand side by its equal, $n$, to find the conditions for $fA = 1$:

(6) $$fA \geq \frac{1}{n} \quad \text{or} \quad n \geq 1$$

The transistor in Fig. 12-33 is a common-base amplifier with no phase reversal, consequently the transformer windings are connected for no phase reversal between $V_o$ and $E_i$.

# PROBLEMS

**12-1** In Fig. 12-1, $A = 500$ and $f = 0.02$. What values are required for $E_i$ for $V_o = 5$ V in each circuit? What is the relationship between $E_i$ found for Fig. 12-1(a) and $E_{if}$ for Fig. 12-1(b)?

## Chap. 12　Problems　409

*12-2*　In Fig. 12-1(b), $E_i$ is held constant at 1 mV and $A = 100$. Plot $A_f$, $V_o$, and $fV_o$ versus $f$, for $f = 0, 0.001, 0.01$, and $0.1$.

*12-3*　What is the degree of feedback in decibels for Example 12-1 for each amplifier?

*12-4*　Identify the sampled output variable, fed-back variable, and stabilized input variable for (a) common-emitter amplifier with emitter resistance, (b) common-collector amplifier, and (c) common-emitter amplifier with collector-to-base resistor. (d) What output-input relationships are stabilized in (a), (b), and (c)?

*12-5*　What changes result in Example 12-2 if $A_{v1}$ is increased to 500?

*12-6*　A distortion voltage of 10% $V_o$ is introduced into both circuits of Figs. 12-7(a) and (b). If $A_{v1} = 500$, $A_{v2} = 1, f = 0.2$, and $V'_o = 2.5$ V what distortion-voltage component is present in the output of each circuit?

*12-7*　Open-loop gain $A_v = 1000$ of an amplifier is reduced to 100 by negative feedback. What is the (a) required feedback fraction, (b) resulting improvement factor, and (c) loop gain or return ratio?

*12-8*　In Fig. 12-8, (a) what output-input relationship is stabilized? (b) Express $f$ in terms of the external feedback elements. (c) Express the stabilized output-input relationship in terms of $f$ and also the feedback elements.

*12-9*　Express $f$ and the stabilized output-input relation in terms of the external feedback elements for the circuits in Figs. 12-11, 12-16, and 12-17.

*12-10*　When $R_E$ is changed to 100 Ω in Fig. 12-15, what changes result in the solutions of Example 12-6?

*12-11*　Employing the results of Problem 12-10 and Example 12-7 find the input-resistance changes for Fig. 12-15 caused by changing $R_E$ to 100 Ω.

*12-12*　Source resistance $R_g$ is increased to 1 kΩ in Example 12-8 and 12-9. What are the new gain and resistance levels?

*12-13*　What changes occur in the solutions of Example 12-11 if $R_E$ is doubled to 1 kΩ?

*12-14*　Feedback of $f = 0.01$ is applied to an amplifier with open-loop gain of 1000, $f_H = 50$ kHz and $f_L = 20$ Hz. What are the resulting (a) improvement factor, (b) voltage gain, (c) upper and lower cutoff frequencies?

*12-15*　$R_F$ is halved to 5000 Ω in Fig. 12-21. What changes result in the solutions of Example 12-13?

**12–16** Revise the results of Example 12-14(a) by selecting an open-loop gain of 200.

**12–17** If $\omega_H = 100$ kHz for each stage in Fig. 12-26, at what frequency will it oscillate? Assuming $A_o = 5$ for each stage, what is the minimum feedback fraction to begin oscillation?

**12–18** In Fig. 12-27(a), $R_L = 3$ k$\Omega$, $V_{CC} = 6$ V, $I_C = 1$ mA, $\beta_o = \beta_F = 200$, $R = 10$ k$\Omega$. (a) Calculate $R_E$ for a gain of 29, assuming the feedback network does not load $R_L$. (b) Calculate $R_B$ to establish $I_C = 1$ mA. (c) Find $C$ for $\omega_r = 10$ krad/sec. (d) Evaluate $R_3$ so that $R_3 \| R_i \| R_B = R$.

**12–19** In the Wien bridge oscillator of Fig. 12-28, $R_1 = R_2 = 10$ k$\Omega$. (a) Find $C_1 = C_2 = C$ for an oscillating frequency of 10 krad/sec. (b) Choose $R_F$ for a gain of 3.

**12–20** $L = 100$ $\mu$H and $\beta_o = 50$ in the Colpitts oscillator of Fig. 12-30. Choose $C_1$ and $C_2$ for oscillation at 100 krad/sec.

**12–21** In Fig. 12-28(a), $R_1 = 10$ k$\Omega$, $R_2 = 10R_1$, $C_1 = 0.0159$ $\mu$F, $C_2 = C_1/10$. Find $f_r$ and the gain requirement on the amplifier.

**12–22** A quartz crystal is modeled by $R_s = 250$ $\Omega$, $C = 0.05$ pF, $C_m = 5$ pF, and $L_s = 5$ H. Find (a) the series resonant frequency, (b) the parallel resonant frequency, and (c) $Q$ of the series model.

**12–23** What capacitance is required for oscillation at 1 kHz, 10 kHz, and 100 kHz in the integrated circuit oscillator of Fig. 12-29.

**12–24** In Fig. 12-33, $L_1 = 100$ $\mu$H. Between what values must $C$ vary to adjust the oscillating frequency between 1 and 2 MHz.

# Chapter 13

13-0 INTRODUCTION .............................413
13-1 CHARACTERISTICS OF THE VACUUM DIODE ..........414
13-2 TRIODE CHARACTERISTICS AND MODEL ............416
13-3 BIASING ....................................423
13-4 VOLTAGE GAIN AND RESISTANCE LEVELS ..........424
13-5 LOW-FREQUENCY RESPONSE .....................429
13-6 HIGH-FREQUENCY RESPONSE.....................431
13-7 PENTODE VACUUM TUBES .......................433
13-8 VACUUM TUBE PHASE INVERTERS.................437
13-9 PUSH-PULL OPERATION ........................439
    PROBLEMS ....................................446

# Vacuum Tubes

## 13-0 Introduction

A vacuum tube is a device that controls the passage of charge carriers through a vacuum by means of an electric field. Electrons are usually the only significant charge carriers, although in some tubes, gas molecules which ionize and become charge carriers are introduced. Electrons are supplied by heating a specially treated surface in order that thermally agitated valence electrons may leave their parent atoms and escape from the surface into the surrounding vacuum. Vacuum tubes are referred to as thermionic devices because they depend on externally supplied heat for operation. Common to all vacuum tubes is a resistance wire called the *heater* or *filament* and an element to emit electrons called the *cathode*. Cathodes coated with oxides are efficient electron emitters at temperatures of 1000 °C and are indirectly heated as in Fig. 12-1(a).

Thoriated tungsten cathodes are heated directly up to temperatures of 2500 °C, to supply higher electron emissions as symbolized in Fig. 12-1(b).

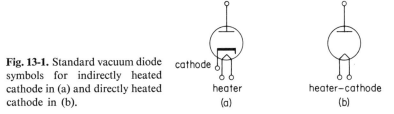

Fig. 13-1. Standard vacuum diode symbols for indirectly heated cathode in (a) and directly heated cathode in (b).

Movement of electrons within the vacuum will be studied by reference to the simplest thermionic device—the vacuum diode.

## 13-1 Characteristics of the Vacuum Diode

Heat energy from the heater element in Fig. 13-2 is imparted to the cathode, forcing electrons to escape into the vacuum environment within the tube. As each electron escapes from the surface it leaves behind a unit positive charge. As more electrons escape, a concentration is built up that reaches a dynamic equilibrium or balance between attraction by the positively charged cathode and the thermal velocities of the electrons. Electrons are continuously leaving the cathode and returning from the cloud to form a dynamic equilibrium.

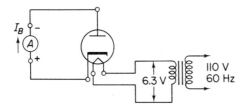

Fig. 13-2. Circuit to demonstrate current flow due to high-energy electrons. Conventional or positive charge movement is shown in the external circuit.

The *anode* or *plate* of a vacuum diode is a cylinder surrounding the cathode cylinder. Some of the higher-energy electrons will reach the anode, and if a return path is available, via a wire between plate and cathode, they will traverse it and a small "Edison effect" current of up to 2.0 mA may be measured in the external plate-to-cathode circuit, as shown in Fig. 13-2. If the anode is made positive with respect to the cathode an electric field is established, attracting electrons from the cloud to the plate. Increasing voltage $E_B$ between plate and cathode will attract an increasing number of electrons so that current through the diode increases with voltage across it. There is a maximum value of $E_B$, above which there is little further increase in current, indicating that all of the electrons in the cloud have been collected. The plate is collecting them as fast as they are being emitted in a condition defined as *saturation*. A test circuit is shown in Fig. 13-3 to sweep the characteristic of the diode. With switch $Sw$ in position 1, rated filament voltage is applied to the heater, giving the characteristic curve in Fig. 13-4(a). The rising portion of the curve shows an increasing current due to an increasing electric field caused by increasing $E_B$. It is on this portion of the characteristic that the diode is usually operated. Saturation is demonstrated by throwing switch $Sw$ to position 2 in Fig. 13-3 to halve the filament voltage.

A negative plate potential forces electrons back into the electron cloud, shrinking it about the cathode. Thus the vacuum diode performs as a rectifier, passing electrons from cathode to plate. We can derive a simple circuit model from the diode characteristic of Fig. 13-4 that will apply only

Sec. 13-1          Characteristics of the Vacuum Diode          415

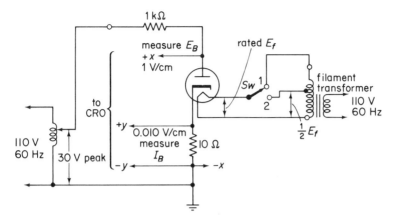

**Fig. 13-3.** Circuit to display vacuum diode $I_B - E_B$ characteristic. Reducing filament voltage $E_f$ allows a display of saturation.

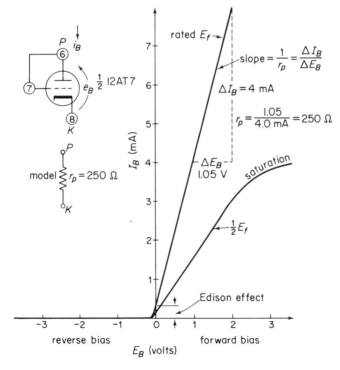

**Fig. 13-4.** Characteristic curve of a diode-connected 12AT7 is swept by the circuit of Fig. 13-3 to show that the diode may be modeled by a resistance of 250 Ω, under forward bias.

under forward bias and when current flow is not emission limited. Since the general equation of a line through the origin is $y = mx$, associate $y$ with $I_B$, slope $m$ with $1/r_p$, and $x$ with $E_B$. Employing total instantaneous values, the current-voltage relationship of a diode can be modeled by plate resistance $r_p$, where

$$i_B = \frac{1}{r_p} e_B \tag{13-1}$$

## 13-2 Triode Characteristics and Model

Placing a metal electrode called a *grid* between cathode and plate introduces a method of influencing the electric field between them. If the grid is positioned close to the cathode and structured to allow electrons to pass through it, a relatively small negative grid voltage can offset or even cancel the field due to the plate voltage. Thus amplification is possible in this three-element vacuum tube known as a *triode*.

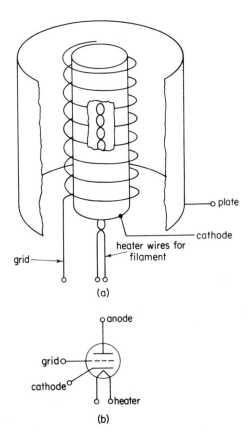

**Fig. 13-5.** Triode structure in (a) and symbol in (b).

Sec. 13-2                    Triode Characteristics and Model    417

A basic triode structure and symbol is given in Fig. 13-5. Once the heater function is understood we assume it is connected properly and omit it from the drawing. Electrical behavior of the triode is then determined by voltages between the plate, grid, and cathode terminals.

Plate characteristic curves may be swept by the circuit of Fig. 13-6(a)

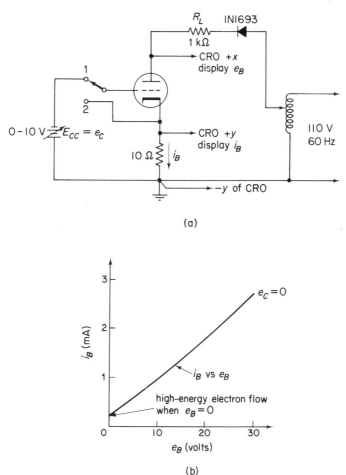

Fig. 13-6. The sweep circuit in (a) displays a plate characteristic curve for $e_C = 0$ in (b).

to measure parameters for a model of the vacuum tube. Grid-to-cathode voltage $e_C$ is the input variable and output characteristic curves show the relationship between plate current $i_B$ and plate-to-cathode voltage $e_B$.

Typical characteristics for a 12AT7 triode are shown in Fig. 13-6(b). The $i_B - e_B$ curve for $e_C = 0$ is *not* identical with the diode-connected 12AT7 curve in Fig. 13-4.

In Fig. 13-4, 0.2 V of $e_B$ caused 1 mA to flow in the tube because the connection between plate and grid brought the plate effectively closer to the cathode. However, in Fig. 13-6(b), $e_B$ must go to 10 V before 1 mA flows because most of the field lines from the plate are terminated on the grid and do not reach the electron cloud.

An idealized set of triode plate characteristics are shown in Fig. 13-7.

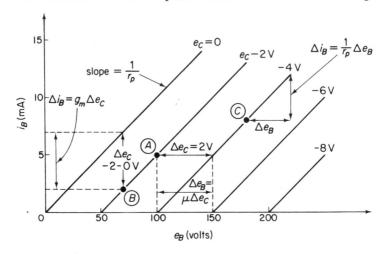

Fig. 13-7. Measurement of tube parameters $\mu$, $g_M$, and $r_p$.

Consider point $A$, which indicates there are 100 V across the tube and 5 mA flowing between plate and cathode, with a bias voltage, $e_C = -2$ V, between grid and cathode. Let $e_C$ be increased to $-4$ V but hold $i_B$ to 5 mA by increasing $e_B$ to 150 V. The amount of increase in $e_B$ of $\Delta e_B = 50$ required to balance the increase in $e_C$ of $\Delta e_C = -2$ V is a measure of the relative effect of plate and grid voltages on the plate current. The *amplification factor* is defined by their ratio:

$$\mu = \frac{\Delta e_B}{\Delta e_C}, \qquad i_B = \text{constant} \tag{13-2}$$

In Fig. 13-7, $\mu$ is seen to be $50/2 = 25$ at $I_B = 5$ mA. Consider point $B$ in Fig. 13-7. If bias voltage $e_C$ is decreased from $-2$ to $0$ $V = \Delta e_C = 2$, and $e_B$ is held constant at 70 V, plate current will change from 2 to 7 mA and $\Delta i_B = 5$ mA. The relationship between plate current change due to a grid voltage change at constant plate voltage is defined as *transconductance* $g_m$, where

$$g_m = \frac{\Delta i_B}{\Delta e_C}, \qquad e_B = \text{constant} \tag{13-3}$$

At point $B$, $g_m = 5 \text{ mA}/2\text{V} = 2500$ $\mu$mho.

*Sec. 13-2*  *Triode Characteristics and Model* 419

The final tube parameter is *plate resistance*, $r_p$, and is defined as the ratio of the change in plate current to the corresponding change in plate voltage, with $e_C$ held constant.

Refer to point $C$ and assume $e_B$ increases from 180 to 220 V = $\Delta e_B$ = 40 V. If $e_C$ is held constant, $i_B$ will increase from 8 to 12 mA for a change of $\Delta i_B = 4$ mA. The definition and value of $r_p$ is

$$r_p = \frac{e_B}{i_B}, \quad e_C = \text{constant}$$
$$= \frac{40}{4 + 10^{-3}} = 10 \text{ k}\Omega, \quad e_C = -2 \text{ V} \qquad (13\text{-}4)$$

Since tube parameters are defined from the same characteristic they are related, as seen from the product of Eqs. (13-3) and (13-4)

$$g_m r_p = \frac{\Delta i_B}{\Delta e_C}\frac{\Delta e_B}{\Delta i_B} = \frac{\Delta e_B}{\Delta e_C} = \mu \qquad (13\text{-}5)$$

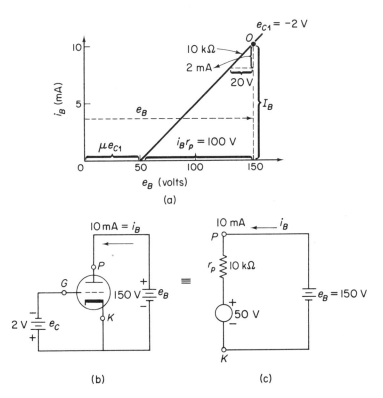

**Fig. 13-8.** Operating point $O$ in (a) describes the tube's terminal current and voltages in (b). An equivalent circuit model is shown in (c).

A dc model for the triode is developed from the idealized triode characteristic in Fig. 13-8(a), where $e_B$ is pictured as the sum of $\mu e_{C1}$ and $i_B r_p$. The tube is modeled at point 0 by (1) dependent voltage generator $\mu e_{C1}$ of $25(2) = 50$ V, plus (2) voltage drop $i_B r_p$ of $10$ mA $\times 10^4 = 100$ V in Fig. 13-8(c).

Assuming the dc bias on a vacuum tube is properly established we can superimpose signal currents and voltages and permit the use of a small-signal model for ac analysis. Assume signal increment $e_c = 1$ V is applied to the grid with a polarity that will cause an incremental increase in plate current of $I_B$ in Fig. 13-9(a). [The dc value of $I_B$ is assumed to be present, although not shown, with the value given in Fig. 13-8(b).] Since $e_B$ cannot change, the change in $e_C$ from $-2$ to $-1$ V caused $i_B$ to rise along the $e_B = 150$ V line in Fig. 13-9(b) from $O$ to $O'$. $i_B$ increases from 10 to 12.5 mA or $\Delta i_B = 2.5$ mA. Distance $AO$ is found from

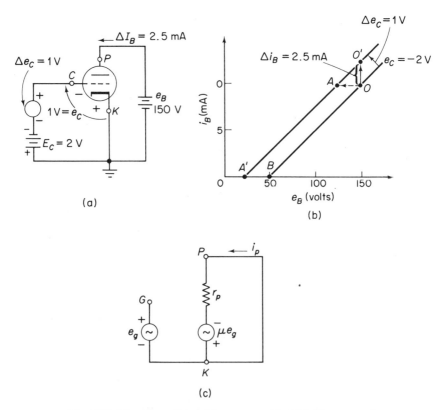

**Fig. 13-9.** The effect of signal increment $e_C$ in (a) is shown graphically in (b). The triode may be replaced by the ac circuit model in (b) for incremental or rms values.

Sec. 13-2          Triode Characteristics and Model    421

$$\text{Slope} = \frac{1}{r_p} = \frac{O'O}{AO} = \frac{\Delta i_B}{AO} \quad \text{or} \quad AO = \Delta i_B r_p$$

But $BA' = OA = \Delta e_C$ and

$$\Delta e_C = \Delta i_B r_p \tag{13-6}$$

Evidently a voltage change in the plate circuit is related to a change in grid voltage by $\mu$ and causes a plate current change of $\Delta i_B$. Replacing delta quantities in Eq. (13-6) by instantaneous quantities yields

$$\mu e_g = i_p r_p \tag{13-7}$$

where $e_g =$ instantaneous grid-to-cathode signal voltage, and
   $i_p =$ instantaneous plate signal current.

A small-signal model of Eq. (13-7) for the triode is shown in Fig. 13-9(c). Since G is positive with respect to $K$, increment $\Delta I_B$ or $i_p$ enters the plate terminal $P$. This fixes the polarity of dependent generator $\mu e_g$ that apparently causes $i_p$.

*Example 13-1.* On the plate characteristics of Fig. 13-10, locate an operating point at $e_B = 175$ V, $e_C = -2$ V. Calculate $\mu$, $r_p$, and $g_m$.

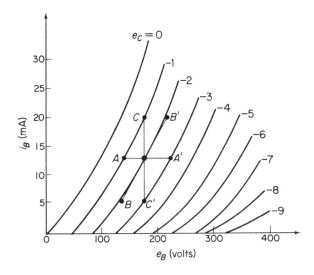

**Fig. 13-10.** Solution to Example 13-1.

*Solution.* Point $O$ is at $e_B = 175$ V, $e_C = -2$ V, and $i_B = 13$ mA in Fig. 13-10. $\mu$ is determined by drawing $AA'$ between two adjacent grid voltage lines, through point $O$. This horizontal line corresponds to a constant $I_B = 13$ mA with length $\Delta e_C = 2$ V. Plate voltages are 140 and 225 V at points $A$ and $A'$, respectively, corresponding to $\Delta e_B = 85$ V. From Eq. (13-2), $\mu = 85/2 = 42$.

Incremental plate resistance $r_p$ is found by drawing line $BB'$ tangent to $e_C = -2$ V at point $O$ between convenient plate currents for a $\Delta i_B$ of 15 mA. Projecting downward from points $B$ and $B'$ to the $e_B$ axis gives $\Delta e_B = 80$ V. The ratio of $\Delta e_B$ to $\Delta i_B$ is 5.3 k$\Omega$.

Transconductance $g_m$ can be found from dashed line $CC'$ drawn vertical (to correspond with constant $e_B$) between $e_C = -1$ V and $e_C = -3$ V for $\Delta e_C = 2$ V. Values of $i_B$ at points $C$ and $C'$ are 22 and 6 mA, respectively, for $\Delta i_B = 16$ mA. The resulting value of $g_m$ from Eq. (13-3) is $g_m = 16 \times 10^{-3}/2 = 8000$ $\mu$mho. We compute the product, $g_m r_p = \mu = (8000 \times 10^{-6})(5.3 \times 10^3) = 42.4$, and check it against the graphical value of 42.

**Example 13-2.** Find the peak value of $V_o$ for the circuit in Fig. 13-11(a). Operating-point and triode characteristics are identical to those in Example 13-1.

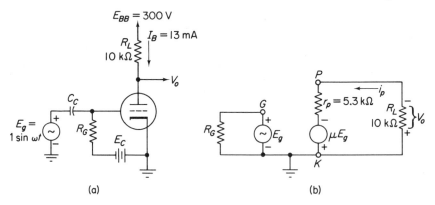

**Fig. 13-11.** Circuit in (a) and model in (b) of a common-cathode triode amplifier.

*Solution.* Draw the model as in Fig. 13-11(b), replacing the tube with its small-signal model. Power supplies $E_C$ and $E_{BB}$ plus $C_C$ are replaced by their zero resistance to signal currents. Since $\mu E_g$ divides between $r_p$ and $R_L$, substitute peak values into

$$V_o = -\frac{R_L}{R_L + r_p}\mu E_g = -42\frac{10^4}{(10 + 5.3)10^3}(1) = 27.4 \text{ V}$$

Voltage gain is expressed from the relation developed in Example 13-2 as

$$\frac{V_o}{E_g} = -\frac{\mu R_L}{r_p + R_L} \tag{13-8}$$

Voltage gain can never exceed $\mu$ and increases with larger values of load and smaller values of plate resistance.

## 13-3 Biasing

The biasing procedure for a vacuum tube is much simpler than it is for a transistor. In Fig. 13-12(a), resistor $R_K$ causes a voltage drop of $E_C =$

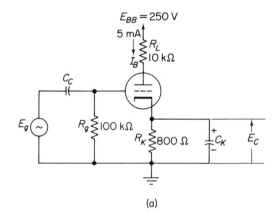

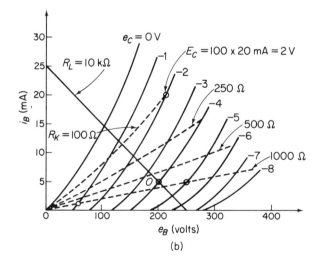

**Fig. 13-12.** Cathode resistor biasing in (a) is analyzed from bias lines in (b).

$I_B R_K$. No current flows in $R_g$, so there is no voltage drop across it. Therefore the grid is at the same potential as the bottom terminal of $R_K$, which is negative with respect to the cathode. The dc load seen by the tube is $R_L + R_K$. Ignoring $R_K$ in the load line for the moment, refer to the load line for $R_L =$

10 kΩ in Fig. 13-12(b). Assume it is desired to operate at point $O$, of $I_B = 5$ mA, $E_C = -4$ V.

Knowing that 5 mA will flow, *if* we establish the $-4$ V bias, we *assume* it flows through $R_K$ and calculate $R_K$ to give a drop of 4 V from

$$R_K = \frac{E_C}{I_B} = \frac{4}{5 \times 10^{-3}} = 800 \ \Omega \qquad (13\text{-}9)$$

Since 800 Ω is below 10% of $R_L$, it is not really necessary to redraw the dc load line with a slope of $-1/10{,}800 \ \Omega$. It is possible to superimpose the bias curve $E_C = I_B R_K$ for various values of $R_K$ on the plate characteristics as in Fig. 13-12(b). When an operating point is chosen on a load line we merely estimate the value of $R_K$ from the nearest intersection between the load line and bias line. If $R_K$ is chosen to be 1 kΩ. in Fig. 13-12(b) an $I_B$ of 5 mA will cause an $E_C$ of $-5$ V, and an $I_B$ of 2 mA will give $E_C = -2$ V. These two points are circled to plot the dashed line of $R_K = 1$ kΩ. Other lines are shown for $R_K = 100$, 250, and 500 Ω.

Capacitor $C_K$ is chosen to have a reactance of $0.1R_K$ at the low-frequency cutoff. $C_K$ bypasses the signal plate currents and effectively shorts out $R_K$ to ac signals. $C_K$ will be large enough so that its charge is not changed by the signal currents and will maintain its average dc potential at $E_C$ to hold the tube at the designated operating point.

## 13-4 Voltage Gain and Resistance Levels

Voltage gain expressions are developed for common-cathode and common-plate amplifiers from the circuit of Fig. 13-13(a). From the model in Fig. 13-13(b), $V_{o1}$ goes negative when $E_i$ goes positive, indicating a 180° phase reversal between grid and plate. No phase reversal occurs between grid and cathode. Write the input-loop equation and multiply both sides by $\mu$.

$$E_g = E_i - I_p R_K \qquad (13\text{-}10\text{a})$$
$$\mu E_g = \mu E_i - I_p(\mu R_K) \qquad (13\text{-}10\text{b})$$

From Eq. (13-10b) we can replace $\mu E_g$ with a generator $\mu E_i$ that depends on input (not grid) signal in series with a resistor $\mu R_K$ as shown in Fig. 13-13 (c). By inspection of the resulting voltage division of $\mu E_i$ we can write the output voltage and gains directly:

$$V_{o2} = \frac{R_K}{r_p + R_L + (\mu + 1)R_K}(\mu E_i) \qquad (13\text{-}11\text{a})$$

and

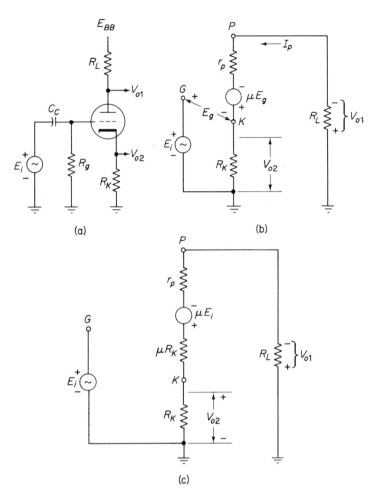

**Fig. 13-13.** The amplifier in (a) is modeled in (b) and simplified in (c).

$$A_{v2} = \frac{\mu R_K}{r_p + R_L + (\mu + 1)R_K} = \frac{V_{o2}}{E_i} \quad (13\text{-}11\text{b})$$

$$V_{o1} = -\frac{R_L}{r_p + R_L + (\mu + 1)R_K}(\mu E_i) \quad (13\text{-}12\text{a})$$

$$A_{v1} = -\frac{\mu R_L}{r_p + R_L + (\mu + 1)R_K} = \frac{V_{o1}}{E_i} \quad (13\text{-}12\text{b})$$

Equation (13-11b) represents the gain of a common plate provided $R_L = 0$. Equation (13-12b) reduces to Eq. (13-8) when $R_K = 0$ due to capacitor bypassing.

Resistance seen by looking into the grid is considered infinite since there is negligible coupling to the electron stream. To find resistance-transformation properties of the plate terminal, apply test voltage $E$ in Fig. 13-14 (a) to write the output loop

$$E = I(r_p + R_K) + \mu E_g \tag{13-13a}$$

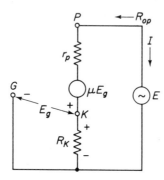

Fig. 13-14. Test circuit to measure output resistance of the plate terminal.

Substituting for $E_g = IR_K$ shows that $R_K$ or any cathode leg resistance is increased by the factor $(\mu + 1)$.

$$R_{op} = \frac{E}{I} = r_p + (\mu + 1)R_K \tag{13-13b}$$

We illustrate another technique of finding output resistance for the cathode follower. If we can work the output-voltage expression from Eq. (13-11a),

$$V_{o2} = \frac{R_K}{r_p + (\mu + 1)R_K} \mu E_i, \qquad R_L = 0 \tag{13-11a}$$

into the form

$$V_o = \frac{R_{\text{load}}}{R_{oK} + R_{\text{load}}} V_{\text{tube}} \tag{13-14}$$

We can pick out $R_{oK}$ by inspection. Since $R_{\text{load}} = R_K$, divide numerator and denominator of Eq. (13-11a) by $(\mu + 1)$:

$$V_{o2} = \frac{R_K}{\dfrac{r_p}{\mu + 1} + R_K} \left( \frac{\mu}{\mu + 1} E_i \right) \tag{13-15}$$

to see that

$$R_{oK} = \frac{r_p}{\mu + 1} \tag{13-16}$$

Sec. 13-4  Voltage Gain and Resistance Levels  427

We conclude that $r_p$ and any resistance in the plate leg will be divided by $(\mu + 1)$. These transformations are illustrated in Examples 13-3 and 13-4.

*Example 13-3.* In the circuit of Fig. 13-15(a), calculate $V_o$ using (a) voltage gain and (b) output resistance.

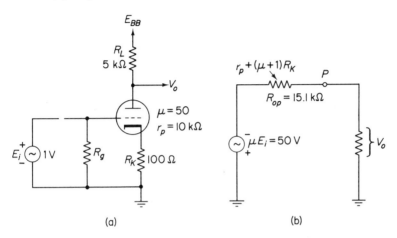

**Fig. 13-15.** Circuit for Example 13-3.

*Solution.* (a) From Eq. (13-12b),

$$A_v = -\frac{50(5000)}{[10 + 10 + 51(.1)]10^3} = -12.4$$

$V_o$ is $A_v E_i = 12.4$ V.
(b) Evaluate $R_{op}$ in Fig. 13-15(b) from Eq. (13-13b):

$$R_{op} = 10 + 51(0.1) = 15.1 \text{ k}\Omega$$

From the voltage division of $\mu E_i$,

$$V_o = \frac{5}{5 + 15.1}(50) = 12.4 \text{ V}$$

*Example 13-4.* Evaluate $V_o$ for the cathode follower in Fig. 13-16(a) both by (a) voltage gain and (b) output resistance.

*Solution.* (a) From Eq. (13-11a),

$$V_o = \frac{50(0.1) \times 10^3}{[10 + 51(0.1)]10^3} = 0.33 \text{ V}$$

(b) Evaluate $R_{oK}$ in Fig. 13-1(b) from Eq. (13-16):

$$R_{oK} = \frac{10000}{51} = 196 \text{ }\Omega$$

**428** *Vacuum Tubes* **Chap. 13**

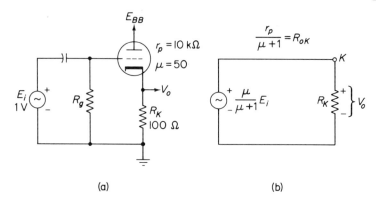

**Fig. 13-16.** Circuit for Example 13-4.

From the voltage division of $\mu/(\mu + 1)E_i$ or Eq. (13-15),

$$V_o = \frac{100}{296}\left(\frac{50}{51}\right)(1) = 0.33 \text{ V}$$

One point that must be emphasized is the effect of a cathode resistor on the permissible value of $E_i$. Without a cathode resistor, $E_i$ equals $E_g$ and the peak allowable value of $E_g$ depends on the bias voltage $E_C$. For example, if $E_C$ were $-3$ V, then $E_g$ must be limited to 3 V on its positive peak so that the grid will not go positive. Or if $E_C$ were biased at $-12$ V and the tube is cut off at $-14$ V, then $E_g$ is limited by its negative excursion to a peak of $-2$ V. It is convenient to develop an expression showing how the maximum permissible value of $E_i$ is determined if we know the maximum allowable value of $E_g$ from our operating-point and tube characteristics. In Fig. 13-13(b),

$$I_p R_K = \frac{R_K}{R_K + R_p + R_L}\mu E_g$$

Substituting for $I_p R_K$ into Eq. (13-10a) gives

$$E_i = E_g\left[1 + \frac{\mu R_K}{r_p + R_K + R_L}\right] \tag{13-17}$$

Voltage gain and output resistance of the common-grid amplifier in Fig.13-17(a) are developed from the outside-loop equation in the model of Fig. 13-17(b):

$$\mu E_g + E_i = I_p(R_L + r_p + R_K) \tag{13-18}$$

Substituting for $E_g = E_i - I_p R_K$, solving for $I_p$, and noting that $V_o = I_p R_L$ yields

Sec. 13-5                                      Low-Frequency Response

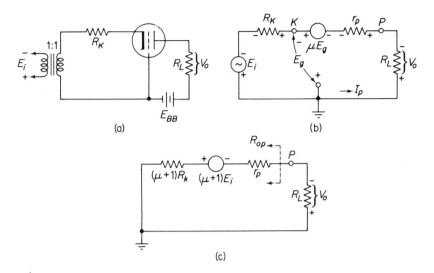

Fig. 13-17. A common-grid amplifier in (a) is modeled in (b) to show output resistance in (c).

$$V_o = I_p R_L = \frac{R_L}{R_L + r_p + (\mu + 1)R_K}(\mu + 1)E_i \qquad (13\text{-}19)$$

Comparing Eq. (13-19) with Eq. (13-14) shows that resistance seen looking into the plate is still given by Eq. (13-13b). A model of the tube as seen by the load is given in Fig. 13-17(c) to agree with Eq. (13-19).

## 13-5 Low-Frequency Response

Coupling capacitor $C_C$ and bypass capacitor $C_K$ are isolated from one another by the grid in Fig. 13-18(a). We assume that $C_K$ is a short circuit and analyze low-frequency circuit behavior due to $C_C$ alone. Assume that input signal $E_i$ is held constant in amplitude. As frequency of $E_i$ is reduced, reactance of $C_C$ will no longer be negligible but will increase, reducing $V_i$ and consequently $V_o$. The low break frequency $\omega_L$ is the reciprocal of the time constant formed in Fig. 13-18(b).

$$\omega_C = \frac{1}{R_G C_C} \qquad (13\text{-}20)$$

Now assume $C_C$ is large enough to be a short circuit at a low frequency where $C_K$ acts as an open circuit. Voltage gain will be expressed by Eq. (13-12b).

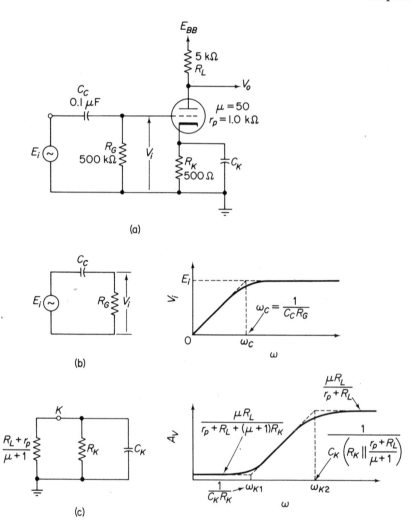

**Fig. 13-18.** The vacuum tube circuit in (a) is modeled at low frequencies by (b) to study effects of $C_C$ alone. In (c) we study the effects of $C_K$ alone.

As frequency of $E_i$ is increased, $C_K$ begins bypassing $R_K$, increasing voltage gain until $C_K$ effectively short circuits $R_K$ with gain expressed by Eq. (13-8). By looking into the cathode terminal we can show the effective resistance presented to $C_K$ in Fig. 13-18(c) and from the resulting time constant, express the low break frequency due to $C_K$ as $\omega_K$:

$$\omega_K = \frac{1}{C_K R} \quad \text{where} \quad R = R_K \| \frac{R_L + r_p}{\mu + 1} \tag{13-21}$$

Sec. 13-6                                  High-Frequency Response   431

*Example 13-5.* (a) Find the low break frequency due to $C_C$ in Fig. 13-18. (b) Calculate $C_K$ to cut off a decade higher.

*Solution.* (a) From Eq. (13-20),

$$\omega_C = \frac{10^6}{0.1 \times 5 \times 10^5} = 20 \text{ rad/sec}$$

(b) Evaluating $R$ from Eq. (13-21),

$$R = 500 \| \frac{15000}{51} = 185 \ \Omega.$$

Substitute for $\omega_K = 200$ rad/sec:

$$C_K = \frac{1}{200(185)} = 27 \ \mu\text{F}$$

## 13-6 High-Frequency Response

Interelectrode or geometric capacitances $C_{gp}$, $C_{gk}$, and $C_{pk}$ are shown on the Norton triode model in Fig. 13-19(a). As the grid goes positive the

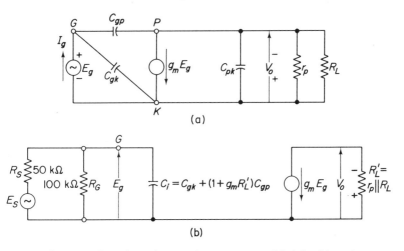

**Fig. 13-19.** Interelectrode capacitances are modeled in (a) and may be simplified by the equivalent model in (b).

plate goes negative and at high frequencies the low reactance of $C_{gp}$ allows negative feedback to reduce voltage gain. A simplified model is derived by summing the currents at node $G$.

$$j\omega C_{gp}(E_g + V_o) + j\omega C_{gk}E_g = I_g \qquad (13\text{-}22)$$

Currents through $C_{gp}$ and $C_{pk}$ will be negligible at the $P$ node and we can substitute in Eq. (13-22) for $V_o = g_m R'_L E_g$ to obtain input impedance of the grid

$$\frac{E_g}{I_g} = \frac{1}{j\omega[C_{gk} + (1 + g_m R'_L)C_{gp}]} = \frac{1}{\omega C_i} \qquad (13\text{-}23)$$

Just as with the transistor, a Miller capacitance $C_i$ is formed and modeled in Fig. 13-19(b) as

$$C_i = C_{gk} + (1 + g_m R'_L)C_{gp} \qquad \text{where } R'_L = r_p \| R_L \qquad (13\text{-}24)$$

*Example 13-6.* Given $R'_L = 10 \text{ k}\Omega$, $g_m = 8000\mu$ mho, $C_{gp} = 4 \text{ pF}$, and $C_{gk} = 2 \text{ pF}$ in the circuit of Fig. 13-19(a). Find the upper cutoff frequency.

*Solution.* From Eq. (13-24) $C_i = 2 + (1 + 8 \times 10^{-3} \times 10^4)4 = 326 \text{ pF}$. $C_i$ forms a time constant with $R_s \| R_G$ in Fig. 13-19(b) to give a break frequency $\omega_H$ of

$$\omega_H = \frac{1}{R \| R_g \times C_i} = \frac{10^{12}}{(50 \| 100)(10^3)(326)} = 92 \times 10^3 \text{ rad/sec}$$

The triode presents an input capacitance only if there is a purely resistive load. A tuned-circuit load presents an impedance which varies with frequency and can be purely resistive, or contain either inductive or capacitive reactances depending on the frequency. To examine possible effects of an impedance load, replace $R'_L$ with $Z_L = (R_L \pm jX)$ in Eq. (13-23) to obtain

$$\frac{E_g}{I_g} = \frac{1}{j\omega[C_{gk} + (1 + g_m R_L)C_{gp}] - g_m C_{gp}(\pm X)} = \frac{1}{Y_{in}} \qquad (13\text{-}25)$$

Admittance $Y_{in}$ is shown in Fig. 13-20 and can be written in the form

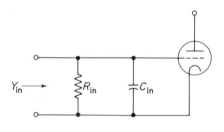

Fig. 13-20. High-frequency equivalent input circuit of a triode.

$$Y_{in} = \frac{1}{R_{in}} + j\omega C_{in}$$

where

$$R_{in} = \frac{1}{-\omega g_m C_{gp}(\pm X)} \qquad (13\text{-}26)$$

$R_{in}$ in Fig. 13-20 may take on three possible ranges.

Sec. 13-7                                            Pentode Vacuum Tubes    433

1. If the load is pure resistance, $X$ is zero and $R_{in}$ is very large or infinite.
2. Should the load have a component of capacitance, $X$ is negative and $R_{in}$ has a positive value.
3. When the load has an inductive component, $X$ is positive and $R_{in}$ becomes negative.

A negative input resistance means that current is being supplied by $R_{in}$ rather than being dissipated. Physically, current is supplied from the plate terminal through $C_{gp}$ to the grid. This feedback current *may* be of a phase and amplitude sufficient to cause sustained oscillation. Thus, as an *LC* load circuit is tuned, a point can be reached where the grid signal voltage loses control to the capacitively-coupled plate terminal. Section 13-7 deals with a vacuum tube structure that reduces the coupling between output and input terminals to almost negligible proportions.

## 13-7 Pentode Vacuum Tubes

A fourth element was introduced into the triode as a screen between grid and plate, resulting in a structure called the *tetrode*. This fourth element was called *a screen grid* because it provided an electrostatic shield between grid and plate. Flux lines from the plate terminate on the screen, and plate signal voltage can influence the screen grid. Connecting a capacitor between screen and cathode will allow these signal variations from the plate to be passed back to the cathode and prevent them from influencing the control grid. The screen is held at a positive potential whose value is usually below the dc plate voltage. Electrons from the cathode experience the electric field from the screen and pass through it to the plate. In the triode as well as the tetrode, electrons strike the plate, impart energy to the plate atoms, and ionize some of them. High-energy electrons from the ionized atoms are emitted from the plate by a phenomenom called *secondary emission*. In the case of the triode, secondary electrons saw a negative space charge and returned to the plate. But a positive screen will collect these electrons and return them through the power supply to the cathode. Plate current is diminished by this process, particularly at low values of plate voltage.

A *suppressor grid* is installed between plate and screen grid to reestablish a potential that is negative with respect to the plate, so secondary electrons will be returned to the plate and not be directed to the screen grid. The suppressor grid is wired directly to the cathode to establish the necessary negative potential. The fifth element results in the name *pentode*. Most modern pentodes have an internal connection between suppressor and cathode, as shown in Fig. 13-21. These additional elements result in a plate resistance so high that the current-source model is employed. Physically the plate is

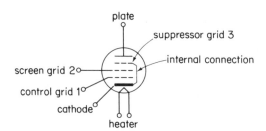

Fig. 13-21. Circuit symbol for a pentode.

so isolated from the grid that plate voltage has hardly any effect on the plate. Plate current is controlled almost entirely by the grid voltage as shown by the plate characteristics in Fig. 13-22. Capacitance between control grid and plate is low and of the order of 0.5 pF.

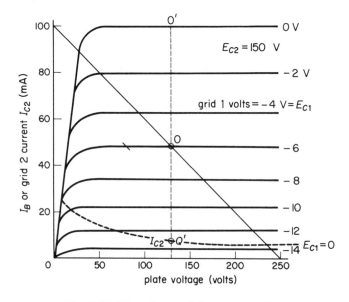

Fig. 13-22. Plate characteristics of a pentode.

Operation of the pentode will be studied by reference to the circuit of Fig. 13-23. $R_K$ carries both dc plate current $I_B$ and dc screen current $I_{C2}$ to cause a bias voltage

$$E_{C1} = (I_B + I_{C2})R_K \qquad (13\text{-}27)$$

The screen grid resistor $R_{SG}$ is determined empirically to deliver the desired value of $E_{C2}$ or, from

$$R_{SG} = \frac{E_{BB} - E_{C2}}{I_{C2}} \qquad (13\text{-}28)$$

Sec. 13-7  Pentode Vacuum Tubes  435

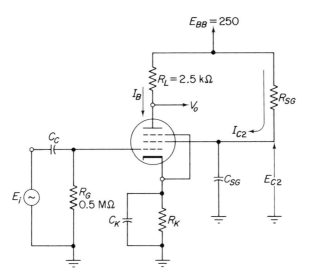

**Fig. 13-23.** Self-biased pentode amplifier.

The dc load line is represented by a loop equation written through $R_L$ and $R_K$:

$$E_{BB} = I_B R_L + V_B + (I_B + I_{C2})R_K \qquad (13\text{-}29)$$

Usually the term $(I_B + I_{C2})R_K$ can be neglected in Eq. (13-29) and $R_L$ makes up the dc loadline.

Capacitor $C_K$ determines low-frequency cutoff $\omega_L$ principally through a time constant in the denominator of

$$\omega_L = \frac{1}{C_K R_K} \qquad (13\text{-}30)$$

$C_{SG}$ is determined empirically by driving the pentode with a signal a decade above $\omega_L$ and varying $C_{SG}$ until the gain drops to 0.707 times the midfrequency value. Multiply this value of $C_{SG}$ by 10 to get a sharp break frequency at $2\omega_L$, or multiply $C_{SG}$ by 100 to let $C_K$ dominate the break frequency at $\omega_L$.

**Example 13-7.** Characteristic curves in Fig. 13-22 apply to the pentode in the circuit of Fig. 13-23. (a) Find $R_K$ to set $E_{C1} = -6$ V, (b) $R_{SG}$ to set $E_{C2} = 150$ V, (c) $C_K$ for $\omega_L = 500$ rad/sec, and (d) $V_o/E_i$.

**Solution.** (a) Assuming $R_K$ does not affect the dc load line locate operating point 0 in Fig. 13-22 at $I_B = 48$ mA, $E_B = 130$ V, $E_{C1} = -6$ V. Locate points $Q'$ and $0'$ that specify $I_B$ and $I_{C2}$ at $E_{C1} = 0$ V and $E_B = 130$ V. Since $I_B$ decreases by 50% when $E_{C1}$ changes from 0 to $-6$ V ($0'$ to 0), $I_{C2}$ will decrease by 50% from 7 mA to 3.5 mA ($Q$ to $Q'$). Find $R_K$ from Eq. (13-7):

$$R_K = \frac{6}{(48 + 3.5)10^{-3}} \cong 120\ \Omega$$

(b) From Eq. (13-28),

$$R_{SG} = \frac{250 - 150}{3.5 \times 10^{-3}} \cong 30\ \text{k}\Omega$$

(c) From Eq. (13-30),

$$C_K = \frac{1}{500(120)} \cong 20\ \mu\text{F}$$

(d) Graphically measure $g_m = 7250\ \mu$mho at the operating point in Fig. 13-22 and evaluate

$$A_v = \frac{V_o}{E_i} = g_m R_L = (7250 \times 10^{-6})(2.5 \times 10^3) \cong 18$$

Input capacitance of a pentode is expressed by

$$C_i = C_{gk} + C_{gs} + C_{gp}(1 + g_m Z_L) \tag{13-31}$$

where  $C_i$ = equivalent input capacitance   Typical values
$C_{gk}$ = grid-to-plate capacitance   3 pF
$C_{gs}$ = grid-to-screen capacitance   3 pF
$C_{gp}$ = grid-to-plate capactance   0.03 pF

Notice that $C_{gp}$ of a pentode is typically 100 times less than that of a triode so we would expect that a pentode's high-frequency response would be 100 times higher than a comparable triode.

**Example 13-8.** The amplifier of Fig. 13-23 is driven by a signal generator with an internal resistance of 20 kΩ. A tube manual lists $C_{gK} = C_{gs} = 3$ pF, $C_{gp} = 0.025$ pF. Find $\omega_H$.

**Solution.** (a) From Ex. 13-7 and Eq. (13-30)

$$C_i = 3 + 3 + 0.025(1 + 18) = 6.5\ \text{pF}$$

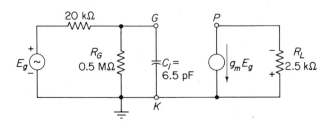

Fig. 13-24. Circuit model for Example 13-8.

From the equivalent input circuit in Fig. 13-24, $C_i$ forms a time constant with $R_G \| 20 \text{ k}\Omega$ so that

$$\omega_H = \frac{10^{12}}{6.5(20\|500)10^3} = 8 \times 10^6 \text{ rad/sec}$$

## 13-8 Vacuum Tube Phase Inverters

A phase inverter converts an input signal into two output signals that are equal and 180° out of phase with one another. It is used primarily to interface the output of a voltage amplifier with the grids of two power output tubes operated in a push-pull arrangement. If $R_K$ and $R_L$ are made equal in Fig. 13-14, $V_{o1}$ will equal $-V_{o2}$ and the circuit will perform as a *cathode-follower* phase inverter. A second tube is added in the phase inverter of Fig. 13-25 to give gain as well as phase inversion. Voltage division of $V_{o1}$

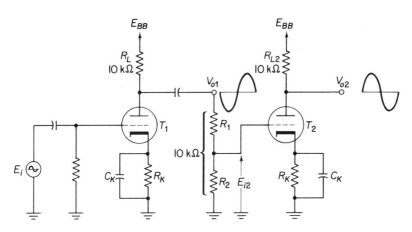

Fig. 13-25. Phase inverter, $V_{o1} = V_{o2}$.

by resistors $R_1$ and $R_2$ must be exactly balanced by the gain of $T_2$ in order for $V_{o1}$ to equal $V_{o2}$. Thus $V_{o1}/E_{i2}$ must equal $V_{o2}/E_{i2} = A_{v2}$, or

$$A_{v2} = \frac{R_1 + R_2}{R_2} \qquad (13\text{-}32)$$

The *self-balancing* phase inverter of Fig. 13-26(a) is modeled in Fig. 13-26(b) to show

$$E_i = E_{g1} + E_{g2} \quad \text{and} \quad E_{g2} = (I_1 - I_2)R_K$$

Substitute into the $I_1$ and $I_2$ loop equation to obtain the ratio of $V_{o1}$ to $V_{o2}$

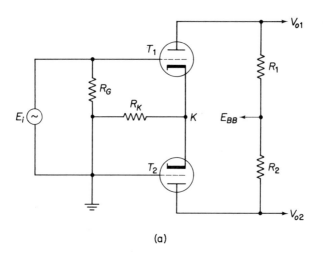

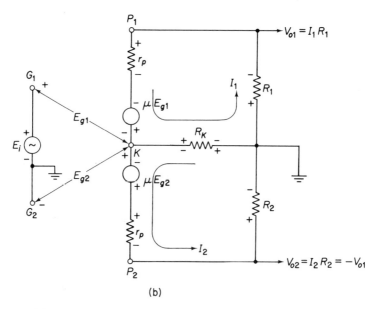

**Fig. 13-26.** Self-balancing phase inverter circuit in (a) and model in (b).

$$\frac{V_{o1}}{V_{o2}} = \frac{R_1}{R_2}\left(1 + \frac{r_p + R_2}{(\mu + 1)R_K}\right) \quad (13\text{-}33)$$

If $R_K$ and $\mu$ are large in Eq. (13-33) and $R_1 = R_2$, then $V_{o1}/V_{o2} = 1$ and is independent of tube parameter variation.

## 13-9 Push-Pull Operation

Relatively large power outputs are obtained from vacuum tubes by large voltage swings. This is the primary difference between transistor and vacuum tube power devices. Speakers are current-driven devices at modest voltages and an output transformer is employed to change high-voltage, low-current output from the vacuum tube to low-voltage, high-current power for the speaker load. The most common arrangenemt is the push-pull amplifier in Fig. 13-27(a). Push-pull operation automatically mitigates against distortion introduced by large-signal voltage swings. For example, on the characteristics of $V_{T1}$ in Fig. 13-27(c), $E_{CC} = 4$ V and $E_{g1}$ is a sinusoidal variation of 8-V peak to peak. The resulting positive and negative excursions of plate current $i_{B1}$ are unequal and result from distortion due to the unequal spacing of the vacuum tubes curves. During this same interval $E_{g2}$ is causing the distorted variation of $i_{B2}$ in Fig. 13-27(b).

The dashed dc load lines in Figs. 13-27(b) and (c) are based on a zero resistance for the primary windings of output transformer $T_2$. The ac load line $R_p$ is drawn for an ac resistance of 300/80 mA = 3.75 k$\Omega$. Tube $V_{T1}$ looks at $R_L$ through $N$ and $N_S$ of the output transformer. This turns ratio should be caluclated to make $R_L$ look like 3.75 k$\Omega$ to $V_{T1}$. Conversely, $V_{T2}$ should be presented with 3.75 k$\Omega$ through the proper ratio of $N$ to $N_S$ and $R_L$. Thus if $R_L$ were a 3.2-$\Omega$ speaker, the turns ratios should be

$$R_p = \left(\frac{N}{N_S}\right)^2 R_L \qquad (13\text{-}34)$$

$$\frac{N}{N_S} = \left(\frac{R_p}{R_L}\right)^{1/2} = \left(\frac{3750}{3.2}\right)^{1/2} \cong 34$$

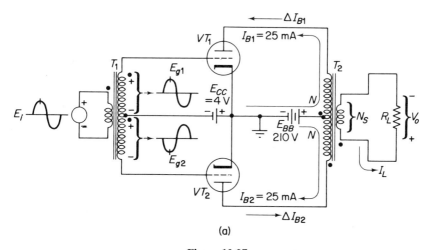

(a)

**Figure 13-27**

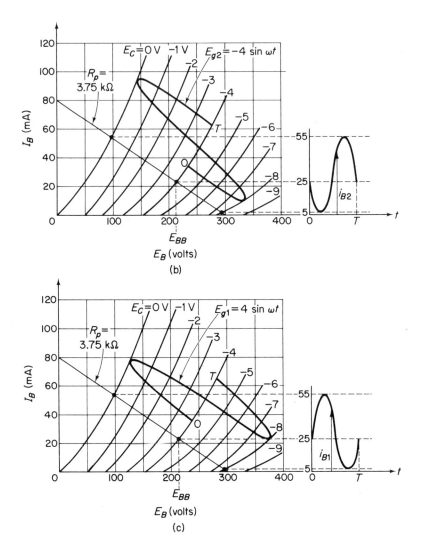

**Fig. 13-27.** Phase-inverting transformer $T_1$ drives the push-pull amplifier in (a). Large signals cause nonsinusoidal plate currents in $VT_2$ and $VT_1$ in (b) and (c), respectively.

However, it is standard paractice to specify these output transformers by the plate-to-plate resistance $R_{p/p}$, which is related to $R_p$ by

$$R_{p/p} = \left(\frac{2N}{N_S}\right)^2 R_L = 4\left(\frac{N}{N_S}\right)^2 R_L = 4R_p \tag{13-35}$$

$R_{p/p}$ is 4 times that seen by each tube or 15 kΩ and we would purchase a 15-kΩ–4-Ω center-tapped output transformer for $T_2$.

Sec. 13-9                                              Push-Pull Operation    441

To understand how push-pull operation mitigates against distortion we can approximate the complex wave forms of $i_{B1}$ and $i_{B2}$ in Fig. 13-28 by the sum of (1) a fundamental sine-wave component, equal in frequency to the complex wave, plus (2) a second harmonic at twice the fundamental's frequency, plus (3) a dc component. Verify this approximation by adding component amplitudes at time $= T/4$.

$$\text{At } t = \frac{T}{4} \quad I_{B1} + i_{b12} + i_{b1} = i_{B1}$$

$$25 + 5 + 25 = 55 \quad \text{check}$$

$$\text{At } t = \frac{T}{4} \quad I_{B2} + i_{b22} + i_{b2} = i_{B2}$$

$$25 + 5 - 25 = 5 \quad \text{check}$$

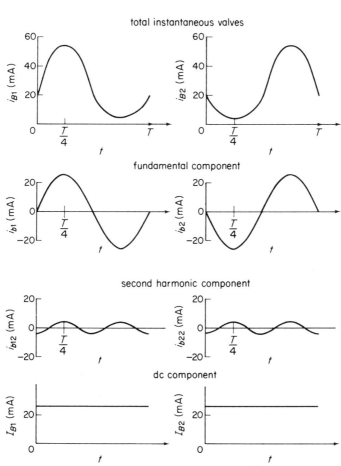

**Fig. 13-28.** Currents $i_{B2}$ and $i_{B2}$ in Fig. 13-27(b) and (c) can be approximated by the sum of three components.

We can now examine separately the action of each component in the push-pull output circuit.

DC COMPONENT. In Fig. 13-29 (a) the dc component plate currents set up fluxes that cancel one another in the output transformer's primary. As far as $T_1$ is concerned, magnetically it is carrying no dc current and this distinct advantage of the push-pull connection reduces transformer size and cost. If a self-biasing resistor $R_K$ is employed for $E_{CC}$ it will carry $I_{B1}$ equal to $I_{B2}$. It is good practice to insert separate cathode-biasing resistors adjusted to match the output dc currents exactly. Of course each must be bypassed with a large capacitor.

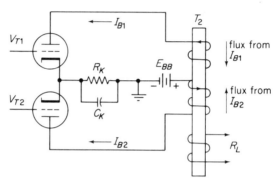

**Fig. 13-29.** DC component currents establish flux linkages that cancel in the core.

SECONDARY HARMONIC COMPONENTS. At time $T/4$ in Fig. 13-29 both second-harmonic current components are at their maximum positive peak. They represent a simultaneous incremental *increase* in the plate currents of both $i_{B1}$ and $i_{B2}$ and thus have the same direction and effect as the dc current components. Both increments pass through $R_K$, and $R_K$ must be heavily bypassed so that no appreciable second-harmonic voltage drop is developed across it. The power supply also carries both second-harmonic currents. At time $T/2$ the currents and flux directions would be reversed in Fig. 13-30

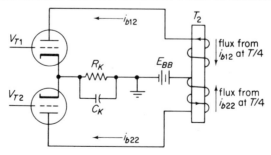

**Fig. 13-30.** Second-harmonic plate current components cancel in the primary of $T_2$.

Sec. 13-9  Push-Pull Operation  443

with the flux canceling out in the primary of $T_2$ so that no second-harmonic voltage would be induced in the secondary. This analysis applies to all even-harmonics and it is fortunate that second-harmonic distortion is the predominant factor in vacuum tube and semiconductor devices.

FUNDAMENTAL COMPONENT. The first-harmonic or fundamental component is a regenerated duplicate of the input signal. At time $T/4$, $i_{b1}$ is at its positive maximum and $i_{b2}$ is at its negative maximum, as shown in Fig. 13-31. Both fundamental plate current components set up fluxes that aid

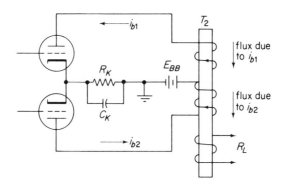

Fig. 13-31. Fundamental plate current components help to establish a secondary output voltage.

one another and generate a voltage in the secondary winding. In one sense $i_{b2}$ is pushing into one primary terminal while $i_{b1}$ is pulling out of the other, hence the name "push-pull." The fundamental current components cancel out in $R_K$ and the power supply. In summary, it is concluded that push-pull operation allows us to drive a vacuum tube with large signals over a nonlinear region on its characteristic but obtain an essentially undistorted output signal. If any odd harmonics are present (third, fifth, etc.) there will be no mitigation against their distortion.

A circuit model for the push-pull amplifier in Fig. 13-32(a) shows cancelation of the rms fundamental components in wire $AA'$. Since $AA'$ carries no current it may be removed to yield the simplified model in Fig. 13-32(b), where it is clear that maximum power transfer occurs when $2r_p = R_{p/p}$. Voltage developed across the primary of $T_2$ is

$$V'_o = 2\mu E_g \frac{R_{p/p}}{2r_p + R_{p/p}} \quad (13\text{-}36)$$

Actual output voltage $V_o$ is found from the turns ratio in Fig. 13-32(c) to be

$$V_o = \frac{N_S}{2N} V'_o \quad (13\text{-}37)$$

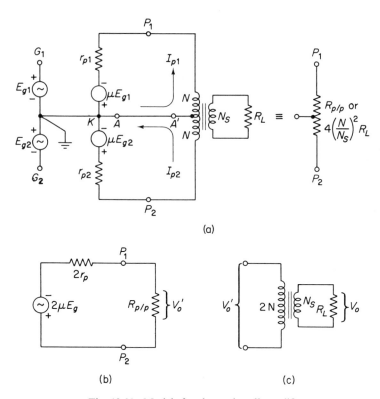

**Fig. 13-32.** Models for the push-pull amplifier.

*Example 13-9.* For a push-pull amplifier circuit $R_L = 3.2\ \Omega$, $\mu = 40$, $r_p = 2.2$ k$\Omega$ at the operating point, and $T_1$ is a 1:2 center-tapped transformer so that $E_i = E_{g1} = -E_{g2} = 2$ V. (a) Select $T_2$ for maximum power transfer and (b) evaluate power output to the load.

*Solution.* (a) $R_{p/p}$ must equal $2(2.2)$ k$\Omega = 4.4$ k$\Omega$ and $T_2$ is selected for a turns ratio from Eq. (13-35) of

$$\frac{2N}{N_S} = \left(\frac{4400}{3.2}\right)^{1/2} = 37$$

and the ac resistance seen by each tube will be $R_{p/p}/4 = 1.1$ k$\Omega$.
(b) From Eq. (13-36),

$$V'_o = 2(40)2\left[\frac{4.4 \times 10^3}{8.8 \times 10^3}\right] = 80\ \text{V}.$$

From Eq. (13-37), $V_o = 80/37 = 2.16$ V and $P_o = (2.16)^2/3.2 \cong 1.5$ W.

*Example 13-10.* Given the push-pull circuit and tube characteristics in Fig. 13-33. Assume $T_1$ is 80% efficient, (a) Find its impedance rating to deliver 4 W to $R_L$. (b) What are the required values of $E_{g1}$ and $E_{g2}$?

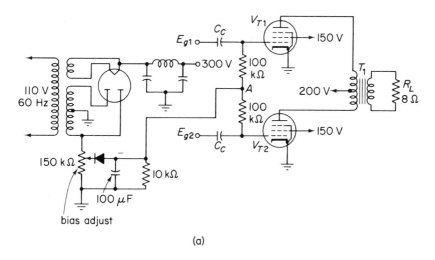

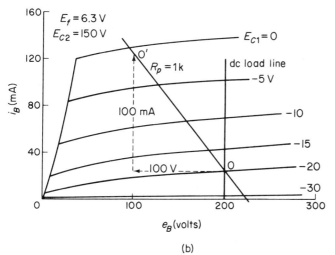

**Fig. 13-33.** Circuit and tube characteristics for Example 13-10.

*Solution.* (a) 5 W must be delivered to the primary of $T_1$ to obtain 4 W at the load. Arbitrarily choose a peak tube voltage swing of 100 V so that the primary must present to each tube an ac load resistance of

$$R_p = \frac{(V'_o)^2}{P} = \left(\frac{100}{2}\right)^2 \frac{1}{5} = 1000 \quad \text{and} \quad R_{p/p} = 4 \text{ k}\Omega$$

$T_1$ must be rated at 8-Ω–4-kΩ, center tapped.

(b) Choose a bias voltage of $-20$ V in Fig. 13-33(b) and locate operating point 0 on the dc load line. Draw ac load line $R_p$ and for a peak tube voltage of 100 V. The grid voltages must swing a peak of 19 V between points 0 and 0'.

Note that a simple rectifier is used in Fig. 13-33 to develop grid bias for the output tubes. The 60-Hz ripple component will be rather large but it will swing the grids of $T_1$ and $T_2$ up and down together. For example, if the grids are both going positive, both plate currents will increase and generate opposing fluxes in the primary of $T_1$. No ripple output will be developed in the secondary.

## PROBLEMS

*13-1* Sketch a circuit to display the current-voltage characteristic of a diode.

*13-2* State the relationships between increments of plate current, plate voltage, and grid voltage that define (a) incremental plate resistance, (b) transconductance, and (c) amplification factor. (d) How are the vacuum tube parameters in (a), (b), and (c) related?

*13-3* Graphically evaluate incremental resistance of the diode in Fig. 13-4 when operated at $E_f/2$.

*13-4* Sketch a sweep circuit to display the plate-current versus plate-voltage characteristic of a triode with a negative bias of 2 V.

*13-5* Locate an operating point in Fig. 13-7 at $i_B = 5$ mA, $e_B = 150$ V, and $e_C = -4$ V. Evaluate $r_p$, $g_m$, and $\mu$ at this point.

*13-6* An operating point is located at $i_B = 5$ mA, $e_B = 250$ V, $e_C = -5$ V in Fig. 13-10. (a) Evaluate the tube parameters at this point. (b) Draw a Thévenin model of the tube when operated at this point.

*13-7* In order to operate the tube in Fig. 13-11 at the operating point of Problem 13-6, (a) what supply voltage is required? (b) Draw the load line for $R_L$ to check your answer. (c) What is the voltage gain $V_o/E_g$?

*13-8* Choose a cathode-bias resistor to establish the operating point (a) in Problem 13-6 or 13-7; (b) shown in Fig. 13-10.

*13-9* If the cathode resistor were changed to 200 Ω in Fig. 13-12(a), locate the new operating point and state the dc voltage level across $C_K$.

*13-10* In the circuit of Fig. 13-13(a), $R_L = 10$ kΩ, $R_E = 1$ kΩ, $r_p = 10$ kΩ, $g_m = 4000$ μmho, and $E_i = 1$ V. (a) Find $V_{o1}$ and $V_{o2}$ from the model in Fig. 13-13(c). Evaluate resistance seen looking into the (b) plate terminal and (c) the emitter terminal.

*13-11* In a triode amplifier $R_E = 0$, $\mu = 20$, $r_p = 10$ kΩ, and the plate load $R_L$ is varied from 0 to 50 kΩ. (a) Plot voltage gain versus $R_L$. (b) On the same graph plot voltage gain versus $R_L$ if a cathode resistor is added with a value of $R_K = 500$ Ω.

*Chap. 13* *Problems* 447

*13–12* In Example 13-3, what changes result in the solutions when $R_L$ is doubled to 10 kΩ?

*13–13* In Example 13-4, what changes result when $r_p$ is halved to 5 kΩ?

*13–14* Choose $C_C$ and $C_K$ in Fig. 13-18(a) for a low-frequency cutoff at 100 rad/sec.

*13–15* In Example 13-6 let $R'_L$ have values of 5 kΩ, 10 kΩ, and 20 kΩ to evaluate corresponding values of $e_i$ and $\omega_H$. Calculate voltage gain from $g_m R_L$ and compare the product of $\omega_H$ and voltage gain for each value of $R'_L$.

*13–16* $E_{BB}$ is reduced to 200 V in Fig. 13-23. What changes will result in Example 13-8?

*13–17* In Example 13-9 and Fig. 13-24, which action extends $\omega_H$ to a higher value, (a) reducing internal generator resistance by half, or (b) reducing $R_L$ by half?

*13–18* In Fig. 13-25, $r_p = 10$ kΩ and $\mu = 20$ for each tube. Select values for $R_1$ and $R_2$ to make a phase inverter.

*13–19* In Fig. 13-26, $r_p = 10$ kΩ, $\mu = 49$, $R_K = 1$ kΩ, and $R_2 = 20$ kΩ. Choose $R_1$ to construct a phase inverter.

*13–20* Revise the solution to Example 13-10 for $R_L = 8$.

*13–21* A 4-Ω speaker replaces the 8-Ω speaker in Fig. 13-33(a). Employing the solutions to Example 13-10 and with a peak grid voltage swing of 19 V, what power will be delivered to the 4-Ω speaker?

# Chapter 14

**14-0** INTRODUCTION ............................. 449
**14-1** PHYSICAL MODEL OF THE JFET ................. 449
**14-2** CHARACTERISTIC CURVES AND PARAMETER
        MEASUREMENTS OF THE JFET ................... 452
**14-3** MEASUREMENT OF PINCHOFF VOLTAGE FOR THE
        JFET ....................................... 456
**14-4** PHYSICAL MODEL OF THE IGFET ............... 458
**14-5** CHARACTERISTIC CURVES AND PARAMETER
        MEASUREMENTS OF THE IGFET ................. 461
**14-6** MAXIMUM RATINGS ........................... 464
**14-7** BIASING FOR DEPLETION-MODE OPERATION ...... 465
**14-8** BIASING FOR ENHANCEMENT-MODE OPERATION .... 467
**14-9** ZERO BIAS SHIFT OF THE OPERATING POINT .... 471
**14-10** LOW-FREQUENCY CIRCUIT MODELS FOR THE FET .. 473
**14-11** VOLTAGE GAIN ............................. 475
**14-12** RESISTANCE LEVELS ........................ 480
**14-13** HIGH-FREQUENCY DEPENDENCE ................ 482
**14-14** LOW-FREQUENCY VOLTAGE GAIN ............... 484
**14-15** HYBRID AND TWO-STAGE AMPLIFIERS .......... 486
**14-16** CHOPPER AMPLIFIERS ....................... 488
**14-17** ANALOG SWITCHING AND COMMUTATION ......... 493
         PROBLEMS .................................. 495

# Field-Effect Transistors and Applications

## 14-0 Introduction

*Field-effect transistors* are named for their operating principle whereby the electric field of a *pn* junction controls resistivity of a conducting channel. FET operation depends on transport of one type of carrier (majority) in the conducting channel, *either* holes or electrons but not both. For this reason the FET is classified as a unipolar device in contrast to the junction transistor, which is a *bipolar* or two-carrier device. Advantages of FETs over junction transistors are (1) high input impedance, (2) lower noise, (3) approximately 25% fewer steps are required in the manufacturing process, permitting higher yield, (4) more devices can be packed into a smaller area for integrated circuit arrays, and (5) no thermal runaway. Disadvantages of the FET include (1) low power operation, (2) low gain, and (3) high on-resistance (corresponding to saturation resistance $R_{sat}$ of the junction transistor).

The FET promises to displace junction transistors principally in (1) the construction of medium- or large-scale integrated circuit arrays such as memory elements for computers and (2) low-power applications which require low noise or high input impedance.

Two of the most important types of field-effect solid-state devices are classified as (1) diffused-junction field-effect transistor (JFET) and (2) insulated-gate (IGFET) or metal-oxide semiconductor (MOSFET) field-effect transistor.

## 14-1 Physical Model of the JFET

Structural features of a JFET are shown in Fig. 14-1 (a). A sample of *p*-type semiconductor material forms a channel for transport of majority carriers (holes), usually from *source* contact to *drain* contact. In many

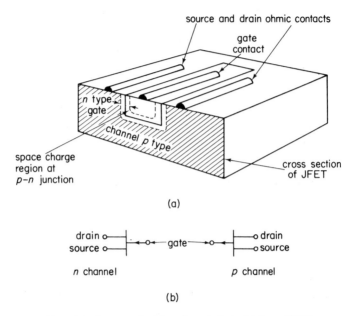

**Fig. 14-1.** Structure in (a) and symbols in (b) for a JFET

JFETs, majority carriers may be transported from drain to source so that connections to drain or source terminals are interchangeable. Diffused into the *p*-type body is a volume of *n*-type material to form a *gate*. A space-charge region, complete with electric field and devoid of carriers, is formed along the junction between *n*-type and *p*-type materials. *n* channel FETs are made with *p*-type diffused gates, and have electron carriers for conduction between source and drain. Since electron mobility is higher than hole mobility, *n*-type JFETs are inherently faster than *p*-type ones. Symbols are shown in Fig. 14-1(b), where a convenient memory aid is to think of *n*-in to associate an arrow pointing *into* the device for the *n* channel.

Physical operation of the JFET will be studied by reference to the conceptually simpler structure in Fig. 14-2(a), where $V_{GS} = 0$ and $V_{DS}$ is small. The space-charge regions are small, and drain current $I_D$ is small because $V_{DS}$ is small. $I_D$ is determined by $V_{DS}$ and the channel's resistivity. As $V_{GS}$ is increased, this reverse bias causes the space-charge regions to penetrate evenly and deeper into the channel. Eventually a value of $V_{GS}$ will be reached where the space-charge regions meet as shown in Fig. 14-2(b). This value of $V_{GS}$ has been designated as *pinchoff* voltage $V_p$. Since no carriers are found in the expanded space-charge region there can be no conduction and drain-current flow ceases. It is convenient to visualize $V_{GS}$ as controlling channel width and consequently channel resistance. Essentially the JFET behaves as a variable resistor in the region where $V_{DS}$ is small.

## Sec. 14-1  Physical Model of the JFET  451

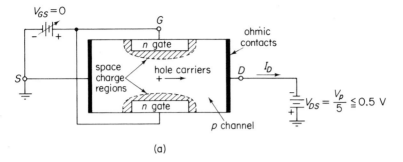

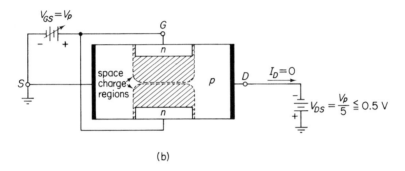

**Fig. 14-2.** JFET model showing control of channel conductivity by $V_{GS}$. Increasing $V_{GS}$ from 0 in (a) expands the space charge regions until the channel is pinched off in (b) at $V_{GS} = V_p$.

There is another physical mechanism that controls channel width. In Fig. 14-3, $V_{GS} = 0$ and $V_{DS}$ is increased to increase $I_D$. We show the bulk resistance of the channel as a series of resistors to indicate how $I_D$ establishes a voltage gradient along the channel. Hole current flow in the $p$ channel establishes voltage drops across each resistor as shown. Measuring from the common-source terminal, it is seen that the drain is more negative with respect to the gate than is the source. The gate's drain end is more reverse biased than is its source end. Consequently, space-charge regions at the drain end will penetrate further into the channel. Drain-current flow is self-limiting and with zero bias (or even with forward bias) current will be limited when the space charge regions nearly meet to *pinch off* the channel.

The minimum value of $V_{DS}$, where the space-charge regions limit drain current from any further increase, is also equal to $V_p$, and this condition is described as *pinchoff*. Since maximum possible drain current flows, the condition is also called *saturation*. Thus there are two basic ways to pinch off the channel; (1) hold $V_{DS}$ at a small value and increase reverse bias $V_{GS}$ to $V_p$, where the space-charge regions meet. (2) Hold $V_{GS} = 0$ (short gate to source

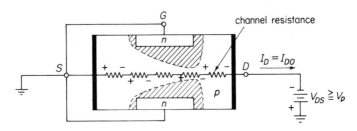

Fig. 14-3. Pinchoff due to $V_{DS}$.

and increase $V_{DS}$ to $V_p$, where the space-charge regions nearly meet. In both cases the electric fields across the space-charge regions are equal and thus the pinchoff voltages must be equal. Of course when $V_{DS}$ is less than $V_p$, and a reverse-bias gate voltage is applied, both will act to pinch off the channel so that the drain-source voltage at pinchoff is

$$V_{DS}(\text{at pinchoff}) + V_{GS} = V_p \qquad (14\text{-}1)$$

Magnitudes only are used in Eq. (14-1).

## 14-2 Characteristic Curves and Parameter Measurements of the JFET

The ordinary bipolar-transistor curve tracer may be converted quite simply to test field-effect transistors by connecting a 1000-Ω resistor between base and emitter terminals of the curve plotter. This converts the base-current step generator to a base-voltage step generator. Since the gate of a reverse-biased JFET draws no current, multiply each current-step value by 1000 Ω to obtain the base-voltage step value. For example, when the base step generator is set to provide 0.2 mA/step, $V_{GS}$ will equal 0.2 V/step. Sweep techniques also may be used to obtain FET characteristic curves in the same fashion as for vacuum tubes or by substituting voltage bias for current bias with BJTs.

Drain characteristics for a $p$-channel JFET in the region below saturation are shown in Fig. 14-4. These $I_D$-$V_{DS}$ characteristics correspond to the circuit of Fig. 14-2(a). $V_{DS}$ is swept over a small voltage range between approximately 0 and 1.5 V for each value of $V_{GS}$. The straight-line curve was taken with an open-gate connection and below $V_{DS} = -0.6$ V this line coincides reasonably with the curve of $V_{GS} = 0$. The slope of the line $V_{GS} = 0$ is called the incremental drain or *on-resistance* $R_o$ and is defined and evaluated by

$$R_o = \left.\frac{\Delta V_{DS}}{\Delta I_D}\right|_{V_{GS}=0} = \frac{0.8}{1 \times 10^{-3}} = 800 \text{ Ω} \qquad (14\text{-}2)$$

Sec. 14-2  Curves and Measurements of the JFET  453

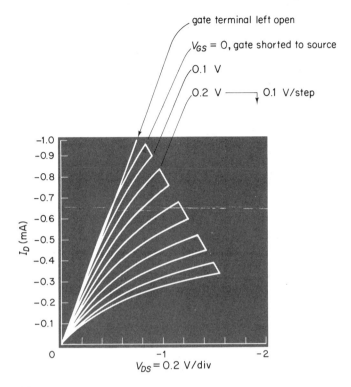

**Fig. 14-4.** Ohmic-region drain characteristics for the $p$-channel transistor in the condition of Fig. 14-2(a) ($V_{DS} < V_p$).

As $V_{GS}$ is increased in steps, $I_D$ decreases for corresponding values of $V_{DS}$ at each step, showing that the channel is being closed toward pinchoff and that the resistance of the channel is increasing. This region of the curves is the *ohmic* or *triode region*.

By increasing the sweep voltage magnitude of $V_{DS}$ we extend the characteristic curves into the pinchoff or saturation region in the $I_D$-$V_{DS}$ characteristics of Fig. 14-5. The curve $V_{GS} = 0$ illustrates pinchoff due to $V_{DS}$ and corresponds to conditions in the circuit of Fig. 14-3. $I_{DO}$ is measured above $V_{DS} = V_p$ along the flat portion of $V_{GS} = 0$. At values of $V_{DS} > V_p$, increasing bias voltage $V_{GS}$ lowers drain current until $I_D = 0$ at $V_{GS} = V_p \cong 2$ V in Fig. 14-5. The dashed line is a locus of $V_{DS}$ (at pinchoff) and the corresponding $V_{GS}$. Their values are related by Eq. (14-1). We conclude from Fig. 14-5 that $V_p$ can be measured with limited accuracy from the characteristic curves.

In the pinchoff region, incremental channel resistance becomes large since the slope of the $I_D$-$V_{DS}$ curves becomes almost flat. To find the incremental channel resistance $r_d$ at any operating point on a characteristic curve we evaluate the slope graphically from

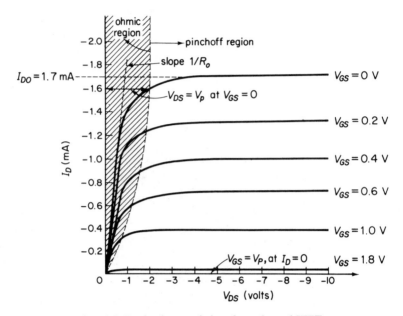

Fig. 14-5. Drain characteristics of a *p*-channel JFET.

$$r_d = \frac{\Delta V_{DS}}{\Delta I_D}\bigg|_{V_{GS}=k} \quad (14\text{-}3)$$

Note that Eq. (14-2) is a special case of Eq. (14-3).

Transconductance $g_m$ of the JFET is a parameter which tells how much drain current will be changed by a change in grid-to-source voltage at a constant value of drain-to-source voltage. We shall be interested primarily in the *maximum transconductance* $G_o$ which occurs in the pinchoff region and when $V_{GS} = 0$. Therefore we evaluate $G_o$ graphically from the drain characteristics according to

$$G_o = \frac{\Delta I_D}{\Delta V_{GS}} \quad \text{with } V_{GS} = 0, V_{DS} = k > V_p \quad (14\text{-}4)$$

There is an approximate relationship between $G_o$ and $R_o$ given by

$$G_o \simeq \frac{1}{R_o} \quad (14\text{-}5)$$

Drain current can be described in terms of $V_{GS}$, $V_p$, and $I_{DO}$ by the equation

$$I_D = I_{DO}\left(1 - \frac{V_{GS}}{V_p}\right)^2 \quad (14\text{-}6)$$

and the magnitude of $I_{DO}$ is expressed in terms of $G_o$ and $V_p$ by

Sec. 14-2  Curves and Measurements of the JFET  455

$$I_{DO} = \frac{G_o V_p}{2} \quad (14\text{-}7)$$

Transconductance may be measured by sweep techniques or curve plotter with a display of $I_D$ versus $V_{GS}$ as in Fig. 14-6. Incremental trans-

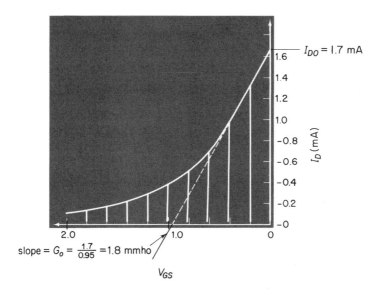

Fig. 14-6. Transconductance $g_m$ and $G_o$ from the transfer characteristic curve of $I_D$ versus $V_{GS}$.

conductance $g_m$ is measured from the slope of the curve at an operating point. $I_{DO}$ is shown as 1.7 mA at $V_{GS} = 0$ and $G_o$ is measured from the slope at $I_{DO}$, $V_{GS} = 0$. Transconductance $g_m$ may also be expressed by

$$g_m = G_o\left(1 - \frac{V_{GS}}{V_p}\right) \quad (14\text{-}8)$$

*Example 14-1.* Determine $R_o$ and $G_o$ from Figs. 14-5 and 14-6, respectively, to verify Eq. (14-5).

*Solution.* Taking a voltage increment of approximately 1 V from Fig. 14-4,

$$R_o = \left.\frac{\Delta V_{DS}}{\Delta I_D}\right|_{V_{GS}=0} = \frac{1}{1.8 \times 10^{-3}} = 0.55 \text{ k}\Omega$$

From Fig. 14-6,

$$G_o = \frac{\Delta I_D}{\Delta V_{GS}} = \frac{1.7 \times 10^{-3}}{0.95} \cong 1800 \ \mu\text{mho}$$

*Example 14-2.* Obtain $G_o$ and $I_{DO}$ from Fig. 14-6 to calculate pinchoff voltage from Eq. (14-7).

**Solution.** From Example 14-1 and Fig. 14-6, $I_{DO} = 1.7$ mA and $G_o = 1.8$ mmho. From Eq. (14-7),

$$V_p = \frac{2I_{DO}}{G_o} = -\frac{2(1.7 \times 10^{-3})}{1.8 \times 10^{-3}} = 1.9$$

**Example 14-3.** Given $G_o = 1.8$ mmho and $V_p = 1.8$ V. Find $I_D$ in the pinchoff region when $V_{GS} = 0.9$ V.

**Solution.** From Eq. (14-7), $I_{DO} = G_o V_p/2 = 1.6$ mA. From Eq. (14-6),

$$I_D = I_{DO}\left(1 - \frac{V_{GS}}{V_p}\right)^2 = 1.6\left(1 - \frac{0.9}{1.8}\right)^2 = 0.4 \text{ mA}$$

$G_o$ may also be measured from the drain characteristic of Fig. 14-5 at $V_{GS} = 0$ in the pinchoff region. For example, along the vertical line of $V_{DS} = -6$ V, a change of $V_{GS}$ from 0 to 0.2 V causes a change in $I_D$ of 0.38 mA, so that

$$G_o = \frac{0.38 \times 10^{-3}}{2} = 1.9 \text{ mmho} \qquad \text{at } V_{DS} = -6 \text{ V}$$

From Eq. (14-3) we can evaluate $r_d$ as *output resistance* of the drain in the pinchoff region. For example, at $V_{DS} = -6$ V and $V_{GS} = 0$ in Fig. 14-5, a change in $V_{DS}$ from $-5$ to $-7$ will increase $I_D$ by approximately 0.02 mA, or

$$r_d = \frac{2}{0.02 \times 10^{-3}} = 100 \text{ k}\Omega$$

Plate resistance of the vacuum tube $r_p$ is analogous to $r_d$. As in the vacuum tube, an amplification factor $\mu$ can be defined from the product of $g_m$ and $r_d$, or

$$g_m r_d = \frac{\Delta I_D}{\Delta V_{GS}} \frac{\Delta V_{DS}}{\Delta I_D} = \frac{\Delta V_{DS}}{\Delta V_{GS}}\bigg|_{I_D = k} = \mu \qquad (14\text{-}9)$$

## 14-3 Measurement of Pinchoff Voltage for the JFET

In Section 14-2 we saw that $V_p$ could be measured from the drain characteristics from two rather arbitrary measurements. That is, the curve of $V_{GS} = 0$ does not break sharply in the transition from ohmic to pinchoff region in Fig. 14-5, and $I_D$ approaches zero very slowly as reverse bias $V_{GS}$ is increased. We can, however, rewrite Eq. (14-6) as

$$(I_D)^{1/2} = \left(1 - \frac{V_{GS}}{V_o}\right)(I_{DO})^{1/2} \qquad (14\text{-}10)$$

Sec. 14-3    Measurement of Pinchoff Voltage for the JFET    457

Above pinchoff on the drain characteristics, extract corresponding values of $I_D$ and $V_{GS}$ at points along a convenient vertical line of $V_{DS} = k > V_p$. Plot $(I_D)^{1/2}$ versus $V_{GS}$ as in Fig. 14-7, and extrapolate to obtain $V_p$ at a horizontal-axis intersection, where $(I_D)^{1/2} = 0$.

A direct and simple method of measuring $V_p$ has been developed by B. Wedlock in his "Direct Determination of the Static Parameters of a Junction Field-Effect Transistor," (in press). In Fig. 14-8(a), voltage between drain

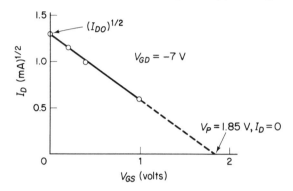

Fig. 14-7. Square root of $I_D$ is plotted against $V_{GS}$ for the JFET in Fig. 14-5. $\bar{V}_p = V_{GS}$ when $(I_D)^{1/2} = 0$ from Eq. (14-10).

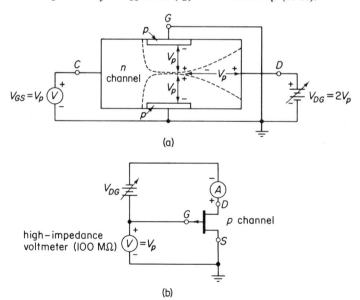

Fig. 14-8. Measure $V_p$ by increasing $V_{DG}$ to $3V_p$ and reading $V_p$ on the voltmeter. Then short the gate and source to read $I_{DO}$ on the milliammeter. n- and p-channel tests are shown in (a) and (b), respectively.

and gate is increased from 0 to $2V_p$. Space-charge regions expand into the channel until they touch at $V_{DG} = V_p$ and when $V_{DG}$ is increased beyond $V_p$, the channel is closed and the increase in $V_{DG}$ is taken up in the channel body near the drain contact. Potential along the channel's center, toward the source, remains at $V_p$. Hence $V_p$ will be measured by voltmeter $V$ for all values of test voltage above several $V_p$.

Manufacturer's data sheets often specify pinchoff voltage at a specific, low value of drain current. Figure 14-9 illustrates the use of a differential

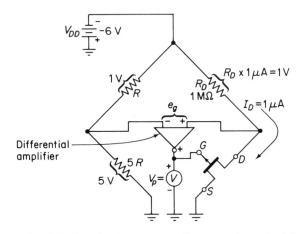

**Fig. 14-9.** Test circuit to measure $V_p$ at a predetermined $I_D$.

amplifier in a bridge network to measure $V_p$ automatically at a specified $I_D$. The inverting (−) input terminal of the differential amplifier will always be at −5 V due to the division of $V_{DD}$ between $R$ and $5R$. Assuming the FET is conducting, its drain will be more positive than −5 V so that the amplifier's noninverting (+) input terminal gives a positive output to increase reverse bias, decreasing drain current. At balance, $e_g$ is close to zero and the drain will be at −5 V. Output bias voltage from the amplifier will be just enough to hold 1-V drop across $R_D$. If $R_D$ is made to equal 1 MΩ, then the circuit balances with $I_D = 1\mu A$ and $V_p$ is read on the voltmeter at $I_D = 1\mu A$.

## 14-4 Physical Model of the IGFET

When channel resistivity is controlled by the electric field through an insulator rather than through a space-charge layer, the device is classified as an *Insulated-Gate Field-Effect Transistor* or *IGFET*. The same device may also be known as a *Metal-Oxide Semiconductor FET* or *MOSFET*. IGFETs are divided into two types as determined by whether the device is conducting (depletion), or nonconducting (enhancement) with zero bias.

Structure of a zero-bias, nonconducting IGFET and its circuit symbol are shown in Fig. 14-10(a). Heavily doped *n*-type drain and source are diffused into a lightly doped *p*-type substrate. Silicon dioxide, $SiO_2$, is diffused over the surface to passivate or protect against contamination. Holes are etched in the silicon for deposit of aluminum contacts to source and drain. An aluminum gate is also deposited on the $SiO_2$ so that it is insulated from the substrate. Gate and substrate form two plates of a capacitor with $SiO_2$ as the dielectric. No carriers are available in the channel for conduction between source and drain in the zero-bias condition of Fig. 14-10(a). The substrate isolates drain and source since the three elements form two back-to-back diodes.

Connecting a voltage between gate and source with the polarity of Fig. 14-10(b) will attract minority electron carriers from the substrate towards the gate to form a conducting *n* channel between source and drain. Increasing the bias increases the number of channel carriers, increasing drain current. Since conduction is enhanced by application of bias, this type of IGFET structure is classified as *enhancement* type.

A complementary device is made with an *n*-type substrate and *p*-type

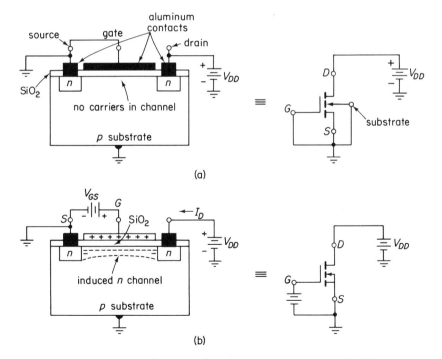

Fig. 14-10. No drain current flows in the *enhancement* type IGFET in (a) until an *n*-channel is induced by the positive gate voltage in (b).

drain and source. Negative bias attracts holes toward the gate to form a *p*-channel, *enhancement-type* IGFET. Drain voltage must be negative with respect to the source to attract channel carriers. Do not confuse the substrate type with the IGFET channel. A *p*-type substrate results in an *n*-channel IGFET. Instead of being connected internally to the source, the substrate may be brought out to an external connection to make a four-terminal device.

Another type of IGFET is constructed to operate in the *depletion mode*, similar to operation of a JFET. That is, the device conducts between drain and source with zero bias. Structure of an *n*-channel, depletion-type IGFET is pictured in Fig. 14-11(a). Drain and source are not isolated by a substrate

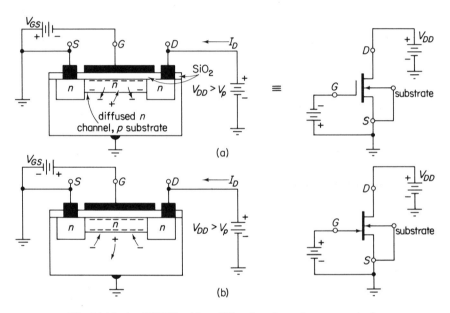

**Fig. 14-11.** An IGFET with a diffused *n*-channel can operate by *depletion* with reverse bias in (a) or by enhancement with forward bias in (b).

as in the enhancement type but are *connected by a conducting channel* as in the JFET. However, unlike the JEFT, $SiO_2$ insulates the gate and *n*-channel to form a capacitor. Making the gate negative with respect to the source repels electrons from the channel into the substrate to increase channel resistance by depleting it of electron carriers. The channel experiences essentially the same pinchoff mechanisms as a JFET and the output-characteristic curves are similar for reverse bias on the gate. In Fig. 14-11(b) a positive voltage is applied to the gate to attract minority electrons from the substrate in to the channel to enhance conduction and increase drain current.

Positive gate potentials are limited by the *gate-to-source breakdown voltage* $V_{(BR)GSS}$ and not by a forward-bias potential as in the JFET. Symbols used by various manufacturers are given in Figs. 4-10 and 4-11.

SUMMARY. If the gate terminal is insulated from drain and source by a dielectric the FET is classified as an IGFET. Either the IGFET is constructed with a diffused channel between source and drain or it is not. *Without a diffused channel* it cannot conduct at zero bias. Operation is possible only in the *enhancement mode*, by forward biasing the gate to induce a channel (+ gate for electron conduction). *With a diffused channel*, conduction occurs at zero bias; charge carriers are present in the channel with reverse bias (− gate for electron conduction) for operation in the *depletion mode*. In this respect, the diffused-channel IGFET operates like a JFET. Enhancement-mode operation is also possible with a diffused-channel IGFET and a forward bias on the gate. Since either electron-conduction (*n* channels induced or diffused) or hole-conduction (*p* channels either induced or diffused) devices may be constructed, four types of IGFETs are available.

## 14-5 Characteristic Curves and Parameter Measurements of the IGFET

Drain-characteristic curves are shown for a typical depletion-type *n*-channel IGFET in Fig. 14-12(a). Switch $Sw$ is a reversing switch in the point-by-point test circuit of Fig. 14-12(b) to allow depletion or enhancement curves.

On data sheets $I_{DO}$ is specified at zero bias. $I_{DO}$ may be listed as $I_{DSS}$ to signify *drain-saturation current with gate and source short circuited.* Saturation means the channel is carrying maximum possible current, in the pinchoff condition, with $V_{DS}$ greater than $V_p$. For enhancement operation a value of drain current $I_{D(on)}$ is specified within the enhancement mode and always at a $V_{DS}$ greater than $V_p$. $I_{DO}$ and $I_{D(on)}$ are shown in Fig. 14-12(a).

*Small-signal forward transfer admittance* $g_m$ (or $y_{fs}$ on data sheets) is the most important transistor parameter. For a depletion type IGFET, $g_m$ is found from Eq. (14-8) and $I_D$ from Eqs. (14-6) and (14-7). All relationships given for the JFET hold approximately for the depletion type IGFET and $V_p$ can be measured by any of the methods discussed except that of Fig. 14-8. A simple circuit to measure $g_m$ is shown in Fig. 14-13(a). $V_{GS}$ and $V_{DS}$ are set to establish operating point 0 in the depletion region of the characteristics in Fig. 14-12. $R_S$ samples drain current and the voltage drop across $R_S$ is applied to $+y$ of a CRO, with a gain of 10 mV/div to display $I_D$ on the vertical axis at 1 mA/div. An oscillator $v_{gs}$ sweeps the gate voltage which is displayed on the *x* axis of the CRO at 1 V/div. The resultant display is shown

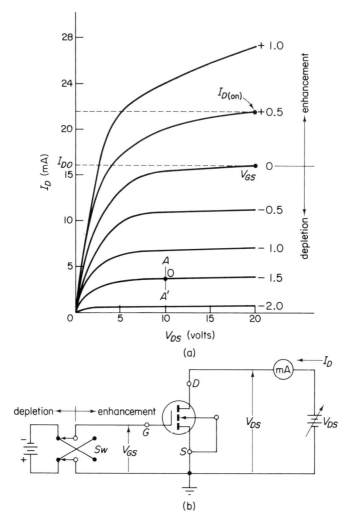

**Fig. 14-12.** Drain-characteristic curves in (a) for a depletion IGFET are plotted from the circuit in (b).

in Fig. 14-13(b) for a peak gate voltage of 0.2 V at 1000 Hz. Transconductance is evaluated graphically from the slope as shown. Reverse polarity of $V_{GS}$ to measure $g_m$ in the enhancement mode.

All of the equations listed thus far apply to both JFET and depletion-type IGFETs. For enhancement-type IGFETs, data sheets specify a threshold gate voltage $V_T$ which is just enough to induce a channel. In order to make the equations approximately applicable to the enhancement IGFET, (1)

Sec. 14-5    Curves and Measurements of the IGFET    463

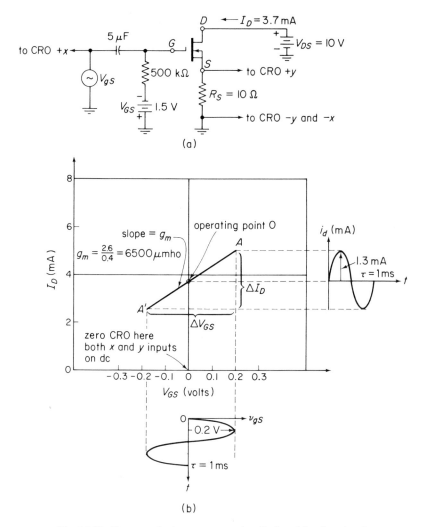

Fig. 14-13. Transconductance $g_m$ may be displayed by the circuit in (a) and measured from the CRO display in (b).

specify a gate voltage of $V_{GS} = 2V_T$ to ensure that a channel is created, and (2) Set $V_{DS} = 0$ (or small) and measure transconductance $G'_o$ or $R'_o$. $R'_o$ is the slope of the $I_D$-$V_{DS}$ characteristic for $V_{GS} = -8$ V $= 2V_T$ in Fig. 14-14, at some *small* value of $V_{DS}$. Practically, the slope is extended by the dashed line for calculation. It is seen that $V_T$ is analogous to $V_p$ and the value of $V_T$ may be used in our equations for $V_p$. $I_{DSS}$ is usually specified as 10 μA at $V_{GS} = V_T$. Holding $V_{GS}$ at $2V_T$, increase $V_{DS}$ to a value of $V_p$ and measure

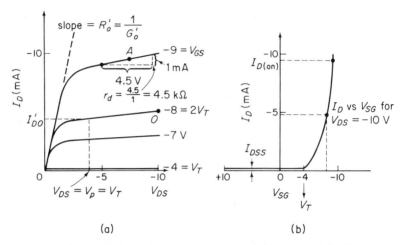

**Fig. 14-14.** Drain and transfer characteristics in (a) and (b) for enhancement-type, $p$-channel IGFET.

$I_{DO'}$. These measurements of $I_{DO'}$, $G_o'$, and $V_T$ may be used for $I_{DO}$, $G_o$, and $V_p$, respectively, in the basic equations of Sections 14.2 and 14.3. Finally $I_{D(on)}$ is shown on the transfer characteristic of Fig. 14-14. This is a typical value as given on a data sheet for an operating point in the enhancement mode.

## 14-6 Maximum Ratings

Maximum voltages and current ratings of FETS are specified in data sheets. Maximum power dissipation is limited by heat-dissipation ability and is typically 200–500 mW. $V_{DS(max)}$ is determined when carriers in the channel are accelerated to an energy where the onset of avalanche multiplication begins. Maximum gate-to-source or gate-to-drain voltage in an IGFET is determined by the breakdown voltage of the insulator under the gate. The insulating $SiO_2$ layer must be thin to allow control by an electric field and therefore necessarily has a low breakdown voltage, of the order of 10–50 V. The gate insulation is so good that serious problems are encountered in the handling of IGFETs. A static voltage can build up on the gate by merely holding the device. Insulation rupture can and does occur through careless or uninformed handling. IGFET leads should always be shorted together with the washer furnished by the manufacturer. Handle the IGFET by the case, never by the leads. When inserting or removing an IGFET, make sure that power is off to avoid transient damage and that the leads are shorted. There should always be a leakage path of high resistance from gate to source, provided by the circuit. Then the short circuit can be safely removed.

Zener diodes may be built into the FET to protect against gate insulation breakdown. When the static charge across the gate-insulator capacitor builds up to the Zener voltage the Zener conducts, preventing further voltage buildup.

## 14-7 Biasing for Depletion-Mode Operation

JFET or depletion-model IGFET biasing is much like a vacuum tube in that a reverse bias is applied between gate and source or grid and cathode. Source resistor $R_S$ in Fig. 14-15(a) is analogous to cathode bias resistor $R_K$

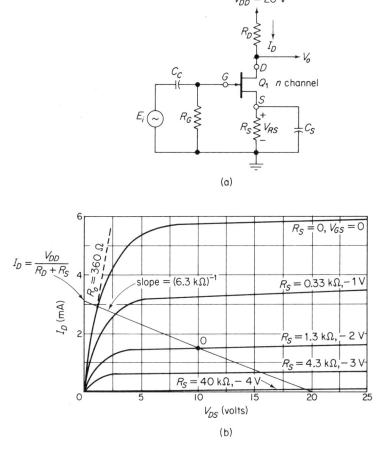

Fig. 14-15. Self-biased, common-source amplifier circuit in (a) and calibrated drain characteristics in (b) determine the operating point.

because conducting drain or plate bias current will set up a voltage drop that can be employed to establish a reverse bias. For example, in Fig. 14-15(a) the $n$-channel JFET conducts electrons but $I_D$ is represented by conventional current flow so that the source end of $R_S$ is driven positive with respect to the reference. Gate resistor $R_G$ functions to allow $V_{RS}$ to act between gate and source.

Since $R_G$ carries no current, it places the gate at ground potential and should be in the megohm range to minimize shunting effect on input resistance. $R_S$ introduces a measure of stabilization against operating-point shift. If for any reason $I_D$ begins to decrease, (1) the voltage drop across $R_S$ decreases, (2) thus reducing reverse bias to (3) increase $I_D$ back toward its original value.

A small value of leakage current, $I_{GSS}$, of the order of 1–10 nA, may flow through $R_G$. $I_{GSS}$ is temperature dependent and can cause a temperature-dependent increase in voltage across $R_G$ by as much as 0.5 V for a 75°C increase in temperature. It may be important to account for the $I_G R_G$ drop when measuring an operating point.

Two constraints are imposed on operating-point drain current by (1) Eq. (14-6) and (2) the bias voltage $V_{RS}$ in the common-source circuit of Fig. 14-15(a), where

$$V_{GS} = V_{RS} = -I_D R_S \quad (14\text{-}11)$$

From these two equations and Eq. (14-7) we can obtain an expression for $I_D$ in terms of $R_S$ and transistor parameters $G_o$ and $V_p$; that is,

$$I_D = \frac{1 - G_o R_S + (1 - 2G_o R_S)^{1/2}}{\dfrac{G_o R_S^2}{V_p}} \quad (14\text{-}12)$$

However, it is far easier to use graphical techniques for establishing or analyzing depletion-mode operating points determined by a source resistor. In Fig. 14-15, as long as we stay in the pinchoff region, each $V_{GS}$ curve is flat and represents a constant-current line. For example, a $V_{GS}$ of $-1$ V will cause a saturation current of approximately $I_D = 3.3$ mA for all $V_{DS} > V_p$ in Fig. 14-15(b). It follows that a drain current of 3.3 mA flowing through an $R_S = V_{GS}/I_D = 1/(3.3 \times 10^{-3}) = 330\ \Omega$ will cause a $V_{GS}$ of $-1$ V. Thus as long as we stay above pinchoff, an $R_S$ of 330 $\Omega$ will *always generate a bias voltage of* $-1$ V *and cause a drain current of 3.3 mA*. For each bias line an approximate value of $R_S$ can be calculated from $V_{GS}$ and $I_D$ and *bias lines of the common-source output characteristic can be calibrated in terms of $R_S$*. Of course $I_D$ does increase slightly with increasing $V_{DS}$, but to keep it simple, read $I_D$ at $V_{DS} = 12.5$ V, approximately in the center of the saturation region. Accordingly, bias lines for Fig. 14-15(b) would be calibrated in terms of $R_S$ as shown in Table 14-1.

Intermediate values of $R_S$ may be interpolated. For example, a source resistor of 2 k$\Omega$ would result in a bias voltage of approximately $V_{GS} = 2.5$ V.

Sec. 14-8            Biasing for Enhancement-Mode Operation      467

Table 14-1

| $V_{GS}$ (V) | $I_D$ at 12.5 V $= V_{DS}$ (mA) | $R_S = V_{GS}/I_D$ (kΩ) |
|---|---|---|
| 1 | 3.3 | 0.33 |
| 2 | 1.5 | 1.3 |
| 3 | 0.7 | 4.3 |
| 4 | 0.1 | 40 |

The interpolation is not in direct proportion since an increase of $R_S$ from 1.3 kΩ to 4.3 kΩ increased $V_{GS}$ by 1 V from $-2$ to $-3$. But to increase $V_{GS}$ another volt to $-4$ V requires an increase in $R_S$ from 4.3 to 40 kΩ, or about 10 times as much. In many practical circuits or test circuits, operating-point location may be found simply by placing a potentiometer in the source leg.

To locate an operating point draw a dc load from the load-line equation written for Fig. 14-15(a):

$$V_{DD} = I_D R_D + V_{DS} + I_D R_S \qquad (14\text{-}13)$$

Intersection of the load line and the bias line locates the operating point.

*Example 14-4.* Given $R_D = 5$ kΩ, and $R_S = 1.3$ kΩ in Fig. 14-15(a). Find the operating point.

*Solution.* Substituting into Eq. (14-13):

$$V_{DD} = 20 = 6300 I_D + V_{DS}$$

This equation is plotted as a straight line between the voltage-axis intercept at $V_{DS} = 20$, $I_D = 0$ and the current-axis intercept at $V_{DS} = 0$, $I_{DS} = V_{DD}/(R_D + R_S) = 20/6300 = 3.2$ mA. Point 0 is shown in Fig. 14-15(b) with $I_D = 1.5$ mA, $V_{DS} = 10$ V at the intersection of the load line with the bias line of $R_S = 1.3$ kΩ.

## 14-8 Biasing for Enhancement-Mode Operation

A forward bias voltage is applied to the gate in Fig. 14-16(a) for operation in the enhancement mode. $R_S$ is included for stability and establishes a reverse bias to partially offset the forward bias caused by $R_1$ and $R_2$. Biasing is simplified by the Thévenin input circuit in Fig. 14-16(b). Since no grid current flows in the IGFET, bias voltage $V_{GS}$ is determined from

$$V_{GS} = I_D R_S - V_G \qquad (14\text{-}14a)$$

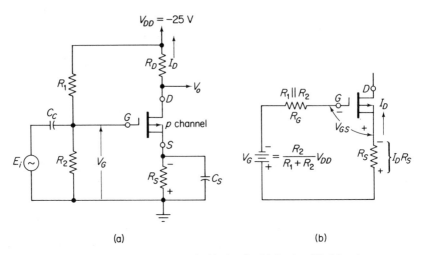

**Fig. 14-16.** Enhancement-mode biasing in (a) is simplified by the dc input model in (b). $V_G < I_D R_S$.

$$V_G = \frac{R_2}{(R_1 + R_2)} V_{DD} \tag{14-14b}$$

A resistor could be added between the junction of $R_1$ and $R_2$ and the gate to increase ac input resistance with no effect on the bias voltage, because no grid current flows.

**Example 14-5.** The circuit of Fig. 14-16(a) is to be operated at $I_D = -5$ mA, $V_{DS} = -10$ V, $V_{GS} = -8$ V and corresponds to point 0 in Fig. 14-14. Given $R_D = 2$ k$\Omega$ and $R_S = 1$ k$\Omega$. Select $R_1$ and $R_2$ to give an ac input resistance of 1 M$\Omega$ and to establish $V_{GS}$ at $-8$ V.

**Solution.** The reverse bias across $R_S$ will be $I_D R_S = 5 \times 10^{-3} \times 10^3 = 5$ V. From Eq. (14-14b), $V_G$ must equal $-13$ V or,

(1) $$\frac{R_2}{(R_1 + R_2)}(-25) = -13$$

The ac input resistance presented to $E_i$ is $R_G$, or

(2) $$R_G = 10^6 = \frac{R_1 R_2}{R_1 + R_2}$$

Solving Eqs. (1) and (2) simultaneously gives $R_1 = 1.9$ M$\Omega$, $R_2 = 2.1$ M$\Omega$.

**Example 14-6.** Given $R_D = 2.3$ k$\Omega$, $R_S = 4$ k$\Omega$, $V_{DD} = 20$ V, $R_1 = 5$ M$\Omega$, and $R_2 = 1.25$ M$\Omega$ in Fig. 14-16(a). Find the operating point on the transistor's characteristics in Fig. 14-17.

Sec. 14-8   Biasing for Enhancement-Mode Operation   469

*Solution.* From Eq. (14-4), $V_G - (1.25)(-20)/6.25 = -4$ V. Eq. (14-14a) may be expressed by

$$V_{GS} = 4000 I_D - 4$$

Assume a value for $V_{GS}$, calculate the resultant $I_D$, and plot this point on Fig. 14-17. Repeat for a smooth bias curve, as illustrated by the following calculations (the last point cannot be plotted).

Table 14-2

| $V_{GS}$ (V) | 0 | 1 | 1.2 | 1.5 | 2.0 | 4.0 |
|---|---|---|---|---|---|---|
| $I_D$ (mA) | 1 | 1.25 | 1.3 | 1.2 | 1.5 | 2.0 |

Draw the load line of $R_D + R_S$ to locate depletion-mode operating point 0 at its intersection with the bias curve in Fig. 14-17.

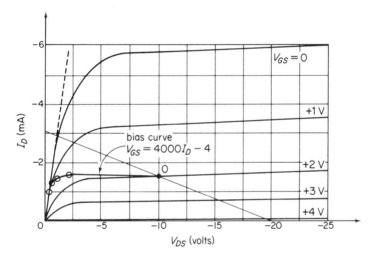

Fig. 14-17. Solution to Example 14-6.

Another method of biasing an FET for operation in the enhancement mode is to connect a resistor between drain and gate. In Fig. 14-18 drain characteristics for an enhancement-type, n-channel IGFET require a positive drain voltage to collect n-channel carriers and a positive gate voltage to induce an n channel. In the circuit of Fig. 14-18(a), no dc current flows through $R_F$ so the potentials are equal at drain and gate. One constraint imposed on the operating point is that

$$V_{GS} = V_{DS} \tag{14-15}$$

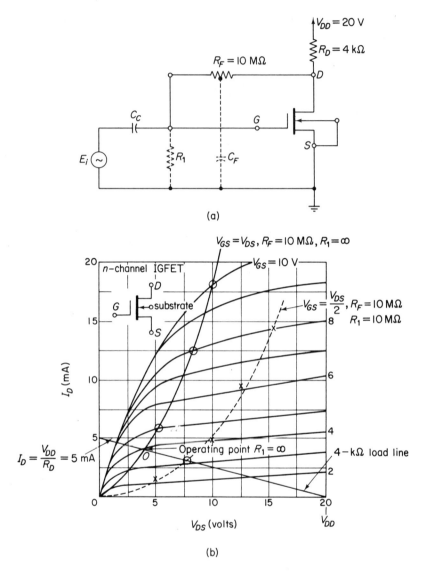

**Fig. 14-18.** Enhancement mode biasing in (a) is analyzed graphically on the drain characteristics in (b).

A second constraint is the load-line equation:

$$V_{DD} = I_D R_D + V_{DS} \tag{14-16}$$

If we plot both curves from Eqs. (14-15) and (14-16), respectively, on the drain characteristic, their intersection will locate the operating point as illustrated in Example 14-7.

Sec. 14-9        Zero Bias Shift of the Operating Point    471

*Example 14-7.* Find the operating point for the enhancement-mode circuit of Fig. 14-18.

*Solution.* (a) Plot the curve of $V_{GS}$ versus $V_{DS}$ on Fig. 14-18(b). Three points are shown at $V_{DS} = V_{GS} = 10$ V, 8 V, and 5 V, respectively. (b) From Eq. (14-16), plot the load line for $V_{DD} = 20$ and $R_D = 4$ k$\Omega$. (c) Locate operating point 0 at the intersection of curves (a) and (b). Point 0 is located at $I_D = 4$ mA, $V_{DS} = V_{GS} = 4$ V. Verify this point by substituting for $I_D$ and $V_{DS}$ into Eq. (14-16).

$$V_{DD} = I_D R_D + V_{DS}$$
$$20 = (4 \times 10^{-3})(4 \times 10^3) + 4 = 20$$

Capacitor $C_F$ in Fig. 14-18(a) may be connected to prevent any ac feedback from drain to gate, and reduces the input impedance to 5 M$\Omega$. This type of biasing stabilizes operating-point variation. For example, if $I_D$ tends to increase, the voltage drop across $R_D$ increases, causing $V_{GS}$ to decrease and reduce drain current.

If it is necessary to change the operating point, add resistor $R_1$ in Fig. 14-18 so

$$V_{GS} = \left(\frac{R_1}{R_1 + R_F}\right) V_{DS} \qquad (14\text{-}17)$$

Gate voltage is lower than $V_{DS}$ and shifts the operating point towards the saturation region. For example, if $R_1 = 10$ M$\Omega$ were connected in Fig. 14-18, the bias constraint would now be

$$V_{GS} = \left(\frac{10}{10+10}\right) V_{DS} = \frac{V_{DS}}{2}$$

$V_{GS} = V_{DS}/2$ is plotted as a dashed line in Fig. 4-18(b) and shows that the operating point will move to $V_{GS} = 3$ V, $I_D = 3$ mA, when $R_1 = 10$ M$\Omega$.

## 14-9 Zero Bias Shift of the Operating Point

There exists an operating point for a FET which is independent of temperature. This characteristic is unique for the FET and is a distinct advantage for applications where extreme operating-point stability is mandatory and large gain is not important. Perhaps the best way to demonstrate and measure this operating point is to photograph the transfer characteristic of an FET with a double exposure under extremes of hot and cold.

In Fig. 14-19 the curve with steepest slope was caused by freezing the transistor with a commercial freezing spray. As temperature was increased the curve appears to rotate clockwise about the point $V_{GS} = -0.9$ V, $I_D = 0.2$ mA.

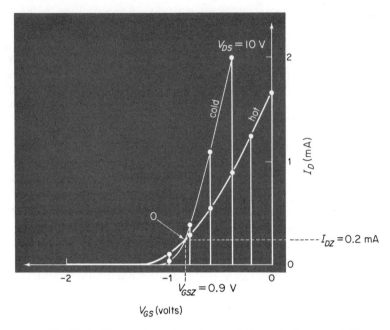

**Fig. 14-19.** Temperature dependence of $I_D$ with the zero drift operating point 0.

Two competing physical mechanisms are responsible for this temperature-dependent behavior. First, carrier mobility decreases with increasing temperature because thermal agitation of molecules in the crystal increases collisions between carriers to reduce the net carrier velocity. For a given $V_{DS}$, drain current decreases by 0.7%/°C rise. However, the decreasing drain current causes the space-charge regions to shrink, which tends to increase $I_D$ at a rate equivalent to 2.2 mV/°C drop in $V_{GS}$. There is some *zero shift value of drain current* $I_{DZ}$ whose 0.7% change equals that due to $V_{GS}$, or

$$0.0022 g_m = 0.007 I_{DZ}$$

and

$$\frac{I_{DZ}}{g_m} = 0.314 \tag{14-18}$$

At $I_{DZ}$ it is informative to obtain an expression for the corresponding zero shift bias voltage $V_{GSZ}$ by writing Eq. (14-6) and (14-8) for the zero shift point,

$$I_{DZ} = I_{DO}\left(1 - \frac{V_{GDZ}}{V_p}\right)^2 \tag{14-6}$$

$$g_{mz} = \frac{2I_{DO}}{V_p}\left(1 - \frac{V_{GSZ}}{V_p}\right) \tag{14-8}$$

Divide Eq. (14-6) by Eq. (14-8) to obtain

$$\frac{I_{DZ}}{g_{mz}} = \frac{V_p - V_{GSZ}}{2} \tag{14-19}$$

Equating Eqs. (14-19) and (14-18) yields

$$|V_{GSZ}| = |V_p| - 0.63 \tag{14-20}$$

In order to predict $I_{DZ}$ from knowledge of $I_{DO}$ and $V_p$, substitute Eq. (14-20) into Eq. (14-6).

$$I_{DZ} = I_{DO}\left(1 - \frac{V_p - 0.63}{V_p}\right)^2 = I_{DO}\left(\frac{0.63}{V_p}\right)^2 \tag{14-21}$$

If $V_p = 0.63$ V then $I_{DZ}$ occurs at $I_{DO}$ so that maximum $g_m$ and gain can be realized. However, a device with $V_p = 3.1$ V requires a bias at $I_{DZ} = I_{DO}/25$ for zero drift and yields a much lower gain due to the lower $g_m$. As was to be expected, the price of stability is gain. For applications where an amplifier will be cycled repeatedly between temperature extremes, the zero shift operating point represents a distinct advantage over other device types, since there is no transient shift.

## 14-10 Low-Frequency Circuit Models for the FET

Considerable simplicity is realized in a low-frequency model for either the JFET or IGFET because of their characteristically high input impedance. The gate terminal is considered to be an open circuit. Any signal input voltage $v$ superimposed on $V_{GS}$ causes a drain current output signal variation $i_d$ to be superimposed on the $I_D$ current. Signal voltage $v$ is related to the signal drain current by the incremental *transconductance* $g_m$. Transconductance may also be represented by the symbol $y_{sf}$, signifying *small-signal forward transfer admittance* in the common-source configuration. A second parameter in the model is *incremental drain resistance $r_d$* or *output resistance*. Since our interest will be concerned primarily with amplifier applications, our models will be limited to the pinchoff region. Parameters $r_d$ and $g_m$ were developed with respect to the output characteristic curves in previous sections and graphical evaluation examples are given for the enhancement-mode IGFET in Fig. 14-20 at various operating points. Parameter $g_m$ may also be expressed by solving Eqs. (14-8) and (14-10) as

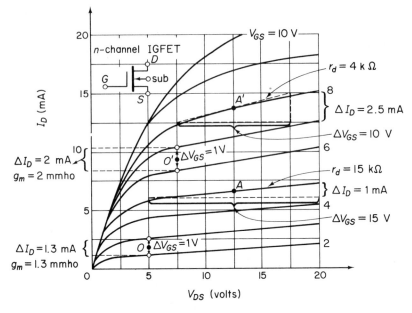

**Fig. 14-20.** Graphical evaluation of $g_m$ and $r_d$.

$$g_m = G_o \left[\frac{I_D}{I_{DO}}\right]^{1/2} \quad (14\text{-}22)$$

Low-frequency incremental models for either JFET or IGFET are presented in Fig. 14-21. Gate $G$ is shown as an open circuit. Input voltage $V$,

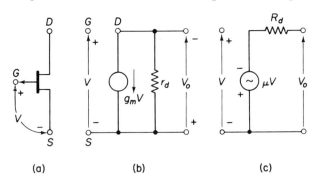

**Fig. 14-21.** The FET in (a) is represented by a Norton model in (b) and a Thévenin model in (c).

appearing between gate and source, excites dependent current generator $g_m V$, which acts between drain and source. A positive-going gate increment will (1) act to close off the $p$ channel and (2) reduce drain current by an increment which will (3) reduce the drain load resistor drop by an increment

Sec. 14-11  Voltage Gain  475

to (4) make the drain negative with respect to the source. This action leads to the phase reversal between gate and drain of 180° and requires the downward direction of $g_m V$ for a positive-going gate signal in Fig. 14-21(b).

A Thévenin equivalent circuit model is drawn in Fig. 14-21(c) employing amplification factor $\mu$. Equation (14-9) shows that $\mu$ may be obtained graphically by taking the ratio of a change in $V_{DS}$ to $V_{GS}$ at a constant $I_D$ around the operating point. Often $\mu$ is difficult to evaluate because the drain characteristic's slope, in pinchoff, is very small. All of the Thévenin relationships for the vacuum tube are applicable to the FET Thévenin model, including the resistance- and voltage-transforming properties. Since the product of $g_m$ and $r_d$ is $\mu$ and the open-circuit output voltages are equal in Figs. 14-21(b) and (c),

$$g_m V r_d = \mu V \qquad (14\text{-}23)$$

## 14-11 Voltage Gain

Practice with circuit models of the FET will be concerned with the Norton model since Thévenin relationships are covered in Chapter 13. A common-source model of Fig 14-16 is represented at low frequencies in Fig. 14-22. Output voltage $V_o$ develops from the product of signal output current $g_m E_i$ and the parallel combination of $r_d$ and $R_{,D}$ or

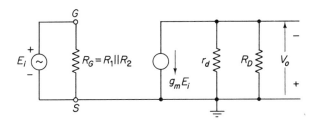

Fig. 14-22. Small-signal model of the common-source amplifier in in Fig. 14-16(a).

$$V_o = -g_m E_i (R_D || r_d) \qquad (14\text{-}24\text{a})$$

Voltage gain $A_v$ is simply

$$A_v = \frac{V_o}{E_i} = -g_m (R_D || r_d) \qquad (14\text{-}24\text{b})$$

In enhancement-type IGFETS, $r_d$ often cannot be neglected. For JFETS and depletion-type IGFETS, $r_d$ is large with respect to $R_L$ so that Eq. (14-24b) may be approximated by

$$A_v = \frac{V_o}{E_i} \cong -g_m R_D \qquad (14\text{-}24\text{c})$$

It is obvious from Eq. (14-24) that voltage gain is a function of $g_m$, which in turn is a function of the operating point via Eq. (14-22) or (14-8). $G_o$ is the maximum value for $g_m$, occurring at zero bias for JFETS and depletion-type IGFETS. Consequently *maximum gain* occurs at

$$A_{v(\max)} = \frac{V_o}{E_i} \cong -G_o R_D \cong -\frac{2I_{DO}}{V_p} R_D \qquad (14\text{-}25a)$$

Adjust $R_D$ and $V_{DD}$ to locate the $Q$ point on the zero bias line so that $V_{DD}$ will divide approximately evenly between $I_{DO}R_D$ and $V_{DS}$. Then the voltage drops across $R_D$ and $V_{DS}$ will both be equal to $I_{DO}R_D$ or $V_{DD} = 2I_{DO}R_D$. Substituting into Eq. (14-25a) gives the maximum gain in terms of $V_{DD}$ for zero bias operation.

$$A_{v(\max)} = -\frac{V_{DD}}{V_p} \qquad \text{zero bias, center load line} \qquad (14\text{-}25b)$$

Equation (14-25) shows that FETs with low $V_p$ and circuits with high supply voltage yield maximum gain. Practically, we will locate the $Q$ point at $\frac{1}{2} I_{DO}$ and estimate $g_m \cong \frac{1}{2} G_o$ for gain calculations. These points are demonstrated in Example 14-8, together with a graphical evaluation of voltage gain.

**Example 14-8.** (a) Calculate voltage gain for the common-source circuit of Fig. 14-16(a), with $R_S = 1.3\,\text{k}\Omega$ ($V_{GS} = -2\,\text{V}$), $R_D = 5\,\text{k}\Omega$, $V_{DD} = 20\,\text{V}$ $G_o = 2800\,\mu$mho, $V_p = 4.2\,\text{V}$. Characteristic curves for the transistor are given in Fig. 14-23. (b) Let $E_{ip} = 1\,\text{V}$ and graphically find $A_V$ for a check.

**Solution.** (a) Draw the dc load line in Fig. 14-23 to locate operating point 0 at $V_{GS} = -2\,\text{V}$. Evaluate $r_d$ at the operating point from $r_d = \Delta V_{DS}/\Delta I_D = 10/(0.2 \times 10^{-3}) = 50\,\text{k}\Omega$. Calculate $g_m$ from Eq. (14-8):

$$g_m = G_o\left(1 - \frac{V_{GS}}{V_p}\right) = 2800\left(1 - \frac{2}{4.2}\right) = 1330\,\mu\text{mho}$$

Voltage gain is found from Eq. (14-24b):

$$A_v = g_m(R_D \,\|\, r_d) = 1300 \times 10^{-6}(5\,\|\,50) \times 10^3 = 6.0$$

(b) In Fig. 14-23 an ac load line with slope equal to the reciprocal of $R_D \,\|\, r_d = 4.55\,\text{k}\Omega$ is drawn through operating point 0. A peak-to-peak voltage swing of $E_i = 1\,\text{V}$ is sketched around $V_{GS} = -2\,\text{V}$ and the peak bias lines of $V_{GS} = -1.5$, $V_{GS} = -2.5$ are drawn as dashed lines to intersect the ac load line. Vertical projections to the $V_{DS}$ axis shows an output voltage swing of approximately 6 V for a gain of 6 with a phase reversal of 180°.

An unbypassed source resistor is introduced into the circuit of Fig. 14-24(a). If output is taken from the drain, we have a common-source configuration and $R_S$ introduces negative feedback to reduce voltage gain. If $R_D$

Sec. 14-11          Voltage Gain    477

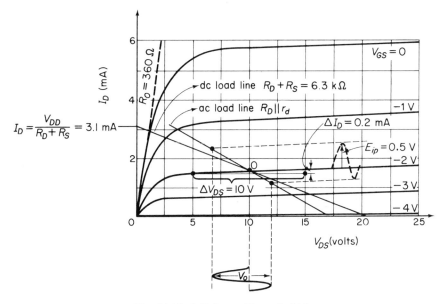

Fig. 14-23. Solution to Example 14-8.

is deleted and output is taken from $R_S$, we have a common-drain amplifier. Our first analysis of this circuit will disregard $r_d$ to simplify and focus attention on a method of transforming current generator $g_m V$ into a generator which depends on $E_i$. Essentially what happens is that $g_m$ will be modified into an equivalent transconductance which depends on $R_S$.

From the input loop of Fig. 14-24(a), $E_i$ feeds $V$ and the drop across $R_S$, or

$$E_i = V + g_m V R_S$$

Solving for $V$ yields

$$V = \frac{E_i}{1 + g_m R_S} \tag{14-26}$$

Now modify the current generator in terms of Eq. (14-27) as shown in Fig. 14-24(b). Voltage gains may be written directly from this figure as the product of output current (in terms of $E_i$) and either $R_S$ or $R_D$.

*Common source:*

$$V_{o2} = -\frac{g_m}{1 + g_m R_S} E_i R_D \tag{14-27a}$$

or

$$A_v = \frac{V_{o2}}{E_i} = -\frac{g_m R_D}{1 + g_m R_S} \tag{14-27b}$$

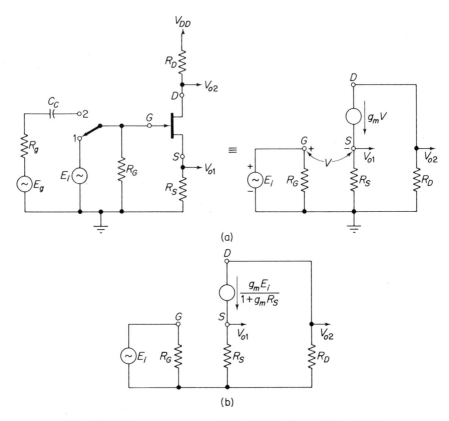

**Fig. 14-24.** Common-source or drain circuit; the model in (a) is simplified in (b). Parameter $r_d$ has negligible effect.

*Source follower:*

$$V_{o1} = \frac{g_m}{1 + g_m R_S} E_i R_S \qquad (14\text{-}28a)$$

$$A_v = \frac{V_{o1}}{E_i} = \frac{g_m R_S}{1 + g_m R_S} \simeq 1 \quad \text{if } g_m R_S \gg 1 \qquad (14\text{-}28b)$$

Input resistance for either source-follower or common-source amplifier in Fig. 14-24(a) is $R_G$. However, if $E_g$ is taken as the reference, then input resistance is the sum of $R_g$ and $R_G$. Voltage gain may be taken with respect to $E_g$ by writing $E_i$ in terms of $E_g$ from the division of $E_g$ between $R_g$ and $R_G$. Throw the switch to point 2 in Fig. 14-24(a) to write

$$E_i = \frac{R_G}{R_g + R_G} E_g \qquad (14\text{-}29)$$

Substitute for $E_i$ from Eq. (14-29) in either Eqs. (14-27) or (14-28) to obtain

Sec. 14-11    Voltage Gain    479

$A_v = V_o/E_g$. Finally, if $r_d$ is large enough to be inconsequential then output resistance is large for either configuration due to the dependent-current generator.

When $r_d$ is not negligible ($r_d < 10R_D$), the effect of its presence is investigated in Fig. 14-25(a). Here the signal drain current $g_m V$ divides

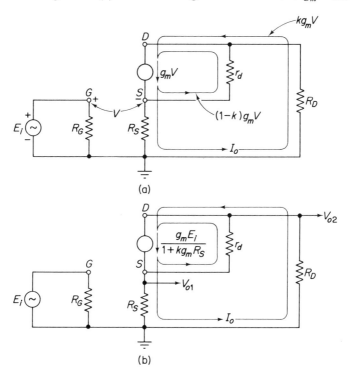

**Fig. 14-25.** Common-source or drain models with $r_d$.

between the branch $r_d$ and the branch $R_D + R_S$. It is convenient to replace $g_m V$ with a generator dependent on $E_i$ by again writing the input loop, noting that $R_S$ does not carry all of $g_m V$ but only a fraction which depends on the relative sizes of $r_d$ and $R_S + R_D$. That is, $R_S$ carries a signal current of $kg_m V$, where

$$k = \frac{r_d}{R_D + R_S + r_d} \tag{14-30}$$

Thus the input loop is $E_i = V + kg_m V R_S$. Solving for $V$,

$$V = \frac{E_i}{1 + kg_m R_S} \tag{14-31}$$

Modify the current generator according to Eq. (14-31) in Fig. 14-25(b).

Output voltages may now be written directly from Fig. 14-25(b) by noting that the actual output signal current $I_o$ results from the division between $r_d$ and $R_S + R_D$ of the modified current generator in terms of $E_i$.

Common source:

$$V_{o2} = -k\left(\frac{g_m E_i}{1 + k g_m R_S}\right) R_D \qquad (14\text{-}32)$$

Source follower:

$$V_{o1} = k\left(\frac{g_m E_i}{1 + k g_m R_S}\right) R_S \qquad (14\text{-}33)$$

Observe that $r_d$ has two effects on voltage gain. First, $r_d$ shunts part of the drain current away from $R_s$ so that $V$ is larger with $r_d$ than it would be without $r_d$. However, the resulting larger drain current is also shunted away from $R_D$ and $R_s$ by $r_d$ to offset the increase.

## 14-12 Resistance Levels

The presence of $r_d$ will affect output resistance because it introduces coupling between drain and source. A test voltage $V$ is connected between source and ground in Fig. 14-26(a). Test voltage $V$ excites the $g_m V$ generator

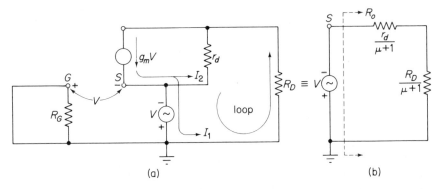

**Fig. 14-26.** The test circuit in (a) shows resistance transformation by the source in (b).

which divides between $R_D$ and $r_d$. Resistance presented to $V$ is $R_o$, as determined by the ratio of $V$ to $I_1$. Writing the indicated loop and node equations:

$$V = I_1 R_D - I_2 r_d \qquad (14\text{-}34a)$$

$$g_m V = I_1 + I_2 \qquad (14\text{-}34b)$$

Solving Eq. (14-34) for $V$ in terms of $I_1$

$$R_o = \frac{V}{I_1} = \frac{r_d + R_D}{1 + g_m r_d} = \frac{r_d + R_D}{1 + \mu} \qquad (14\text{-}35)$$

As shown in Fig. 14-26(b), looking into the *source* we see each resistance apparently divided by one plus the amplification factor.

Apply test voltage $E$ in Fig. 14-27(a) to measure output resistance,

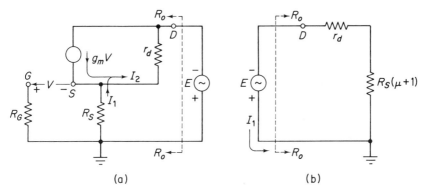

Fig. 14-27. FET resistance transformation at the drain.

$R_o$, seen looking into the *drain terminal*. Three equations may be written

$$E = R_s I_1 + I_2 r_d \qquad (14\text{-}36a)$$

$$I_2 = g_m V + I_1 \qquad (14\text{-}36b)$$

$$V = I_1 R_s \qquad (14\text{-}36c)$$

Solving for the ratio of $E$ to $I_1$,

$$R_o = \frac{E}{I_1} = (1 + \mu)R_s + r_d \qquad (14\text{-}37)$$

Equation (14-37) indicates that we see $r_d$ unchanged and source leg resistance multiplied by $\mu + 1$, as modeled in Fig. 14-27(b).

The final investigation in this section is the effect of feedback resistor $R_F$ between drain and gate. From the model of Fig. 14-28,

$$g_m E_i = I + I_f \qquad (14\text{-}38)$$

Assuming that $E_i$ is small with respect to $V_o$, the current through $R_F$ is $I_f \cong V_o/R_f$, and current $I$ is $I = V_o/R_L$, where $R_L = R_D \parallel r_d$. Substituting for $I$ and $I_f$ in Eq. (14-38) gives a gain expression:

$$g_m E_i = \frac{V_o}{R_L} + \frac{V_o}{R_F} \qquad (14\text{-}39)$$

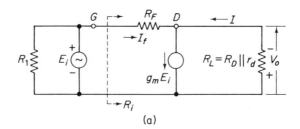

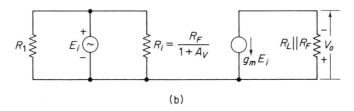

**Fig. 14-28.** The circuit of Fig. 14-18 is modeled in (a) and simplified in (b).

$$A_v = \frac{V_o}{E_i} = -g_m(R_L \parallel R_F) \qquad (14\text{-}40)$$

Input resistance is developed by writing the outside-loop equation of Fig. 14-28 as $E_i - I_f R_F + V_o = 0$.

Substituting for $V_o$ from Eq. (14-40) and solving for the ratio of $E_i$ to $I_f$ gives

$$R_i = \frac{E_i}{I_f} = \frac{R_F}{1 + A_v} \qquad (14\text{-}41)$$

## 14-13 High-Frequency Dependence

When the signal frequencies are high enough, gate voltage increments will cause changes in the stored charges of the space-charge region that can be modeled by capacitance in Fig. 14-29(a). Subscripts denote capacitor terminals and are $C_{gs}$, gate to source, $C_{gd}$, gate to drain, and $C_{ds}$, drain to source. Typical values are 1–10 pF for $C_{gs}$ and $C_{gd}$, and 0.1–2 pF for $C_{ds}$. $C_{ds}$ is usually neglected since its effects are masked by the other two capacitances. $C_{gd}$ is listed in data sheets as $C_{rss}$, *reverse transfer capacitance*. If we place an ac short circuit across $R_D$ in Fig. 14-29 and measure the input capacitance we would see $C_{gs}$ in parallel with $C_{gd}$ or a capacitance equal to their sum. This measurement is specified on data sheets as $C_{iss}$, *small-signal, common-source, short-circuit input capacitance.* The relationships are summarized by $C_{rss} = C_{gd}$, and $C_{iss} = C_{gs} + C_{gd}$.

$C_{gd}$ introduces coupling between drain and gate and is analogous

Sec. 14-13            High-Frequency Dependence    483

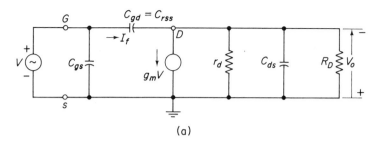

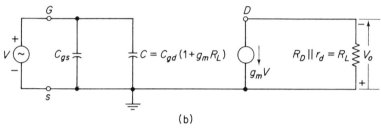

**Fig. 14-29.** The high-frequency model of Fig. 14-22 in (a) is simplified by the equivalent Miller capacitance in (b).

to the Miller effect of $C_\mu$ in a bipolar transistor and $C_{gp}$ in a vacuum tube. An equivalent high-frequency input model of Fig. 14-29(a) is developed by discarding $C_{ds}$ and writing the outside-loop equation:

$$V - I_f \frac{1}{j\omega C_{gd}} + V_o = 0$$

But $V_o = g_m V R_L$, and $R_L = R_D || r_d$. Solving for the ratio of $V$ to $I_f$ yields

$$\frac{V}{I_F} = \frac{1}{j\omega C_{gd}(1 + g_m R_L)} = \frac{1}{j\omega C} \qquad (14\text{-}42)$$

where

$$C = C_{gd}(1 + g_m R_L)$$

The total capacitance $C_T$ between gate and source may be expressed by either

$$C_T = C_{gs} + C_{gd}(1 + g_m R_L) \qquad (14\text{-}43a)$$

or

$$C_T = C_{iss} + C_{rss} g_m R_L \qquad (14\text{-}43b)$$

*Example 14-9.* Manufacturer's data is shown for the transistor and operating point of Fig. 14-30(a). Find the low-frequency voltage gain $A_v = V_o/E_g$. (b) locate the upper break frequency $\omega_H$.

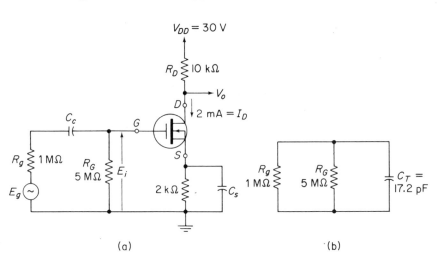

Fig. 14-30. Circuit in (a) and high-frequency time constant in (b) for Example 14-9. Transistor parameters are $y_o = 50$ μmhos, $y_{fs} = g_m = 2000$ μmhos, $C_{iss} = 5$ pF, $C_{rss} = 1$ pF.

**Solution.** (a) Evaluate $r_d$ from $r_d = 1/y_o = 1/(50 \times 10^6) = 20$ kΩ. Evaluate $R_L$ from $R_L = R_D \| r_d = (10 \| 20) \times 10^3 = 6.6$ kΩ. $R_S$ has no effect on gain because it is bypassed, so that

$$\frac{V_o}{E_i} = -g_m R_L = (-6.6 \times 10^3)(2000 \times 10^{-6}) = -13.2$$

Since

$$\frac{E_i}{E_g} = \frac{R_G}{R_g + R_G} = \frac{5}{1+5} = 0.83$$

then

$$A_v = \frac{V_o}{E_g} = \frac{V_o}{E_i}\frac{E_i}{E_g} = (-13.2)(0.83) = -11.0$$

From Eq. (14-43b), $C_T = C_{iss} + C_{rss}g_m R_L = 5 + (1)(13.2) = 17.2$ pF. Draw the equivalent input model in Fig. 14-30(b) and evaluate $\omega_H$ from the time constant.

$$\omega_H = \frac{1}{T} = \frac{1}{C_T(R_g \| R_G)} = \frac{1}{(17.2 \times 10^{-12})(0.83 \times 10^6)} = 7 \times 10^4 \text{ rad/sec}$$

## 14-14 Low-Frequency Voltage Gain

*Low-frequency dependence* will be analyzed by reference to Fig. 14-30, where coupling capacitor $C_c$ and bypass capacitor $C_s$ affect voltage gain. Considering $C_s$ to be a short circuit, $C_c$ will reduce gain when its reactance increases at low frequencies. More of $E_g$ will then drop across $C_c$, reducing the available driving-gate voltage. Employ the impedance-transforming properties of the FET to find the effective resistance presented to $C_c$ in Fig. 14-31(a). Considering $C_c$ to be a short circuit, find the effective resistance

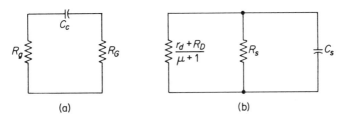

**Fig. 14-31.** Low-frequency time constants for the coupling capacitor in (a) and bypass capacitor in (b).

presented to $C_s$ in Fig. 14-31(b). The time constant of each circuit is expressed by

$$T_s = C_s \left( R_S \left\| \frac{r_d + R_D}{1 + \mu} \right. \right) \tag{14-44}$$

and

$$T_c = C_c(R_g + R_G) \tag{14-45}$$

Actual lower break frequency $\omega_L$ results from both time constants:

$$\omega_L \cong \frac{1}{T_s} + \frac{1}{T_c} \tag{14-46}$$

It is good practice to make either $C_c$ or $C_s$ determine $\omega_L$ and make the other capacitor cause a break 1 decade lower. The choice is based on economics as illustrated in Example 14-10.

*Example 14-10.* Refer to the circuit of Fig. 14-30. (a) Calculate magnitudes for $C_c$ and $C_s$ independently to cause a break frequency at 200 rad/sec, by assuming that the other capacitor is a short circuit. (b) Based on economics, increase one capacitor by a factor of 10, causing it to break a decade lower. (c) Use Eq. (14-46) to calculate the resulting break frequency due to both capacitors.

*Solution.* (a) From Eq. (14-45),

$$\omega_L = \frac{1}{T_c} = \frac{1}{C_c(R_g + R_G)} = 200$$

$$C_c = \frac{1}{(200)(1+5)10^6} = 830 \text{ pF}$$

From Eq. (14-44),

$$\omega_L = \frac{1}{T_s} = \frac{1}{C_s\left(R_S \left\|\frac{r_d + R_D}{\mu + 1}\right.\right)}$$

$$C_s = \frac{1}{200\left[2 \left\|\left(\frac{20+10}{41}\right)\right.\right]10^3} = \frac{1}{200(2\|0.73)10^3} = 9.3\ \mu F$$

where $\mu = g_m r_d = (2000 \times 10^{-6})(20 \times 10^3) = 40$.

(b) Increasing $C_s$ from 9.3 to 93 $\mu$F would be more expensive than increasing $C_c$ from 830 to 8300 pF. Therefore retain $C_s$ at 9.3 $\mu$F to begin rolloff at 200 rad/sec, and increase $C_c$ to 8300 pF to cut in at 20 rad/sec.

(c) From Eq. (14-47) $\omega_L \cong \frac{1}{T_s} + \frac{1}{T_c} = 200 + 20 = 220$ rad/sec.

## 14-15 Hybrid and Two-Stage Amplifiers

Hybrid amplifiers are constructed to combine the advantages of FET and bipolar transistors. Such applications usually employ an FET as a high-input impedance first stage, followed by a bipolar transistor to offset the FET's low gain. Several applications will be considered by analyzing typical examples.

*Example 14-11.* Figure 14-32(a) shows a circuit where high-input impedance is contributed by $Q_1$ and low-output impedance by $Q_2$.

BIASING  Operating-point drain current of $Q_1$ will be practically constant (in pinchoff) whether $Q_2$ is connected or not. We can therefore take a Thévenin equivalent of $Q_1$ and $Q_2$ to calculate base bias current $I_B$ in Fig. 14-32(b). From $I_B$ and $\beta_F$, $Q_2$'s operating point can be calculated. Since $R_D$ carries both $I_D$ and $I_B$, $V_{DS}$ of $Q_1$ can be calculated.

SMALL-SIGNAL ANALYSIS  From Fig. 14-32(c), voltage gain of $Q_1$ is

(1) $\quad \dfrac{V_{o1}}{E_i} = g_m(R \| R_i) \quad$ where $R = R_D \| r_D$ and $R_i = r_{\pi 2} + (\beta_0 + 1)R_L$

Voltage gain of emitter follower $Q_2$ is approximately 1 so overall gain is given by Eq. (1).

*Example 14-12.* A direct-coupled cascode circuit in Fig. 14-33(a) exhibits high input impedance, moderate gain, large output-swing capability and isolation between output and input terminals. Voltage gain of $Q_1$ is seen from Fig. 14-33(b) to be low because of the low input impedance of $Q_2$. Thus $V_{o1}$ will be small and negligible energy will be coupled back through $C_{gd}$.

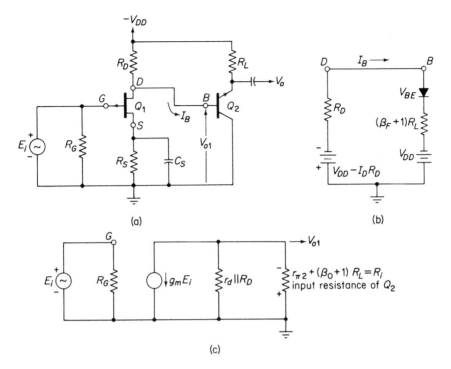

**Fig. 14-32.** The common-source to common-collector amplifier in (a) is anlyzed from the dc model in (b) and the ac model in (c).

(1) $$\frac{V_{o1}}{E_i} \cong \frac{g_{m1}}{g_{m2}} = A_{v1}$$

Since the dc collector current of $Q_2$ is approximately equal to the drain current of $Q_1$, a typical value for Eq. (1) may be found. Assume $I_C = I_D = 1$ mA. Typically $g_{m1} = 2000$ $\mu$mho and $g_{m2} = 40000$ $\mu$mho so that $A_{V1} \cong 1/20$. Voltage gain of the common-base connected $Q_2$ would be high. From Fig. 14-33(c) assume $R_L = 20$ k$\Omega$, so that $A_{V2} = g_{m2}R_L = (0.040)20 \times 10^3 = 800$ and total voltage gain is $(800)(1/20) = 40$. Even though $Q_2$ has high gain, coupling back through $C_\mu$ is bypassed to ground through $C_B$. Thus $R_L$ could be replaced by a tuned circuit and no neutralization would be required.

An application is examined in Example 14-13, where one FET serves as a load for another.

*Example 14-13.* In Fig. 14-34(a), $Q_2$ serves as a load for $Q_1$. Both bias voltages, $V_{GS}$, equal zero and the dc drain currents are equal because they are in

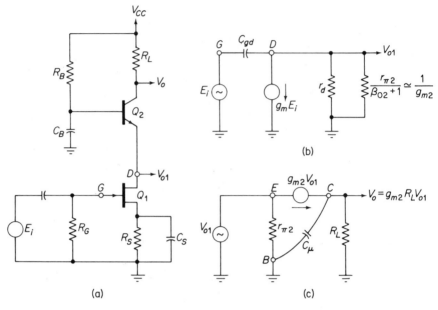

**Fig. 14-33.** Cascoded FET and BJT in (a). Models for voltage gain are shown in (b) for $Q_1$ and in (c) for $Q_2$.

series. If transistor characteristics are matched, $V_{DS}$ of each transistor will equal $V_{DD}/2$. Graphically we can show the dc of load of $Q_2$ as a load line on the characteristics of $Q_1$. That is we pretend $Q_2$ is a nonlinear resistor which varies according to its $I_D - V_{DS}$ characteristic at $V_{GS} = 0$, and draw the load line as in Fig. 11-34(b). Operating point 0 is located at the $V_{GS} = 0$ intersection and operating path $AA'$ shows that a small-signal variation yields a relatively large output-voltage variation. The slope of $AA'$ is $r_{d2}$ of $Q_2$.

A small-signal model is shown in Fig. 14-34(c). Taking a Thévenin equivalent of the current generator $g_m E_i$ and $r_{d1}$ gives the simpler model of Fig. 14-34(d), where we write output voltage by inspection:

(1) $$V_o = \frac{r_{d2}}{r_{d1} + r_{d2}} \mu E_i$$

Assuming matched characteristics so that $r_{d1} = r_{d2}$, voltage gain will be $V_o/E_i = \mu/2$.

## 14-16 Chopper Amplifiers

There are many practical problems encountered when attempting to amplify dc signals or very low-frequency signals. For example, a 100-mV signal level may remain steady for 1 hr and then increase at the rate of 1 mV/h. To amplify these levels by a factor of 100 is a complex engineering problem

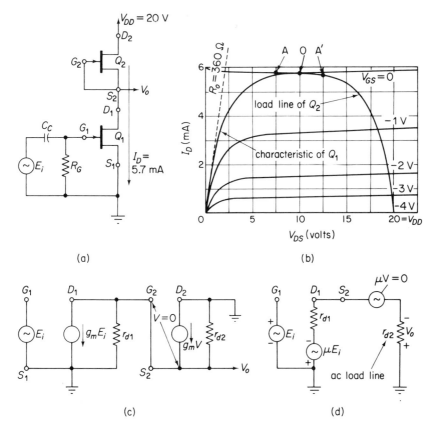

**Fig. 14-34.** Two FETs are cascoded in (a). Their operating point is located graphically in (b). The small-signal model in (c) is simplified in (d).

because of equal voltage changes due to variations in environment and power supply voltages. The task is considerably simplified by *chopping* or *modulating* the dc signal and converting it to an ac signal whose amplitude is directly proportional to the original dc level. A stable ac amplifier builds up the ac signal level to an output ac voltage which is rectified and filtered to present a precise, amplified version of the original dc voltage. Recovery of the amplified dc voltage is often referred to as *demodulation*.

Field-effect transistors are almost ideal choppers. Their advantages in this application will be considered by comparison with electromechanical choppers.

ELECTROMECHANICAL CHOPPERS

A relay functions as a shunt chopper in Fig. 14-35(a) and is driven by chopper voltage $e_c$. Chopper voltage $e_c$ may be either sinusoidal or a

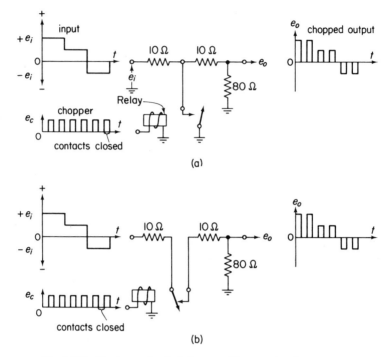

**Fig. 14-35.** Electromechanical chopper. Shunt and series types are shown in (a) and (b).

square wave with a frequency that is at least 10 times higher than the highest frequency variation of the input voltage. When $e_c = 0$, the relay is deenergized with contacts open and $e_o$ is related to $e_i$ by the voltage division

$$e_o = \frac{80 e_i}{10 + 10 + 80} = 0.8\, e_i$$

When $e_c$ energizes the relay, its contacts ground node $A$ and $e_o$ equals zero. Contact resistance for relays with mercury-wetted contacts may be on the order of a few milliohms. However, any resistance $r_s$ in the contacts or contact leads will raise point $A$ above ground and $e_o$ will not be zero when contacts are closed. The resulting small output voltage is called *offset voltage* and is approximately

$$e_{\text{offset}} = \frac{r_s}{r_s + 10} e_i \tag{14-47}$$

In the series chopper of Fig. 14-34(b) there is no offset voltage when the relay's contacts are open. With contacts closed, $e_o$ is related to $e_i$ exactly as in the shunt arrangement.

Sec. 14-16                                   Chopper Amplifiers   491

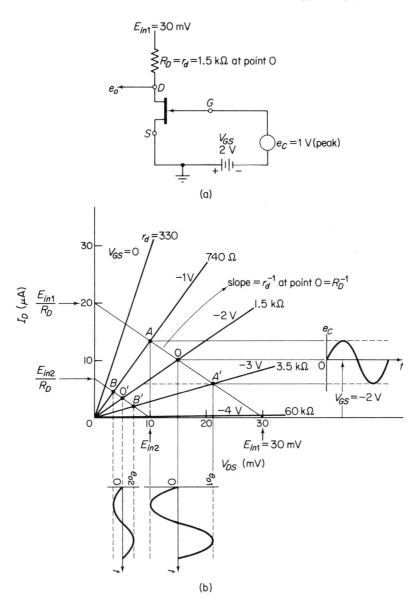

**Fig. 14-36.** Low-level FET shunt chopper in (a) operates in the ohmic region of (b).

Electromechanical choppers are (1) limited to chopping frequencies of about 400 Hz, (2) subject to wear because of moving parts, (3) susceptible to contact pitting with reactive loads, and (4) unable to withstand severe mechanical shock. They do have the advantage of extremely low *on* resistance through closed contacts and extremely high *off* resistance with open contacts.

Low-level signals may be chopped by FETs without the offset voltage inherent in vacuum tubes and bipolar transistors. Sine-wave drive is considered in the circuit of Fig. 14-36(a), where a dc input voltage of 30 mV is chopped by $e_c$. Operating point 0 is located in the ohmic region at $V_{GS} = -2$ V, on the load line in Fig. 14-36(b). Chopper voltage $e_c$ generates a peak output voltage of $e_{o1} = 5$ mV for an input voltage of $E_{in1} = 30$ mV.

Assume the chopper voltage remains constant, but $e_i$ changes abruptly to 10 mV $= E_{in2}$. On the characteristic curves of Fig. 14-36(b) the dc load line shifts to the left. A new operating point is established at 0', and a new operating path along B-0'-B'. Output voltage $e_{o2}$ is now reduced to a peak swing of 2.5 mV and reflects directly the reduction in $E_{in}$.

It can be seen from the geometry of the triangle, origin-B-A-A'-B'-origin, that the same load line with different values of $E_{in}$ will always form similar triangles. Thus as $E_{in}$ is reduced from 30 mV to zero, $e_o$ will reduce *in direct proportion* to zero. Consequently the amplified and detected version of $e_o$ will be directly related to $E_{in}$.

SUMMARY. For sinusoidal drive (1) the gate is operated at a fixed bias, (2) operation is restricted to the ohmic region with large values of load resistances unless input voltages are at a low level, (3) chopper drive is adjusted so that the gate is not driven to $V_p$ or into forward conduction.

*Square-wave drive* is defined when $V_{GS}$ is varied between zero and values equal to or greater than $V_p$. $V_{GS}$ is varied by the chopping voltage $e_c$ and must be restricted against driving the FET into forward conduction. A typical application is given in the series chopper of Fig. 14-37. When chopper voltage $e_c$ equals zero, the FET is conducting or on. $R_s$ is small enough so that it

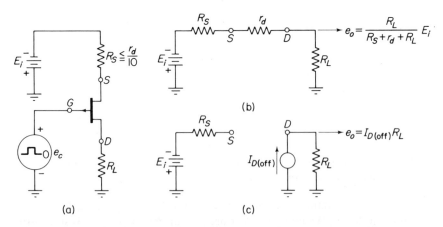

Fig. 14-37. The series FET chopper in (a) is represented by its *on*-circuit model ($e_c = V_{GS} = 0$) in (b), and its *off*-circuit model ($e_c = V_{GS} = V_p$) in (c).

does not introduce a significant bias voltage in the *on* circuit model of Fig. 14-37(b). When $V_{GS}$ is driven beyond $V_p$ by a negative-going $e_c$, the FET is cut off in Fig. 14-37(c) and only the small drain-leakage current $I_{D(\text{off})}$ will flow though $R_L$. Since $I_{D(\text{off})}$ is of the order of picoamperes, offset voltage $I_{D(\text{off})}R_L$ will be of the order of 0.5 $\mu$V.

## 14-17 Analog Switching and Commutation

An analog switching circuit passes an analog signal or blocks its transmission periodically. When several analog signals are transmitted in sequence along a common transmission link, the process is called *commutation, sequential switching, or multiplexing*. The circuit of Fig. 14-37(a) is not capable of transmitting large negative input voltages that will cause a forward bias to hold the FET on. Since an analog signal may be continuously variable and of both polarities, a modification is made to the series chopper as shown in Fig. 14-38. This electronic commutator functions as a single-pole-single-throw switch which connects and disconnects $e_i$ from $R_L$ by the on-off action of $Q_2$.

When $e_c$ turns $Q_1$ off, +10 V reverse biases diode $D$. If $e_i$ goes positive, $V_{GS}$ remains zero so that $Q_2$ is on and $e_i$ is connected across $R_S$, $r_d$ of $Q_2$, and $R_L$ to generate an output voltage of

$$e_o = \frac{R_L}{R_S + r_d + R_L} e_{\text{in}} \tag{14-48}$$

Values of $e_i$ should not exceed +10 V or $D$ will turn on and the resulting drop across $R$ will tend to turn off $Q_2$. If $e_i$ goes negative, source and gate terminals remain at the same potential unless $e_i$ is increased to a value which will break down the diode.

When $e_c$ turns $Q_1$ on, it connects 15 V to the diode. For all positive values of $e_i$ and negative values up to $-15$ V diode $D$ conducts and clamps the gate to about $-14$ V. If we restrict $e_i$ to values between + or $-10$ V, $V_{GS}$ will be reverse biased from 24 to 4 V, respectively, and $Q_2$ will remain off. Effectively $e_i$ is disconnected from $R_L$. In summary, $e_i$ may be a high-frequency analog signal which can be transmitted in bursts by the off-on action of $Q_2$ under direction of chopper voltage $e_c$.

The chopper of Fig. 14-38 may be used as an analog switch in the basic sequential switching arrangement of Fig. 14-39. The rotary switch or electronic equivalent steps continuously. Whenever the switch wiper is on an odd numbered contact $-e_c$ is applied to chopper 1 to (1) turn off $Q_1$, which (2) turns on $Q_2$ to (3) connect analog input 1 to the output bus. Meanwhile $+e_c$ holds $Q_2$ off in chopper 2, disconnecting analog input 2 from the output bus.

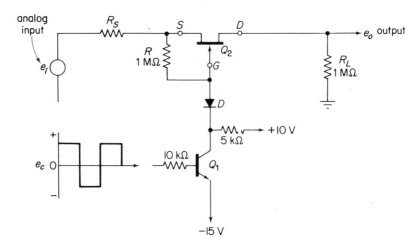

**Fig. 14-38.** Analog switch network $Q_1$ and $Q_2$ connects or disconnects input to output under control of $e_c$.

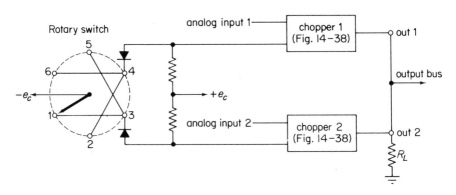

**Fig. 14-39.** Two analog inputs are connected sequentially to the output bus.

Whenever the rotary switch wiper is on an even-numbered terminal, chopper 2 transmits analog input 2 to the output bus. Analog input 1 is disconnected from the output bus by chopper 1.

At the receiving end a similar circuit separates the two signals into different channels by (1) synchronizing the receiving rotary switch with the sending switch, (2) connecting incoming signals from the output bus in parallel with both analog inputs so that (3) chopper 1 connects the input to out 1 signals and (4) chopper 2 connects the same input to out 2 during reception of 2 signals. To summarize: A sequential switching arrangement and chopper connects several input signals in sequence to a single transmission line and can pick off each input signal as it is received to switch it into a separate channel for recovery.

## PROBLEMS

**14-1** From Fig. 14-4, evaluate $r_d$ from the slope of each curve where it intersects the vertical line of $V_{DS} = 0.8$ V. Plot $r_d$ versus $V_{GS}$ for $V_{GS}$ between 0 and $-0.9$ V to show how the FET may be employed as a voltage-controlled resistor.

**14-2** A JFET has $I_{DO} = 1.7$ mA and $V_p = 3$ V. Plot the output characteristics from Eq. (14-6) for $V_{GS} = 0$, 0.5, 1.0, and 3.0 V for values of $V_{DS}$ between $V_p$ and 10 V.

**14-3** Calculate pinchoff voltage from measured values of (a) $I_{DO} = 1.8$ mA, $G_o = 1800$ mho, (b) $I_D = 0.36$ mA, $I_{DO} = 1.7$ mA, $V_{GS} = 1$ V.

**14-4** Find $I_D$ in the pinchoff region from the data of Example 14-3 for $V_{GS} = 0$, 0.3, 1.2, and 1.8 V.

**14-5** Graphically evaluate $g_m$ at $V_{DS} = -5$ V, $V_{GS} = 0.6$ V from the characteristics in Fig. 14-5.

**14-6** Given $I_{DO} = 1.7$ mA in Fig. 14-5 for $V_{DS} = -7$ V. Plot $(I_{DO})^{1/2}$ versus $V_{GS}$ for $V_{GS} = 0$, 0.2, 0.4, and 1.0 V. Extrapolate the curve to measure $V_p$ and compare with Fig. 14-7.

**14-7** Sketch a sweep circuit to display $g_m$ of a $p$-channel JFET on a CRO.

**14-8** Estimate the value of $r_d$ and $g_m$ at $V_{DS} = -10$ V, $V_{GS} = -8$ V for the $p$-channel IGFET in Fig. 14-14.

**14-9** Calibrate the bias lines in Fig. 14-5 with corresponding source resistor values for operation in the pinchoff region.

**14-10** $R_D = 4.7$ k$\Omega$ and $R_s = 1.3$ k$\Omega$ in Fig. 14-15(a). Find the operating point.

**14-11** Change the operating point in Example 14-5 to $I_D = -3$ mA, $V_{GS} = -7$ V, and $V_{DS} = -16$ V by redesigning $R_1$ and $R_2$.

**14-12** $R_D$ is increased to 6 k$\Omega$ in Example 14-6 find the new operating point.

**14-13** $R_1 = 5$ M$\Omega$ and $R_F = 10$ M$\Omega$ in Fig. 14-18. Find the operating point.

**14-14** $R_s$ is changed to 330 $\Omega$, $V_{DD}$ to 25 V, and $R_D$ to 4.2 k$\Omega$ in Example 14-8. (a) Verify that the operating-point bias voltage is $V_{GS} = -1$ V and $V_{DS} = 10$ V. (b) Revise the solutions to Example 14-8 accordingly.

**14-15** From the characteristics of Fig. 14-5, $V_p = 1.85$ V, $I_{DO} = 1.7$ mA, and $G_o = 1800$ mho. (b) Evaluate the zero bias shift operating point $I_{DZ}$ and $V_{GSZ}$. (b) Find $g_m$ at this operating point. (c) Repeat (a) and (b) for the FET in Fig. 14-23, where $I_{DO} = 5.6$ mA, $V_p = 4.2$ V.

**14-16** Assuming the operating point remains constant in Fig. 14-24(a), $R_S = 500\ \Omega$, $r_d$ is negligible, $e_i = 1$ mV, and $g_m = 2500\ \mu$mho, (a) plot $V_{o2}$ versus $R_D$ for $1\ \text{k}\Omega < R_D < 10\ \text{k}\Omega$. (b) Evaluate $V_{o1}$.

**14-17** Assume $r_d = 10\ \text{k}\Omega$ and, using the data in Problem 14-16, plot $V_{o2}$ versus $R_D$ over the same range of values and compare results.

**14-18** Using the transistor data for $g_m$ and $r_d$ in Problems 14-16 and 14-17, calculate output resistance seen looking into the drain terminal.

**14-19** In Fig. 14-18(a), $R_1 = 10\ \text{M}\Omega$, $C_F$ is removed, $r_d = 20\ \text{k}\Omega$, and $g_m = 1350\ \mu$mho. Find the input resistance presented to $E_i$.

**14-20** $R_g$ is reduced to $1\ \text{k}\Omega$ in Fig. 14-30(a) and $R_D$ is increased to $20\ \text{k}\Omega$. Find the upper break frequency and gain $V_o/E_g$.

**14-21** Choose coupling and bypass capacitors for the circuit of Fig. 14-30 to give a lower break frequency of 100 Hz.

**14-22** Derive an expression for $I_E$ of $Q_2$ from Fig. 14-32(b) in terms of $R_D$, $I_D$, $R_L$, $V_{BE}$, and $\beta_{F2}$.

**14-23** In the circuit of Fig. 14-34(a), $g_m = 2500\ \mu$mho and $r_d = 50\ \text{k}\Omega$. What is the voltage gain $V_o/E_i$?

**14-24** If $R_D$ is changed to $1\ \text{k}\Omega$ in Fig. 14-36(a) and all other conditions are unchanged, sketch the graphical development of $e_{o1}$ and $e_{o2}$.

**14-25** $R_S = 50\ \Omega$, $r_d = 10\ \text{k}\Omega$, and $R_L = 10\ \text{k}\Omega$ in Fig. 14-37(a), $E_i = 2$ V and $I_{D(\text{off})} = 5 \times 10^{-12}$/A. Find the values of $e_o$ when under the *on* and *off* conditions of the FET.

# Chapter 15

**15-0** INTRODUCTION ............................... 499
**15-1** CHARACTERISTIC CURVES OF THE UJT ............ 500
**15-2** PARAMETER MEASUREMENTS OF THE UJT .......... 505
**15-3** TEMPERATURE STABILIZATION OF $V_P$ .............. 508
**15-4** THE UNIJUNCTION AS A RELAXATION OSCILLATOR .. 510
**15-5** SAWTOOTH GENERATOR ....................... 516
**15-6** TIMING AND VOLTAGE LEVEL SENSING WITH THE
 UJT ........................................ 518
**15-7** INTRODUCTION TO THE SILICON CONTROLLED
 RECTIFIER (SCR) ............................. 520
**15-8** PULSED GATE OPERATION ..................... 523
**15-9** PHASE CONTROL ............................. 525
**15-10** SCR APPLICATIONS ........................... 529
 PROBLEMS .................................. 530

# The Unijunction Transistor and Silicon-Controlled Rectifier

## 15-0 Introduction

Another field-effect device is the *Unijunction Transistor* (UJT) which functions principally to discharge a capacitor in applications for timing

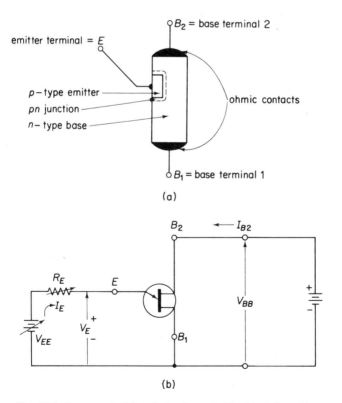

Fig. 15-1. Structure in (a) and circuit symbol in (b) of the unijunction transistor.

circuits or triangular wave-shape generators. Significant differences between the UJT and FET are in the nomenclature and construction, as indicated in Fig. 15-1. In its simplest form the UJT is constructed from a base bar of high-resistivity (lightly doped) *n*-type material. Ohmic contacts are brought out from the base as $B_1$ and $B_2$ in Fig. 15-1(a). The *p*-type material is diffused into the base body to form an emitter with its accompanying *pn* junction and space-charge region. The emitter is located physically closer to $B_2$, both in the structure and on the circuit *symbol* of Fig. 15-1(b).

## 15-1 Characteristic Curves of the UJT

Operation of the UJT will be described by considering its characteristic curves and parameters. *Interbase resistance* $R_{BB}$ represents bulk resistance of the base material measured between $B_1$ and $B_2$. $R_{BB}$ may be measured with an ohmmeter, bridge, or by sweep techniques as in Fig. 15-2. Since $B_1$ and $B_2$ are ohmic contacts, the value of $R_{BB}$ is unaffected by the polarity of $V_{BB}$. From the $I_{B2}$ versus $V_{BB}$ characteristic of Fig. 15-2(b), measure the slope from origin to point $V_{BB} = 5$ V, $I_{B2} = 0.7$ mA to obtain

$$R_{BB} = \frac{V_{BB}}{I_{B2}} = \frac{5}{0.7 \times 10^{-3}} = 7.1 \text{ k}\Omega$$

Typical values for $R_{BB}$ range between 4.0 and 12.0 k$\Omega$.

A simplified model of the UJT can be constructed from the measurement of $R_{BB}$ by placing a resistor between $B_2$ and $B_1$. This resistor is to be tapped at point $A$, where a diode is connected to model the emitter-base junction as shown in Fig. 15-3. The tap at point $A$ conveniently divides $R_{BB}$ into two resistors. With $I_E = 0$ (by adjusting $V_{EE} = 0$) a current $I_{B2}$ will flow which will depend only on the value of $V_{BB}$ and $R_{BB}$, or

$$I_{B2} = \frac{V_{BB}}{R_{BB}} = \frac{V_{BB}}{R_1 + R_2}, \quad I_E = 0 \qquad (15\text{-}1)$$

$I_{B2}$ will establish a voltage drop across $R_1$ with the polarity shown in Fig. 15-3 and a value of $V_A$, where

$$V_A = \frac{R_1}{R_{BB}} V_{BB}, \quad I_E = 0 \qquad (15\text{-}2)$$

Of course, $V_A$ established a reverse bias across the emitter diode so that the diode acts as an open circuit. The coefficient of $V_{BB}$ in Eq. (15-2) represents the voltage division of $V_{BB}$ between $R_2$ and $R_1$ and defines the *intrinsic standoff ratio* ($\eta$), where

## Characteristic Curves of the UJT

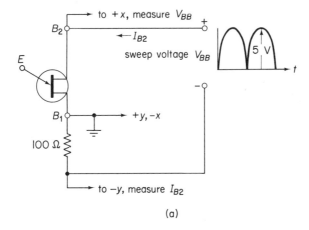

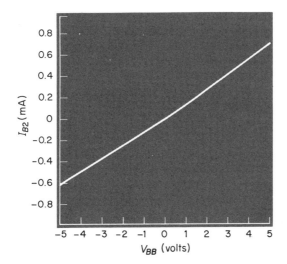

**Fig. 15-2.** Measuring circuit and display of interbase resistance $R_{BB}$. The emitter terminal is left open during this measurement. (a) Sweep circuit to measure interbase resistance $R_{BB}$. (b) Plot of $I_{B2}$ versus $V_{BB}$. Slope is the reciprocal of $R_{BB}$.

$$\eta = \frac{R_1}{R_1 + R_2} = \frac{R_1}{R_{BB}} \qquad (15\text{-}3)$$

A characteristic curve for the input circuit is obtained by varying $V_E$ and measuring $I_E$. Let $V_E$ be increased in Fig. 15-3 to a value equal to $V_A$ in Eq. (15-2). Diode $D$ has 0 V across it and will not conduct. $V_E$ must be increased beyond $V_A$ by approximately $V_D = 0.5$ V in order for the diode to just begin conduction. A special name is assigned to that value of $V_E$ which

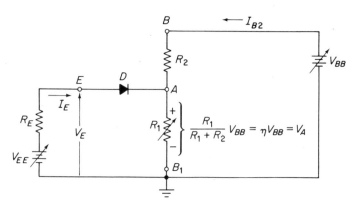

**Fig. 15-3.** Circuit model of the UJT.

just barely turns on the diode. It is *peak-point emitter voltage* ($V_P$), where

$$V_P = V_D + \eta V_{BB} \tag{15-4}$$

Once $V_E$ is raised to $V_P$, a minimum value of emitter current is necessary to begin unijunction transistor action. This minimum current value is defined at $V_P$ as *peak-point current* ($I_P$) and is normally 1 μA. $V_D$ is measured when the diode carries $I_P$ by the test circuit of Fig. 15-4. Note that $V_{EE}$ and $R_E$

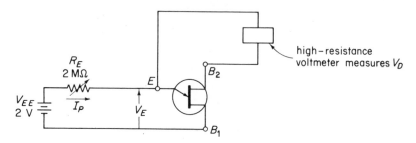

**Fig. 15-4.** A method of measuring $V_D$.

form an equivalent constant current source of 1 μA, which forward biases the emitter diode. Since no current flows through $R_2$, the voltage at $B_2$ equals that at point $A$. Connecting a high-resistance voltmeter between $B_2$ and $E$ measures the voltage between $E$ and $A$ or the drop $V_D$ at 10 μA. Another method of measuring $V_D$ is to sweep the emitter diode's characteristic with a curve tracer by the test circuit of Fig. 15-5(a). In Fig. 15-5(b) (assuming that $R_1$ is a maximum of 5 kΩ then 1 μA of emitter current will cause a drop of only 5 mV), the voltage $V_E$ is approximately 0.30 V when $I_E$ is 1 μA. Thus $V_E \cong V_D = 0.30$ V.

*Unijunction transistor action* begins when the voltage between emitter and base exceeds $V_P$ and $I_P$ begins to flow. Holes from the emitter p-type

## Sec. 15-1      Characteristic Curves of the UJT      503

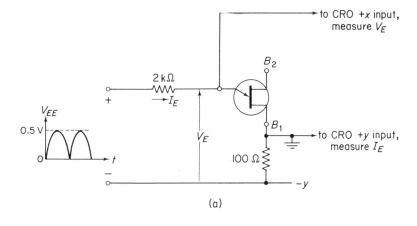

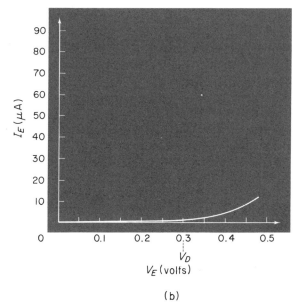

**Fig. 15-5.** The sweep circuit in (a) measures $I_E$ versus $V_E$ to obtain the forward conducting diode voltage $V_D$ in (b). $V_D$ is measured at 1 or 10 μA to be 0.3 and 0.45 V, respectively.

material are injected by the forward bias of $V_P$ into the base material, where they proceed toward $B_1$ due to the voltage gradient across $R_1$. For each hole injected by the emitter, an electron is injected by $B_1$ into the base. The base region comprising $R_1$ becomes flooded with extra charge carriers so that the resistance of $R_1$ becomes smaller. Refer to Fig. 15-3 to see that point $A$ becomes less positive and acts to increase the forward bias across $D$. This action is regenerative. That is, once $R_1$ begins to decrease it initiates an action which causes $R_1$ to decrease further. In addition, emitter current $I_E$ increases while $V_E$ decreases because (1) we assume $V_{EE}$ is constant, (2) as $R_1$ goes down $I_E$ must go up, and (3) as $R_1$ goes down $R_E$ must necessarily absorb more of the division of $V_{EE}$ between $R_E$ and $R_1$. This action is shown in the sweep circuit and characteristic curve of Fig. 15-6. $R_1$ will not decrease

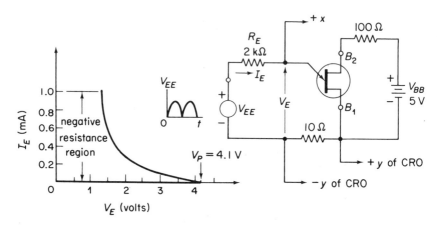

Fig. 15-6. Emitter characteristic curve shows that $I_E$ decreases once $V_E$ is increased to $V_P$. This is the characteristic of a negative resistance.

indefinitely because eventually the carrier concentrations increase to a point where recombinations counteract the effect of carriers being injected. Negative resistance characterizes the region where $I_E$ increases as $V_E$ decreases.

The point where $R_1$ reaches its minimum value locates the *valley point*. The valley point marks the lowest value of emitter-base-1 voltage and is

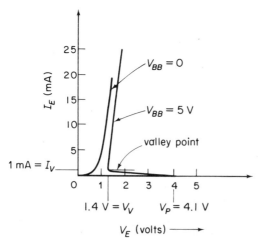

Fig. 15-7. Reducing the vertical scale in Fig. 15-6 shows the valley point for $V_{BB} = 5$ V. The curve at $V_{BB} = 0$ shows emitter diode characteristics.

designated as valley voltage $V_V$. The corresponding valley current is $I_V$. Any further increase in emitter current, beyond the valley point is accompanied by an increase in emitter voltage so that the UJT emitter characteristic exhibits a positive resistance characteristic and enters its *saturation region*. An emitter characteristic curve shows a valley point in Fig. 15-7 for $V_{BB} = 5$ V, together with another emitter characteristic curve at $V_{BB} = 0$.

## 15-2 Parameter Measurements of the UJT

From the measurement of $V_P = 4.1$ V in Fig. 15-7 and $V_D = 0.3$ V in Fig. 15-5 we can calculate the intrinsic standoff ratio from Eq. (15-4), where

$$V_P = V_D + \eta V_{BB},$$
$$4.1 = 0.3 + \eta(5), \text{ and}$$
$$\eta = 3.8/5 = 0.76.$$

From Eq. (15-3) and Fig. 15-2(b), calculate $R_1$. Since $R_{BB} = 7.1$ k$\Omega$ and $\eta = R_1/R_{BB}$,

$$R_1 = \eta R_{BB} = (0.76)(7.1) = 5.4 \text{ k}\Omega$$

Finally, calculate $R_2$ from

$$R_2 = R_{BB} - R_1 = 7.1 - 5.4 = 1.7 \text{ k}\Omega$$

Two other UJT parameters are of interest: *saturation resistance $R_s$* and *emitter saturation voltage $V_{E(\text{sat})}$*. Above the valley point, emitter current is limited (in the transistor) only by the *saturation resistance $R_s$*. $R_s$ is found from the slope of the emitter characteristic above the valley point. Emitter saturation voltage $V_{E(\text{sat})}$ gives the forward voltage drop between emitter and base 1 in the saturation region. By convention $V_{E(\text{sat})}$ is measured at $I_E = 50$ mA and $V_{BB} = 10$ V.

Measurement of both $R_s$ and $V_{E(\text{sat})}$ is illustrated in Fig. 15-8.

When comparing Fig. 15-8 with Fig. 15-7 it is seen that $V_P$ is different, because $V_{BB}$ is different. The effect of interbase voltage $V_{BB}$ on $V_P$ should be almost directly proportional; that is, if $V_{BB}$ is doubled ($R_{BB}$ will not change) then $V_P$ must be doubled to start UJT action. A simple method of demonstrating the direct increase of $V_P$ with $V_{BB}$ is to make multiple exposures of the emitter characteristic at different values of $V_{BB}$. However, the measurements of $V_P$ will fall on top of one another unless we employ a technique illustrated in Fig. 15-9. A quadruple exposure is employed in the following sequence. (1) Connect emitter and base 1 of the UJT to the collector and emitter

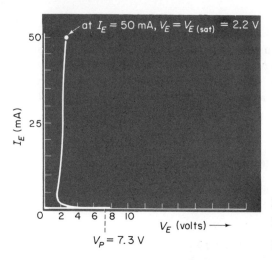

**Fig. 15-8.** When $V_{BB} = 10$ V and $I_E = 50$ mA, $V_E = V_{E(\text{sat})} = 2.2$ V. $R_s$ is the slope above the valley point of 0.8 V/50 mA $\geq$ 16 ohms.

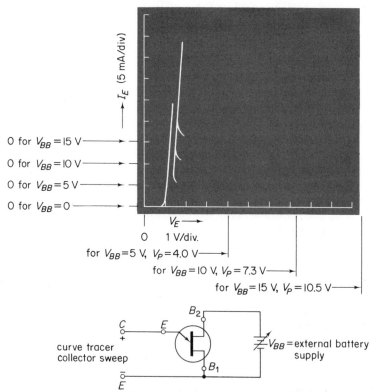

**Fig. 15-9.** Emitter characteristics for various $V_{BB}$ are shown in (a) as generated by the circuit in (b), (a) Variation of peak-point emitter voltage $V_P$ with interbase voltage $V_{BB}$. (b) Test circuit for (a).

terminal of a curve plotter or to a sweep circuit and CRO. Apply a positive sweep to the UJT emitter (*npn* position) and dial the plotter to read $I_E$ on the vertical, $V_E$ on the horizontal. (2) Set $V_{BB} = 0$ and photograph or sketch the emitter-diode characteristic. (3) Increase $V_{BB}$ to 5 V. Employ the vertical position control to raise the $I_E = 0$ reference by one scale division and

Table 15-1

| $V_{BB}$ (volts) | $V_P$ (volts) |
|---|---|
| 5 | 4.0 |
| 10 | 7.3 |
| 15 | 10.5 |

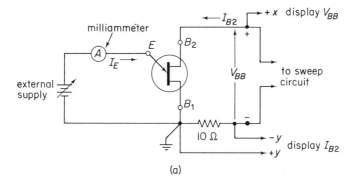

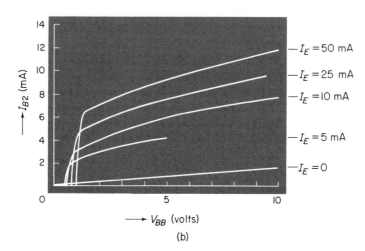

**Fig. 15-10.** The sweep circuit in (a) measures interbase characteristics in (b).

photograph the $V_{BB} = 5$ V curve in Fig. 15-9(a). (3) Repeat the procedure in (2) for $V_{BB} = 10$ V. (4) Turn on the background reticule illumination and repeat (2) for $V_{BB} = 15$ V. As shown in Fig. 15-9(a), $V_P$ clearly increases with increasing $V_{BB}$, as in Table 15-1.

*Interbase characteristics* are measured by (1) varying interbase voltage $V_{BB}$, (2) holding $I_E$ at a constant value, and (3) reading the resulting variation in interbase current $I_{B2}$ as shown in Fig. 15-10.

In Fig. 15-10(b) the lowest linear characteristic is a plot of $I_{B2}$ versus $V_{BB}$ for $I_E = 0$ and its slope's reciprocal represents $R_{BB}$. Calculating $R_{BB} = V_{BB}/I_{B2} = 10/1.4$ mA $= 7.1$ kΩ shows reasonable agreement with $R_{BB}$ calculated from Fig. 15-2(b). Focus attention on the top interbase characteristic curve of Fig. 15-10 corresponding to $I_E = 50$ mA. $V_{BB}$ must rise to a voltage of approximately 1 V, which is more positive than $V_D + I_E R_s$, before an appreciable amount of electrons are attracted to $B_2$. Above $V_{BB} = 1$ V the slope of $I_{B2}$ versus $V_{BB}$ is approximately equal to $1/R_2$, from

$$R_2 = \frac{\Delta V_{BB}}{\Delta I_{B2}} = \frac{10}{6 \text{ mA}} = 1.6 \text{ k}\Omega, \qquad V_{BB} > 1 \text{ V}$$

The UJT parameter $I_{B2(\text{mod})}$ is measured at $I_E = 50$ mA, $V_{BB} = 10$ V and is seen from Fig. 15-10 to be $I_{B2(\text{mod})} = 12$ mA. $I_{B2(\text{mod})}$ is specified on data sheets to describe a point in the saturation region.

### 15-3 Temperature Stabilization of $V_P$

Temperature dependence of peak-point voltage $V_P$ is illustrated in Fig. 15-11, where a reference emitter characteristic is shown at 25°C, with $V_P$ equal to 4.1 V. Raise the reference curve by 3 div and heat the transistor to see $V_P$ shift to the left. In Fig. 15-11 the UJT was heated to over 200°C to show how $V_P$ dropped to 2.5 V.

In any application of the UJT it is usually a fundamental requirement that $V_P$ be constant. From Eq. (15-4), temperature dependence of $V_P$ depends on temperature dependence of $V_D$, $\eta$, and $V_{BB}$. Take the derivative with respect to temperature $T$ of Eq. (15-4) and set it equal to zero so that the shift of $dV_P/dT$ will be zero. Assuming $\eta$ is not temperature dependent, we have

$$\frac{dV_P}{dT} = \frac{dV_D}{dT} + \eta \frac{dV_{BB}}{dT} = 0 \tag{15-5}$$

or

$$\frac{dV_D}{dT} = -\eta \frac{dV_{BB}}{dT} \tag{15-6}$$

Temperature dependence of the diode is approximately $-2.2$ mV/°C so that Eq. (15-6) becomes

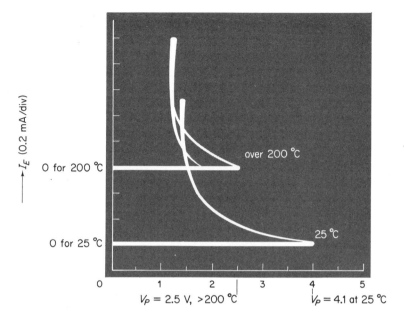

**Fig. 15-11.** Peak-point voltage $V_p$ decreases with an increase in temperature. $V_{BB} = 5$ V.

$$0.0022 = \eta \frac{dV_{BB}}{dT} \tag{15-7}$$

Interbase resistance $R_{BB}$ increases with increasing temperature because of decreasing mobility in the crystal lattice at the rate of approximately 0.8%/°C. It has been found that adding a resistor $R_T$ in series with base 2 will allow temperature compensation by balancing the negative and positive temperature coefficients of $V_D$ and $R_{BB}$, respectively. By inspection of Fig. 15-12 we can write $V_{BB}$ in terms of $R_{BB}$ from the voltage division of $V_B$

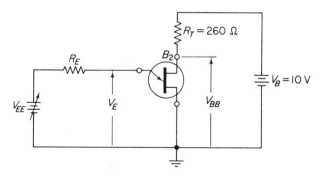

**Fig. 15-12.** Temperature-compensating resistor $R_T$ mitigates against change of $V_P$ with temperature change.

$$V_{BB} = \frac{R_{BB}}{R_T + R_{BB}} V_B \qquad (15\text{-}8)$$

Differentiating both sides of Eq. (15-8) with respect to temperature (assuming $R_T$ and $V_B$ are constants),

$$\frac{dV_{BB}}{dT} = \left[\frac{(R_T + R_{BB}) - (R_{BB})}{(R_T + R_{BB})^2}\right] \frac{dR_{BB}}{dT} V_B \qquad (15\text{-}9)$$

Stipulate the $R_T \ll R_{BB}$ and substitute $0.008 R_{BB}$ for the differential of $R_{BB}$ with respect to temperature

$$\frac{dV_{BB}}{dT} = 0.008 \frac{R_T}{R_{BB}} V_B \qquad (15\text{-}10)$$

Substituting for $dV_{BB}/dT$ from Eq. (15-7) into Eq. (15-10) yields

$$0.0022 = 0.008 \frac{R_T}{R_{BB}} \eta V_B$$

or

$$R_T = 0.28 \frac{R_{BB}}{\eta V_B} \quad \text{for} \quad \frac{dV_P}{dT} = 0 \qquad (15\text{-}11)$$

Equation (15-11) is only approximate and gives a starting point to select $R_T$. For critical applications $R_T$ should be selected empirically. Typically $R_T$ is in the order of a few hundred ohms.

## 15-4 The Unijunction as a Relaxation Oscillator

Introducing capacitor $C$ between emitter and ground in Fig. 15-13 applies an exponentially increasing voltage $V_E$ to the emitter as soon as switch $Sw$ is closed. As long as $V_E$ remains below $V_P$ the UJT is off, because its emitter diode is reverse biased. $V_E$ increases at a rate determined by $R_E$ and

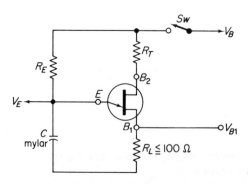

Fig. 15-13. Basic unijunction relaxation oscillator.

$C$ toward supply voltage $V_B$. When $V_E$ reaches a value equal to $V_P$, interbase resistance $R_1$ abruptly decreases. At this point, the voltage on $C$ is equal to $V_P$ and the capacitor sees the (1) low incremental diode resistance, (2) low saturation resistance $R_{sat}$, and (3) low resistance of $R_L$. $C$ will discharge quickly through this low-resistance path toward a voltage approximately equal to $V_V$. If certain restrictions are met by $R_E$ the UJT will not enter saturation, but will turn off, leaving $C$ charged to the valley voltage.

The sequence will repeat indefinitely with (1) the UJT off while $C$ charges from $V_V$ to $V_P$ and (2) the UJT turning on briefly while $C$ discharges through $R_L$ from $V_P$ to $V_V$. $R_E$ will be restricted to an upper limit imposed by the need to furnish $I_P$ at the peak point in order to trigger the UJT. This value of $R_{E(max)}$ is expressed by

$$R_{E(max)} = \frac{V_B - V_P}{I_P} \tag{15-12}$$

As long as $R_E < R_{E(max)}$, the UJT will trigger and $C$ will discharge through the emitter. A lower limit is imposed on $R_E$ to prevent the UJT from attaining a stable operating point in the saturation region. That is operation must be restricted to the negative resistance region when the UJT turns on. Emitter current must not drop below the valley current $I_V$, or

$$R_{E(min)} = \frac{V_B - V_V}{I_V} \tag{15-13}$$

In Fig. 15-14 load lines are drawn based on Eqs. (15-11) and (15-12). The emitter-current axis is grossly distorted to show the principles. In practice

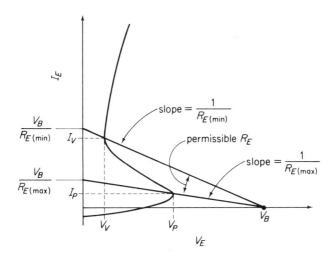

**Fig. 15-14.** Emitter characteristic load lines show the permissable range of $R_E$. The vertical axis is not to scale.

select a resistor at least $3R_{E(\min)}$ to ensure stable operation. Actual evaluation of the limits for $R_E$ are found by reference to data in specification sheets. The following examples lead to calculations of $R_{E(\min)}$ and $R_{E(\max)}$ which are typical, and show a wide latitude for choosing $R_E$. There is approximately a 1000 : 1 ratio between maximum and minimum values of $R_E$.

**Example 15-1.** From the transistor characteristics in Figs. 15-8 and 15-10, and with $V_{BB} = 10$ V, (a) list the values of $V_P$, $V_V$, and $I_V$. (b) Evaluate $R_{BB}$ graphically. (c) Calculate the value of intrinsic standoff ratio $\eta$ (assume $V_D = 0.4$ V).

**Solution.** (a) From Fig. 15-8, $V_P = 7.3$ V, $V_V = 1.8$ V, $I_V = 3$ mA. (b) From Fig. 15-10(b), calculate $R_{BB}$ from the slope of the straight line at $I_E = 0$ and from Eq. (15-1),

$$R_{BB} = \frac{V_{BB}}{I_{B2}} = \frac{10}{1.8 \text{ mA}} = 5.5 \text{ k}\Omega$$

(c) From Eq. (15-4),

$$V_P = V_D + \eta V_{BB}$$
$$7.3 = 0.4 + \eta 10$$
$$\eta = 0.69$$

**Example 15-2.** Employ the data from Example 15-1 to find a value for temperature-stabilizing resistor $R_T$. $V_B = 10$ V $\cong V_{BB}$.

**Solution.** From Eq. (15-11),

$$R_T = 0.28 \frac{R_{BB}}{\eta V_B} = \frac{0.28(5500)}{(0.69)10} = 220 \text{ }\Omega$$

**Example 15-3.** From the data in Example 15-1, calculate values of $R_{E(\max)}$ and $R_{E(\min)}$ for use in the circuit of Fig. 15-13. $V_B = 10$ V and $I_P = 0.5$ $\mu$A.

**Solution.** From Eq. (15-12),

$$R_{E(\max)} = \frac{V_B - V_P}{I_P} = \frac{10 - 7.3}{0.5 \times 10^{-6}} = 5.4 \text{ M}\Omega.$$

From Eq. (15-13),

$$R_{E(\min)} = \frac{V_B - V_V}{I_B} = \frac{10 - 1.8}{3 \times 10^{-3}} = 2.7 \text{ k}\Omega$$

The minimum actual value of $R_E$ should be two to three times $R_{E(\min)}$ to meet differences between transistors.

FREQUENCY OF OSCILLATION. Frequency of oscillation is dependent on (1) the time $T$ for capacitor $C$ to charge from $V_V$ to $V_P$; (2) turn-on time of the UJT, $t_{on}$; plus (3) *emitter-voltage fall time* or turn-off time $t_f$. Turn-on represents the time required for carrier injection to lower $R_1$ and cause an abrupt increase in both $I_E$ and $I_{B2}$. It is of the order of 1 μsec. Turn-off time occurs while $V_E$ discharges to approximately $V_V$, and the number of holes injected into the base decreases. $R_1$ increases to reverse bias the diode, again reducing hole injection and the UJT turns off due to this regenerative effect. The magnitude of $t_f$ depends primarily on $C$ and $R_L$ and is given in data sheets by a plot of $t_f$ versus $C$. Typical values for $t_f$ are 1 μsec at $C = 0.001$ μF and 180 μsec for $C = 10$ μF. Time $t_{on}$ is shorter than $t_f$ so $t_{on}$ can usually be neglected.

Charging time $T$ is found by reasoning from Fig. 15-13 that $C$ charges from a low voltage of $V_V$ exponentially toward $V_B$ or voltage across capacitor $C$ is $V_C$, where

$$V_C = V_V + (V_B - V_V)(1 - e^{-t/(R_E C)}) \tag{15-14}$$

The capacitor stops charging when $V_C$ rises to $V_P$ in the time $T$. At time $T$, $V_P$ is given by Eq. (15-4), or

$$V_P = V_C = V_D + \eta V_{BB} = V_V + (V_B - V_V)(1 - e^{-t/(R_E C)})$$

Solving for $T$ yields

$$T = R_E C \ln\left(\frac{V_B - V_V}{V_B - V_D - \eta V_{BB}}\right) \tag{15-15}$$

Assuming that $V_{BB} \cong V_B$ and $V_D \ll V_B$ simplifies Eq. (15-15) to

$$T = R_E C \ln \frac{1}{1-\eta} \cong 2.3 R_E C \log \frac{1}{1-\eta} \tag{15-16}$$

Note that if $\eta = 0.63$, Eq. (15-16) reduces to

$$T = R_E C \quad \text{for} \quad \eta = 0.63 \tag{15-17a}$$

The period of oscillation $T_o$ is the sum of $T$ and $t_f$ and is

$$T_o = T + t_f \tag{15-17b}$$

Where $t_f$ is small with respect to $T$ and $\eta = 0.63$ the frequency of oscillation is given in hertz by

$$f = \frac{1}{T_o} \cong \frac{1}{R_E C} \tag{15-18}$$

*Example 15-4.* Given $R_T = 100$ Ω, $R_E = 10$ kΩ, $t_f = 7$ μsec, $C = 0.1$ μF in the oscillator of Fig. 15-13. Employ the transistor data in Examples 15-1 to 15-3 to find the frequency of oscillation with $V_B = 20$ V.

*Solution.* Select Eq. (15-15) and given $V_B = 20$ V, $V_V = 1.8$ V, $V_D = 0.4$ V, find $V_{BB}$ from the voltage divider.

$$V_{BB} = \left(\frac{5500}{5500 + 100 + 50}\right)(20) = 19.5 \text{ V}$$

Substituting into Eq. (15-15),

$$T = (10^4)(0.1 \times 10^{-6}) \ln\left(\frac{20 - 1.8}{20 - 0.4 - (0.69)(19.5)}\right)$$

$$= 1.09 \text{ msec}$$

Discard $t_f$ because it is small with respect to $T$, so $T_o = T$ and from Eq. (15-18),

$$f = \frac{1}{T} = \frac{1000}{1.09} = 920 \text{ Hz}$$

An oscillator was built with components selected to illustrate Example 15-4. The circuit and wave shapes are shown in Fig. 15-15. $V_E$ is seen to be a ramp with a minimum at $V_V \cong 2$ V, and a maximum at approximately 14 V. Verify the maximum value as $V_P$, where

$$V_P = V_D + \eta V_{BB} = 0.4 + (0.69)(19.5) = 14 \text{ V}$$

Frequency of oscillation is found from the period of $V_E$ where for one cycle, $T_o = (0.2 \text{ msec/div})(5 \text{ div}) = 1$ msec and

$$f = \frac{1}{T_o} \cong 1 \text{ kHz.}$$

The time for $t_f$ begins at $V_{E(\max)}$ and ends at $V_{E(\min)}$. In Fig. 15-15(a), the time scale is too large for a measurement but shows that $t_f$ is unimportant in this particular circuit.

$V_{B1}$ is a pulse which begins when the UJT triggers and lasts for an interval equal to $t_f$. The magnitude of $V_{B1}$ depends on the discharge current from $C$ through the emitter and also upon the value of $I_{B2}$ during $t_f$. Both $I_{B2}$ and $I_E$ (discharge) flow through $R_L$ to develop output voltage $V_{B1}$. $I_{B2}$ is usually smaller than the $I_E$ discharge so for a given supply voltage $V_{B1}$ depends primarily on $C$ and $R_L$. Relationships for both $V_{B1}$ and $V_{B2}$ are involved and it is advisable to build the circuit and measure their values. Tables for values of $V_{B1}$ as a function of $R_L$, $C$, and $\eta$ are available in the literature but the student is advised to make his own experimentally for the sake of experience. For example, $V_{B1}$ in Fig. 15-15 is equal to 8 V.

*Example 15-5.* From Fig. 15-15, estimate (a) the peak increment of $I_{B2}$ which flows through $R_L$ during $t_f$, (b) the total peak current flowing through $R_L$ during $t_f$, and (c) peak current delivered by the capacitor.

## Sec. 15-4    The Unijunction as a Relaxation Oscillator    515

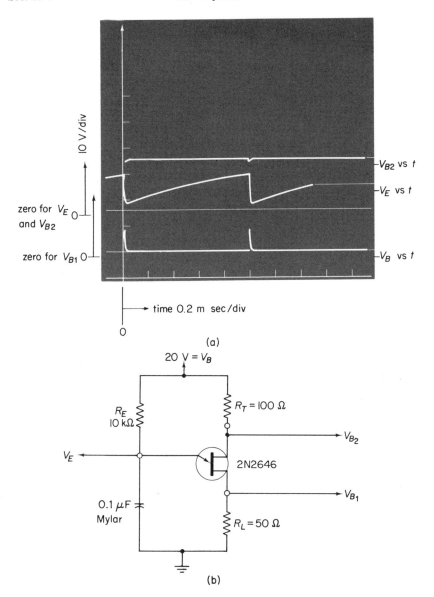

**Fig. 15-15.** UJT relaxation oscillator and voltage wave shapes for Example 15-4.

*Solution.* (a) From the plot of $V_{B2}$ versus $t$ measure a negative-going increment of approximately 1 V during $t_f$. This 1-V increment is the increase in $I_{B2}$ or $\Delta I_{B2}$ which flows through $R_T$, or

$$\Delta I_{B2} = \frac{\Delta V_{B2}}{R_T} = \frac{1}{100} = 10 \text{ mA}$$

The dc value of $I_{B2}$ (while the UJT is timing) is almost negligible at

$$I_{B2} = \frac{V_1}{R_{BB} + R_T + R_L} = \frac{20}{5650} = 3.5 \text{ mA}$$

(b) From the peak value of $V_{B1}$ which measures 8 V in Fig. 15-15(a), calculate a total current peak $I$ through $R_L$ of

$$I = \frac{V_{B1(peak)}}{R_L} = \frac{8}{50} = 160 \text{ mA}$$

(c) Neglect the small current furnished through $R_E$. Peak emitter current, $I_{E(peak)}$ is then

$$I_{E(peak)} = I - \Delta I_{B2} - I_{B2} = 160 - 13 \cong 150 \text{ mA}$$

## 15-5 Sawtooth Generator

The emitter-voltage wave shape in Fig. 15-15(a) is not suitable for a *ramp* or *sawtooth generator* because of its poor linearity. This circuit can be modified to improve linearity by charging the capacitor with a constant current. Replace $R_T$ with three diodes in series which will serve to forward bias *pnp* transistor $Q_1$ in Fig. 15-16(a). The collector current of $Q_1$ is the actual capacitor charging current but since $I_C \cong I_E$ we shall base our analysis on the fact that $I_E$ charges $C$, until the UJT turns on.

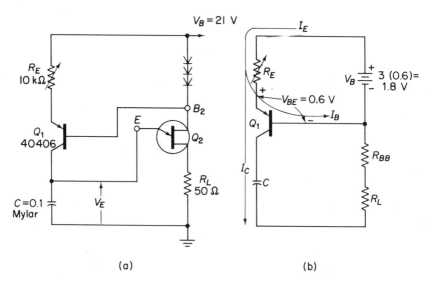

**Fig. 15-16.** Circuit in (a) and constant current generator in (b) for a high-linearity sawtooth generator.

## Sec. 15-5  Sawtooth Generator

It is simple to calculate $I_E$ from the model in Fig. 15-16(b). As long as any $I_{B2}$ flows in the UJT the diodes are on and supply a constant voltage of 1.8 V to $R_E$ and the emitter junction drop of $V_{BE} = 0.6$ V. Therefore $I_E$ is simply calculated from

$$I_E = \frac{(N-1)0.6}{R_E} \tag{15-19}$$

where $N$ = number of diodes between $B_2$ and $V_B$. From Eq. (15-19) it is evident that $I_E$ will be constant. Capacitor $C$ will be charged from $V_V$ to $V_P$ by $I_E$ in a time $T$ equal to

$$T = \frac{C(V_P - V_V)}{I_E} \tag{15-20}$$

Equation (15-20) is derived as follows. The charge on a capacitor equals the product of its voltage and capacitance:

$$Q = CV$$

But $Q$ results from a constant current $I_E$ over a period of time $T$. Substituting for $Q = I_E T$,

$$I_E T = CV \tag{15-21}$$

Since $R_E$ is variable, substitute Eq. (15-19) for $I_E$ into Eq. (15-20):

$$T = \frac{C(V_P - V_V)R_E}{(N-1)(0.6)} \tag{15-22}$$

Assuming $t_f$ is small with respect to $T$, we take the reciprocal of $T$ to find the frequency of oscillation in hertz:

$$f = \frac{1}{T} = \frac{(N-1)(0.6)}{CR_E(V_P - V_V)} \tag{15-23}$$

Since $C$ and $R_E$ are not changed between the circuits of Figs. 15-15 and 15-16 the peak value of $V_{B1}$ will not change. There will be no output pulse at $V_{B2}$ because the diode voltages will remain constant, even when the UJT turns on during $t_f$. Also, $V_V$ is the same since $V_B$ has been increased in Fig. 15-16 to compensate $V_{BB}$ for the diode drops. $V_E$ will increase almost linearly, even for low-frequency ramps.

*Example 15-6.* Calculate the frequency of oscillation for the circuit in Fig. 15-16. Transistor data is given in Example 15-4.

*Solution.* From Example 15-4 and Fig. 15-15(a), $V_V = 1.8$ V, $\eta = 0.69$, and $V_P$ is calculated from

$$V_{BB} = [V_B - 3(0.6)]\frac{R_{BB}}{R_{BB} + R_L} = (21 - 1.8)\frac{(5500)}{(5500 + 50)} = 19.2 \text{ V}$$

$V_P = V_D + \eta V_{BB} = 0.4 + (0.69)(19.2) = 13.7$ V. Substituting into Eq. (15-23) yields

$$f = \frac{(N-1)(0.6)}{CR_E(V_P - V_V)} = \frac{2(0.6)}{(0.1 \times 10^{-6})(10^4)(13.7 - 1.8)} = 100 \text{ Hz}$$

It is interesting to increase the value of $C$ in Fig. 15-16 to 10 $\mu$F in order to make a low-frequency ramp generator. If ordinary electrolytic capacitors are employed $Q_2$ will not fire when $R_E$ is increased much above 50 k$\Omega$. This observation leads to a simple method of measuring capacitor-leakage current. For example, if $R_E$ is 240 k$\Omega$, $I_E$ will be 5 $\mu$A. If the capacitor's leakage current is 5 $\mu$A the charge from $C$ will leak away as fast as it is furnished by $Q_1$. $R_E$ can be easily calibrated in terms of $I_E$. Hence increase $R_E$ until the circuit stops oscillating and read the leakage current from the calibration on $R_E$.

## 15-6 Timing and Voltage-Level Sensing with the UJT

The UJT performs well in applications involving the need for a timer. Only slight modifications to the basic oscillator circuit allow a control pulse to be generated after a precise time interval following an event. Refer to Fig. 15-17, where switch $Sw$ and relay $RY$ form a timer. With $Sw$ on "reset timer," no voltage is applied to the UJT and a discharge path for $C$ is provided through diode $D$, resistor $R_D$, and contacts 1 and 2 of $RY$. When switch $Sw$ is thrown to "start timing":

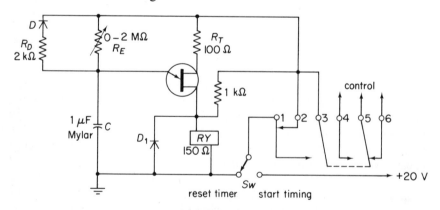

Fig. 15-17. UJT timer.

1. $I_E = 0$, since the capacitor is discharged and a small current $I_{B2}$ of

$$I_{B2} = \frac{V_B}{R_{BB} + R_T + RY}$$

flows through $RY$.

2. $C$ charges through $R_E$ for a time interval to $V_P$ and discharges a large current through the winding of $RY$ to operate the relay.

RELAY OPERATION. Contacts 1, 2, and 3 are make-before-break contacts. That is, the battery supply on contact 1 is applied to contact 3 (to lock the relay through resistor $R$ in the operated position) before contact 2 breaks from 1. Thus the relay needs only enough pulse energy to make contact 3 touch 1, then $R$ completes the relay operation. $RY$ is now locked through its own contacts and will remain locked until switch $SN$ is thrown to "reset timer," whereupon the battery is removed from the circuit to release the relay. Diode $D_1$ gives a path for discharging energy in the magnetic field of $RY$ to protect the UJT from transients. Contacts 4, 5, and 6 are transfer contacts which control any sequence to be initiated after the time interval.

For long time intervals we would increase $C$, $R_E$, or both. The magnitude of $C_E$ is limited primarily by the cost of large-capacity, low-leakage capacitors. If $R_E$ is made too large (typically greater than 5 MΩ) the current through it will not be sufficient to furnish peak-point current $I_P$. In addition, a large time constant means that the voltage across $C$ will approach $V_P$ very slowly. As a matter of fact the closer $V_C$ gets to $V_P$ the slower $V_C$ approaches $V_P$. For this reason it is distinctly advantageous to replace $R_E$ and $R_T$ with the diode-transistor, constant-current charging network of Fig. 15-16. This will result in a more linear charging voltage of $V_C$ and improve the time interval's repeatabily.

Figure 15-17 can also illustrate the application of the UJT as a voltage sensor. For example, let capacitor $C$ be replaced with an automobile battery. Substitute a 50-Ω adjustable resistor for $R_E$ to adjust battery charging current. Since $V_P$ is approximately 14 V, the battery will charge to 14 V and turn on the UJT to stop charging. Of course a 10-Ω resistor should be added in series with the emitter to limit emitter current when the UJT fires.

The relay used for control purposes by the UJT has disadvantages inherent in any mechanical device when compared with an electronic device, such as (1) slower speed of operation, (2) subject to mechanical wear, and (3) limited resistance to shock and vibration. We therefore turn our attention to another type of electronic device, which can replace the relay and act as a companion device to the UJT.

## 15-7 Introduction to the Silcon-Controlled Rectifier (SCR)

The *silicon-controlled rectifier* is a semiconductor device that operates in a manner analogous to a thyratron. Output terminals of both devices are the anode and cathode which exercise control through two stable states: (1) the *forward conduction or "on" state* characterized by high current through, and low voltage drop across, cathode and anode; and (2) the *forward blocking or "off" state* characterized by a low leakage current through, and high voltage drop across, cathode and anode. Thus cathode and anode terminals function essentially as two terminals of a switch that are either short circuited together or form an open circuit. A third terminal, the grid, exercises partial control over either device in that the grid can drive the SCR from its *off* state to its *on* state but then immediately loses its ability to exercise any further control over the SCR. It will help to think of anode and cathode as an output circuit partially controlled by an input circuit consisting of grid and cathode.

A physical explanation of operation is undertaken from the construction details in Fig. 15-18(a), where principle features of the SCR are seen to be (1) four alternate layers of semiconductor material between cathode and anode, *npnp*, labeled *cathode, control, blocking,* and *anode layers,* (2) three

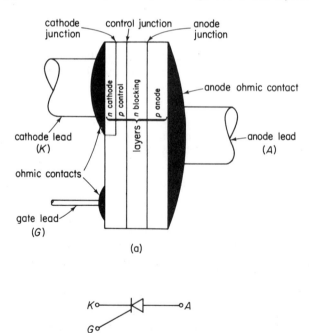

**Fig. 15-18.** Structure and symbol of a silicon-controlled rectifier. (a) Basic structure of an SCR. (b) Circuit symbol for the SCR.

*pn* junctions between cathode and anode, labeled *cathode, control,* and *anode junctions,* and (3) one junction between adjacent cathode and gate leads.

Within limits, the SCR will conduct only leakage currents for either polarity of voltage between cathode and anode. With anode negative, cathode positive, and gate lead open the SCR is in a *reverse-blocking* condition defined when both anode and cathode junctions are reverse biased. If the reverse anode-cathode voltage is increased beyond a specified maximum *reverse-breakdown* voltage rating, avalanching occurs and the *I-V* characteristic curve is similar to that of a diode in reverse breakdown.

With anode positive, cathode negative, and no gate voltage, only the control junction is reverse biased in Fig. 15-19. To examine the reverse leakage current, we employ the two-transistor analog of an SCR in Fig. 15-19(b), and write the current-node equations

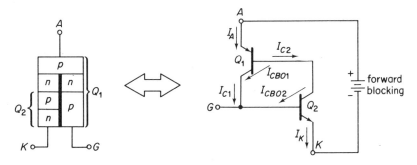

Fig. 15-19. Two-transistor analog of an SCR.

$$I_A = I_{C1} + I_{C2} \tag{15-24a}$$

$$I_{C1} = \beta_{F1} I_{C2} + (\beta_{F1} + 1) I_{CBO1} \tag{15-24b}$$

$$I_{C2} = \beta_{F2} I_{C1} + (\beta_{F2} + 1) I_{CBO2} \tag{15-24c}$$

Solving these three equations for $I_A$ in terms of leakage currents and $\beta_F$ yields

$$I_A = \frac{(\beta_{F1} + 1)(\beta_{F2} + 1)(I_{CBO1} + I_{CBO2})}{1 - \beta_{F1} \beta_{F2}} \tag{15-25}$$

At low values of emitter current, $\beta_{F1}$ and $\beta_{F2}$ approach zero and $I_A \cong I_{CBO1} + I_{CBO2}$. $I_A$ will remain at this low value unless some action is taken external to the SCR to increase emitter current. Once emitter current is increased, $\beta_{F1}$ and $\beta_{F2}$ begin to increase, causing the denominator of Eq. (15-25) to approach zero and $I_A$ to become large, limited only by the external circuit. This describes the *on* state.

There are four ways to increase emitter current and initiate a turn-on sequence:

1. *Voltage change:* Sudden application of a forward blocking voltage can cause a charging current through the junction capacitances sufficient to initiate turn-on.
2. *Carrier generation:* Free holes and electrons can be created by application of heat or radiant (light) energy and initiate turn-on. The *light activated SCR,* or *LASCR,* is turned on by light received through a translucent window in the SCR.
3. *High voltage:* When the forward blocking sweep voltage in Fig. 15-20(a) exceeds the forward breakover voltage, leakage current

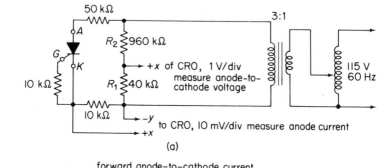

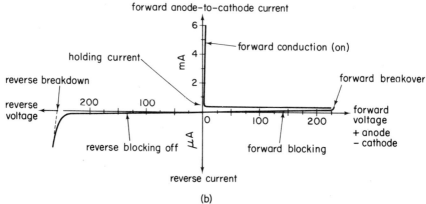

**Fig. 15-20.** Sweep circuit in (a) and characteristic curves in (b) for the SCR.

through the control junction increases $\beta_{F1}$ and $\beta_{F2}$ and the SCR goes abruptly into *forward conduction.* Current is then limited only by the 50-k$\Omega$ resistor in the characteristic of Fig. 15-20(b). When the reverse blocking sweep voltage exceeds the *reverse breakdown* voltage, both cathode and anode junctions are reverse biased and the SCR exhibits a diode breakdown characteristic.

4. *Transistor action:* Injecting carriers into the gate terminal is the most common and reliable method of initiating turn-on. There is a minimum gate voltage and minimum gate-current requirement that can be measured simply by the circuit of Fig. 15-21. Adjust $R_G$

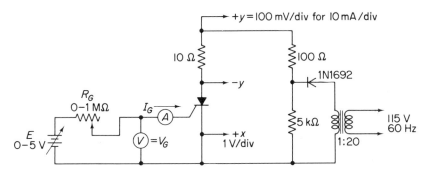

**Fig. 15-21.** Measurement of minimum gate current and voltage.

to zero and increase $E$ until the forward conduction characteristic appears on the CRO. The value of $E$ just before turn-on is the minimum gate voltage requirement. Increase $R_G$ to 1 MΩ and increase $E$ from 0 until the SCR turns on. Measure minimum gate current just before turn-on.

Once the SCR switches on into forward conduction, no further control is exercised over anode current by the gate. As long as anode current exceeds a minimum value specified by a *holding current* rating $I_{HX}$, the SCR will conduct. Typically $I_{HX}$ is 1 mA and in Fig. 15-21 the diode extinguishes the SCR once each cycle during the reverse blocking sweep, by reducing anode current to zero.

## 15-8 Pulsed Gate Operation

The SCR is commonly triggered by applying a pulse to its gate from a UJT. For example, $V_{B1}$ in Fig. 15-15 furnishes a pulse with peak amplitude of 8 V. The gate pulse must satisfy restrictions imposed by peak allowable gate current, voltage, and power ratings exemplified in Fig. 15-22. A typical minimum firing-point combination of $V_{G(\min)} = 0.8$ V and $I_{G(\min)} = 20$ μA is plotted to show the wide latitude in allowable gate-pulse amplitude. As long as the gate-pulse voltage peak is between 0.8 and 6.0 V and gate current is between 20 μA and 1.6 A the SCR will be turned on.

The forward gate-cathode diode characteristic may be measured by a conventional sweep circuit to obtain the typical display in Fig. 15-22. We can

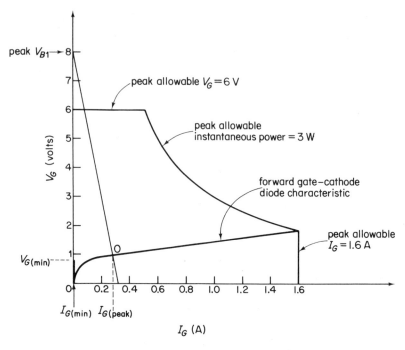

**Fig. 15-22.** Maximum gate ratings and gate circuit load lines.

describe pulsed-gate operation along this characteristic by focusing attention on the pulse measurement across $R_L$ in Fig. 15-15. Since its peak value is 8 V we could halve $R_L$ from 50 to 25 Ω and still ensure that the peak value of $V_G$ would exceed $V_{G(\min)}$. A Thévenin equivalent of the UJT output during this pulse is shown in Fig. 15-23(a). We take the open circuit voltage $V_{B1}$ (meas-

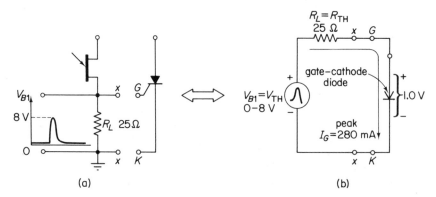

**Fig. 15-23.** Thévenin UJT output circuit in (a) drives the gate in (b) to solve graphically for gate current in Fig. 15-22.

ured before the SCR gate is connected) as the open-circuit Thévenin voltage in Fig. 15-23(b).

A load line for $R_L = 25\ \Omega$ is plotted on the gate characteristic of Fig. 15-22 at the peak $V_{B1} = 8$ V and peak gate current = 280 mA is measured at peak operating point 0. The load line actually begins as a point at the origin when $V_{B1} = 0$, sweeping up the characteristic to a peak excursion at point 0 when $V_{B1} = 8$ V. The load line then returns over the same path as the pulse decays to zero.

One final precaution must be observed with the UJT trigger circuit. $R_L$ must small enough so that its voltage drop is less than the minimum gate-firing voltage. Assuming a margin of safely, assign $V_{G(\min)} = 0.25$ V; then in the circuit of Fig. 15-13 (with the UJT off) $R_L$ must be chosen so that

$$\frac{R_L}{R_L + R_{BB} + R_T} V_B \leq V_{G(\min)} = 0.25 \text{ V} \qquad (15\text{-}26)$$

*Example 15-7.* With $V_B = 20$ V, $R_T = 100\ \Omega$, $R_{BB} = 5.5$ k$\Omega$, and $R_L = 25\ \Omega$, what gate voltage is applied to an SCR?

*Solution.* From Eq. (15-26),

$$V_G = \frac{25(20)}{5625} = 0.09 \text{ V}$$

## 15-9 Phase Control

Average or effective voltage across a load can be varied by controlling the firing angle of an SCR. The 60-Hz line voltage in Fig. 15-24(a) is rectified and equals the load voltage across $R_{\text{load}}$ when the SCR is on. If the firing angle of the SCR is delayed by angle $\alpha$, the average load voltage, $E_{\text{dc}}$, will depend on the averaged crosshatched area in Fig. 15-24(b) and is dependent on $\alpha$ by

$$E_{\text{dc}} = \frac{1}{\pi} \int_\pi^\alpha E_p \sin\theta = \frac{E_p}{\pi}(1 + \cos\alpha) \qquad (15\text{-}27\text{a})$$

For example, when $\alpha = 0°$, $E_{\text{dc}} = 2E_p/\pi$ for a maximum value equal to the average of a fully rectified wave. The root-mean-square (rms) load voltage can be expressed in terms of $\alpha$ by

$$E_{\text{rms}} = E_p \sqrt{\frac{2(\pi - \alpha) + \sin 2\alpha}{4\pi}} \qquad (15\text{-}27\text{b})$$

The ratios of $E_{\text{rms}}$ to $E_p$ and $E_{\text{dc}}$ to $E_p$ are plotted in Fig. 15-25 as a function of firing angle where it is clear that a lamp load would dim as $\alpha$ exceeds 40°.

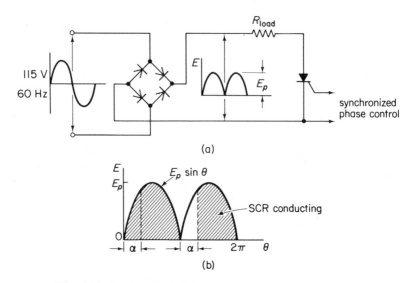

**Fig. 15-24.** Synchronized phase control in (a) controls firing angle α in (b).

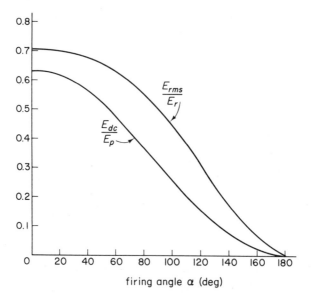

**Fig. 15-25.** Dependence of average or rms load voltage on firing angle α.

Synchronization of the trigger pulse is accomplished by driving the SCR and a UJT from the same ac power line. In Fig. 15-26(a) we assume $R_E$ is large with respect to $R_T + R_{BB} + R_L$ and can express UJT supply voltage $V_{bb}$ in terms of line voltage $E_p$

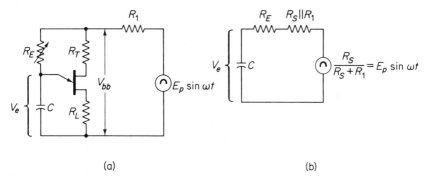

**Fig. 15-26.** AC supply for a UJT in (a) and Thévenin model in (b) to find $V_e$.

$$V_{bb} = \frac{R_S}{R_1 + R_S} E_p \sin \omega t \qquad (15\text{-}28)$$

where $R_S = R_T + R_{BB} + R_L$. Neglecting $V_D$ in Eq. (15-4), peak-point emitter voltage is

$$V_p = \eta V_{bb} \qquad (15\text{-}29)$$

Voltage across the capacitor $V_e$ is given in terms of $E_p$ by

$$V_e = \frac{1}{1 + j\omega \tau} V_{bb} \qquad (15\text{-}30)$$

where $\tau = RC$ and $R = R_E + R_1 \parallel R_2$, and
$\omega$ = line voltage frequency in radians.

Assuming $C$ is fixed we can vary $R_E$ and consequently $\tau$ in Eq. (15-30). The UJT will fire when $V_e$ equals or just exceeds $V_p$. How $V_e$ is varied in amplitude and phase in order to vary $\alpha$ will be investigated in Examples 15-8 and 15-9.

**Example 15-8.** Assume $R_E$ is varied in Fig. 15-26 so that $\tau = 1/\omega$, $\tau = 0.1/\omega$, and $\tau = 10/\omega$, and $\omega = 2\pi(60)$ rad/sec. What is the instantaneous equation for $V_e$ in terms of $V_{bb}$ for each time constant?

**Solution.** Equation (15-30) can be rewritten in terms of a magnitude and phase angle as

$$V_e = \frac{1}{\sqrt{1 + (\omega \tau)^2}} V_{bb} \underline{/-\tan^{-1} \omega \tau} \qquad (15\text{-}31)$$

Substituting for $\tau = 1/\omega$, gives $\omega \tau = 1$ and, from Eq. (15-31),

$$V_e = 0.707 V_{bb} \underline{/-45°} \quad \text{at } \tau = \frac{1}{\omega} = 2.65 \text{ msec}$$

Repeating the substitutions for $\tau = 0.1/\omega$ and $\tau = 10/\omega$,

$$V_e \cong V_{bb} \underline{/-6°} \quad \text{at } \tau = \frac{0.1}{\omega} = 0.265 \text{ msec}$$

$$V_e = 0.1 V_{bb} \underline{/-84°} \quad \text{at } \tau = \frac{10}{\omega} = 26.5 \text{ msec}$$

***Example 15-9.*** In Example 15-8, Assume $\eta = 0.6 V_{bb}$ and $V_{bb}$ has a 10-V peak. ($V_{bb}$ is a fully rectified wave with a period of $1/60 = 16.7$ msec). Plot the voltage wave of $V_{bb}$ for $\frac{1}{2}$ cycle. (b) Plot $V_e$ for each time constant on the same graph and measure the resulting values of $\alpha$.

***Solution.*** At the intersection of each dashed $V_e$ curve with the solid $\eta V_{bb}$ curve in Fig. 15-27 we measure $\alpha$ to be 6°, 102°, and 172° at $\tau = 0.1\omega$, $1/\omega$, and $10/\omega$, respectively.

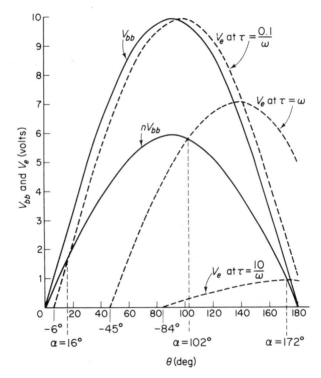

**Fig. 15-27.** Solution to Example 15-9.

From Examples 15-8 and 15-9 we see that $\alpha$ not only depends on $T$ but also on $\eta$. To fire the UJT early in the cycle we need an $RC$ time constant approximately $\frac{1}{10}$ the power-line frequency's period.

Sec. 15-10           SCR Applications    529

## 15-10 SCR Applications

Variable ac control is featured by the lamp dimmer or ac motor control (universal and shaded pole) in the circuit of Fig. 15-28. $R_1$ is chosen to protect the UJT and the 2-k$\Omega$ resistor in series with potentiometer $R_E$ establishes $\tau \cong 0.265$ msec when $R_E = 0$, for a minimum firing angle (maximum SCR conduction) of $\alpha \cong 10°$.

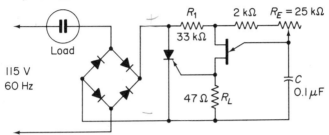

**Fig. 15-28.** Variable ac control.

The SCR can replace the relay in the timer of Fig. 15-17 and eliminate maintenance, or can be arranged to trigger from the sawtooth generator in Fig. 15-16.

A dc control is illustrated in Fig. 15-29, where a Zener diode is added to protect the UJT in case the SCR does not fire. $R_E$ allows variation in $\alpha$ and load voltage. UJT supply voltage is equal to $V_Z = V_B$ and, assuming negligible charge accumulates on $C$ while $E_p$ is rising toward $V_Z$, timing is found from Eqs. (15-15) or (15-17a).

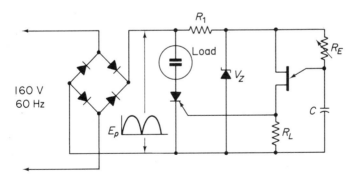

**Fig. 15-29.** Variable dc control.

## PROBLEMS

**15-1** How will the display look in Fig. 15-2 if connections to the $y$ amplifier of the CRO are reversed?

**15-2** Sketch the resultant CRO display for Fig. 15-5(a) if the polarity of sweep voltage $V_{EE}$ is reversed.

**15-3** Identify on a sketch the negative and positive resistance regions of the $I_E - V_E$ characteristic curve at $V_{BB} = 5$ V in Fig. 15-7.

**15-4** What are the valley voltage and current measurements in Fig. 15-8?

**15-5** Estimate saturation resistance from the characteristics in Fig. 15-9.

**15-6** Calculate the intrinsic standoff ratio for each characteristic curve in Fig. 15-9(a). Assume $V_D = 0.3$ V.

**15-7** Calculate $R_{BB}$ and $R_2$ from the proper characteristic curves in Fig. 15-10.

**15-8** In Fig. 15-13, $V_B = 20$ V, $\eta = 0.75$, $V_D = 0.3$ V, $I_P = 1$ μA, $I_V = 3$ mA, $V_V = 2$ V, and $R_{BB} = 8.0$ kΩ. Find (a) $R_T$ for minimum temperature dependence of $V_p$, (b) maximum and minimum values of $R_E$, (c) the oscillation period from Eq. (15-16), assuming $t_f$ is negligible, at maximum and minimum values of $R_E$ with $C = 0.1$ μF, (d) the frequency of oscillation at maximum and minimum $R_E$.

**15-9** Repeat Problem 15-8, changing $\eta$ to 0.63 and using Eq. (15-17a) to compare the differences, particularly the oscillation frequencies at $R_{E(\max)}$ and $R_{E(\min)}$.

**15-10** Obtain measurements from Fig. 15-15 to calculate $V_p$ and $\eta$ by assumming $V_D$ is negligible.

**15-11** Redesign the circuit of Fig. 15-15 for oscillation at (a) 10 kHz and (b) 10 Hz.

**15-12** $R_E$ is increased to 1 MΩ in Fig. 15-16 and $C$ is increased to 10 μF. Find the frequency of oscillation.

**15-13** One of the diodes is removed from the circuit in Problem 5-12. What is the new frequency of oscillation, assuming $V_p$ does not change?

**15-14** A 100-μF electrolytic capacitor replaces $C$ in Fig. 15-16. $R_E$ is increased to 300 kΩ and oscillation stops or becomes erratic. What is the leakage current of the capacitor?

**15-15** Assuming $V_{B1(\text{peak})} = 6$ V and $R_L = 10$ Ω, what peak gate current will flow as evaluated from the characteristics of Fig. 15-22?

**15-16** An SCR is to be driven by the pulse across $R_L$ in Fig. 15-16. What gate voltage is applied to the SCR, when the UJT is off, if $R_{BB} = 8$ kΩ?

*15-17* Verify, by calculation, the curves in Fig. 15-25 by assuming $E_p = 1$ V in Eqs. (15-27a) and (15-27b).

*15-18* $R_{BB} = 5$ kΩ in Fig. 15-28. Find the value of $R_E$ required to set $T = 1/\omega$. Refer to Example 15-9 to see that firing angle α will be 102°.

# Bibliography

Applications Engineering Dept., *Semiconductor Power Circuits Handbook*, Motorola Semiconductor Products Inc., Phoenix, Arizona, 1968.

Applications Engineering Dept., *Unijunction Transistor Times and Oscillators*, AN-294, Motorola Semiconductor Products Inc., Phoenix, Arizona, 1968.

Bergerson, Thor B., *Field Effect Transistors in Chopper and Analog Switching Circuits*, AN-220, Motorola Semiconductor Products Inc., Phoenix, Arizona, 1967.

Coughlin, Robert F., et al., *Laboratory Manual in Transistors and Semiconductor Devices*, Prentice-Hall Inc., Englewood Cliffs, N.J., 1970.

Fitchen, Franklin C., *Transistor Circuit Analysis and Design*, 2nd Edition, D. Van Nostrand Inc., Princeton, New Jersey, 1967.

Gray, Paul E., *Introduction to Electronics*, John Wiley & Sons, Inc., New York, 1967.

Gray, Paul E. and Searle, Campbell L., *Electronic Principles, Physics, Models, and Circuits*, John Wiley & Sons, Inc., New York, 1969.

*RCA Linear Integrated Circuit Fundamentals*, Radio Corporation of America, Harrison, N.J., 1966.

*RCA Transistor Manual*, Radio Corporation of America, Harrison, N.J., 1967.

Robertson, John J., *Tuned Amplifier Design with an Emitter-Coupled Integrated RF Amplifier*, AN-203, Motorola Semiconductor Products Inc., Phoenix, Arizona, 1967.

*SCR Manual*, 4th Edition, General Electric Co., Auburn, N.Y., 1967.

Searle, Campbell L., et. al., *Elementary Circuit Properties of Transistors*, John Wiley & Sons, Inc., New York, 1964.

*The Controlled Rectifier, Volume I*, International Rectifier Corporation, El Segundo, California, 1966.

*Transistor Manual*, General Electric Company, Syracuse, New York, 1964.

Wedlock, Bruce D. and Roberge, James K., *Electronic Components and Measurements*, Prentice-Hall Inc., Englewood Cliffs, N.J., 1969.

Welling, Brent, *An Integrated Circuit RF-IF Amplifier*, Motorola Semiconductor Products Inc., Phoenix, Arizona, 1968.

Zinder, David A., *Unijunction Trigger Circuits for Gated Thyristors*, AN-413, Motorola Semiconductor Products Inc., Phoenix, Arizona, 1968.

# Appendix 1

Courtesy RCA

## SILICON MOS TRANSISTORS
## N-Channel Depletion Types

For Amplifier Mixer & Oscillator
Applications in Military & Industrial
VHF Communications Equipment
Operating up to 250 MHz

TO-72

RCA-3N128 and 3N143* are N-channel depletion-type silicon field-effect transistors utilizing the MOS construction. The 3N128 is intended primarily for VHF amplifier service in military and industrial applications. It also is extremely well suited for use in dc and low-frequency amplifier applications requiring a transistor having high power gain, very high input impedance, and low gate leakage.

**Maximum Ratings,** *Absolute-Maximum Values:*

| | | |
|---|---|---|
| DRAIN-TO-SOURCE VOLTAGE, $V_{DS}$ | +20 max. | V |
| GATE-TO-SOURCE VOLTAGE, $V_{GS}$: | | |
|   Continuous dc | +1, -8 max. | V |
|   Peak ac | ±15 max. | V |
| DRAIN CURRENT, $I_D$ (PULSED) | 50 | mA |
|   Peak duration ≤ 20 ms, duty factor ≤ 0.15 | | |
| TRANSISTOR DISSIPATION, $P_T$: | | |
|   At Ambient {up to 25°C | 400 | mW |
|   Temperatures {above 25°C ... Derate at 2.67 mW/°C | | |
| AMBIENT TEMPERATURE RANGE: | | |
|   Storage and Operating | -65 to +175 | °C |
| LEAD TEMPERATURE (During Soldering): | | |
|   At distances not closer than 1/32 inch to seating surface for 10 seconds maximum | 265 max. | °C |

### APPLICATIONS

- VHF amplifiers, mixers, converters and if-amplifiers in communication receivers.
- High impedance timing circuits
- Detectors, oscillators, frequency multipliers, phase splitters, pulse stretchers and current limiters
- Electrometer amplifiers
- Voltage-controlled attenuators
- High impedance differential amplifiers

### DEVICE FEATURES

- Low noise figure — 3.5 dB typ. at 200 MHz
- High VHF amplifier gain — 18 dB typ. at 200 MHz
- Low input capacitance — 5.5 pF typ.
- High transconductance — 7500 $\mu$mho typ.
- High input resistance — $10^{14}$ $\Omega$ typ.

**ELECTRICAL CHARACTERISTICS:** (At $T_A = 25°C$)

Measured with Substrate Connected to Source Unless Otherwise Specified.

| CHARACTERISTIC | SYMBOL | CONDITIONS | 3N128 MIN | 3N128 TYP | 3N128 MAX | 3N143 MIN | 3N143 TYP | 3N143 MAX | UNITS |
|---|---|---|---|---|---|---|---|---|---|
| Forward Transconductance | $g_{fs}$ | $V_{DS} = 15$ V, $V_{GS} = 0$, f = 1 kHz | - | 10,000 | - | - | - | - | μmho |
| | | $V_{DS} = 15$ V, $I_D = 5$ mA, f = 1 kHz | 5,000 | 7,500 | 12,000 | 5,000 | 7,500 | 12,000 | μmho |
| Magnitude of Forward Transadmittance | $\|y_{fs}\|$ | $V_{DS} = 15$ V, $I_D = 5$ mA, f = 200 MHz | 5,000 | 7,500 | - | - | - | - | μmho |
| Gate Leakage Current | $I_{GSS}$ | $V_{DS} = 0$, $V_{GS} = -8$ V $T_A = 25°C$ | - | 0.1 | 50 | - | 0.1 | 1000 | pA |
| | | $V_{DS} = 0$, $V_{GS} = -8$ V $T_A = 125°C$ | - | - | 5 | - | - | 100 | nA |
| Small-Signal Short-Circuit Input Capacitance | $C_{iss}$ | $V_{DS} = 15$ V, $I_D = 5$ mA, f = 0.1 to 1 MHz | - | 5.5 | 7 | - | 5.5 | 7 | pF |
| Small-Signal Short-Circuit Reverse Transfer Capacitance* | $C_{rss}$ | $V_{DS} = 15$ V, $I_D = 5$ mA, f = 0.1 to 1 MHz | - | 0.12 | 0.20 | - | 0.12 | 0.20 | pF |
| Small-Signal, Short-Circuit Output Capacitance | $C_{oss}$ | $V_{DS} = 15$ V, $I_D = 5$ mA, f = 0.1 to 1 MHz | - | 1.4 | - | - | 1.4 | - | pF |
| Gate Leakage Resistance | $R_{GS}$ | $V_{DS} = 0$, $V_{GS} = -8$ V | - | $10^{14}$ | - | - | $10^{14}$ | - | Ω |
| Drain-to-Source Channel Resistance | $r_{DS}(on)$ | $V_{DS} = 0$, $V_{GS} = 0$, f = 1 kHz | - | 200 | - | - | 200 | - | Ω |
| Gate-to-Source Cutoff Voltage | $V_{GS}(off)$ | $V_{DS} = 15$ V, $I_D = 50$ μA | -2 | -3.5 | -8 | -2 | -3.5 | -8 | V |
| Drain-to-Source Cutoff Current | $I_D(off)$ | $V_{DS} = 20$ V $V_{GS} = -8$ V | - | - | 50 | - | - | 50 | μA |
| Zero-Bias Drain Current** | $I_{DSS}$ | $V_{DS} = 15$ V, $V_{GS} = 0$ | 5 | 15 | 25 | 10 | 20 | 50 | mA |
| Input Conductance | $g_{is}$ | $V_{DS} = 15$ V, $I_D = 5$ mA, f = 1 kHz | - | - | - | - | - | 10 | μmho |
| Output Conductance | $g_{os}$ | $V_{DS} = 15$ V, $I_D = 5$ mA, f = 1 kHz | - | - | - | - | - | 1,000 | μmho |
| Power Gain Maximum Available Gain Maximum Usable Gain (Neutralized) see Fig.1 | $G_{PS}$ | $V_{DS} = 15$V, $I_D = 5$ mA, f = 200 MHz | 15 13.5 | 20 16 | - - | - - | - - | - - | dB dB |
| Power Gain (Conversion (See Fig.3) | $G_{PS}(c)$ | $V_{DS} = 15$ V, $I_D = 1$ mA, $f_{in} = 200$ MHz, $f_{out} = 30$ MHz | - | - | - | 10 | 13.5 | - | dB |
| Noise Figure (see Figs. 1 & 2) | NF | $V_{DS} = 15$ V, $I_D = 5$ mA, f = 200 MHz | - | 3.5 | 5 | - | - | - | |

\* Three-Terminal Measurement: Source Returned to Guard Terminal.  
\*\* Pulse Test: Pulse Duration 20 ms max. Duty Factor $\leq 0.15$.

## TYPICAL CHARACTERISTICS

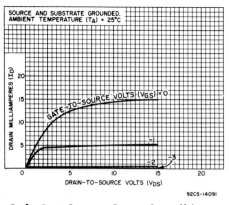

**Fig.1 - Drain Current vs Drain-to-Source Voltage.**

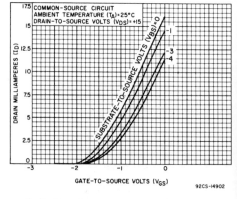

**Fig.2 - Drain Current vs Gate-to-Source Voltage.**

Appendix 537

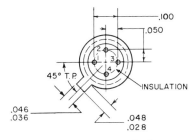

Dimensions in Inches

## DIMENSIONAL OUTLINE
## JEDEC TO-72

### TERMINAL DIAGRAM

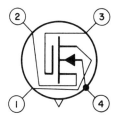

1 - Drain
2 - Source
3 - Insulated Gate
4 - Bulk (Substrate) and Case

### OPERATING CONSIDERATIONS

The flexible leads of the 3N128 and 3N143 are usually soldered to the circuit elements. As in the case of any high-frequency semiconductor device, the tips of soldering irons should be grounded, and appropriate precautions should be taken to protect the devices against high electric fields.

These devices should not be connected into or disconnected from circuits with the power on because high transient voltages may cause permanent damage to the devices.

# Appendix 2

## Silicon N-P-N & P-N-P Power Transistors for Audio Amplifier Applications

Courtesy RCA

RCA-40406 through 40411 are diffused-junction, silicon n-p-n and p-n-p transistors intended for a variety of uses in audio amplifiers. Giving high-quality performance economically, these 6 devices have power dissipation ratings of 1 to 150 watts. Supply voltages for these types range from 50 volts for the 40406 and 40407, to 90 volts for the 40408–40410.

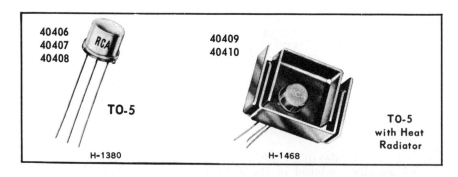

40406 40407 40408 — TO-5 — H-1380

40409 40410 — TO-5 with Heat Radiator — H-1468

**40406 & 40407**
- $V_{CEO}(sus) = -50$ V max. (40406)
- $V_{CEO}(sus) = 50$ V max. (40407)
- Type 40406 is P-N-P complement of type 40407
- 1 watt dissipation rating
- TO-5 package

**40408**
- $V_{CEO}(sus) = 90$ V max.
- 1 watt dissipation rating
- N-P-N type in JEDEC TO-5 package

**40409 & 40410**
- $V_{CER}(sus) = 90$ V max. (40409)
- $V_{CER}(sus) = -90$ V max. (40410)
- Type 40410 is P-N-P complement of type 40409
- 3 watt free-air dissipation rating
- TO-5 package with heat radiator

Appendix 539

| MAXIMUM RATINGS<br>Absolute-Maximum Values: | 40406 | 40407 | 40408 | 40409 | 40410 | UNITS |
|---|---|---|---|---|---|---|
| DC Collector-to-Emitter Sustaining Voltage:<br>With Base Open, $V_{CEO}(sus)$ ............<br>With $R_{BE} = 100\ \Omega$, $V_{CER}(sus)$ ........... | −50<br>— | 50<br>— | 90<br>— | —<br>90 | —<br>−90 | V<br>V |
| DC Emitter-to-Base Voltage:<br>With Collector Open, $V_{EBO}$ ............. | −4 | 4 | 4 | 4 | −4 | V |
| DC Collector Current, $I_C$ ................. | −0.7 | 0.7 | 0.7 | 0.7 | −0.7 | A |
| DC Base Current, $I_B$ .................... | −0.2 | 0.2 | 0.2 | 0.2 | −0.2 | A |
| Transistor Power Dissipation $(P_T)$:<br>At Free Air Temperatures up to 25° C.......<br>At Free Air Temperatures up to 50° C.......<br>At Case Temperatures up to 25° C.........<br>At Other Temperatures................ | 1<br>—<br>—<br> | 1<br>—<br>—<br>See Fig.1 | 1<br>—<br>—<br> | —<br>3<br>—<br> | —<br>3<br>—<br>See Fig.2 | W<br>W<br>W |
| Operating Junction Temperature Range | ←——————— −65 to +200 ———————→ | | | | | °C |

### DISSIPATION DERATING CURVE FOR TYPES 40406, 40407, AND 40408

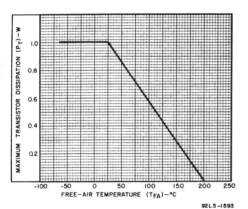

Fig. 1

### DISSIPATION DERATING CURVE FOR TYPES 40409 AND 40410

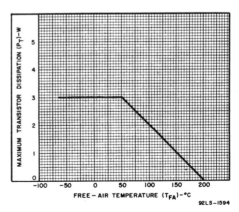

Fig. 2

## ELECTRICAL CHARACTERISTICS

| Characteristic | TEST CONDITIONS ||||||  LIMITS ||||||||||||
| --- | --- | --- | --- | --- | --- | --- | --- | --- | --- | --- | --- | --- | --- | --- | --- | --- | --- | --- |
|  | $V_{CB}$ | $V_{CE}$ | $V_{EB}$ | $I_C$ | $I_B$ | $T_C$ | 40406 || 40407 || 40408 || 40409 || 40410 || 40411 ||
|  | Volts ||| mA || °C | Min. | Max. | Min. | Max. | Min. | Max. | Min. | Max. | Min. | Max. | Min. | Max. |
| $I_{CEO}$ | 40ᵃ |  |  |  |  | 25 | -1μA | 1μA |  |  |  |  |  |  |  |  |  |  |
|  | 80 |  |  |  |  | 25 |  |  |  |  |  | 1μA |  |  |  |  |  |  |
|  | 40ᵃ |  |  |  |  | 150 | -10μA | 100μA |  |  |  |  |  |  |  |  |  |  |
|  | 80 |  |  |  |  | 150 |  |  |  |  |  | 250μA |  |  |  |  |  |  |
| $I_{CER}$ᵇ | 80ᵃ |  |  |  |  | 25 |  |  |  |  |  | 1μA |  |  |  | -1μA |  | 500μA |
|  | 80ᵃ |  |  |  |  | 150 |  |  |  |  |  |  |  | 100μA |  | -100μA |  | 2mA |
| $I_{CBO}$ | 10 |  |  |  |  |  |  | 0.25 μA |  |  |  |  |  |  |  |  |  |  |
| $I_{EBO}$ |  |  | 4ᵃ |  |  |  |  | -1mA |  | 1mA |  | 1mA |  | 1mA |  | -1mA |  | 5mA |
| $V_{CEO}$(sus) |  |  |  | 100ᵃ |  |  | -50 V |  | 50 V |  | 90 V |  |  |  |  |  |  |  |
| $V_{CER}$(sus)ᵇ |  |  |  | 100ᵃ |  |  |  |  |  |  |  |  | 90 V |  | -90 V |  |  |  |
|  |  |  |  | 200 |  |  |  |  |  |  |  |  |  |  |  |  | 90 V |  |
| $V_{CE}$(sat) |  |  |  | 150ᵃ | 15 |  |  |  |  |  |  | 1.4 V |  | 1.4 V |  | -1.4 V |  |  |
|  |  |  |  | 4A | 400 |  |  |  |  |  |  |  |  |  |  |  |  | 0.8 V |
| $V_{BE}$ |  | -10 |  | -0.1 |  |  |  | -0.8 V |  |  |  |  |  |  |  |  |  |  |
|  |  | 10 |  | 1 |  |  |  |  |  |  |  | 0.8 V |  |  |  |  |  |  |
|  |  | 4 |  | 10 |  |  |  |  |  |  |  | 1V |  |  |  |  |  |  |
|  |  | 4ᵃ |  | 150ᵃ |  |  |  |  |  |  |  |  |  | 1 V |  | -1 V |  |  |
|  |  | 4 |  | 4A |  |  |  |  |  |  |  |  |  |  |  |  |  | 1.2 V |
| $h_{FE}$ |  | -10 |  | -0.1 |  |  | 30 | 200 |  |  |  |  |  |  |  |  |  |  |
|  |  | 10 |  | 1 |  |  |  |  | 40 | 200 |  |  |  |  |  |  |  |  |
|  |  | 4 |  | 10 |  |  |  |  |  |  | 40 | 200 |  |  |  |  |  |  |
|  |  | 4 |  | 150 |  |  |  |  |  |  |  |  | 50 | 250 |  |  |  |  |
|  |  | -4 |  | -150 |  |  |  |  |  |  |  |  |  |  | 50 | 250 |  |  |
|  |  | 4 |  | 4A |  |  |  |  |  |  |  |  |  |  |  |  | 35 | 100 |
| $h_{fe}$ᶜ |  | 10 |  | 50 |  |  |  |  |  | 6 |  |  |  |  |  |  |  |  |
| $f_T$ |  | 4ᵃ |  | 50ᵃ |  |  | ◄──────── 100 MHz (Typ) ────────► ||||||||||
|  |  | 4 |  | 4A |  |  |  |  |  |  |  |  |  |  |  |  |  | 800 kHz (Typ) |
| $\theta_{J\text{-}C}$ |  |  |  |  |  |  | 35° C/W |  | 35° C/W |  | 35° C/W |  |  |  |  |  |  | 1.17° C/W |
| $\theta_{J\text{-}FA}$ |  |  |  |  |  |  | 175°C/W |  | 175°C/W |  | 175°C/W |  | 50° C/W |  | 50° C/W |  |  |  |
| $C_{ob}$ᵈ | 10 |  |  |  |  |  |  | 15 pF |  |  |  |  |  |  |  |  |  |  |
| PRTᵉ |  |  |  | 40 | 5A |  |  |  |  |  |  |  |  |  |  |  |  | 1 sec |

ᵃ Negative for types 40406 & 40410

ᵇ $R_{BE} = 100\ \Omega$

ᶜ $F = 20$ MHz

ᵈ $F = 1$ MHz, $I_E = 0$

ᵉ Power rating test at 200 watts

## Appendix 541

### TYPICAL OPERATION CHARACTERISTICS FOR TYPES 40406 & 40410

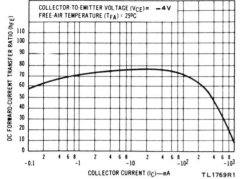

Fig. 3

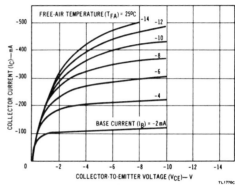

Fig. 4

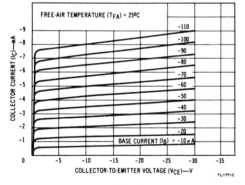

Fig. 5

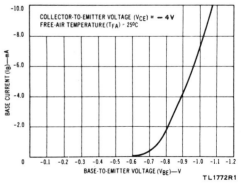

Fig. 6

## TYPICAL OPERATION CHARACTERISTICS FOR TYPES 40407, 40408, & 40409

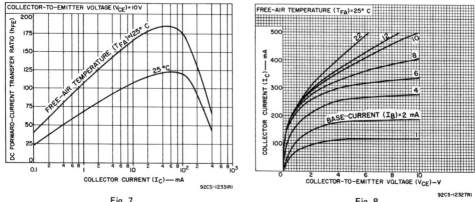

Fig. 7

Fig. 8

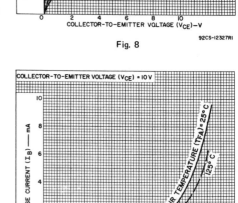

Fig. 9

Fig. 10

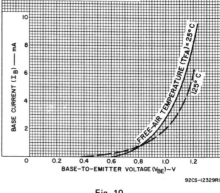

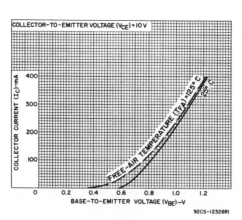

Fig. 11

Appendix 543

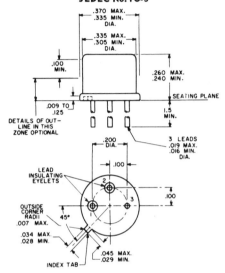

Fig. 12

# Appendix 3

Courtesy RCA

## SILICON N-P-N GENERAL-PURPOSE TYPES FOR INDUSTRIAL AND COMMERCIAL APPLICATIONS

RCA-2N3055* is a silicon n-p-n transistor useful for power-switching circuits, for series-and shunt-regulator driver and output stages, and for high-fidelity amplifiers.
*Formerly Dev. Type Nos TA2402A and TA2403A, respectively.

## 2N3055

- High dissipation capability--115W $V_{CEV}$(sus) = 100 V min.
- Low saturation voltage, $V_{CE}$(sat) = 1.1 V (at $I_C$ = 4A)
- Maximum area-of-operation curves for DC and pulse operation
- Now possible to determine maximum operating conditions for operation free from second breakdown

### MAXIMUM RATINGS

*Absolute-Maximum Values:*

| | | 2N3055 | |
|---|---|---|---|
| COLLECTOR-TO-BASE VOLTAGE | $V_{CBO}$ | 100 | V |
| COLLECTOR-TO-EMITTER VOLTAGE: | | | |
| With $-1.5$ V ($V_{BE}$) of reverse bias | $V_{CEV}$(sus) | 100 | V |
| With external base-to-emitter resistance ($R_{BE}$) = 10 Ω | $V_{CER}$(sus) | — | V |
| = 100 Ω | $V_{CER}$(sus) | 70 | V |
| With base open | $V_{CEO}$(sus) | 60 | V |
| EMITTER-TO-BASE VOLTAGE | $V_{EBO}$ | 7 | V |
| COLLECTOR CURRENT | $I_C$ | 15 | A |
| BASE CURRENT | $I_B$ | 7 | A |
| TRANSISTOR DISSIPATION: | $P_T$ | | |
| At case temperatures up to 25°C | | 115 | W |
| | | — | W |
| At free-air temperatures up to 25°C | | — | W |
| | | — | W |
| At temperatures above 25°C, See Figs. | | 2, 4, & 6 | |
| TEMPERATURE RANGE: | | | |
| Storage & Operating (Junction) | | ←−65 to 200→ | °C |
| LEAD OR PIN TEMPERATURE (During soldering): | | | |
| At distance ≥ 1/32" from seating plane for 10 s max. | | 235 | °C |

2N3055
JEDEC TO-3

H-1339

## ELECTRICAL CHARACTERISTICS
*Case Temperature $(T_C)$ = 25°C, Unless Otherwise Specified*

| Characteristics | Symbol | TEST CONDITIONS | | | | | | | LIMITS Type 2N3055 | | Units |
|---|---|---|---|---|---|---|---|---|---|---|---|
| | | DC Collector Volts | | DC Emitter or Base Volts | | DC Current milliamperes | | | | | |
| | | $V_{CB}$ | $V_{CE}$ | $V_{EB}$ | $V_{BE}$ | $I_C$ | $I_E$ | $I_B$ | Min. | Max. | |
| Collector-Cutoff Current | $I_{CBO}$ | 30 | | | | | 0 | | — | — | μA |
| | $I_{CEV}$ | | 90<br>100 | | -1.5<br>-1.5 | | | | —<br>— | —<br>5.0 | mA |
| At $T_C$ = 150°C | $I_{CEV}$ | | 30<br>60 | | -1.5<br>-1.5 | | | | —<br>— | —<br>10.0 | mA |
| Emitter-Cutoff Current | $I_{EBO}$ | | | 4<br>7<br>7 | | 0<br>0<br>0 | | | —<br>—<br>— | —<br>—<br>5.0 | μA<br>mA<br>mA |
| DC Forward-Current Transfer Ratio | $h_{FE}$ | | 10<br>4<br>4 | | | 150[a]<br>500<br>4A[a] | | | —<br>—<br>20 | —<br>—<br>70 | |
| Collector-to-Base Breakdown Voltage | $BV_{CBO}$ | | | | | 0.1 | 0 | | — | — | V |
| Emitter-to-Base Breakdown Voltage | $BV_{EBO}$ | | | | | 0<br>0<br>0 | 0.1<br>1<br>5 | | —<br>—<br>7 | —<br>—<br>— | V |
| Collector-to-Emitter Sustaining Voltage:<br>With base open | $V_{CEO}$(sus) | | | | | 100[a]<br>100<br>200 | | 0<br>0<br>0 | —<br>—<br>60 | —<br>—<br>— | V |
| With base-emitter junction reverse biased | $V_{CEV}$(sus) | | | | -1.5 | 100 | | 100 | — | — | V |
| With external base-to-emitter resistance<br>($R_{BE}$) = 10 Ω<br>= 100 Ω<br>= 100 Ω | $V_{CER}$(sus) | | | | | 100[a]<br>100<br>200 | | | —<br>—<br>70 | —<br>—<br>— | V |
| Base-to-Emitter Voltage | $V_{BE}$ | | 4<br>4 | | | 500<br>4A[a] | | | —<br>— | —<br>1.8 | V |
| Base-to-Emitter Saturation Voltage | | | | | | 150 | | 15 | — | — | V |
| Collector-to-Emitter Saturation Voltage | $V_{CE}$(sat) | | | | | 150<br>500<br>4A[a] | | 15<br>50<br>400 | —<br>—<br>— | —<br>—<br>1.1 | V |
| Small-Signal, Forward Current Transfer Ratio (At 20 MHz) | $h_{fe}$ | | 10 | | | 50 | | | — | — | |
| Gain-Bandwidth Product | $f_T$ | | | | | 200<br>1 A | | | —<br>800 | —<br>— | kHz |
| Output Capacitance | $C_{ob}$ | 10 | | | | | 0 | | — | — | pF |
| Input Capacitance | $C_{ib}$ | | | 0.5 | | 0 | | | — | — | pF |
| Power Rating Test | PRT | | 39 | | | 3 A | | | — | 1[b] | s |
| Thermal Resistance:<br>Junction-to-Case | $\theta_{J-C}$ | | | | | | | | —<br>— | 1.5<br>— | °C/W<br>°C/W |
| Junction-to-Free Air | $\theta_{J-FA}$ | | | | | | | | —<br>— | —<br>— | °C/W<br>°C/W |

[a] Pulsed; pulse duration = 300 μs, duty factor = 1.8 %.  [b] At 115 W.

## 546 Appendix

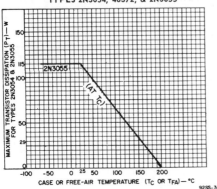

Fig. 1

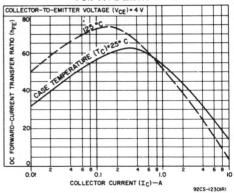

Fig. 2

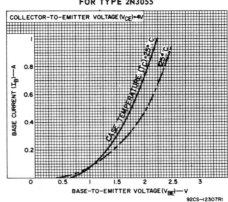

Fig. 3

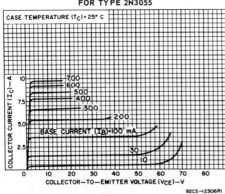

Fig. 4

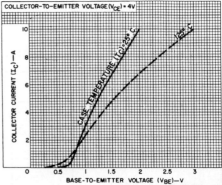

Fig. 5

*Appendix* 547

### FOR TYPE 2N3055
### JEDEC No. TO-3

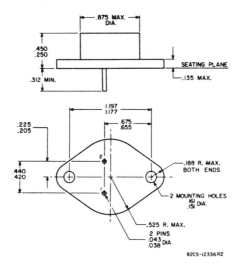

Dimensions in Inches

**TERMINAL CONNECTIONS**
FOR TYPE 2N3055

Pin 1 - Base
Pin 2 - Emitter
Case, Flange - Collector

# Appendix 4

Courtesy of Motorola Semiconductor Products Inc.

## 2N2646  2N2647

**SILICON ANNULAR† PN UNIJUNCTION TRANSISTORS**

PN UNIJUNCTION TRANSISTORS

... designed for use in pulse and timing circuits, sensing circuits and thyristor trigger circuits. These devices feature:

JANUARY 1967 — DS 2500

(Replaces DS 6514)

- Low Peak Point Current — 2 μA max
- Low Emitter Reverse Current — 200 nA max
- Passivated Surface for Reliability and Uniformity

**MAXIMUM RATINGS** ($T_A$ = 25°C unless otherwise noted)

| Characteristic | Symbol | Rating | Unit |
|---|---|---|---|
| RMS Power Dissipation* | $P_D$ | 300* | mW |
| RMS Emitter Current | $I_e$ | 50 | mA |
| Peak Pulse Emitter Current** | $i_e$ | 2** | Amp |
| Emitter Reverse Voltage | $V_{B2E}$ | 30 | Volts |
| Interbase Voltage | $V_{B2B1}$ | 35 | Volts |
| Operating Junction Temperature Range | $T_J$ | -65 to +125 | °C |
| Storage Temperature Range | $T_{stg}$ | -65 to +150 | °C |

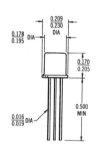

APPROX. WEIGHT .015 OZ.

\* Derate 3.0 mW/°C increase in ambient temperature. The total power dissipation (available power to Emitter and Base-Two) must be limited by the external circuitry.

\*\* Capacitor discharge — 10 μF or less, 30 volts or less.

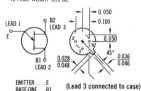

EMITTER....E
BASE-ONE...B1
BASE-TWO...B2

(Lead 3 connected to case)

TO-18 PACKAGE
(Except for lead position)

†Annular Semiconductors Patented by Motorola Inc.

*Appendix* 549

# 2N2646  2N2647

**ELECTRICAL CHARACTERISTICS** ($T_A = 25°C$ unless otherwise noted)

| Characteristic | | Symbol | Min | Typ | Max | Unit |
|---|---|---|---|---|---|---|
| Intrinsic Standoff Ratio ($V_{B2B1} = 10$ V) (Note 1) | 2N2646<br>2N2647 | $\eta$ | 0.56<br>0.68 | —<br>— | 0.75<br>0.82 | — |
| Interbase Resistance ($V_{B2B1} = 3$ V, $I_E = 0$) | | $R_{BB}$ | 4.7 | 7.0 | 9.1 | K ohms |
| Interbase Resistance Temperature Coefficient ($V_{B2B1} = 3$ V, $I_E = 0$, $T_A = -55°C$ to $+125°C$) | | $\alpha R_{BB}$ | 0.1 | — | 0.9 | %/°C |
| Emitter Saturation Voltage ($V_{B2B1} = 10$ V, $I_E = 50$ mA) (Note 2) | | $V_{EB1(sat)}$ | — | 3.5 | — | Volts |
| Modulated Interbase Current ($V_{B2B1} = 10$ V, $I_E = 50$ mA) | | $I_{B2(mod)}$ | — | 15 | — | mA |
| Emitter Reverse Current ($V_{B2E} = 30$ V, $I_{B1} = 0$) | 2N2646<br>2N2647 | $I_{EO}$ | —<br>— | 0.005<br>0.005 | 12<br>0.2 | $\mu$A |
| Peak Point Emitter Current ($V_{B2B1} = 25$ V) | 2N2646<br>2N2647 | $I_P$ | —<br>— | 1.0<br>1.0 | 5.0<br>2.0 | $\mu$A |
| Valley Point Current ($V_{B2B1} = 20$ V, $R_{B2} = 100$ ohms) (Note 2) | 2N2646<br>2N2647 | $I_V$ | 4.0<br>8.0 | 6.0<br>10 | —<br>18 | mA |
| Base-One Peak Pulse Voltage (Note 3, Figure 3) | 2N2646<br>2N2647 | $V_{OB1}$ | 3.0<br>6.0 | 5.0<br>7.0 | —<br>— | Volts |

## NOTES

1. Intrinsic standoff ratio, $\eta$, is defined by equation:
$$\eta = \frac{V_P - V_{(EB1)}}{V_{B2B1}}$$
Where $V_P$ = Peak Point Emitter Voltage
$V_{B2B1}$ = Interbase Voltage
$V_{(EB1)}$ = Emitter to Base-One Junction Diode Drop
($\sim 0.5$ V @ $10$ $\mu$A)

2. Use pulse techniques: PW $\approx 300$ $\mu$s duty cycle $\leq 2\%$ to avoid internal heating due to interbase modulation which may result in erroneous readings.

3. Base-One Peak Pulse Voltage is measured in circuit of Figure 3. This specification is used to ensure minimum pulse amplitude for applications in SCR firing circuits and other types of pulse circuits.

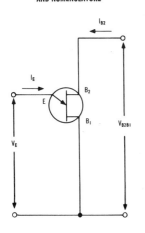

FIGURE 1 — UNIJUNCTION TRANSISTOR SYMBOL AND NOMENCLATURE

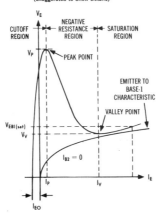

FIGURE 2 — STATIC EMITTER CHARACTERISTIC CURVES
(Exaggerated to Show Details)

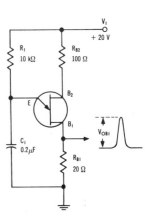

FIGURE 3 — $V_{OB1}$ TEST CIRCUIT
(Typical Relaxation Oscillator)

# Appendix 5

Courtesy of Motorola Semiconductor Products Inc.

## 2N1595 thru 2N1599

### for 0-1.6 AMPERE (RMS), 50-400 VOLT APPLICATIONS up to 125°C MAXIMUM JUNCTION TEMPERATURE

**SILICON CONTROLLED RECTIFIERS**
PNPN

**1.6 AMPERES RMS**
**50 thru 400 VOLTS**

MARCH, 1964
DS 6503

- † Annular Construction
- All Welded Construction
- Low Forward Voltage Drop
- Uniform Gate Characteristics
- TO-5 Package Ideal for Printed Circuit Applications

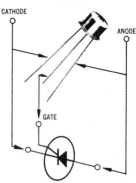

**MAXIMUM RATINGS\*** ($T_J = 125°C$ unless otherwise noted)

| Characteristic | Symbol | Rating | Unit |
|---|---|---|---|
| Peak Forward Voltage | PFV | 600 | Volts |
| Peak Reverse Blocking Voltage\* <br> 2N1595 <br> 2N1596 <br> 2N1597 <br> 2N1598 <br> 2N1599 | $V_{ROM(rep)}$\* | <br> 50 <br> 100 <br> 200 <br> 300 <br> 400 | Volts |
| Peak Forward Blocking Voltage \* <br> ($R_{GC}$ = 1000 ohms) 2N1595 <br> 2N1596 <br> 2N1597 <br> 2N1598 <br> 2N1599 | $V_{FOM}$ | <br> 50 <br> 100 <br> 200 <br> 300 <br> 400 | Volts |
| Forward Current RMS (All Conduction Angles) | $I_f$ | 1.6 | Amps |
| Peak Surge Current (One Cycle, 60 cps, $T_J$ = -65 to +125°C) | $I_{FM(surge)}$ | 15 | Amps |
| Peak Gate Power | $P_{GM}$ | 0.1 | Watt |
| Average Gate Power | $P_{G(AV)}$ | 0.01 | Watt |
| Peak Gate Current | $I_{GM}$ | 0.1 | Amp |
| Peak Gate Voltage - Forward <br> Reverse | $V_{GFM}$ <br> $V_{GRM}$ | 10 <br> 10 | Volts |
| Operating Temperature Range | $T_J$ | -65 to +125 | °C |
| Storage Temperature Range | $T_{stg}$ | -65 to +150 | °C |

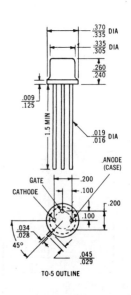

TO-5 OUTLINE

## 2N1595 thru 2N1599

**ELECTRICAL CHARACTERISTICS** ($T_J = 25°C$ unless otherwise noted, $R_{GC} = 1000$ ohms)

| Characteristics | Symbol | Min | Typ | Max | Unit |
|---|---|---|---|---|---|
| Peak Reverse Blocking Current (@ rated $V_{ROM}$, $T_J = 125°C$) | $I_{ROM}$ | — | — | 1000 | μA |
| Peak Forward Blocking Current (@ rated $V_{FOM}$, $T_J = 125°C$) | $I_{FOM}$ | — | — | 1000 | μA |
| Forward On Voltage ($I_F = 1$ Adc) | $V_F$ | — | 1.1 | 2.0* | Volts |
| Gate Trigger Current (Anode Voltage = 7 V, $R_L = 12 \Omega$) | $I_{GT}$ | — | 2.0 | 10.0* | mA |
| Gate Trigger Voltage (Anode Voltage = 7 V, $R_L = 12 \Omega$) (Anode Voltage = 7 V, $R_L = 12 \Omega$, $T_J = 125°C$) | $V_{GT}$ $V_{GNT}$ | — 0.2 | 0.7 — | 3.0* — | Volts |
| Holding Current (Anode Voltage = 7 V) | $I_{HX}$ | — | 5.0 | — | mA |
| Turn-on Time ($I_{GT} = 10$ mA, $I_F = 1$ A) ($I_{GT} = 20$ mA, $I_F = 1$ A) | $t_{on}$ | — — | 0.8 0.6 | — — | μ sec |
| Turn-off Time ($I_F = 1$ A, $I_R = 1$ A, dv/dt = 20 V/μsec, $T_J = 125°C$) | $t_{off}$ | — | 10 | — | μ sec |

*JEDEC Registered Values

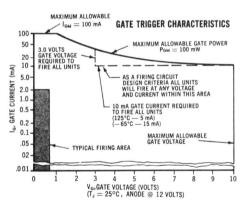

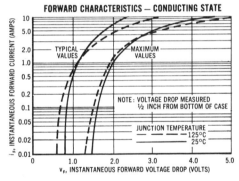

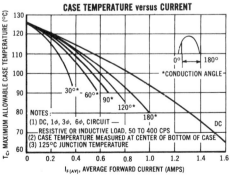

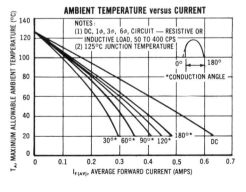

# Index

## A

Acceptor atoms, 6, 7
Active mode, 39, 43, 46
Admittance, triode, 432–433
Alpha:
  cutoff frequency, 188, 190
  large-signal, 48, 51
  small-signal, 168
Aluminum impurity, 6
Ambient, 240
Amplification factor, 418–420
Amplification, graphical analysis, 85–89
Amplifier:
  classification, 235–236
  chopper, 488–493
  common-base, 166–169
  common-cathode, 422–426
  common-collector, 159–160
  common-drain, 477–480
  common-emitter, 145–158
  common-emitter to common-base cascode, 227–230
  common-grid, 428–429
  common-plate, 424–426
  common-source, 475–480
  complementary (See complementary amplifier)
  Darlington, 221–227

Amplifier (*cont.*):
  design example, 285–290
  differential (See differential amplifier)
  feedback (See feedback amplifiers)
  hybrid, 486–488
  operational, 295
  oscillation in, 396–398
  power (See power amplifiers)
  quasicomplementary, 290–296
  tuned circuit, 342–348, 350–356
Amplitude modulation, 316–320
Analog switching, 493–494
Anode, vacuum tube, 414
Antenna, 329
Atom, doping, 6–7
  pentavalent, 6
  tetravalent, 4–5
  trivalent, 6
Automatic gain control, 314
Avalanche, breakdown, 27, 32, 237–239
  multiplication factor, 238

## B

Balanced modulator, 320
Bandwidth, 193, 331–334, 348–350
Barkhausen criterion, 398
Base, 39

554  *Index*

Base-spreading resistance $r_x$, 142
Beta, cutoff frequency, 188, 190
   large-signal, 50, 59
   frequency dependence, 189–191
   measurement, 60, 68, 137
   small-signal, 60
Biasing, 97–141
   collector-base resistor, 126–130
   complementary amplifier 270–272
   constant current, 309–310
   diode, 12, 133–135
   emitter, 129–131
   emitter resistor, 103–108
   field-effect transistor, 465–473
   integrated circuits, 134–136
   push-pull, 442, 446
   quasicomplementary amplifier, 293–396
   single battery, 98–101
   stability, 112–126
   vacuum tube, 423–424
   voltage, 12
Bipolar device, 449
Bleeder current, 100, 271–273
Blocking junction, 521
Boltzmann's constant, 15, 143
Bootstrapping, 132
   common-emitter, 210–211
   complementary output, 279–285
   emitter-follower, 207–210
   quasicomplementary amplifier, 295–296
Bond, covalent, 4
Breakdown (See avalanching)
Break frequency (See cutoff frequency)
Bridge circuit, 402

## C

Capacitance:
   input, bipolar transistor, 187–189
      differential, 350–352
      diffusion, 187
      field-effect transistor, 482–483
      interelectrode, 431
      triode, 432
      pentode, 436

Capacitance (*cont.*):
   output, common-base, 187
   space-charge, 11, 14
Carrier frequency, 318–321
Cascode, CE to CB, 227–230
Cathode, semiconductor diode, 13
   vacuum tube, 413–416
Cathode-follower (See common-plate amplifier)
Cathode-ray oscilloscope, 21–27
Characteristic curve, 14
Charge carriers, 1, 5, 7
Chopper, electromechanical, 489–490
   field-effect transistor, 491–493
Circuit notation convention, 41
Clamp circuit, 32
Class A power amplifiers, 248–260
Class B power amplifiers, 261–264, 269–296, 440–446
Clipping, 86
Collector, 39
Colpitts oscillator, 403–406
Contact potential, 11
Common-base amplifier, 46–50, 166–170
Common-cathode amplifier, 422–426
Common-collector amplifier, 159–161
Common-drain amplifier, 477–480
Common-emitter configuration, 50–56, 145–158
Common-emitter, high-frequency cutoff, 191–194
Common-plate amplifier, 424–426
Common-mode rejection ratio, 304–308
Common-mode voltage, 305
Common-source amplifier, 475–480
Common-grid amplifier, 428–429
Communication circuits, 299–325
Commutation, 493
Comparator, 321–325
Complementary amplifier, 269–295
   biasing, 270–273
   bootstrapping, 279–282
   circuit analysis, 273–278
   crossover distortion, 269–270
   design example, 285–290
   driver, 282–285
   power relationships, 274–276

Index 555

Complementary amplifier (*cont.*):
  quasi type, 290–295
Complementary output stage, 269–281
Complex variable, 33
Composite transistor, 292–294
Conductance, 330
Constant current gain control, 311–313
  biasing, 309
Contact potential, 11
Control, voltage gain, 311–314
Control circuits, 299–325
Convection, 243
Conventional current, 43
Correction curves, 72
Covalent bond, 4
CRO (See cathode-ray oscilloscope)
Crossover distortion, 269–270, 290
Crystal oscillator, 405–406
Cubic lattice, 4
Current axis intercept, 77
Current flow, semiconductor, 7–10, 41
Current gain, alpha, 48, 68
  beta, 50, 60
  graphical analysis, 90–92
Current gain-bandwidth product, 188
Current-sensing resistor, 22–24
Current source, bias, 128, 309–310
  drive, 213
  components, 114–128
  model, 131
Cutoff region, 61–62
Cutoff frequency control, 216
Cutoff frequency, 177–203, 196–205
  definition, 177–179
  lower, 179–184, 289–290, 296
  measurement, 189, 199–203
  upper, 187–198

# D

Darlington amplifier, 222–227
DC restorer, 32
Decade, 184–185
Decibels, feedback, 362–363
  gain, 171–173
Demodulation, 489

Demonstration ac-dc load line, 92
  cutoff operation, 61
  hot probe, 10
  operating point location, 93
  operating point stability, 137
  transistor parameters, 199–203
Depletion region, 12, 26, 32
Detector, zero crossing, 323–325
Differential amplifier, 299–325
  applications, 310–324
    comparator, 321–325
    frequency shift, 321
    gain control, 311–314
    modulator, 316–321
    multiplier, 314–316
    phase inverter, 310
  biasing, 300–301, 309
  common-mode rejection ratio,
    304–308
  differential output voltage, 308
Differential output voltage, 300, 308–309
Differential input voltage, 300, 304–308
Diffusion capacitance, 187
Diffusion current, 9, 10
Digital voltmeter, 321
Diode, applications, 30–32
  bias, 12, 13, 22
  bulk resistance, 16, 27
  characteristic curves, 12, 23
  circuit model, 15–17
  equation, 14–17, 20
  ideal, 16, 21
  junction resistance, 17, 27
  ohmmeter test, 19
  operating point, 24, 28
  rectifying properties, 25
  reverse breakdown, 25
  reverse saturation, 14, 15, 24
  vacuum tube, 413–416
Dissipation derating curve, 241
Distortion, 108
  crossover, 269–270, 290
  feedback reduction, 368–371
  non-linear, 231
  second harmonic, 441–443
Donor atoms, 6, 7
Doping, 6, 7

556    *Index*

Doubler, voltage, 31–32
Drift current, 8, 9
Drain, 449
Drain resistance, 473–474
Driver stage, 282–287

## E

Edison effect, 414
Efficiency, Class A power amplifier, 252-253
   Class B power amplifier, 264
   complementary output, 275
   transformer-coupled, 255
Electrons, valence, 4
Emitter, 39
Emitter saturation voltage, 505
Envelope, 319, 321
Equivalent circuit (See model)

## F

Feedback, 359–409
   decibels, 362–363
   design example, 391–395
   distortion, 368–371
   frequency response, 390–391
   fundamentals, 363–365
   gain stabilization, 363–368
   negative, 359–395
   multistage amplifiers, 371–388
   oscillators, 396–409
   positive, 359, 396–408
   resistance levels, 389–391
   single-stage amplifiers, 365–368
Feedback amplifier, 365–395
   distortion, 368–371
   feedback network loading, 372–375
   input resistance, 375, 380
   output resistance, 389–390
   oscillation, 396
   parallel-parallel, 381–385
   parallel-series, 371–375
   series-parallel, 375–381
   series-series, 385–389
   single stage, 365–368
FET (See Field-effect transistor)

Field-effect transistor:
   characteristic curves, 452–455, 461–465
   depletion, 458–464
   enhancement, 458–464
   handling precaution, 464
   biasing, 465–473
   circuit models, 473–480
   frequency dependence, 482–486
Filament, 413–416
Filter, 329
Firing angle, 525–526
Forward bias, 12
Forward conduction, 520, 522
Frequency, cutoff (See cutoff frequency)
Frequency, dependence of $\beta$, 189–191
Frequency, deviation from resonance, 331–333
Frequency response, 177–178, 390–391, 429–433
Frequency shifting, 321
Fundamental frequency, 441–443

## G

Gain:
   common-mode, 305
   current, 305
   decibel, 171–173
   voltage, 83, 88
      common-base, 166–170
      common-cathode, 422–426
      common-collector, 159, 161
      common-drain, 477–480
      common-emitter, 145–158, 179–183, 191–194
      common-mode, 305
      common-plate, 424–426
      complementary, 277, 292
      feedback, 360–363
      stabilization, 363
      resonance, 342
Gain–bandwidth product, 194, 344
Gain control, 311–314
Gallium, 6
Gas tube, 413
Gate, 521

Germanium, properties, 5
Graphical analysis, 75–93
　ac loadline, 88–93
　dc load line, 75–83
　input resistance, 83–84
　operating point, 82–83
　voltage amplification, 85–88
Grid, 416
Grounded-emitter (See common emitter)

## H

Harmonic distortion, second, 441–443
Hartley oscillator, 406–408
Header, 240
Heater (See filament)
Heat radiation, 243
Heat sinking, 242–245
High-linearity sawtooth generator, 516–518
Holding current, 523
Hole, 4, 6
Hot-probe experiment, 10
Hybrid amplifier, 486–488
Hybrid-pi model:
　frequency dependence, 189–191
　high frequency, 187–189
　limitation, 148
　low frequency, 142–145
　parameter measurement, 199–203
　window, 149

## I

$I_{CBO}$, 47, 61, 109
$I_{CEO}$, 53, 56, 61, 109
Idle, current, 270
IF transformer, 347
IGFET, 449, 458
Impedance, input, 217
　transformation, 161–166
　transformer, 337–342
　tuned circuit, 331–333
Improvement factor, 369
Incremental resistance, 27, 29, 273
Incremental voltage, 27
Inductance, 334–336

Interelectrode capacitance, 431
Intrinsic semiconductor, 1–5
Input impedance, 217
Input resistance:
　common-base, 167–168
　common-collector, 159
　common-emitter, 154, 212
　feedback, 389–391
　graphical analysis, 83–85
　incremental, 85
　large signal, 262
　measurement, 171, 199–202
　open loop, 373, 377–381
Integrated circuit:
　amplifiers, 299
　FET, 449
　internal feedback, 350
　oscillator, 403
Interbase resistance, 500–505
Intermediate frequency, 321
Intrinsic standoff ratio, 500, 505–507
Ionization, 6–7

## J

JFET (See field-effect transistor)
Junction, anode, 521
　bias, 12, 43
　blocking, 521
　capacitance, 11, 187
　cathode, 521
　collector, 39
　diode, 12–21
　emitter, 39
　$pn$, 10–12, 39–41
Junction transistor, 39

## K

Kinetic energy, 4
Kelvin, degree, 15

## L

Lamp dimmer, 529
Large signal:
　amplifiers (See power amplifiers)

558  Index

Large signal (*cont.*):
  current gain, 276–277
  input resistance, 262
LASCR, 522
Leakage current:
  components, 114–128
  $I_{CBO}$, 53, 56, 61, 109
  $I_{CEO}$, 47, 61, 109
  SCR, 521–522
Linear mixing, 231–233
Linear multiplier, 314–316
Load line, ac, 88–93
  dc, 75–89, 284, 435
  definition, 29
  demonstration, 92
Local, oscillator, 321
Loop gain, 361
Low-frequency cutoff:
  bypass capacitor, 182–186
  coupling capacitor, 179–181

# M

Majority carriers, 7, 11–14
Maximum output to dissipation ratio:
  class A, 253
  class A transformer coupled, 255
  class B, 264
  complementary output, 275
Maximum output voltage, 284–285
Measurement:
  $\alpha_F$, 65
  $\beta_o$, 60, 68, 70
  $\beta_F$, 68, 137
  characteristic curves:
    common-base, 64–66
    common-emitter, 56–63, 67
    diode, 21–30
    IGFET, 461–465
    JFET, 453–457
    SCR, 522–523
    triode, 417–418
    unijunction, 501–507
    vacuum diode, 415
  current gain, 59–61, 68
  drain resistance, 473
  Edison effect, 414

Measurement (*cont.*):
  firing voltage, 523
  hybrid-pi parameters, 199–204
  $I_{CBO}$, 47
  $I_{CEO}$, 52–53, 58
  inductance, 334–336
  input resistance, 67, 87, 170–171
  operating point, 82, 99
  output resistance, 170–171, 426
  pinch off voltage, 456–458
  $R_{sat}$, 62
  transconductance, 67, 474
  valley current, 504
  valley voltage, 504
  zero bias point, 472
Meter protection, 30
Miller capacitance, 432
  effect, 191–196
  FET, 483
Minority carriers, 11–14
Mixer, linear, 231–233
Model, 15, 18
  bipolar transistor:
    heat flow, 241
    high-frequency, 187–188
    hybrid-pi (See hybrid-pi model)
  diode:
    junction, 16–19
    vacuum, 415
    Zener, 33
  FET, 449–552, 473–480
  inductor, 334–335
  pentode, 436
  SCR, 521
  small signal, 141
  triode, 419–420, 431
  UJT, 502
Modulating frequency, 318–320
Modulation, amplitude, 316–320
MOSFET, 449, 458
Motor speed control, 529
Multiplier, linear, 314–316
Multiplexing, 493

# N

Narrow-bandwidth, 329–330

## Index

Negative feedback (See feedback)
Negative resistance, triodes, 433
   UJT, 504
Neutralization, 350
Non-linear, 14
Norton's theorem, 376, 383
Notation, standard circuit, 41–43
$n$-type, 7

## O

Offset voltage, 313–114
   chopper amplifier, 490
Ohmic contacts, 9
Open-loop gain, 373
Operating modes, 39, 43–46, 61–64
Operating path, 29, 84–87
Operating point, 58, 75, 79–83, 98, 105
Operating point stability, 108
On-resistance, 452, 491
Oscillator, amplitude, 399, 403
   Colpitts, 405–406
   crystal, 405–406
   feedback, 398–399
   Hartley, 406–408
   phase shift, 399–403
   tuned circuit, 403–408
   Wien bridge, 402
Oscillation, in amplifiers, 219
   in triodes, 433
Oscilloscope (See CRO)
Output characteristic curve, 57
Output resistance, 161, 215
   measurement, 170–171, 426
Oxide coated cathode, 413

## P

Parallel resonance (See resonance)
Peak point, current, 502
   voltage, 502–509
Peak-reading voltmeter, 31
Peak-to-peak voltmeter, 31
Pentode, 433–437
Permeability, 336–337
Phase control, SCR, 525–528

Phase inverter, 45
   cathode-follower, 437
   differential amplifier, 310
   self-balancing, 437
   vacuum tube, 437–438
Phase-shift oscillators, 399–403
Piezoelectric effect, 405–406
Pinch off, 450
Plate, 414
Plate resistance, 419
Plate-to-plate resistance, 440
$p$-$n$ junction, 10–12
Positive feedback (See feedback)
Potential barrier, 11
Power amplifier, 235–264
   classification, 235–236
   class A, 248–260
   class B, 261–264
   current limitations, 236–237
   direct-coupled, 248–253
   heat sinking, 242–246
   limitations, 236–240
   thermal limitations, 239–240
   thermal resistance, 240–242
   transformer-coupled, 253–259
   voltage limitations, 237–239
Power, average, 246–247
   fundamentals review, 243–246
   instantaneous, 246
$p$-type, 7
Push-pull operation, 439–446

## Q

Q point (See operating point)
Q meter, 335–336
Quality factor Q, 334–336
Quasicomplementary amplifier, 290–296
Quiescent point (See operating point)

## R

Ramp generator, 516
Ramp voltage, 321
Reach-through voltage, 237
Recombination, 5
Rectifier, diode, 25

Rectifier, meter, 30
Reference voltage, 321, 322
Regulator, voltage, 33
Relaxation oscillator, 510–516
Relay operation, 489–490, 519
Resistance, semiconductor, 5
Resistance transformation:
    BJT, 104, 127, 155, 159–166
    bootstrapping, 105, 209
    Darlington, 224–227
    FET, 480–482
    vacuum tube, 426–427
Resonance, 329–334, 340–341
    crystal, 406
Return ratio, 361
Reverse bias, 13, 18, 43
Reverse current, 17, 19
Reverse resistance, 17, 19
$R_{sat}$, 59, 62

## S

Sampling resistor, 57
Saturation region, 59, 61–64, 65
    FET, 451
    UJT, 505
    vacuum tube, 414–415
Sawtooth generator, 516–518
SCR (See silicon controlled rectifier)
Secondary emission, 433
Second harmonic distortion, 441–443
Selectivity, 333
Self-bias, 423, 465–471
Semiconductor:
    current flow, 7–10
    doping, 6–7
    intrinsic, 3–6
    material, 5
Sensitivity, 22
Sequential switching, 493
Screen grid, 433
Side band, 318
Side frequency, 318–319
Silicon controlled rectifier:
    applications, 529
    characteristics, 520–523
    pulsed gate operation, 523–525

Silicon controlled rectifier (*cont.*):
    phase control, 525–528
Silicon, properties of, 4, 5
Single side-band transmission, 320
Slug, 337
Solar cell, 308
Soldering precautions, 240
Source, constant current, 43, 114–128
    constant voltage, 43
    terminal, FET, 449
Space charge region, 12
    FET, 450–452
Stability, 97
    coefficient, 119
    factors, 119
Stabilization:
    gain, 363
    operating point:
        collector-base, resistor, 123–124
        collector-base and emitter resistor, 126–129
        emitter resistor, 115–123
        feedback, 120
Standard letter symbols, 41–43
Superposition, 30
    majority and minority current, 54, 115–129
    stability analysis, 112–129
Sustaining voltage, 238
Sweep measurements (See measurement, characteristic curves)
Symbols (See notation)
Symmetry, 300–302
Synchronization, 526

## T

Tank circuit, 329
Tapped inductor, 338–342
Temperature dependence, 108–112
    beta, 109
    charge density, 4, 5
    drain current, 472
    $I_{CBO}$, 48, 108, 111
    operating point, 137
    peak-point voltage, 509

*Index* 561

Temperature dependence (*cont.*):
  transistor parameters, 108–109
  $V_{BE}$, 108, 112, 271
Temperature, junction, 241
Tetrode, 437
Thermal resistance, 240–245
Thermal velocity, 8
Thermistor, 321–322
Thermistor, biasing, 133–134
Threshold voltage, 16, 35
Thévenin's theorem, 97, 100–101, 131, 163–164
Thyratron (See silicon controlled rectifier)
Time constant, 179, 180, 182
  high-frequency, 191–193
  short-circuit, 184–186
Timers, 518–519
Transconductance:
  BJT, 142
  FET, 454–455, 473–474
  gain, 359, 364, 385–388
  measurement, 67, 474
  vacuum tube, 418–420
Transducer, 216
Transfer characteristic, 67, 68
Transformation, resistance (See resistance transformation)
  impedance, 183
Transformer, audio, 258–259
  coupling, 253–258
  ideal, 337
Transistor, 39
  action, 46–50
  active region, 59
  alpha (See alpha)
  beta (See beta)
  bipolar, 39
  circuit configurations, 44
    symbols, 40
  common-base, 46–56, 156–170
  common-collector, 159–161
  common-emitter, 50–56, 145–155
  construction, 40
  current control, 57
  cutoff region, 61
  Field-effect (See Field-effect transistor)

Transistor (*cont.*):
  hybrid-pi model (See hybrid-pi)
  insulated gate (IGFET), 458–475
  junction, (JFET), 449–458
  majority carrier currents, 46
  minority carrier currents, 46
  operating point, 58
  parameter correction factors, 71
  power dissipation, 250–251
  recombination current, 47–52
  reverse saturation current, 47–49
  saturated region, 43–49, 59, 62
  space-charge region, 39
  symbols (See notation)
  transconductance (See transconductance)
  unijunction (See unijunction transistor)
Transresistance gain, 359, 368, 381–385
Transverse cutoff frequency, 205
Triode, 416–422
Tuned circuits, 321, 329–355
Tuned circuit, oscillators, 403–407
Tuning cores, 337
Turns ratio, 339, 348

## U

Unijunction transistor:
  action, 502–505
  applications, 510–520
  characteristic curves, 500–505
  parameter measurement, 505–508
  temperature stabilization, 508–510
  relaxation oscillator, 510–516
Unipolar device, 449
Upper cutoff frequency:
  common-emitter, 191–193, 196–197
  common-base, 197–199
  common-collector, 197–199

## V

Vacuum tubes, 413–415
Valence electron, 4
Valley current, 505

Valley voltage, 504–505
Voltage axis intercept, 77
Voltage doubler (See peak-to-peak voltmeter)
Voltage gain (See gain)
Voltage gain format, 147
Voltage-level sensing, 518–519
Voltage source components, 114–127
Voltage regulator, 33
Volt-ampere characteristic (See characteristic curves)

## W
Wein-bridge oscillator, 402

## Z
Zero bias shift, FET, 471–473
Zener, breakdown mechanism, 32
  diode, 32–35
  model, 33
  regulator, 33
  resistance, 32
  voltage, 32
Zero-crossing detector, 323–325